ABOUT THE AUTHOR

Jordan Stoyanov was born in Bulgaria and his alma mater is Moscow State University, where he studied mathematics and earned degrees in probability theory as one of the scientific sons of Albert Shiryaev.

He has published a series of papers—on his own and with co-authors—on a range of contemporary stochastic problems, and wrote five books—which have been published in several different languages and countries. He has also been an invited speaker at several international forums, and has served on editorial boards of professional journals.

He has been a visiting professor at universities in Europe and America, has taught standard and non-standard courses, and has supervised a number of students for MSc and PhD degrees. He spent the last 15 years in the UK holding a position at Newcastle University.

He recently retired, and now enjoys life and hobbies, among them research, travels, seminar talks, meeting good people and seeing extra-ordinary places all over the world. He values a sense of humor, enjoys difficult things, and accepts the challenge of sometimes adopting an unusual position—whether in everyday life or professionally.

Over the last three decades the idea of using counterexamples in probability has been well-received by people involved in learning, teaching, and research. Stoyanov's contributions in this field have won praise from both his peers and students.

Counterexamples in PROBABILITY

THIRD EDITION

Courtesy of Professor A. T. Fomenko of Moscow University.

Counterexamples in PROBABILITY

THIRD EDITION

Jordan M. Stoyanov

Formerly at:

Newcastle University, Newcastle upon Tyne, United Kingdom
Universidade Federal do Rio de Janeiro, Rio de Janeiro, Brasil
Northern Kentucky University, Highland Heights, Kentucky, USA
Université Joseph Fourier, Grenoble, France
Miami University, Oxford, Ohio, USA
Queen's University, Kingston, Ontario, Canada
Università di Padova, Padova, Italy
Concordia University, Montréal, Québec, Canada
Sofia University "St. Kliment Ohridski", Sofia, Bulgaria
Bulgarian Academy of Sciences, Sofia, Bulgaria

DOVER PUBLICATIONS, INC.
Mineola, New York

Bibliographical Note

This Dover edition, first published in 2013, is a revised, corrected, and amended republication of the second edition of the work, originally published by John Wiley & Sons, Inc., Chichester and New York, in 1997 [First edition: 1987].

The publisher joins the author in thanking Venelin Chernogorov for his work in preparing the final files for publication. The illustrations are reprinted here with the generous permission of Prof. A. T. Fomenko.

Library of Congress Cataloging-in-Publication Data

Stoianov, Iordan, author.

Counterexamples in probability / Jordan M. Stoyanov, formerly at: Newcastle University, Newcastle upon Tyne, United Kingdom; Universidade Federal do Rio de Janeiro, Rio de Janeiro, Brasil; Northern Kentucky University, Highland Heights, Kentucky, USA; Université Joseph Fourier, Grenoble, France; Miami University, Oxford, Ohio, USA; Queen's University, Kingston, Ontario, Canada; Università di Padova, Padova, Italy; Concordia University, Montréal, Québec, Canada; Sofia University "St. Kliment Ohridski", Sofia, Bulgaria; Bulgarian Academy of Sciences, Sofia, Bulgaria.—Third edition.

pages cm

Summary: "While most mathematical examples illustrate the truth of a statement, counterexamples demonstrate a statement's falsity. Enjoyable topics of study, counterexamples are valuable tools for teaching, learning, and research. The definitive book on the subject in regards to probability and stochastic processes, this third edition features the author's revisions and corrections plus a substantial new appendix. 2013 edition"—Provided by publisher.

Includes bibliographical references and index.

ISBN-13: 978-0-486-49998-7 (pbk.)

ISBN-10: 0-486-49998-7 (pbk.)

1. Probabilities. 2. Stochastic processes. I. Title.

QA273.S7535 2013

519.2—dc23

2013030644

Manufactured in the United States by LSC Communications
4500056942
www.doverpublications.com

To *Lyubomira, Viktoria, Rosalie and Maya*

Contents

Preface to the Third Edition

The Wiley editions, 1987 and 1997, and the Russian editions, 1999 and 2012, of the book received reasonable attention. Apart from some 25 reviews in scientific journals and several citations, the book was used effectively for courses and seminars in probability and also for comprehensive PhD exams at universities in USA. I was invited to universities in Europe and America to deliver special lectures on counterexamples and their rôle in teaching and research.

Leaving positive reactions aside, I was more concerned by letters and messages containing different sort of complaints.

Some colleagues were angrily asking why I had not used counterexamples from their papers and had not included the papers in the References. I apologized to all of them promising to correct this in the next edition.

Other colleagues were not happy about the Index and references for the examples because I had not provided the pages thus leaving the reader to do his/her own search in a book, journal, etc. Others, on the contrary, shared with me that exactly when 'digging' in books they not only found the desired details but also discovered, e.g., how wonderful the books by A. Rényi, K.L. Chung, A.N. Shiryaev, P. Billingsley and L.C.G. Rogers & D. Williams are.

Complaints came from readers who wanted to cite examples from the 2nd edition of my book in their papers, and needed to include the Mathematical Reviews data. But while the 1st edition was reviewed (see MR 89f:60001), strangely enough, the 2nd was not. Ask Mathematical Reviews to explain this mystery.

I received letters and messages from readers, mainly students, who liked the book in many ways but were unable to pay Wiley a price of £200 for a copy on demand or to Amazon a price of $250 and even more. I answered invariably: do not buy it, there will be a new edition with a reasonable price.

Finally, here is an episode that happened during the Joint Annual Meeting of AMS and MAA, January 1997, San Diego, California. I was there with a fresh copy of the 2nd Wiley edition, just published. Among those who showed a keen interest to my book was John Grafton, Senior Editor at Dover Publications, New York. Returning the book to me after perusing it for half an hour he said: "One day I will publish this book at Dover!" That day arrived.

Meanwhile, I accumulated a large amount of new material on old and new topics. In order to reflect these developments I prepared for the Dover edition a special Appendix containing key words followed by references. The main text of the 2nd Wiley edition was checked again, slightly revised, corrected and updated. As before, I followed the rule: "Always pick the lightest item that still fits in the bag!"

Despite some difficult years in the past, I have been a lucky man! My unusual life started from Divotino/Pernik (Bulgaria) and my later studies, work and travels brought me via Sofia and Moscow to Montreal, Paris, New York, Tokyo, Rio de Janeiro, Taipei and London. In one or another way, I was inspired in my work and life by remarkable people. I will only mention here a few names: Ivan Tiufekchiev, Albert N. Shiryaev, Constance Van Eeden, Fortunato Pesarin, Alain Le Breton, Paulo Ribenboim, Bart Braden, Masaaki Kijima and Gwo Dong Lin. I would like to use this opportunity to express my deep gratitude to all of them for their invariable support and precious help.

It would also be absolutely necessary to mention here the names of great contributors to Modern Probability. It was in the late 70s and beginning of 80s last century that I discussed this project with A.N. Kolmogorov (1903–1987), B.V. Gnedenko (1912–1995), D.G. Kendall (1918–2007), K. Itô (1915–2008) and J.L. Doob (1910–2004). Their interest, suggestions and encouraging comments were more than stimulating for my work. What they expressed could be summarized briefly as follows: "It is nice to see that a young and enthusiastic mathematician from Bulgaria, graduate of Moscow State University, is determined to complete an ambitious and original project and publish a book which will be welcomed by everybody in the area of probability."

The material in this book is very diverse. However, the readers, as well as their needs and goals, are also different. Those who find intriguing facts and statements are expected to work out the necessary details in any particular case. It will enhance better understanding of the subject, keep their eyes open and prepare their minds for future challenges.

The pleasure of knowing, using and constructing counterexamples in probability can only be compared with the pleasure of walking horizontally, thinking vertically and finding beautiful "items" between.

As always, comments and suggestions from readers are very welcome.

The layout and LaTeX typesetting of the present edition are due to the superb skills of Venelin Chernogorov. I am very grateful to him.

At the very end (but not at the end of the World!), I will not miss the chance to tell the readers that I am sending the material to Dover Publications at a unique historical moment that is recorded by an extended use of the number 12:

12.12.12 at 12:12 (GMT).

December 2012
Newcastle upon Tyne, U.K.

Jordan Stoyanov (Dancho)
stoyanovj@gmail.com

Preface to the Second Edition

A large amount of newly collected and created material and the lively interest in the first edition of this book (CEP-1) motivated me towards the second edition (CEP-2). Actually, I have never stopped looking for new counterexamples or thinking about how to achieve completeness and clarity as far as possible in this work.

My strategy was to keep the best from CEP-1, replace some examples by new and more attractive ones and add entirely new examples taken from recent publications or invented especially for CEP-2. Thus the reader will find several original topics well supplementing the material in CEP-1.

Among the topics essentially extended are independence/dependence/exchangeability properties of sets of random events and random variables, characterization of probability distributions, the moment problem, martingales and limit theorems. Clearer interpretations of many statements and improvements in presentation have been made in all sections. The text of CEP-2 is more compact. However, much material has remained unused in order to keep the book a reasonable size. The Index, Supplementary Remarks and the References have been updated and extended accordingly.

My work on CEP-2 took a long time and, as always, my enthusiasm was based on my strong belief about the importance of the role of counterexamples to everyone teaching or learning probability theory. Additional stimuli came from the positive reactions of so many colleagues in so many countries. Like many others I experienced difficulties during this time and first had to solve the problem of how to survive in this changing and unpredictable world. I now use this opportunity to express sincere thanks to many colleagues and friends for their attention and support during my visits to several universities in The Netherlands, Great Britain, Russia, Italy, Canada, USA, France and Spain. In particular, large portions of CEP-2 were prepared when I was visiting Queen's University (Kingston, Ontario) and Miami University (Oxford, Ohio). The last stages of this work were undertaken during a recent visit to Université Joseph Fourier (Grenoble) and in Sofia just before my trip to Kentucky.

I am very grateful for my collaboration with John Wiley & Sons (Chichester). The attention, the patience and the help of Helen Ramsey and Jenny Smith were much appreciated. My thanks go to them and to all the staff at Wiley.

Finally, I hope that you, the reader, will benefit from this edition and my belief that new counterexamples will be created as an essential part of the further development of probability theory. As before, any new suggestions are welcome!

July/August 1996
Europe/America

Jordan Stoyanov

Preface to the First Edition

General comments. We have used the term *counterexample* in the same sense as generally accepted in mathematics. Three previous books related to counterexamples: on analysis (Gelbaum and Olmsted 1964), on topology (Steen and Seebach 1978) and on graph theory (Capobianco and Molluzzo 1978), have been and still are popular among mathematicians. The present book is a collection of counterexamples covering important topics in the field of probability theory and stochastic processes.

It is not only traditional theorems, proofs and illustrative examples, but also counterexamples, which reflect the power, the width, the depth, the degree of non-triviality and the beauty of the theory.

If we have found necessary and sufficient conditions for some statement or result, then any change in the conditions implies that the result is false and accordingly the statement has to be modified. Our attention is focused on interesting questions concerning: (a) the necessity of some sufficient conditions; (b) the sufficiency of certain necessary conditions; (c) the validity of a statement which is the converse to another statement. However, we have included some useful and instructive examples which can be interpreted as counterexamples in a generalized sense.

Purpose of the book. The present book is intended to serve as a supplementary source for many courses in the field of probability theory and stochastic processes. The topics dealt with in the book, and the level of counterexamples, are chosen in such a way that it becomes a multi-purpose book. Firstly, it can be used for any standard course in probability theory for undergraduates. Secondly, some of the material is suitable for advanced courses in probability theory and stochastic processes, assuming that the students have had a course in measure theory and function theory. Thirdly, young researchers and even professionals will find the book useful and may discover new and strange results. The wide variety of content and detail in the discussions of the counterexamples may also help lecturers and tutors in their teaching.

It should be noted that some of the examples considered in the book give the reader an opportunity to become more familiar with standard results in probability and stochastic processes and to develop a better understanding of the subject. However, there exist some examples which are more difficult and their mastering requires a considerable amount of additional work.

Content and structure of the book. The present book includes a relatively large number of counterexamples. Their choice was not easy. We have tried to include a variety of counterexamples concerning different topics in probability theory and stochastic processes. Though we have avoided trivial examples, we have nonetheless included some which cover elementary matters. Pathological examples have been completely avoided. The examples which are most useful and interesting fall in between these two categories.

The material of the book is divided into 4 chapters and 25 sections. Each section begins with short introductory notes giving basic definitions and main results. Then we present the counterexamples related to the main results, the motivation for questions and the counter-statements. Some notions and results are given and analysed in the counterexamples themselves. All counterexamples are named and numbered for the convenience of readers.

The counterexamples range over various degrees of difficulty. Some are elementary and well known counterexamples and can be classified as a part of a probabilistic folklore. Also the style of presentation needs to vary. Some of the counterexamples are only briefly described to economize on space and to provide the reader with a chance for independent work.

Readers of the book are assumed to be familiar with the basic notions and results in probability theory and stochastic processes. Some references are given to textbooks and lecture notes which provide the necessary background to the subject.

At the end of the book, *Supplementary Remarks* are included providing references and some additional explanations for the majority of the counterexamples. For most of the examples we have given at least one relevant early reference. Many of the counterexamples originate from individual probabilists and statisticians and we have cited them fully. Other sources are also indicated where the reader can find new counterexamples, ideas for such examples or some questions whose answers would lead to interesting and useful counterexamples. The *Supplementary Remarks* give readers the opportunity for further work.

Note about references. References Dudley (1972) and (Dudley 1976) indicate a paper or book published by Dudley in 1972 or 1976 respectively. For convenience we have devised abbreviated names for the principal journals in the field of probability theory, stochastic processes and mathematical statistics. In all other cases standard international abbreviations are used.

History of the book. The book is a result of 16 years of my study in the field of probability theory and stochastic processes. I started to collect counterexamples in 1970 when I was a student at Moscow University and later it became an intriguing preoccupation. As a result I increased the number of counterexamples to 500 or so. Many of the counterexamples or different versions of them belong to other authors. Some new and fresh counterexamples were created by colleagues and friends especially for this book. During the preparation of the book I have been guided by my own experience in lecturing on these topics in several European and Canadian universities and in giving special seminars in recent years for students of Sofia University.

The international character of the book is obvious. It is not only my opinion that the present book is an example, not a counterexample, of a successful collaboration and friendship among mathematicians from different countries.

Acknowledgements. The selection and presentation of the material in the book, aimed at covering the wide field of probability theory and stochastic processes, has not been an easy task. I was grateful for the opportunity to discuss the project with my many colleagues and friends whose advice and valuable suggestions were extremely helpful. I wish to express my thanks to all of them.

My special thanks are addressed to my teachers Prof. B. V. Gnedenko, Prof. Yu. V. Prohorov and Prof. A. N. Shiryaev for their attention, general and specific suggestions and encouragement. Among colleagues and friends I have to mention N. V. Krylov, R. Sh. Liptser, A. A. Novikov, Yu. M. Kabanov, S. E. Kuznetsov, A. M. Zubkov, O. B. Enchev and S. D. Gaidov with whom I had very useful discussions on several concrete topics.

My thanks are directed to all colleagues who were so kind as to send me their specific suggestions. The names of these colleagues are included in the list of references.

I use the opportunity to express my special grateful to Prof. A. T. Fomenko for providing five of his extraordinary drawings especially for this book.

I wish to thank Prof. D. G. Kendall for his interest to my work and for his constructive suggestions and encouragement.

The comments of the anonymous referees and the editor helped me to improve both the content and the style of the presentation. I express my appreciation to them.

Finally I should like to thank the collaborators of John Wiley & Sons (Chichester) for their patience and for their precise and excellent work. It is my pleasure to mention the names of Charlotte Farmer and Ian McIntosh.

Suggestions and comments from readers are most welcome and will, if appropriate, be reflected in any subsequent editions of the book.

June 1986, Sofia *Jordan Stoyanov*

Basic Notation and Abbreviations

$(\Omega, \mathcal{F}, \mathbf{P})$	— probability space
$\sigma(\mathcal{C})$	— σ-field generated by the class $\mathcal{C}$
$\mathbf{E}X, \mathbf{V}X$	— expectation and variance of the r.v. X
$\mathbf{E}[X\|\mathcal{D}]$	— conditional expectation given σ-field $\mathcal{D}$
$\mathbb{R}^n$	— n-dimensional Euclidean space
$\mathcal{B}^n$	— Borel σ-field in $\mathbb{R}^n$
$\mathbb{N}$	— set of all natural numbers: $\mathbb{N} = \{1, 2, \ldots\}$
$\mathbb{R}^+$	— set of non-negative numbers: $\mathbb{R}^+ = [0, \infty)$
$I_B, I(B)$	— indicator function of the set B
$\mathcal{N}(a, \sigma^2)$	— normal distribution with parameters a and σ^2
Φ	— the standard normal d.f.
$F_1 * F_2$	— convolution of the d.f.s F_1 and F_2
$:=$	— equal by definition
$x \vee y$	— $x \vee y := \max\{x, y\}$
$x \wedge y$	— $x \wedge y := \min\{x, y\}$
$\Rightarrow, \iff$	— logical implications
iff	— if and only if
r.v.	— random variable
d.f.	— distribution function
ch.f.	— characteristic function
i.i.d.	— independent and identically distributed
a.s.	— almost surely
i.o.	— infinitely often
u.a.n.	— uniform asymptotic negligibility
$\xrightarrow{\text{d}}$	— convergence in distribution
$\xrightarrow{\text{P}}$	— convergence in probability
$\xrightarrow{\text{L}^r}$	— convergence in L^r-sense
$\xrightarrow{\text{w}}$	— weak convergence
$\xrightarrow{\text{a.s.}}$	— almost sure convergence
$\xrightarrow{\text{v}}$	— convergence in variation

CLT	— central limit theorem
WLLN	— weak law of large numbers
SLLN	— strong law of large numbers
IFR	— increasing failure rate
IFRA	— increasing failure rate average
NBU	— new and better than used

Chapter 1

Classes of Random Events and Probabilities

Courtesy of Professor A. T. Fomenko of Moscow University

SECTION 1. CLASSES OF RANDOM EVENTS

Let Ω be an arbitrary non-empty set. Its elements, denoted by ω, will be interpreted as outcomes (results) of some experiment. As usual, we use $A \cup B$ and $A \cap B$ (as well AB) to represent the union and the intersection of any two subsets A and B of Ω respectively. Also, A^c is the complement of $A \subset \Omega$. In particular, $\Omega^c = \emptyset$ where $\emptyset$ is the empty set.

The class $\mathcal{A}$ of subsets of Ω is called a **field** if it contains Ω and is closed under the formation of complements and finite unions, that is if:

(a) $\Omega \in \mathcal{A}$;
(b) $A \in \mathcal{A} \Rightarrow A^c \in \mathcal{A}$;
(c) $A_1, A_2 \in \mathcal{A} \Rightarrow A_1 \cup A_2 \in \mathcal{A}$.

Taking into account the so-called de Morgan laws, $(A_1A_2)^c = A_1^c \cup A_2^c$ and $(A_1 \cup A_2)^c = A_1^c A_2^c$, we easily see that (c) can be replaced by the condition

(c′) $A_1, A_2 \in \mathcal{A} \Rightarrow A_1A_2 \in \mathcal{A}$.

Thus $\mathcal{A}$ is closed under finite intersections.

The class $\mathcal{F}$ of subsets of Ω is called a **σ-field** if it is a field and if it is closed under the formation of countable unions, that is if:

(d) $A_1, A_2, \ldots, \in \mathcal{F} \Rightarrow \bigcup_{n=1}^{\infty} A_n \in \mathcal{F}$.

Again, as above, condition (d) can be replaced by

(d′) $A_1, A_2, \ldots, \in \mathcal{F} \Rightarrow \bigcap_{n=1}^{\infty} A_n \in \mathcal{F}$

and clearly the σ-field $\mathcal{F}$ is closed under countable intersections.

Recall that the elements of any field or σ-field are called **random events** (or simply, **events**). Other classes of events, such as the semi-field, D-system, and product of σ-fields, will be defined and compared with each another in the examples below.

Any textbook on probability theory contains a detailed presentation of all these basic ideas (see Kolmogorov 1956; Breiman 1968; Gihman and Skorohod 1974/1979; Chung 1974; Neveu 1965; Chow and Teicher 1978; Billingsley 1995; Shiryaev 1995). The examples given in this section concern some of the properties of different classes of random events and examine the relationship between notions which seem to be close to one another.

1.1. A class of events which is a field but not a σ-field

Let $\Omega = [0, \infty)$ and $\mathcal{F}_1$ be the class of all intervals of the type $[a, b)$ or $[a, \infty)$ where $0 \le a < b < \infty$. Denote by $\mathcal{F}_2$ the class of all finite sums of intervals of $\mathcal{F}_1$. Then $\mathcal{F}_1$ is not a field, and $\mathcal{F}_2$ is a field but not a σ-field.

Take arbitrary numbers a and b, $0 < a < b < \infty$. Then $A = [a, b) \in \mathcal{F}_1$. However, $A^c = [0, a) \cup [b, \infty) \neq \mathcal{F}_1$ and thus $\mathcal{F}_1$ is not a field.

It is easy to see that: (i) the finite union of finite sums of intervals (of $\mathcal{F}_1$) is again a sum of intervals; (ii) the complement of a finite sum of intervals is also a sum of intervals. This means that $\mathcal{F}_2$ is a field. However, $\mathcal{F}_2$ is not a σ-field because, for example, the set $A_n = [0, 1/n) \in \mathcal{F}_1$ for each $n = 1, 2, \ldots$, and the intersection $\bigcap_{n=1}^{\infty} A_n = \{0\}$ does not belong to $\mathcal{F}_1$.

Let us look at two additional cases.

(a$_1$) Let $\Omega = \mathbb{R}^1$ and $\mathcal{F}$ be the class of all finite sums of intervals of the type $(-\infty, a]$, $(b, c]$ and (d, ∞). Then $\mathcal{F}$ is a field. But the intersection $\bigcap_{n=1}^{\infty}(b - 1/n, c]$ is equal to $[b, c]$ which does not belong to $\mathcal{F}$. Hence the field $\mathcal{F}$ is not a σ-field.

(a$_2$) Let Ω be any infinite set and $\mathcal{A}$ the collection of all subsets $A \in \Omega$ such that either A or its complement A^{c} is finite. Then it is easy to see that $\mathcal{A}$ is a field but not a σ-field.

1.2. A class of events can be closed under finite unions and finite intersections but not under complements

Let $\Omega = \mathbb{R}^1$ and the class $\mathcal{A}$ consist of intervals of the type (x, ∞), $x \in \Omega$. Then using the notations $u = x \wedge y := \min\{x, y\}$ and $v = x \vee y := \max\{x, y\}$ we have:

$$(x, \infty) \cup (y, \infty) = (u, \infty) \in \mathcal{A}$$
$$(x, \infty) \cap (y, \infty) = (v, \infty) \in \mathcal{A}.$$

However, $(x, \infty)^{\mathrm{c}} = (-\infty, x] \notin \mathcal{A}$.

1.3. A class of events which is a semi-field but not a field

Let Ω be an arbitrary set. A non-empty class $\mathcal{J}$ of subsets of Ω is called a *semi-field* if $\Omega \in \mathcal{J}$, $\emptyset \in \mathcal{J}$, $\mathcal{J}$ is closed under the formation of finite intersections, and the complement of any set in $\mathcal{J}$ is a finite sum of disjoint sets of $\mathcal{J}$. It is easy to see that any field of subsets of Ω is also a semi-field. However, the following simple examples show that the converse is not true.

(a$_1$) Let $\Omega = [-\infty, \infty)$ and $\mathcal{J}_1$ contain Ω, $\{\infty\}$ and all intervals of the type $[a, b)$ where $-\infty < a \leq b \leq \infty$. Then $\emptyset \in \mathcal{J}_1$, $\Omega \in \mathcal{J}_1$, $[a_1, b_1 \cap [a_2, b_2) = [a_1 \vee a_2, b_1 \wedge b_2) \in \mathcal{J}_1$ and $[a, b)^{\mathrm{c}} = [-\infty, a) \cup [b, \infty)$. So $\mathcal{J}_1$ is a semi-field. Obviously $\mathcal{J}_1$ is not a field.

(a$_2$) Take $\Omega = \mathbb{R}^1$ and denote by $\mathcal{J}_2$ the class of all subsets of the form AB $(= A \cap B)$ where A is a closed and B is an open set in Ω. Then again, $\mathcal{J}_2$ is a semi-field but not a field.

1.4. A σ-field of subsets of Ω need not contain all subsets of Ω

Recall that the set $A \in \Omega$ is called a *co-finite* set if its complement A^{c} is finite. Let $\mathcal{F}_1$ consist of the finite and co-finite subsets of Ω. Then $\mathcal{F}_1$ is a field. It is a σ-field iff Ω is finite.

Further, the set $A \in \Omega$ is called a *co-countable* set if A^c is countable. Let $\mathcal{F}_2$ consist of the countable and the co-countable subsets of Ω. Then it is easy to check that $\mathcal{F}_2$ is a σ-field.

Suppose now that Ω is uncountable. Then Ω contains a subset A such that A and A^c are both uncountable. This shows that in general a σ-field of Ω need not contain all subsets of Ω and need not be closed under the formation of arbitrary uncountable unions.

1.5. Every σ-field of events is a D-system, but the converse does not always hold

A system $\mathcal{D}$ of subsets of a given set Ω is called a D-*system* (Dynkin system) in Ω if the following three conditions hold: (i) $\Omega \in \mathcal{D}$; (ii) $A, B \in \mathcal{D}$ and $A \subset B \Rightarrow B \setminus A \in \mathcal{D}$; (iii) $A_n \in \mathcal{D}, n = 1, 2, \ldots$ and $A_1 \subset A_2 \subset \ldots \Rightarrow \bigcup_{n=1}^{\infty} A_n \in \mathcal{D}$.

It is obvious that every σ-field is a D-system, but the converse may not be true, as can be seen in the following example.

Take $\Omega = \{\omega_1, \omega_2, \ldots, \omega_{2n}\}, n \in \mathbb{N}$. Denote by $\mathcal{D}_e$ the collection of all subsets $D \in \Omega$ consisting of an even number of elements. Conditions (i), (ii) and (iii) above are satisfied, and hence $\mathcal{D}_e$ is a D-system. However, if $n > 1$ and we take $A = \{\omega_1, \omega_2\}$ and $B = \{\omega_2, \omega_3\}$, we see that $A \in \mathcal{D}_e$, $B \in \mathcal{D}_e$ and $AB = \{\omega_2\} \notin \mathcal{D}_e$. Thus $\mathcal{D}_e$ is not even a field and hence not a σ-field.

Note that a D-system $\mathcal{D}$ is a σ-field iff the intersection of any two sets in $\mathcal{D}$ is again in $\mathcal{D}$ (see Dynkin 1965; Bauer 1996).

1.6. Sets which are not events in the product σ-field

Given two arbitrary sets Ω_1 and Ω_2, we denote their product by $\Omega_1 \times \Omega_2$: $\Omega_1 \times \Omega_1 := \{(\omega_1, \omega_2)\} : \omega_1 \in \Omega_1, \omega_2 \in \Omega_2$. For any set $A \in \Omega_1 \times \Omega_2$ we denote by A_{ω_1} the section of A at ω_1: $A_{\omega_1} = \{\omega_2 \in \Omega_2 : (\omega_1, \omega_2) \in A\}$. Analogously, $A_{\omega_2} = \{\omega_1 \in \Omega_1 : (\omega_1, \omega_2) \in A\}$.

A rectangle in $\Omega_1 \times \Omega_2$ is a subset of the form

$$A_1 \times A_2 = \{(\omega_1, \omega_2) : \omega_1 \in A, \omega_2 \in A_2\},\ A_1 \in \Omega_1,\ A_2 \in \Omega_2.$$

$A_1 \times A_2$ is called a *measurable rectangle* (with respect to $\mathcal{F}_1$ and $\mathcal{F}_2$) if $A_1 \in \mathcal{F}_1$ and $A_2 \in \mathcal{F}_2$ where $\mathcal{F}_1$ and $\mathcal{F}_2$ are σ-fields of subsets of Ω_1 and Ω_2 respectively. The measurable rectangles form a semi-field of subsets in $\Omega_1 \times \Omega_2$. Thus the field generated by the measurable rectangles consists of all finite sums of disjoint measurable rectangles. The σ-field generated by this field is denoted by $\mathcal{F}_1 \times \mathcal{F}_2$ and is called the *product* σ-field of $\mathcal{F}_1$ and $\mathcal{F}_2$.

Let us note the following result (see Neveu 1965; Kingman and Taylor 1966). For every measurable set A in $(\Omega_1 \times \Omega_2, \mathcal{F}_1 \times \mathcal{F}_2)$ and every fixed $\omega_1 \in \Omega_1$ and $\omega_2 \in \Omega_2$, the sections A_{ω_1} and A_{ω_2} are measurable sets in $(\Omega_2, \mathcal{F}_2)$ and $(\Omega_1, \mathcal{F}_1)$ respectively.

However, the converse is not true. To see this, let Ω be any uncountable set and

$\mathcal{F}$ the smallest σ-field of subsets of Ω containing all one-point elements. Then the diagonal $D = \{(\omega, \omega) : \omega \in \Omega\}$ of $\Omega \times \Omega$ does not belong to the product σ-field $\mathcal{F} \times \mathcal{F}$, although all its sections belong to $\mathcal{F}$. In other words, for each $\omega \in \Omega$, the section $D_\omega \in \mathcal{F}$ and is an event but $D \notin \mathcal{F} \times \mathcal{F}$ and is not an event.

1.7. The union of a sequence of σ-fields need not be a σ-field

Let $\mathcal{F}_1, \mathcal{F}_2, \ldots$ be a sequence of σ-fields of subsets of the set Ω. Then their intersection $\bigcap_{n=1}^{\infty} \mathcal{F}_n$ is always a σ-field and it is natural to ask whether the union $\bigcup_{n=1}^{\infty} \mathcal{F}_n$ is a σ-field. We shall now show that the answer to this question is negative.

Consider the set $\Omega = \{\omega_1, \omega_2, \omega_3\}$ and the following two classes of its subsets: $\mathcal{F}_1 = \{\emptyset, \{\omega_1\}, \{\omega_2, \omega_3\}, \Omega\}$, $\mathcal{F}_2 = \{\emptyset, \{\omega_2\}, \{\omega_1, \omega_3\}, \Omega\}$. Then $\mathcal{F}_1$ and $\mathcal{F}_2$ are fields and hence σ-fields. Obviously the intersection $\mathcal{F}_1 \cap \mathcal{F}_2 = \{\emptyset, \Omega\}$, the trivial σ-field. However, the union

$$\mathcal{F} = \mathcal{F}_1 \cup \mathcal{F}_2 = \{\emptyset, \{\omega_1\}, \{\omega_2\}, \{\omega_2, \omega_3\}, \{\omega_1, \omega_3\}, \Omega\}$$

is not a field, and hence not a σ-field because the element $\{\omega_1\} \cup \{\omega_2\} = \{\omega_1, \omega_2\} \notin \mathcal{F}$.

SECTION 2. PROBABILITIES

Let Ω be any set and $\mathcal{A}$ be a field of its subsets. We say that $\mathbf{P}$ is a **probability** on the measurable space $(\Omega, \mathcal{A})$ if $\mathbf{P}$ is defined for all events $A \in \mathcal{A}$ and satisfies the following axioms.

(a) $\mathbf{P}(A) \geq 0$ for each $A \in \mathcal{A}$; $\mathbf{P}(\Omega) = 1$.

(b) $\mathbf{P}$ is finitely additive. That is, for any finite number of pairwise disjoint events $A_1, \ldots, A_n \in \mathcal{A}$ we have

$$\mathbf{P}\left(\bigcup_{i=1}^{n} A_i\right) = \sum_{i=1}^{n} \mathbf{P}(A_i).$$

(c) $\mathbf{P}$ is continuous at $\emptyset$. That is, for any events $A_1, A_2, \ldots \in \mathcal{A}$ such that $A_{n+1} \subset A_n$ and $\bigcap_{n=1}^{\infty} A_n = \emptyset$, it is true that

$$\lim_{n \to \infty} \mathbf{P}(A_n) = 0.$$

Note that conditions (b) and (c) are equivalent to the next one (d).

(d) $\mathbf{P}$ is σ-additive (countably additive), that is

$$\mathbf{P}\left(\bigcup_{n=1}^{\infty} A_n\right) = \sum_{n=1}^{\infty} \mathbf{P}(A_n)$$

for any events $A_1, A_2, \ldots \in \mathcal{A}$ which are pairwise disjoint.

According to the Carathéodory theorem (see Kolmogorov 1956; Loève 1977; Shiryaev 1995), if $\mathbf{P}_0$ is a σ-additive probability on $(\Omega, \mathcal{A})$ and $\mathcal{F} = \sigma(\mathcal{A})$ denotes the smallest σ-field generated by the field $\mathcal{A}$, then there is a unique probability measure $\mathbf{P}$ on $(\Omega, \mathcal{F})$ which is an extension of $\mathbf{P}_0$ in the sense that $\mathbf{P}(A) = \mathbf{P}_0(A)$ for $A \in \mathcal{A}$. In this case we also say that $\mathbf{P}_0$ is a restriction of $\mathbf{P}$ over $\mathcal{A}$ and write $\mathbf{P}|\mathcal{A} = \mathbf{P}_0$.

The ordered triplet $(\Omega, \mathcal{F}, \mathbf{P})$ is called a **probability space** if:

Ω is any set of points called elementary events (outcomes);
$\mathcal{F}$ is a σ-field of subsets of Ω; the elements of $\mathcal{F}$ are events;
$\mathbf{P}$ is a probability on $\mathcal{F}$, that is $\mathbf{P}$ satisfies conditions (a), (b) and (c) above, or, equivalently, (a) and (d).

Thus we have described the axiomatic system which is generally accepted in probability theory. This system was suggested by A. N. Kolmogorov in 1933 (see Kolmogorov 1956).

In this section we present a few examples characterizing some of the properties of probability measures. The important notion of conditional probability is introduced and treated in Example 2.4.

2.1. A probability measure which is additive but not σ-additive

Let Ω be the set of all rational numbers r of the unit interval [0,1] and $\mathcal{F}_1$ the class of the subsets of Ω of the form $[a, b]$, $(a, b]$, (a, b) or $[a, b)$ where a and b are rational numbers. Denote by $\mathcal{F}_2$ the class of all finite sums of disjoint sets of $\mathcal{F}_1$. Then $\mathcal{F}_2$ is a field. Let us define the probability measure $\mathbf{P}$ as follows:

$$\begin{aligned}
&\mathbf{P}(A) = b - a, && \text{if } A \in \mathcal{F}_1, \\
&\mathbf{P}(B) = \sum_{i=1}^{n} P(A_i), && \text{if } B \in \mathcal{F}_2, \text{ that is } B = \textstyle\sum_{i=1}^{n} A_i,\ A_i \in \mathcal{F}_1.
\end{aligned}$$

Consider two disjoint sets of $\mathcal{F}_2$, say

$$B = \sum_{i=1}^{n} A_i \quad \text{and} \quad B' = \sum_{j=1}^{m} A'_j$$

where $A_i, A'_j \in \mathcal{F}_1$ and all A_i, A'_j are disjoint. Then $B + B' = \sum_{k=1}^{m+n} C_k$ where either $C_k = A_i$ for some $i = 1, \ldots, n$, or $C_k = A'_j$ for some $j = 1, \ldots, m$. Moreover,

$$\begin{aligned}
\mathbf{P}(B + B') = \mathbf{P}\left(\sum_k C_k\right) = \sum_k \mathbf{P}(C_k) = \sum_{i,j} (\mathbf{P}(A_i) + \mathbf{P}(A'_j)) \\
\mathbf{P}(A_i) + \sum_j \mathbf{P}(A'_j) = \mathbf{P}(B) + \mathbf{P}(B')
\end{aligned}$$

and hence **P** is an additive measure.

Obviously every one-point set $\{r\} \in \mathcal{F}_2$ and $\mathbf{P}(\{r\}) = 0$. Since Ω is a countable set and $\Omega = \sum_{i=1}^{\infty}\{r_i\}$, we get

$$\mathbf{P}(\Omega) = 1 \neq 0 = \sum_{i=1}^{\infty} \mathbf{P}(\{r_i\}).$$

This contradiction shows that **P** is not σ-additive.

2.2. The coincidence of two probability measures on a given class does not always imply their coincidence on the σ-field generated by this class

Let Ω be a set and $\mathcal{C}$ a class of events such that $A, B \in \mathcal{C} \Rightarrow AB \in \mathcal{C}$ (that is, $\mathcal{C}$ is closed under intersection). Denote by $\mathcal{F} = \sigma(\mathcal{C})$ the σ-field generated by $\mathcal{C}$. Let $\mathbf{Q}_1$ and $\mathbf{Q}_2$ be two probabilities on the measurable space $(\Omega, \mathcal{F})$. The following result is well known (see Breiman 1968):

$$\mathbf{Q}_1 = \mathbf{Q}_2 \text{ on } \mathcal{C} \;\Rightarrow\; \mathbf{Q}_1 = \mathbf{Q}_2 \text{ on } \mathcal{F}.$$

It is not surprising that results of this kind depend essentially on the structure of the class $\mathcal{C}$. By an example we show the importance of the hypothesis that $\mathcal{C}$ is closed under intersection by an example.

Take $\Omega = \{a, b, c, d\}$ and two measures $\mathbf{Q}_1$ and $\mathbf{Q}_2$ defined as follows:

$$\begin{aligned} \mathbf{Q}_1(a) = \mathbf{Q}_1(d) = \mathbf{Q}_2(b) = \mathbf{Q}_2(c) &= \tfrac{1}{6}, \\ \mathbf{Q}_1(b) = \mathbf{Q}_1(c) = \mathbf{Q}_2(a) = \mathbf{Q}_2(d) &= \tfrac{1}{3}. \end{aligned}$$

Let $\mathcal{F}$ be the class of all subsets of Ω and $\mathcal{C} = \{a \cup b, d \cup c, a \cup c, b \cup d\}$. Here and below $x \cup y$ denotes the two-element set $\{x, y\}$. Then it is easy to check that $\mathbf{Q}_1 = \mathbf{Q}_2$ on $\mathcal{C}$. For example,

$$\begin{aligned} \mathbf{Q}_1(d \cup c) &= \mathbf{Q}_1(d) + \mathbf{Q}_1(c) = \tfrac{1}{6} + \tfrac{1}{3} = \tfrac{1}{2} \\ \mathbf{Q}_2(d \cup c) &= \mathbf{Q}_2(d) + \mathbf{Q}_2(c) = \tfrac{1}{3} + \tfrac{1}{6} = \tfrac{1}{2} \end{aligned}$$

and thus $\mathbf{Q}_1(d \cup c) = \mathbf{Q}_2(d \cup c)$. Analogously, $\mathbf{Q}_1(\cdot) = \mathbf{Q}_2(\cdot)$ for all remaining elements of $\mathcal{C}$. However, it is evident from the definition of $\mathbf{Q}_1$ and $\mathbf{Q}_2$ that the equality $\mathbf{Q}_1(\cdot) = \mathbf{Q}_2(\cdot)$ does not hold for all elements of $\mathcal{F}$; for example, it is false for each of a, b, c and d. The reason for this is that $\mathcal{C}$, as taken, is not closed under intersection.

2.3. On the validity of the Kolmogorov extension theorem in $(\mathbb{R}^\infty, \mathcal{B}^\infty)$

Recall that the probability measures in the space $\mathbb{R}^n$, $n > 1$ are constructed in the following way: first for elementary sets (rectangles of the type $(a, b]$), then for sets

$A = \sum(a_i, b_i]$, and finally, by using the Carathéodory theorem (see Loève 1977; Shiryaev 1995), for sets in $\mathcal{B}^n$.

A similar construction can be used for the space $(\mathbb{R}^\infty, \mathcal{B}^\infty)$. Indeed, let $C_n(B) = \{x \in \mathbb{R}^\infty : (x_1, \dots, x_n) \in B\}$, $B \in \mathcal{B}^n$ denote a cylinder set in $\mathbb{R}^\infty$ with base $B \in \mathcal{B}^n$. It is natural to take the cylinder sets as elementary sets in $\mathbb{R}^\infty$ with their probabilities defined by the probability measure on the sets of $\mathcal{B}^\infty$.

Suppose **P** is a probability measure on $(\mathbb{R}^\infty, \mathcal{B}^\infty)$. For $n = 1, 2, \dots$ we put

$$P_n(B) = \mathbf{P}(C_n(B)), \qquad B \in \mathcal{B}^n.$$

Thus we obtain a sequence of probability measures $P_1, P_2, \cdots$ defined respectively on $(\mathbb{R}^1, \mathcal{B}^1)$, $(\mathbb{R}^2, \mathcal{B}^2)$, ... For $n = 1, 2, \dots$ and $B \in \mathcal{B}^n$ the following *consistency* (or *compatibility*) *property* holds:

$$P_{n+1}(B \times \mathbb{R}^1) = P_n(B). \tag{1}$$

We now formulate a fundamental result.

Kolmogorov theorem. *Let $P_1, P_2, \dots$ be a sequence of probability measures respectively on $(\mathbb{R}^1, \mathcal{B}^1)$, $(\mathbb{R}^2, \mathcal{B}^2)$, ... satisfying the consistency property (1). Then there is a unique probability measure* **P** *on $(\mathbb{R}^\infty, \mathcal{B}^\infty)$ such that its restriction on $\mathcal{B}^n$ coincides with P_n, that is,* $\mathbf{P}(C_n(B)) = P_n(B)$, $B \in \mathcal{B}^n$, $n = 1, 2, \dots$.

The proof of this theorem can be found in many textbooks (see Kolmogorov 1956; Doob 1953; Loève 1977; Neveu 1965; Feller 1971; Billingsley 1995; Shiryaev 1995). Let us note that it uses several specific properties of Euclidean spaces. However, this theorem may fail in general (without any hypotheses on the topological nature of measurable spaces and on the structure of the family of measures $\{P_n\}$). This is seen from the following example.

Consider the space $\Omega = (0, 1]$. (Clearly Ω is not complete.) We shall construct a sequence of σ-fields $\mathcal{F}_1, \mathcal{F}_2, \dots$ and a sequence of probability measures $\{P_n\}$ where P_n is defined on $(\Omega, \mathcal{F}_n)$. Let $\mathcal{F} = \sigma(\cup \mathcal{F}_n)$ be the smallest σ-field containing all $\mathcal{F}_n$. Then we shall show that there is no probability measure **P** on $(\Omega, \mathcal{F})$ such that its restriction $\mathbf{P}|\mathcal{F}_n$ on $\mathcal{F}_n$ coincides with P_n, $n = 1, 2, \dots$.

For $n = 1, 2, \dots$ define the function $h_n(\omega) = 1$ if $0 < \omega < 1/n$ and $h_n(\omega) = 0$ if $1/n \le \omega \le 1$. Let $\mathcal{C}_n = \{A \in \Omega : A = \{\omega : h_n(\omega) \in B\},\ B \in \mathcal{B}^1\}$ and $\mathcal{F}_n = \sigma\{\mathcal{C}_1, \dots, \mathcal{C}_n\}$ be the smallest σ-field containing the sets $\mathcal{C}_1, \dots, \mathcal{C}_n$. Clearly $\mathcal{F}_1 \subset \mathcal{F}_2 \subset \dots$. On the measurable space $(\Omega, \mathcal{F}_n)$ define a probability measure P_n as follows:

$$P_n[\omega : (h_1(\omega), \dots, h_n(\omega)) \in B^n] = \begin{cases} 1, & \text{if } (1, \dots, 1) \in B^n \\ 0, & \text{otherwise} \end{cases}$$

where $B^n \in \mathcal{B}^n$. It is easy to see that the family $\{P_n\}$ is consistent: if $A \in \mathcal{F}_n$ then $P_{n+1}(A \times \mathbb{R}^1) = P_n(A)$.

Suppose now that there exists a probability **P** on the measurable space $(\Omega, \mathcal{F})$ such that $\mathbf{P}|\mathcal{F}_n = P_n$. If so, then for $n = 1, 2, \dots$

$$\mathbf{P}[\omega : h_1(\omega) = \dots = h_n(\omega) = 1] = P_n[\omega : h_1(\omega) = \dots = h_n(\omega) = 1] = 1. \tag{2}$$

However, $\{\omega{:}h_1(\omega) = \ldots = h_n(\omega) = 1\} = (0, 1/n) \downarrow \emptyset$, which contradicts (2) and the requirement for the set function $\mathbf{P}$ to be σ-additive (or, equivalently, to be continuous at the 'zero' set $\emptyset$).

2.4. There may not exist a regular conditional probability with respect to a given σ-field

Let $(\Omega, \mathcal{F}, \mathbf{P})$ be a probability space and $\mathcal{F}_1$ a σ-field such that $\mathcal{F}_1 \subset \mathcal{F}$. Recall that the conditional probability $\mathbf{P}(A|\mathcal{F}_1)$ is defined $\mathbf{P}$-a.s. as an $\mathcal{F}_1$-measurable function of ω such that

$$\mathbf{P}(AB) = \int_B \mathbf{P}(A|\mathcal{F}_1)\, \mathrm{d}\mathbf{P}(\omega) \quad \text{for each } B \in \mathcal{F}_1.$$

The conditional probability $\mathbf{P}(A|\mathcal{F}_1)$, $A \in \mathcal{F}$ is said to be *regular* if there exists a function $\mathbf{P}(A, \omega)$, $A \in \mathcal{F}$, $\omega \in \Omega$, which satisfies the following two properties:

(i) $\mathbf{P}(A, \omega) = \mathbf{P}(A|\mathcal{F}_1)$ $\mathbf{P}$-a.s. for an arbitrary $A \in \mathcal{F}$;
(ii) for fixed ω, $\mathbf{P}(\cdot, \omega)$ is a probability measure on $\mathcal{F}$.

If condition (ii) is satisfied and condition (i) holds for all ω (not only for $\mathbf{P}$-almost all ω), then $\mathbf{P}(A|\mathcal{F}_1)$ is called a proper regular conditional probability. (In terms of distributions we speak about regular and proper regular conditional distributions.) Regular conditional probabilities exist in many cases, but proper regular conditional probabilities do not always exist, as can be seen below.

Let $(\Omega, \mathcal{F}, \lambda)$ be a probability space with $\Omega = [0, 1]$, $\mathcal{F}$ the σ-field of the Lebesgue measurable sets in [0,1] and λ the Lebesgue measure. It is well known that in the interval [0,1] there is a non-measurable (in Lebesgue sense) set, say N, such that its outer measure is $\lambda^*(N) = 1$ and its inner measure is $\lambda_*(N) = 0$ (for details see Halmos 1974).

Define a new σ-field $\tilde{\mathcal{F}}$ which is generated by $\mathcal{F}$ and the set N. Thus $\tilde{\mathcal{F}}$ consists of sets of the form $NB_1 \cup N^c B_2$ where $B_1, B_2 \in \mathcal{F}$. Define also the measure $\mathbf{P}$ on the measurable space $([0, 1], \tilde{\mathcal{F}}, \mathbf{P})$ by

$$\mathbf{P}(NB_1 \cup N^c B_2) = \tfrac{1}{2}[\lambda(B_1) + \lambda(B_2)].$$

It is easy to check that $\mathbf{P}$ is well defined and defines a probability on $\tilde{\mathcal{F}}$, so the triplet $([0, 1], \tilde{\mathcal{F}}, \mathbf{P})$ is a probability space. For every $B \in \mathcal{F}$ we have

$$\mathbf{P}(NB_1 \cup N^c B_2) = \mathbf{P}(B) = \lambda(B)$$

and hence $\mathbf{P}$ coincides with λ on $\mathcal{F}$, that is $\mathbf{P}|\mathcal{F} = \lambda$. Moreover,

$$\mathbf{P}(N) = \tfrac{1}{2}.$$

Now we shall prove the following statement: on the probability space $([0, 1], \tilde{\mathcal{F}}, \mathbf{P})$ there is no regular conditional probability $\mathbf{P}(A|\mathcal{F})$, $A \in \tilde{\mathcal{F}}$ with respect to the σ-field $\mathcal{F}$.

Suppose such a probability exists: that is, there is a function, say $\mathbf{P}(A, \omega)$, which satisfies the above conditions (i) and (ii). If so, then for any Borel (and Lebesgue) set A, $\mathbf{P}(A, \omega) = 1_A(\omega)$. Therefore if A is a one-point set, $A = \{\omega\}$, then $\mathbf{P}(\{\omega\}, \omega) = 1$. Now take the set N. From the definition of a conditional probability and the equality $\mathbf{P}(N) = \frac{1}{2}$ we get

$$\tfrac{1}{2} = \mathbf{P}(N) = \int_\Omega \mathbf{P}(N, \omega)\lambda\,(\mathrm{d}\omega).$$

On the other hand, if $\mathbf{P}(\cdot, \omega)$ is a measure for each ω, then

$$\mathbf{P}(N, \omega) \geq \mathbf{P}(\{\omega\}, \omega) = 1 \text{ for all } \omega \in N \Rightarrow \mathbf{P}(N, \omega) = 1 \text{ for all } \omega \in N.$$

Consider the set $C = \{\omega{:}\mathbf{P}(\{\omega\}, \omega) = 1\}$. Since $\mathbf{P}(\cdot, \omega)$ is a Borel function in ω, then the set C is Borel measurable with $\mathbf{P}(C) = 1$. Let $D = \{\omega : \mathbf{P}(N, \omega) = 1\}$. It is clear that D is Borel-measurable and $D \supset CN$, which implies that $D \cup C^c \supset N$. However, the set $D \cup C^c$ is Borel and covers the (non-measurable!) set N which has $\lambda^*(N) = 1$. Therefore $\mathbf{P}(D \cup C^c) = 1$ and $\mathbf{P}(D) = 1$. In other words, for almost all ω we get $\mathbf{P}(N, \omega) = 1$, which implies the following equality

$$\int_\Omega \mathbf{P}(N, \omega)\lambda\,(\mathrm{d}\omega) = 1.$$

However, this contradicts the relation $\int_\Omega \mathbf{P}(N, \omega)\lambda\,(\mathrm{d}\omega) = \frac{1}{2}$ obtained above.

Therefore a regular conditional probability need not always exist.

Let us note that in this counterexample the role of the non-measurable set N is essential. Recall that the construction of N relies on the axiom of choice. Using a weakened form of the axiom of choice, Solovay (1970) derived several interesting results concerning the measurability of sets, measures and their properties.

General results on the existence of regular conditional probabilities can be found in the works of Pfanzagl (1969), Blackwell and Dubins (1975) and Faden (1985).

SECTION 3. INDEPENDENCE OF RANDOM EVENTS

Let $(\Omega, \mathcal{F}, \mathbf{P})$ be a probability space. The events A and B of $\mathcal{F}$ are said to be **independent** (with respect to **P**) if

$$\mathbf{P}(AB) = \mathbf{P}(A)\mathbf{P}(B).$$

More generally, two classes of events (for example fields, σ-fields), say $\mathcal{A}_1$ and $\mathcal{A}_2$, $\mathcal{A}_1, \mathcal{A}_1 \in \mathcal{F}$ are called independent if any two events A_1 and A_2 where $A_1 \in \mathcal{A}_1$, $A_2 \in \mathcal{A}_2$ are independent.

The concept of independence of two events or two classes of events can be extended to any finite number of events or classes. We say that the events $A_1, \ldots, A_n \in \mathcal{F}$ are **mutually independent** if the following relation (product rule)

$$(1) \qquad \mathbf{P}(A_{i_1} A_{i_2} \ldots A_{i_k}) = \mathbf{P}(A_{i_1})\mathbf{P}(A_{i_2}) \ldots \mathbf{P}(A_{i_k})$$

is satisfied for all k and $i_1, i_2, \ldots, i_k$ where $k = 2, \ldots, n$ and $1 \leq i_1 < i_2 < \ldots < i_k \leq n$. Thus for the mutual independence of n events all $2^n - n - 1$ relations (1) must be satisfied. If at least one relation does not hold, the events are dependent. If all the relations (1) fail to hold, we say that the events $A_1, \ldots, A_n$ are *totally dependent*. If the product rule (1) holds only for $k = 2$, the events are *pairwise independent*. Finally, if (1) holds for all k, $2 \leq k \leq m$ for some $m \leq n$, we have a set of n events which are *m-wise independent* (pairwise independent if $m = 2$ and mutually independent if $m = n$).

When considering the independence/dependence properties of collections of random events it is natural to speak about the product rule (1) at *level* k, that is, that (1) holds or does not hold for any of the $\binom{n}{k}$ possible combinations (k-tuples) of events. Thus we can characterize each level k, $k = 2, \ldots, n$, as being independent or dependent. Some interesting (and even unusual) possibilities will be illustrated in the examples below.

It is obvious how to define the independence of a finite number of classes of events.

If $A, B \in \mathcal{F}$ and $\mathbf{P}(B) > 0$ we denote by $\mathbf{P}(A|B)$ the *conditional probability* of A given B and put

$$\mathbf{P}(A|B) = \mathbf{P}(AB)/\mathbf{P}(B).$$

The independence of two events can easily be expressed through conditional probabilities. Another notion, that of conditional independence, is considered in one of the examples.

The examples included in this section aim to help the reader understand the meaning of the fundamental notion of independence more clearly.

3.1. Random events with a different kind of dependence

In a Bernoulli scheme with a parameter p we shall consider two events which, according to the value of p, are either independent or dependent.

Let $H = \{\text{heads}\}$ and $T = \{\text{tails}\}$ be the outcomes at tossing a coin with $\mathbf{P}(H) = p$, $\mathbf{P}(T) = 1 - p$, $0 \leq p \leq 1$. Toss the coin three times independently and consider the events $A = \{\text{at most one tails}\}$ and $B = \{\text{all tosses are the same}\}$. Obviously $A = \{HHH, HHT, HTH, THH\}$, $B = \{HHH, TTT\}$. Hence

$$\mathbf{P}(A) = p^3 + 3p^2(1-p), \quad \mathbf{P}(B) = p^3 + (1-p)^3, \quad \mathbf{P}(AB) = p^3.$$

It is easy to see that the product rule

$$\mathbf{P}(AB) = \mathbf{P}(A)\mathbf{P}(B)$$

holds in the trivial cases $p = 0$ and $p = 1$ and in the symmetric case $p = \frac{1}{2}$. Hence the events A and B are independent if $p = 0$, or $p = 1$, or $p = \frac{1}{2}$. For all other values of p in the interval [0,1], A and B are dependent events.

3.2. The pairwise independence of random events does not imply their mutual independence

It is natural to start with the first ever known examples showing the difference between the mutual and pairwise independence of random events.

The two examples (i) and (ii) below, first presented by Bohlmann (1908) and Bernstein (1928), were created in a period of active studies in probability theory and its establishment as a rigorous branch of mathematics.

(i) (Bohlmann 1908). Suppose we have at our disposal 16 capsules with no difference between them. In each capsule we insert three small balls labelled a, b, c and each ball is either white or black. The capsules are put in an urn, mixed well, and we choose randomly one capsule. We open this capsule to see what is inside, that is what is the outcome of our experiment. We are interested in the property denoted by $(\alpha_1, \alpha_2, \alpha_3)$ where $\alpha_j = 1$ if a white ball is at position j and $\alpha_j = 0$ if that ball is black, $j = 1, 2, 3$. The question is: what kind of dependence exist between α_1, α_2 and α_3?

Clearly, this original and illuminating description is equivalent to considering an urn with 16 capsules marked (inside) as follows: three capsules by 111; three capsules by 100; three capsules by 010; three capsules by 001, and each of the marks 110, 101, 011 and 000 is used just once among the remaining four capsules. We choose one capsule at random and consider the following events:

$$A_j = \{\text{"1" at } j\text{th position}\},\ j = 1, 2, 3$$

(equivalently $A_j = \{\alpha_j = 1\}, j = 1, 2, 3$).

We easily find that $\mathbf{P}(A_1) = \frac{1}{2}, \mathbf{P}(A_2) = \frac{1}{2}, \mathbf{P}(A_3) = \frac{1}{2}$ and then

$$\mathbf{P}(A_1A_2) = \tfrac{1}{4},\ \mathbf{P}(A_1A_3) = \tfrac{1}{4}, \mathbf{P}(A_2A_3) = \tfrac{1}{4}$$

implying that the events A_1, A_2, A_3 are (at least) pairwise independent.

However

$$\mathbf{P}(A_1A_2A_3) = \tfrac{3}{16} \neq \tfrac{1}{2}\tfrac{1}{2}\tfrac{1}{2} = \mathbf{P}(A_1)\mathbf{P}(A_2)\mathbf{P}(A_3)$$

and hence these events are not mutually independent.

(ii) (Bernstein 1928). Suppose a box contains four tickets labelled 112, 121, 211, 222. Choose one ticket at random and consider the events $A_1 = \{1 \text{ occurs in the first place}\}$, $A_2 = \{1 \text{ occurs in the second place}\}$ and $A_3 = \{1 \text{ occurs in the third place}\}$. Obviously $\mathbf{P}(A_1) = \mathbf{P}(A_2) = \mathbf{P}(A_3) = \frac{1}{2}$ and

$$\mathbf{P}(A_1A_2) = \mathbf{P}(A_1A_3) = \mathbf{P}(A_2A_3) = \tfrac{1}{4}.$$

This means that the three events A_1, A_2, A_3 are pairwise independent. However,

$$\mathbf{P}(A_1A_2A_3) = 0 \neq \tfrac{1}{8} = \mathbf{P}(A_1)\mathbf{P}(A_2)\mathbf{P}(A_3)$$

and hence these events are not mutually independent.

(iii) Consider the six permutations of the letters a, b, c as well as the triplets (a, a, a), (b, b, b) and (c, c, c). Let Ω consist of these nine triplets as points, and let each have probability $\frac{1}{9}$. Define the events $A_k = \{\text{the } k\text{th place is occupied by the letter } a\}$, $k = 1, 2, 3$. Then obviously

$$\mathbf{P}(A_1) = \mathbf{P}(A_2) = \mathbf{P}(A_3) = \tfrac{1}{3},$$
$$\mathbf{P}(A_1A_2) = \mathbf{P}(A_1A_3) = \mathbf{P}(A_2A_3) = \tfrac{1}{9}$$

and hence the events A_1, A_2, A_3 are pairwise independent. However, they are not mutually independent, since $A_1A_2 \subset A_3$, which implies that

$$\mathbf{P}(A_1A_2A_3) = \tfrac{1}{9} \neq \tfrac{1}{27}.$$

The same idea can be generalized as follows. Let Ω contain $n! + n$ points, namely the $n!$ permutations of the symbols $a_1, \dots, a_n$ and the n repetitions of the same symbol a_k, $k = 1, \dots, n$. Suppose that each of the permutations has probability $1/[n^2(n-2)!]$ while each of the repetitions has probability $1/n^2$. Then it is not difficult to check that the events $A_k = \{a_1 \text{ occurs at the } k\text{th place}\}$, $k = 1, \dots, n$, are pairwise independent, but no three of them are mutually independent.

(iv) Let A_1, A_2, A_3 be independent events each of probability $\frac{1}{2}$ and put $A_{ij} = (A_i \triangle A_j)^c$ where $\triangle$ denotes the symmetric difference of two sets: $A_i \triangle A_j = A_iA_j^c + A_i^cA_j$ or, equivalently, $A_i \triangle A_j = (A_i \setminus A_j) \cup (A_j \setminus A_i)$. (In particular, we could consider the following simple experiment: three symmetric coins numbered 1, 2, 3 are tossed; then $A_i = \{\text{coin } i \text{ falls heads}\}$, $A_{ij} = \{\text{coins } i \text{ and } j \text{ agree}\}$.) Then the events A_{12}, A_{13}, A_{23} are not mutually independent, though they are independent in pairs.

(v) Let $\mathcal{L}$ be the set of all n^3 three-letter words s of a language and all words are equally likely. Define the events A, B and C as follows:

$$A = \{s \in \mathcal{L}\colon s \text{ begins with a specified letter, say } x\},$$
$$B = \{s \in \mathcal{L}\colon s \text{ has the letter } x \text{ in the middle}\},$$
$$C = \{s \in \mathcal{L}\colon s \text{ has two of its letters the same}\}.$$

Then A, B and C are pairwise but not mutually independent.

3.3. The relation $\mathbf{P}(ABC) = \mathbf{P}(A)\mathbf{P}(B)\mathbf{P}(C)$ does not always imply the mutual independence of the events A, B, C

(i) Let two dice be tossed, Ω = all ordered pairs ij, $i, j = 1, \dots, 6$ and each point of Ω has probability $\frac{1}{36}$. Consider the events:

$$A = \{\text{first die} = 1,\ 2 \text{ or } 3\},$$
$$B = \{\text{first die} = 3,\ 4 \text{ or } 5\},$$
$$C = \{\text{the sum of two faces is } 9\}.$$

Obviously we have $AB = \{31,32,33,34,35,36\}$, $AC = \{136\}$, $BC = \{36,45,54\}$, $ABC = \{36\}$. Then $\mathbf{P}(A) = \frac{1}{2}$, $\mathbf{P}(B) = \frac{1}{2}$, $\mathbf{P}(C) = \frac{1}{9}$ and

$$\mathbf{P}(ABC) = \tfrac{1}{36} = \tfrac{1}{2}\tfrac{1}{2}\tfrac{1}{9} = \mathbf{P}(A)\mathbf{P}(B)\mathbf{P}(C).$$

Nevertheless the events A, B, C are not mutually independent, since

$$\begin{aligned}\mathbf{P}(AB) &= \tfrac{1}{6} \neq \tfrac{1}{4} = \mathbf{P}(A)\mathbf{P}(B),\\ \mathbf{P}(AC) &= \tfrac{1}{36} \neq \tfrac{1}{18} = \mathbf{P}(A)\mathbf{P}(C),\\ \mathbf{P}(BC) &= \tfrac{1}{12} \neq \tfrac{1}{18} = \mathbf{P}(B)\mathbf{P}(C).\end{aligned}$$

In other words, independence at level 3 does not imply independence at level 2.

(ii) Let $\Omega = \{1,2,3,4,5,6,7,8\}$ where each outcome has probability 1/8. Consider the events $B_1 = \{1,2,3,4\}$, $B_2 = B_3 = \{1,5,6,7\}$. Then $\mathbf{P}(B_1) = \mathbf{P}(B_2) = \mathbf{P}(B_3) = \frac{1}{2}$, $B_1B_2B_3 = \{1\}$ and thus $\mathbf{P}(B_1B_2B_3) = \frac{1}{8} = \frac{1}{2}\cdot\frac{1}{2}\cdot\frac{1}{2} = \mathbf{P}(B_1)\mathbf{P}(B_2)\mathbf{P}(B_3)$. However, the events B_2 and B_3 are not independent and hence the three events are not mutually independent.

(iii) Let the space Ω be partitioned into five events, say A_1, A_2, A_3, A_4, A_5, such that $\mathbf{P}(A_1) = \mathbf{P}(A_2) = \mathbf{P}(A_3) = 15/64$, $\mathbf{P}(A_4) = 1/64$, $\mathbf{P}(A_5) = 18/64$. Define three new events, namely $B = A_1 \cup A_4$, $C = A_2 \cup A_4$, $D = A_3 \cup A_4$. Then $\mathbf{P}(B) = \mathbf{P}(C) = \mathbf{P}(D) = 1/4$, $\mathbf{P}(BCD) = 1/64$: that is, $\mathbf{P}(BCD) = \mathbf{P}(B)\mathbf{P}(C)\mathbf{P}(D)$. However, $\mathbf{P}(BC) = \mathbf{P}(A_4) = 1/64 \neq 1/16 = \mathbf{P}(B)\mathbf{P}(C)$ and hence the events B, C, D are not mutually independent.

3.4. A collection of $n+1$ dependent events such that any n of them are mutually independent

(i) A symmetric coin is tossed independently n times. Consider the events $A_k = \{\text{heads at the } k\text{th tossing}\}$, for $k = 1,\dots,n$ and $A_{n+1} = \{\text{the sum of the heads in these } n \text{ tossings is even}\}$. Then obviously

$$\mathbf{P}(A_1) = \frac{1}{2},\dots,\mathbf{P}(A_n) = \frac{1}{2},\ \mathbf{P}(A_{n+1}) = \frac{1}{2^n}\left[\binom{n}{0} + \binom{n}{2} + \cdots\right] = \frac{2^{n-1}}{2^n} = \frac{1}{2}.$$

It is easy to see that the conditional probability $\mathbf{P}(A_{n+1}|A_1 \dots A_n) = 1$ if n is even, and 0 if n is odd. This implies that the equality

$$\mathbf{P}(A_1 \dots A_n A_{n+1}) = \mathbf{P}(A_1)\cdots\mathbf{P}(A_n)\mathbf{P}(A_{n+1})$$

is impossible because the right-hand side is $2^{-(n+1)}$ and the left-hand side $\mathbf{P}(A_1 \dots A_n A_{n+1}) = \mathbf{P}(A_1 \dots A_n)\mathbf{P}(A_{n+1}|A_1 \dots A_n) = 2^{-n}$ if n is even, and 0 if n is odd. Therefore $A_1,\dots,A_n,A_{n+1}$ cannot be a collection of mutually independent events.

Now take any n of these events. If we have chosen $A_1, \ldots, A_n$, they are independent, since for any $A_{i_1}, \ldots, A_{i_k}$, $2 \le k \le n$ we have $\mathbf{P}(A_{i_1} \ldots A_{i_k}) = \mathbf{P}(A_{i_1}) \ldots \mathbf{P}(A_{i_k})$. It remains to consider the choice of n events including A_{n+1} and $n-1$ events taken from $A_1, \ldots, A_n$, for example $A_2, A_3, \ldots, A_n, A_{n+1}$. For their mutual independence it suffices to check that

$$(1) \qquad \mathbf{P}(A_{i_1} \ldots A_{i_m} A_{n+1}) = \mathbf{P}(A_{i_1}) \ldots \mathbf{P}(A_{i_m}) \mathbf{P}(A_{n+1})$$

where $1 \le m \le n-1$, $i_1, \ldots, i_m$ are among $2, \ldots, n$. We have $\mathbf{P}(A_{i_1}) = \ldots = \mathbf{P}(A_{i_m}) = \mathbf{P}(A_{n+1}) = \frac{1}{2}$ and thus the right-hand side of (1) is $2^{-(m+1)}$. Further,

$$\begin{aligned} \mathbf{P}(A_{i_1} \ldots A_{i_m} A_{n+1}) &= \mathbf{P}(A_{i_1} \ldots A_{i_m}) \mathbf{P}(A_{n+1} | A_{i_1} \ldots A_{i_m}) \\ &= 2^{-m} 2^{-1} = 2^{-(m+1)}. \end{aligned}$$

Thus (1) is satisfied and therefore any n events among the given $n+1$ events are mutually independent. In other words, the dependent $n+1$ events $A_1, \ldots, A_{n+1}$ are n-wise independent.

We can conclude that if we have $n+1$ events and any n of them are mutually independent, this does not always imply that the given events are mutually independent. Clearly this is a generalization of the Bernstein example (see Example 3.2(ii)).

(ii) We are given $n+1$ points in the plane, say $M_1, \ldots, M_{n+1}$, which are in a general position (no three of them lie in a straight line). Join up the points in pairs and obtain $\binom{n+1}{2}$ segments. Then we put a pointer to each of the segments by tossing a symmetric coin $\binom{n+1}{2}$ times: if we consider the segment $M_i M_j$, $i < j$, and the result of the tossing is heads, we put a pointer from M_i to M_j; if tails, the pointer goes from M_j to M_i. Consider $n+1$ events $A_1, \ldots, A_{n+1}$, where

$$A_k = \{\text{the number of pointers going to } M_k \text{ is even}\}, \; k = 1, \ldots, n+1.$$

Then for each k, $2 \le k \le n$ and any $1 \le i_1 < i_2 < \ldots < i_k \le n+1$, the events $A_{i_1}, A_{i_2}, \ldots, A_{i_k}$ are mutually independent. However $A_1, \ldots, A_{n+1}$ are dependent and so we have another collection of $n+1$ dependent events which are n-wise independent.

3.5. Collections of random events with 'unusual' independence/dependence properties

Let us describe a few probability models and collections of random events with specific properties.

(i) Suppose that the sample space of an experiment is $\Omega = \{1, 2, 3, 4, 5, 6, 7, 8\}$ with probabilities $p_k = \mathbf{P}(\{k\})$ defined as follows:

$$p_1 = \alpha, \; p_2 = p_3 = p_4 = \tfrac{7-16\alpha}{24}, \; p_5 = p_6 = p_7 = \tfrac{1+8\alpha}{24}, \; p_8 = \tfrac{1}{8}$$

where α is an arbitrary number in the interval $\left(0, \frac{7}{16}\right)$. Consider the events

$$A_1 = \{2,5,6,8\},\ A_2 = \{3,5,6,8\},\ A_3 = \{4,6,7,8\}.$$

We easily find that $\mathbf{P}(A_1) = \mathbf{P}(A_2) = \mathbf{P}(A_3) = \frac{1}{2}$ and then

$$\mathbf{P}(A_1A_2) = \mathbf{P}(A_1A_3) = \mathbf{P}(A_2A_3) = \tfrac{1+2\alpha}{6},\ \mathbf{P}(A_1A_2A_3) = \tfrac{1}{8}.$$

Hence the events A_1, A_2, A_3 are independent at level 3 for any value of the parameter $\alpha \in \left(0, \frac{7}{16}\right)$. If $\alpha = \frac{1}{4}$ they are independent at level 2 and this is the only case when these three events are mutually independent.

(ii) Let $\Omega = \{1,2,3,4,5,6\}$ with $p_1 = \frac{1}{16}$, $p_2 = p_3 = p_4 = p_5 = \frac{7}{48}$, $p_6 = \frac{17}{48}$. Consider the following events:

$$A_1 = \{1,2,3,4\},\ A_2 = \{1,2,3,5\},\ A_3 = \{1,2,4,5\},\ A_4 = \{1,3,4,5\}.$$

Then $\mathbf{P}(A_1) = \mathbf{P}(A_2) = \mathbf{P}(A_3) = \mathbf{P}(A_4) = \frac{1}{2}$ and we find further that

$$\mathbf{P}(A_iA_j) = \tfrac{17}{48},\ \text{all } i<j;\ \mathbf{P}(A_iA_jA_l) = \tfrac{5}{24},\ \text{all } i<j<l;\ \mathbf{P}(A_1A_2A_3A_4) = \tfrac{1}{16}.$$

Therefore these four events are independent at level 4 but they are dependent at level 2 and dependent at level 3.

(iii) Take a sample space Ω containing $|\Omega| = 16$ outcomes denoted by 1,2,...,16 each having the same probability $\frac{1}{16}$. Consider the events:

$$\begin{aligned} A &= \{2,3,4,5,6,9,13,16\}, & B &= \{4,7,8,10,11,13,14,16\}, \\ C &= \{4,6,7,8,10,11,13,14\}, & D &= \{3,4,5,6,9,10,15,16\}. \end{aligned}$$

Then $\mathbf{P}(A) = \mathbf{P}(B) = \mathbf{P}(C) = \mathbf{P}(D) = \frac{1}{2}$ and since $ABCD = \{4\}$ we have

$$\tfrac{1}{16} = \mathbf{P}(ABCD) = \mathbf{P}(A)\mathbf{P}(B)\mathbf{P}(C)\mathbf{P}(D)$$

and hence the product rule is satisfied at level 4. Further, $ABC = \{4, 13\}$ implying that

$$\tfrac{1}{8} = \mathbf{P}(ABC) = \mathbf{P}(A)\mathbf{P}(B)\mathbf{P}(C)$$

and similarly the product rule holds for any of the remaining five possible triplets of events. It turns out, however, that the product rule does not hold for any $6 = \binom{4}{2}$ possible pairs of events. In particular, $CD = \{4, 6, 10\}$ and

$$\tfrac{3}{16} = \mathbf{P}(CD) \neq \mathbf{P}(C)\mathbf{P}(D) = \tfrac{1}{4}.$$

Thus the events A, B, C, D are independent at level 4, independent at level 3 and (completely) dependent at level 2.

(iv) Suppose the space Ω consists of $|\Omega| = 12$ outcomes, say 1,2,...,12 with the following probabilities:

$$p_1 = \tfrac{1}{16},\ p_2 = p_3 = p_4 = p_5 = \tfrac{1}{24},$$
$$p_6 = p_7 = p_8 = p_9 = p_{10} = p_{11} = \tfrac{5}{48},\ p_{12} = \tfrac{7}{48}.$$

Define the events B_1, B_2, B_3, B_4 as follows:

$$B_1 = \{1,2,3,4,6,7,8\}, \quad B_2 = \{1,2,3,5,6,9,10\},$$
$$B_3 = \{1,2,4,5,7,9,11,\}, \quad B_4 = \{1,3,4,5,8,10,11\}.$$

Standard reasoning leads to the following conclusion: the events B_1, B_2, B_3, B_4 are independent at level 4, dependent at level 3 and independent at level 2. (The details are left to the reader.)

3.6. Is there a relationship between conditional and mutual independence of random events?

The random events $A_1, A_2, \dots, A_n$ are called *conditionally independent* given event B with $\mathbf{P}(B) > 0$ if

$$\mathbf{P}(A_1 A_2 \dots A_n | B) = \mathbf{P}(A_1|B)\mathbf{P}(A_2|B)\dots\mathbf{P}(A_n|B).$$

We want to examine the relationship between the two concepts mutual independence and conditional independence.

(i) Suppose we have at our disposal two coins, say a and b. Let p_a and p_b, $p_a \neq p_b$, be the probabilities of heads for a and b respectively. Select a coin at random and toss it twice. Consider the events $A_1 = \{\text{heads at the first tossing}\}$, $A_2 = \{\text{heads at the second tossing}\}$ and $B = \{\text{coin } a \text{ is selected}\}$. Then $\mathbf{P}(A_1 A_2|B) = p_a p_b$, $\mathbf{P}(A_1|B) = p_a$, $\mathbf{P}(A_2|B) = p_a$. Hence $\mathbf{P}(A_1 A_2|B) = \mathbf{P}(A_1|B)\mathbf{P}(A_2|B)$, and the events A_1 and A_2 are conditionally independent given B. However,

$$\mathbf{P}(A_1 A_2) = \tfrac{1}{2}p_a^2 + \tfrac{1}{2}p_b^2, \quad \mathbf{P}(A_1) = \tfrac{1}{2}(p_a + p_b), \quad \mathbf{P}(A_2) = \tfrac{1}{2}(p_a + p_b)$$

and since $p_a \neq p_b$ the equality $\mathbf{P}(A_1 A_2) = \mathbf{P}(A_1)\mathbf{P}(A_2)$ is not satisfied.

Therefore the events A_1 and A_2 are not independent, despite their conditional independence.

(ii) A symmetric coin is tossed twice. Consider the events $A_k = \{\text{heads at } k\text{th tossing}\}$, $k = 1, 2$ and $B = \{\text{at least one tails}\}$. Then $\mathbf{P}(A_1) = \mathbf{P}(A_2) = \frac{1}{2}$, $\mathbf{P}(A_1 A_2) = \frac{1}{4}$ and hence the events A_1 and A_2 are independent. Further, it is easy to see that $\mathbf{P}(A_1|B) = \mathbf{P}(A_2|B) = \frac{1}{3}$. However $\mathbf{P}(A_1 A_2|B) = 0$ and (1) fails to hold. Therefore the independent events A_1 and A_2 are not conditionally independent given B.

The final conclusion is that there is no relationship between conditional independence and mutual independence, that is neither one of these properties implies the other. (See also Example 7.14.)

3.7. Independence type conditions which do not imply the mutual independence of a set of events

Suppose the random events $A_1, A_2, \dots, A_n$ satisfy the conditions

$$(1) \qquad \mathbf{P}(A_k) = p_k, \quad \mathbf{P}(A_1 A_2 \dots A_k) = p_1 p_2 \dots p_k, \quad k = 1, 2, \dots, n$$

which could be called independence-type conditions. In (1) $p_1, \dots, p_n$ are arbitrary numbers in the interval (0,1).

Obviously, if $n = 2$ and (1) is satisfied, this is merely the definition of the independence of two random events A_1 and A_2. We ask the following question: does (1) imply, in the general case when $n \geq 3$, that the given events are mutually independent? Of course, it is clear that (1) is much less than the standard condition for mutual independence. Thus we can expect that the answer to this question is negative. Let us illustrate the truth of this with the following example, considering for simplicity the case $n = 3$.

Suppose A_1, A_2, A_3 are random events such that

$$\begin{aligned} &\mathbf{P}(A_1 A_2 A_3) = \mathbf{P}(A_1 A_2 A_3^c) = \tfrac{1}{8}, \qquad \mathbf{P}(A_1 A_2^c A_3) = \mathbf{P}(A_1^c A_2 A_3) = \tfrac{1}{8} - \varepsilon, \\ &\mathbf{P}(A_1 A_2^c A_3^c) = \mathbf{P}(A_1^c A_2 A_3^c) = \tfrac{1}{8} + \varepsilon, \quad \mathbf{P}(A_1^c A_2^c A_3) = \tfrac{1}{8} + 2\varepsilon, \\ &\mathbf{P}(A_1^c A_2^c A_3^c) = \tfrac{1}{8} - 2\varepsilon \end{aligned}$$

where $0 < \varepsilon \leq \frac{1}{6}$. We can easily check that

$$\mathbf{P}(A_1) = \mathbf{P}(A_2) = \mathbf{P}(A_3) = \tfrac{1}{2}, \quad \mathbf{P}(A_1 A_2) = \tfrac{1}{4}, \quad \mathbf{P}(A_1 A_2 A_3) = \tfrac{1}{8}$$

and thus the conditions in (1) hold. For the mutual independence of A_1, A_2, A_3 the equalities $\mathbf{P}(A_1 A_3) = \mathbf{P}(A_1)\mathbf{P}(A_3)$ and $\mathbf{P}(A_2 A_3) = \mathbf{P}(A_1)\mathbf{P}(A_3)$ must also be satisfied. However,

$$\mathbf{P}(A_1 A_3) = \tfrac{1}{4} - \varepsilon \neq \tfrac{1}{4} = \mathbf{P}(A_1)\mathbf{P}(A_3).$$

Hence the independence-type conditions (1) are satisfied for the events A_1, A_2 and A_3, but these events are not mutually independent.

3.8. Mutually independent events can form families which are strongly dependent

Choose a number x at random in the interval [0,1] and consider the expansions of x in bases $2, 3, \dots$. Denote by $\mathcal{A}_k$, $k = 2, 3, \dots$, the family of sets $A_m^{(k)}$, $m = 1, 2, \dots$, containing all points x whose nth digit in the expansion in base k is equal to zero. Then for every fixed k the events $A_m^{(k)}$, $m = 1, 2, \dots$, are mutually independent. This is easily checked, but for details see Neuts (1973) or Billingsley (1995).

We want to know whether the families $\mathcal{A}_k$, $k = 2, 3, \dots$, are independent. To see this, take the events

$$A_1^{(2)}, A_1^{(3)}, A_1^{(4)}, \dots$$

which are representatives of the families $\mathcal{A}_2, \mathcal{A}_3, \mathcal{A}_4, \ldots$ respectively. On the one hand, for any $n > 2$,

$$\mathbf{P}\left(\bigcap_{k=2}^{n} A_1^{(k)}\right) = \mathbf{P}\left(x < \frac{1}{n}\right) = \frac{1}{n}$$

because the first digit in the number base k is 0 iff $x < 1/k$ for $k = 2, 3, \ldots, n$. However, on the other hand,

$$\prod_{k=1}^{n} \mathbf{P}(A_1^{(k)}) = \prod_{k=2}^{n} \frac{1}{k} = \frac{1}{n!} \neq \frac{1}{n} = \mathbf{P}\left(\bigcap_{k=2}^{n} A_1^{(k)}\right).$$

Therefore the families $\mathcal{A}_k$, $k = 2, 3, \ldots$, are not independent, although they are generated by mutually independent events.

3.9. Independent classes of random events can generate σ-fields which are not independent

Let $(\Omega, \mathcal{F}, \mathbf{P})$ be the standard probability space: $\Omega = [0, 1]$, $\mathcal{F}$ is the Borel σ-field $\mathcal{B}_{[0,1]}$ generated by the subsets of Ω and $\mathbf{P}$ is the Lebesgue measure. Consider the following two classes of random events: $\mathcal{A}_1 = \{A_{11}, A_{12}\}$ and $\mathcal{A}_2 = \{A_2\}$ where

$$A_{11} = \left[0, \tfrac{1}{4}\right) \cup \left[\tfrac{1}{2}, \tfrac{3}{4}\right), \quad A_{12} = \left[0, \tfrac{1}{3}\right) \cup \left[\tfrac{2}{3}, 1\right), \quad A_2 = \left[0, \tfrac{1}{2}\right).$$

Then $\mathbf{P}(A_{11}) = \frac{1}{2}$, $\mathbf{P}(A_{12}) = \frac{2}{3}$, $\mathbf{P}(A_2) = \frac{1}{2}$. Hence

$$\mathbf{P}(A_{11}A_2) = \tfrac{1}{4} = \tfrac{1}{2} \cdot \tfrac{1}{2} = \mathbf{P}(A_{11})\mathbf{P}(A_2),$$
$$\mathbf{P}(A_{12}A_2) = \tfrac{1}{3} = \tfrac{2}{3} \cdot \tfrac{1}{2} = \mathbf{P}(A_{12})\mathbf{P}(A_2).$$

Therefore the classes $\mathcal{A}_1$ and $\mathcal{A}_2$ are independent.

It is easy to see that the σ-fields $\sigma(\mathcal{A}_1)$ and $\sigma(\mathcal{A}_2)$ generated by $\mathcal{A}_1$ and $\mathcal{A}_2$ are not independent. E.g. if $A_1 = A_{11}A_{12}$, then $\mathbf{P}(A_1) = \frac{1}{3}$, $A_1 A_2 = \left[0, \frac{1}{4}\right)$ and

$$\mathbf{P}(A_1 A_2) = \tfrac{1}{4} \neq \tfrac{1}{3} \cdot \tfrac{1}{2} = \mathbf{P}(A_1)\mathbf{P}(A_2).$$

A similar example can be given in the discrete case. It is enough to take e.g. the sample space $\Omega = \{1, 2, 3, 4\}$ with equally likely outcomes and two classes $\mathcal{A}_1$ and $\mathcal{A}_2$ where $\mathcal{A}_1$ contains one of the outcomes of Ω and $\mathcal{A}_2$ contains two of them. A simple calculation leads to a conclusion like that presented above.

Let us note finally that $\sigma(\mathcal{A}_1)$ and $\sigma(\mathcal{A}_2)$ would be independent if each of $\mathcal{A}_1$ and $\mathcal{A}_2$ were a π-system, i.e. $\Omega \in \mathcal{A}_i$ and $\mathcal{A}_i$, $i = 1, 2$, is closed under intersection.

SECTION 4. DIVERSE PROPERTIES OF RANDOM EVENTS AND THEIR PROBABILITIES

Here we introduce and analyse some other properties of random events and probabilities. The corresponding definitions are given in the examples themselves. This section is a natural continuation and an extension of the ideas treated in the previous sections.

4.1. Probability spaces without non-trivial independent events: totally dependent spaces

Let $(\Omega, \mathcal{F}, \mathbf{P})$ be a probability space. Recall that the events $A, B \in \mathcal{F}$ are non-trivial and independent if $0 < \mathbf{P}(A) < 1$, $0 < \mathbf{P}(B) < 1$ and $\mathbf{P}(AB) = \mathbf{P}(A)\mathbf{P}(B)$. One might think that every probability space contains non-trivial independent events. However, this conjecture is false.

(i) Let Ω be a finite set, $\Omega = \{\omega_1, \dots, \omega_n\}$ and

$$\mathbf{P}(\{\omega_1\}) = 1 - (n-1)\varepsilon, \quad \mathbf{P}(\{\omega_2\}) = \mathbf{P}(\{\omega_3\}) = \dots = \mathbf{P}(\{\omega_n\}) = \varepsilon$$

where ε is an irrational number, $0 < \varepsilon < (n-1)^{-1}$. Suppose there exists a pair A, B of non-trivial independent events. We have the following three possibilities: (1) $\omega_1 \notin A$, $\omega_1 \notin B$; (2) $\omega_1 \notin A$, $\omega_1 \in B$, or conversely; (3) $\omega_1 \in A$, $\omega_1 \in B$. We can easily verify that the independence condition is not satisfied in any of the cases (1), (2) or (3). For example, consider case (2). Here A contains some k outcomes taken from $\omega_2, \dots, \omega_n$ and B consists of ω_1 and some l outcomes taken from $\omega_2, \dots, \omega_n$. Then the intersection AB contains elements taken only from $\omega_2, \dots, \omega_n$. Let their number be m, $m < k$. We obtain the following equality:

$$m\varepsilon = [1 - (n-1)\varepsilon + l\varepsilon]k\varepsilon.$$

It follows that $\varepsilon = (k-m)/[k(n-1-l)]$, which contradicts the assumption that ε is irrational.

Similar reasoning can be used in cases (1) and (3). Therefore, in this example non-trivial independent events do not exist. Moreover, it can be shown that more than two non-trivial events also do not exist. Notice that here Ω is a finite set.

(ii) In case (i) Ω was a finite set. Let now Ω be a countably infinite set, $\Omega = \{\omega_1, \omega_2, \dots\}$, and let

$$\mathbf{P}(\{\omega_k\}) = 2^{-k!},\ k = 2, 3, \dots, \quad \mathbf{P}(\{\omega_1\}) = \varepsilon \text{ with } \varepsilon = 1 - \sum_{k=2}^{\infty} \mathbf{P}(\{\omega_k\}).$$

Note that the latter infinite series is convergent and its sum ε is a number in (0,1) and it is crucial for further reasoning that ε is an irrational number (in fact, ε is also transcendental; ε is a Liouville number). It can be shown that any finite or infinite collection of arbitrarily composed random events is totally dependent.

(iii) In cases (i) and (ii) above we have described probability spaces with total dependence of their events, no matter how they are defined. In such a case we use the term *totally dependent probability space*. Notice, however, that in (i) and (ii) Ω is a discrete set and the probability measure $\mathbf{P}$ is purely discrete. Hence there are purely discrete probability spaces which are totally dependent. This immediately leads to the question: is it possible for a non-purely discrete probability space to be totally dependent?

Recall that 'non-purely discrete' means that $\mathbf{P}$ is not just a sum of 'atoms', as in cases (i) and (ii) above. Now we assume that there is a subset $\Omega_c \subset \Omega$ with $\mathbf{P}(\Omega_c) > 0$ and such that the restriction $\mathbf{P}|\Omega_c$ of $\mathbf{P}$ on Ω_c is non-atomic: $\mathbf{P}(\{\omega\}) = 0$ for each $\omega \in \Omega_c$. Let $\mathbf{P}(\Omega_c) = c$ where obviously $0 < c \leq 1$. Let us clarify if such a space can be totally dependent. For this we need the following result known as the Lyapunov theorem (Rudin 1973): For any b, $0 \leq b \leq c$ there is a subset (event) $D \subset \Omega_c$ such that $\mathbf{P}(D) = b$.

Let now p be a fixed number, $0 < p < c$. As a consequence of the above cited result we can find three events, say D_1, D_2, D_3, which are pairwise disjoint and such that

$$\mathbf{P}(D_1) = p^2, \ \mathbf{P}(D_2) = p(1-p), \ \mathbf{P}(D_3) = p(1-p)$$

(the measure of $D_1 \cup D_2 \cup D_3$ is $p - p^2 < c$). Define the events

$$A = D_1 \cup D_2 \text{ and } B = D_1 \cup D_3.$$

Obviously $\mathbf{P}(A) = p$, $\mathbf{P}(B) = p$ and since $AB = D_1$ where $\mathbf{P}(D_1) = p^2$, we get

$$\mathbf{P}(AB) = \mathbf{P}(A)\mathbf{P}(B)$$

and A and B are non-trivial events.

Therefore a non-purely discrete probability space (the measure $\mathbf{P}$ has a 'continuous' part) cannot be totally dependent.

Notice that the examples of Bernstein, Bohlmann and their inverses (Example 3.2) are purely discrete. They all can be realized on probability spaces which are non-purely discrete, that is, on spaces with at least a partially 'continuous' part.

4.2. On the Borel–Cantelli lemma and its corollaries

Let $\{A_n, n \geq 1\}$ be a sequence of events in the probability space $(\Omega, \mathcal{F}, \mathbf{P})$. Define the event $A^* = \bigcap_{n=1}^{\infty} \bigcup_{k=n}^{\infty} A_k$. Then $A^* = \{A_n \text{ i.o.}\}$: that is, infinitely many A_n occur (i.o. means infinitely often). The following result (the **Borel–Cantelli lemma**) can be found in almost all textbooks on probability theory.

(a) If $\sum_{n=1}^{\infty} \mathbf{P}(A_n) < \infty$, then $\mathbf{P}[A_n \text{ i.o.}] = 0$.
(b) If $\sum_{n=1}^{\infty} \mathbf{P}(A_n) = \infty$ and $A_1, A_2, \ldots$ are independent, then $\mathbf{P}[A_n \text{ i.o.}] = 1$.

We show by an example that in general the converse of (a) is not true, and that the independence condition in (b) is essential.

Let $\Omega = [0,1]$, $\mathcal{F} = \mathcal{B}_{[0,1]}$ and let $\mathbf{P}$ be the Lebesgue measure. Consider the following sequence of events: $A_n = [0, 1/n]$, $n = 1, 2, \ldots$. Then obviously we have $A_n \downarrow$ in n as $n \to \infty$, $[A_n \text{ i.o.}] = \bigcap_{n=1}^{\infty} A_n = \emptyset$, so that $\mathbf{P}[A_n \text{ i.o.}] = 0$. However, $\sum_{n=1}^{\infty} \mathbf{P}(A_n) = \sum_{n=1}^{\infty}(1/n) = \infty$. It follows that the converse of (a) is not true. Looking at (b), we see that the condition $\sum_{n=1}^{\infty} \mathbf{P}(A_n) = \infty$ does not imply that $\mathbf{P}[A_n \text{ i.o.}] = 1$ and thus the independence of $A_1, A_2, \ldots$ is essential.

4.3. When can a set of events be both exhaustive and independent?

Let $(\Omega, \mathcal{F}, \mathbf{P})$ be a probability space and $\{A_i, i \in \Lambda\}$ a non-empty set of events. (Λ denotes some non-empty index set.) This set is called *independent* if for any $k \geq 2$ and any subset $A_{i_k}, \ldots, A_{i_k}$, where $i_1, \ldots, i_k \in \Lambda$,

$$\mathbf{P}\left(\bigcap_{j=1}^{k} A_{i_j}\right) = \prod_{j=1}^{k} \mathbf{P}(A_{i_j}).$$

The set is called *exhaustive* if

$$\mathbf{P}\left(\bigcup_{i \in \Lambda} A_i\right) = 1.$$

The following question arises naturally: is it possible for the set $\{A_i\}$ to be both exhaustive and independent? The answer will be given for two cases: when the set $\{A_i\}$ consists of a finite or of an infinite number of events.

(i) Let the index set Λ be finite. Suppose $\{A_i, i \in \Lambda\}$ is an independent set. Then so is the set $\{A_i^c, i \in \Lambda\}$, and

$$\mathbf{P}\left(\bigcup_{i \in \Lambda} A_i\right) = 1 - \mathbf{P}\left(\bigcap_{i \in \Lambda} A_i^c\right) = 1 - \prod_{i \in \Lambda} \mathbf{P}(A_i^c). \tag{1}$$

Obviously, if for all $i \in \Lambda$, $\mathbf{P}(A_i^c) > 0$, then the right-hand side of (1) becomes less than 1 and $\{A_i, i \in \Lambda\}$ cannot be exhaustive. However, if for some i we have $\mathbf{P}(A_i^c) = 0$, this means that $\mathbf{P}(A_i) = 1$ and $A_i = \Omega$. Therefore in this trivial case only (compare Example 4.1) the finite set of independent events can be exhaustive. Of course, a finite set $\{A_i, i \in \Lambda\}$ can be exhaustive without being independent.

(ii) Here the index set $\Lambda = \mathbb{N}$. We shall construct two different sets of independent events such that one of them is exhaustive and the other is not.

Choose at random a number $x \in [0, 1]$. Let A_i be the event that the ith bit in the binary expansion of x is zero. It is easy to check that $A_1, A_2, \ldots$ are independent and moreover $\mathbf{P}(A_i) = \frac{1}{2}$ for each i. Thus $\sum_{i=1}^{\infty} \mathbf{P}(A_i) = \infty$ and, according to the Borel–Cantelli lemma (see Example 4.2), $\mathbf{P}(\bigcup_{i=1}^{\infty} A_i) = 1$. Hence the set $\{A_i, i \geq 1\}$ is both independent and exhaustive.

Consider now another set $\{B_i, i \geq 1\}$ defined by

$$B_i = \bigcap_{j=r}^{s} A_j \qquad \text{where } r = \tfrac{1}{2}i(i-1)+1, \quad s = \tfrac{1}{2}i(i+1).$$

B_1 is the event that the first bit in the binary expansion of x is zero, B_2 that the second and the third bits are zero, B_3 that the next three bits are zero, and so on. Since $\mathbf{P}(B_i) = 2^{-i}$, we have $\sum_{i=1}^{\infty} \mathbf{P}(B_i) < \infty$ and $\mathbf{P}(\bigcup_{i=1}^{\infty} B_i) < 1$. Hence $\{B_i, i \geq 1\}$ is a set of independent events which, however, is not exhaustive.

4.4. How are independence and exchangeability related?

Let us consider a finite collection of random events $\mathcal{A}_n = \{A_1, \ldots, A_n\}$, $n \geq 2$ in probability spaces. $\mathcal{A}_n$ is said to be *exchangeable* (also symmetrically dependent) if the probability $\mathbf{P}(A_{i_1} \ldots A_{i_k})$ is the same for all possible choices of k events, $k \geq 1$, $1 \leq i_1 < i_2 < \ldots < i_k \leq n$. In other words, there are numbers $p_1, p_2, \ldots, p_{k-1}$, all in (0,1), such that

$$\mathbf{P}(A_j) = p_1 \text{ for all } j; \ \mathbf{P}(A_i A_j) = p_2 \text{ for all } i < j;$$
$$\mathbf{P}(A_i A_j A_l) = p_3 \text{ for all } i < j < l \text{ etc.}$$

Like the independence property we can introduce the term *exchangeability at level k* for a fixed k meaning that $\mathbf{P}(A_{i_1} \ldots A_{i_k})$ is the same for all choices of just k events from $\mathcal{A}_n$ regardless of what happens at levels higher than k, and lower than k. It turns out the collection $\mathcal{A}_n$ can be such that exchangeability property does not hold for others. Thus $\mathcal{A}_n$ is *totally exchangeable* (or simply *exchangeable*) if it obeys this property at all levels k, $k = 1, 2, \ldots, n-1$ (for $k = n$ we have only one event, namely $A_1 A_2 \ldots A_n$).

It is easy to see that if $\mathcal{A}_n$ is exchangeable at level 1 ($\mathbf{P}(A_1) = \ldots = \mathbf{P}(A_n) = p_1$) and $\mathcal{A}_n$ is mutually independent, then obviously $\mathcal{A}_n$ is totally exchangeable (now $\mathbf{P}(A_i A_j) = p_1^2$, all $i < j$; $\mathbf{P}(A_i A_j A_l) = p_1^3$, all $i < j < l$ etc.). If, however, $\mathcal{A}_n$ is mutually independent but there are different numbers among $\mathbf{P}(A_1), \ldots, \mathbf{P}(A_n)$, then $\mathcal{A}_n$ is not exchangeable at all.

We can return back to Example 3.5 and derive additional conclusions about the validity of the exchangeability property (total or partial, only at some levels).

Let us turn to another example.

Suppose we have at our disposal 192 cards on which in a special way numbers are written such that: 110 cards are marked by a 'triplet' (each of 123, 124, 125, 134, 135, 145, 234, 235, 245, 345 is written 11 times); 30 cards are marked by a 'quartet' (each of 1234, 1235, 1245, 1345, 2345 is written six times); six cards are marked by the 'quintet' 12345; the remaining 46 cards are blank. All 192 cards are put into a box and well mixed. We are interested in the following five events:

$$A_i = \{\text{randomly chosen card contains the number } i\}, \; i = 1, 2, 3, 4, 5.$$

It is easy to check that for all possible indices i, j, l, s we have:

$$\mathbf{P}(A_1) = \mathbf{P}(A_2) = \mathbf{P}(A_3) = \mathbf{P}(A_4) = \mathbf{P}(A_5) = \tfrac{1}{2};$$
$$\mathbf{P}(A_iA_j) = \tfrac{17}{64}, \; i < j; \quad \mathbf{P}(A_iA_jA_l) = \tfrac{23}{192}, \; i < j < l;$$
$$\mathbf{P}(A_iA_jA_lA_s) = \tfrac{1}{16}, \; i < j < l < s; \quad \mathbf{P}(A_1A_2A_3A_4A_5) = \tfrac{1}{32}.$$

Thus we arrive at the following two conclusions for these five events, namely: (a) they are dependent at level 2, dependent at level 3, independent at level 4 and independent at level 5; (b) they are totally exchangeable.

The final conclusion is that these two properties of random events, independence and exchangeability, are not related.

4.5. A sequence of random events which is stable but not mixing

Let $(\Omega, \mathcal{F}, \mathbf{P})$ be a probability space and $\{A_i, i \geq 1\}$, $A_n \in \mathcal{F}$ a sequence of events such that for every $B \in \mathcal{F}$,

$$\lim_{n\to\infty} \mathbf{P}(A_nB) = \lambda\mathbf{P}(B)$$

where λ is a constant not depending on B, $0 < \lambda < 1$. Then $\{A_n\}$ is called a *mixing sequence* with density λ (see Rényi 1970). In this case it is usual to speak about mixing in the sense of ergodic theory (see Doukhan 1994).

The mixing property can be extended as follows. The sequence $\{A_n\}$ is called a stable sequence of events if for any $B \in \mathcal{F}$ the following limit exists

$$\lim_{n\to\infty} \mathbf{P}(A_nB) = \mathbf{Q}(B).$$

According to Rényi (1970), $\mathbf{Q}$ is a measure on $\mathcal{F}$ which is absolutely continuous with respect to $\mathbf{P}$. The Radon–Nikodym derivative $\mathrm{d}\mathbf{Q}/\mathrm{d}\mathbf{P} = \alpha(\omega)$ exists and for every $B \in \mathcal{F}$, $\mathbf{Q}(B) = \int_B \alpha(\omega)\,\mathrm{d}\mathbf{P}$. Here $0 \leq \alpha(\omega) \leq 1$ with probability 1. The r.v. α is called a (local) density of the stable sequence $\{A_n\}$.

If $\alpha = \lambda =$ constant a.s., $0 < \lambda < 1$, clearly the stable sequence $\{A_n\}$ is mixing and has density λ. However, if α is not a constant, the stable sequence $\{A_n\}$ cannot be mixing. Let us illustrate this statement by an example.

In the probability space $(\Omega, \mathcal{F}, \mathbf{P})$ let $B_1 \in \mathcal{F}$, $0 < \mathbf{P}(B_1) < 1$ and $B_2 = B_1^{\mathrm{c}}$. Consider two spaces, $(\Omega, \mathcal{F}, \mathbf{P}_1)$ and $(\Omega, \mathcal{F}, \mathbf{P}_2)$ where

$$\mathbf{P}_1(A) = \mathbf{P}(A|B_1), \quad \mathbf{P}_2(A) = \mathbf{P}(A|B_2) \;\text{ for each }\; A \in \mathcal{F}.$$

Suppose that $\{A'_n\}$ is a mixing sequence in $(\Omega, \mathcal{F}, \mathbf{P}_1)$ with density λ_1 and $\{A''_n\}$ a mixing sequence in $(\Omega, \mathcal{F}, \mathbf{P}_2)$ with density λ_2 where $0 < \lambda_1 < \lambda_2 < 1$. Put $A_n = A'_n B_1 + A''_n B_2$. Then for every $B \in \mathcal{F}$ we have

$$\mathbf{P}(A_n B) = \mathbf{P}(B_1)\mathbf{P}_1(A'_n B) + \mathbf{P}(B_2)\mathbf{P}_2(A''_n B)$$

and hence

$$\lim_{n\to\infty} \mathbf{P}(A_n B) = \mathbf{Q}(B) \text{ where } \mathbf{Q}(B) = \lambda_1 \mathbf{P}(BB_1) + \lambda_2 \mathbf{P}(BB_2).$$

Define the r.v. $\alpha = \alpha(\omega)$ as follows: $\alpha(\omega) = \lambda_1$ if $\omega \in B_1$, and $\alpha(\omega) = \lambda_2$ if $\omega \in B_2$. Then

$$\mathbf{Q}(B) = \int_B \alpha(\omega)\, \mathrm{d}\mathbf{P}.$$

It follows that the sequence $\{A_n\}$ is stable but not mixing, since its density is not constant but takes two different values with positive probabilities.

As noted by Rényi (1970), in a similar way we can construct a stable sequence of events such that its density has an arbitrarily prescribed discrete distribution.

Chapter 2

Random Variables and Basic Characteristics

Courtesy of Professor A. T. Fomenko of Moscow University.

SECTION 5. DISTRIBUTION FUNCTIONS OF RANDOM VARIABLES

Let $F(x)$, $x \in \mathbb{R}^1$ be a function satisfying the conditions:

(a) F is non-decreasing, that is $x_1 < x_2 \Rightarrow F(x_1) < F(x_2)$;
(b) F is right-continuous and has left-hand limits at each $x \in \mathbb{R}^1$, that is $F(x+) := \lim_{u \downarrow x} F(u) = F(x)$;
(c) $\lim_{x \to -\infty} F(x) = 0, \ \lim_{x \to \infty} F(x) = 1$.

Any function F satisfying conditions (a), (b) and (c) above is said to be a **distribution function** (d.f.).

Now let $(\Omega, \mathcal{F}, \mathbf{P})$ be a probability space. Denote by $\mathcal{B}^1$ the Borel σ-field of the real line $\mathbb{R}^1 = (-\infty, \infty)$. Recall that any measurable function $X : (\Omega, \mathcal{F}) \mapsto (\mathbb{R}^1, \mathcal{B}^1)$ is called a *random variable* (r.v.). By the equality $P_X(B) = \mathbf{P}(X^{-1}(B))$, $B \in \mathcal{B}^1$ we define a probability measure on $\mathcal{B}^1$. Using the properties of the probability $\mathbf{P}$ (see the introductory notes to Section 3 we can easily show that the function

$$F_X(x) = P_X((-\infty, x]), \quad x \in \mathbb{R}^1$$

satisfies the above conditions (a), (b) and (c) and hence F_X is a d.f. In such a case we say that F_X is the d.f. of the r.v. X.

If there is a countable set of numbers $x_1, x_2, \ldots$ (finite or infinite) such that $F_X(x_n) - F_X(x_n-) := p_n > 0$, $\sum_n p_n = 1$, then the d.f. F_X is called *discrete*. The probability measure P_X is also discrete, that is P_X is concentrated at the points $x_1, x_2, \ldots$, called *atoms*, and $P_X(\{x_n\}) = F_X(x_n) - F_X(x_n-) > 0$, $\sum_n P_X(\{x_n\}) = 1$. The set $\{p_1, p_2, \ldots\}$ is called a discrete probability distribution and X a discrete r.v. with values in the set $\{x_1, x_2, \ldots\}$ and with a distribution $\{p_1, p_2, \ldots\}$. Clearly $\mathbf{P}[X = x_n] = p_n$, $n = 1, 2, \ldots$.

The d.f. F_X is said to be *absolutely continuous* if there is a non-negative and integrable function $f(x)$, $x \in \mathbb{R}^1$ such that

$$F_X(x) = \int_{-\infty}^{x} f(u)\, du \quad \text{for all } x \in \mathbb{R}^1.$$

Here f is called a *probability density function* (simply, *density*) of the d.f. F_X as well as of the r.v. X and of the measure P_X.

Let us note that there are measures whose d.f.s are continuous but have their points of increase on sets of zero Lebesgue measure. Such measures and distributions are called singular. They will not be treated here. The interested reader is referred to the books by Feller (1971), Rao (1984), Billingsley (1995) and Shiryaev (1995).

Now we shall define the multi-dimensional d.f. For the n-dimensional random vector $(X_1, \ldots, X_n)$, consider the function

$$G(x_1, \ldots, x_n) = \mathbf{P}[X_1 \le x_1, \ldots, X_n \le x_n], \qquad (x_1, \ldots, x_n) \in \mathbb{R}^n.$$

It is easy to derive the following properties of G:

(a_1) $G(x_1, \dots, x_n)$ is non-decreasing in each of its arguments;
(b_1) $G(x_1, \dots, x_n)$ is right-continuous in each of its arguments;
(c_1) $G(x_1, \dots, x_n) \to 0$ as $x_j \to -\infty$ for at least one j;
$G(x_1, \dots, x_n) \to 1$ as $x_j \to \infty$ for all $i = 1, \dots, n$;
(d) if $a_i \le b_i$, $i = 1, \dots, n$ and

$$\Delta_{a_i,b_i} G(x_1, \dots, x_n) \\ = G(x_1, \dots, x_{i-1}, b_i, x_{i+1}, \dots, x_n) - G(x_1, \dots, x_{i-1}, a_i, x_{i+1}, \dots, x_n)$$

then

$$\Delta_{a_1,b_1} \Delta_{a_2,b_2} \dots \Delta_{a_n,b_n} G(x_1, \dots, x_n) \ge 0.$$

Any function G satisfying conditions (a_1), (b_1), (c_1) and (d) is called an n-*dimensional d.f.* Actually G is the d.f. of the given random vector.

Analogously to the one-dimensional case we can define the notion of a discrete multi-dimensional d.f. Further, we say that the d.f. G and the random vector $(X_1, \dots, X_n)$ are absolutely continuous with density $g(x_1, \dots, x_n)$, $(x_1, \dots, x_n) \in \mathbb{R}^n$ if

$$G(x_1, \dots, x_n) = \int_{-\infty}^{x_1} \dots \int_{-\infty}^{x_n} g(u_1, \dots, u_n) \, du_1 \dots du_n$$

for all $(x_1, \dots, x_n) \in \mathbb{R}^n$. Here g is non-negative and its integral over $\mathbb{R}^n$ is 1.

The *marginal d.f.* $G_i(x_i) = \mathbf{P}[X_i \le x_i]$ is obtained from G in an obvious way, putting $x_j = \infty$ for $j \ne i$. If we integrate $g(x_1, \dots, x_n)$ in the arguments $x_1, \dots, x_{i-1}, x_{i+1}, \dots, x_n$ each in $\mathbb{R}^1$, we obtain the function $g_i(x_i)$ which is the marginal density of X_i, $i = 1, \dots, n$.

We say that the r.v.s X_1 and X_2 are *independent* if

$$\mathbf{P}[X_1 \in B_1, X_2 \in B_2] = \mathbf{P}[X_1 \in B_1]\mathbf{P}[X_2 \in B_2]$$

for any Borel sets B_1 and B_2 (that is, $B_1, B_2 \in \mathcal{B}^1$). Analogously to the case of random events we can introduce the notion of mutual and pairwise independence of a finite collection of r.v.s. If $X_1, \dots, X_n$ are n r.v.s with d.f.s $F_1, \dots, F_n$ respectively, and $F(x_1, \dots, x_n)$ is their joint d.f., then these variables are *mutually independent* (simply independent) iff

$$F(x_1, \dots, x_n) = F_1(x_1) \dots F_n(x_n), \quad x_1, \dots, x_n \in \mathbb{R}^1.$$

In terms of the corresponding densities the independence of the r.v.s $X_1, \dots, X_n$ is expressed in the form

$$f(x_1, \dots, x_n) = f_1(x_1) \dots f_n(x_n), \quad x_1, \dots, x_n \in \mathbb{R}^1.$$

Let us now define the unimodality property for an absolutely continuous d.f. F: $F(x)$, $x \in \mathbb{R}^1$ is said to be *unimodal* with its *mode* (or vertex) at the point $x_0 \in \mathbb{R}^1$ if $F(x)$ is convex for $x < x_0$ and concave for $x > x_0$.

For a detailed description of the properties of one-dimensional and multi-dimensional d.f.s we refer the reader to the works of Feller (1971), Chow and Teicher (1978), Laha and Rohatgi (1979) and Shiryaev (1995).

5.1. Equivalent random variables are identically distributed but the converse is not true

Consider two r.v.s X and Y on the same probability space $(\Omega, \mathcal{F}, \mathbf{P})$ and suppose they are equivalent, that is $\mathbf{P}\{\omega{:}X(\omega) \neq Y(\omega)\} = 0$. Hence

$$F_X(x) = \mathbf{P}\{\omega{:}X(\omega) \leq x\} = \mathbf{P}\{\omega{:}Y(\omega) \leq x\} = F_Y(x) \text{ for each } x \in \mathbb{R}^1.$$

Thus X and Y are identically distributed. In such a case we use the following notation: $X \stackrel{\mathrm{d}}{=} Y$.

To see that the converse is not true, take the r.v. X which is absolutely continuous and symmetric with respect to the origin. Let $Y = -X$. Then the symmetry of X implies that $F_X = F_Y$. Further, as a consequence of the absolute continuity of X, we obtain

$$\mathbf{P}\{\omega : X(\omega) = Y(\omega)\} = \mathbf{P}\{\omega : X(\omega) = -X(\omega)\} = \mathbf{P}\{\omega : X(\omega) = 0\} = 0.$$

Therefore $X \stackrel{\mathrm{d}}{=} Y$, however X and Y are not equivalent.

The same conclusion can be drawn if X is any discrete r.v. which is symmetric with respect to 0 and such that $\mathbf{P}\{X = 0\} < 1$. (The last condition excludes the trivial case.) This means that X takes a finite or infinite number of values $\ldots, -x_2, -x_1, x_0 = 0, x_1, x_2, \ldots$ with probabilities $p_0 = \mathbf{P}\{X = 0\} < 1, p_j = \mathbf{P}\{X = x_j\} = \mathbf{P}\{X = -x_j\}, j = 1, 2, \ldots, p_0 + 2\sum_j p_j = 1$.

5.2. If X, Y, Z are random variables on the same probability space, then $X \stackrel{\mathrm{d}}{=} Y$ does not always imply that $XZ \stackrel{\mathrm{d}}{=} YZ$

Let X and Y be r.v.s (defined, perhaps, on different probability spaces). It is well known that if $X \stackrel{\mathrm{d}}{=} Y$ and $g(x)$, $x \in \mathbb{R}^1$ is a $\mathcal{B}^1$-measurable function, then $g(X)$ and $g(Y)$ are also r.v.s and $g(X) \stackrel{\mathrm{d}}{=} g(Y)$. This fact could suggest the following conjecture. If X, Y and Z are defined on the same probability space, then

$$X \stackrel{\mathrm{d}}{=} Y \Rightarrow XZ \stackrel{\mathrm{d}}{=} YZ \quad \text{for any r.v. } Z.$$

A simple example will show that in general this is not true. Let the r.v. X have a symmetric distribution and let $Y = -X$. Then $X \stackrel{\mathrm{d}}{=} Y$. Now take $Z = Y$, that is $Z = -X$. Then the equality $XZ \stackrel{\mathrm{d}}{=} YZ$ is impossible because $XZ = -X^2$ and $YZ = (-X)(-X) = X^2$. It suffices to note that all values of XZ are non-positive while those of YZ are non-negative. The trivial case $\mathbf{P}\{X = 0\} = 1$ is of no interest.

5.3. Different distributions can be transformed by different functions to the same distribution

Suppose ξ is a r.v. in $\mathbb{R}^1$ and $g_1(x) \neq g_2(x)$, $x \in \mathbb{R}^1$ are Borel-measurable functions. Then $g_1(\xi)$ and $g_2(\xi)$ are r.v.s with different distributions, i.e. $g_1(\xi) \overset{\mathrm{d}}{\neq} g_2(\xi)$ (except trivial cases involving symmetry-type properties).

Further, if ξ_1 and ξ_2 are r.v.s and $g(x)$, $x \in \mathbb{R}^1$ is a Borel-measurable function, then $\xi_1 \overset{\mathrm{d}}{\neq} \xi_2$ implies that $g(\xi_1) \overset{\mathrm{d}}{\neq} g(\xi_2)$ (again except easy cases).

These two facts make the possibility to describe explicitly two r.v.s ξ_1 and ξ_2 and two Borel-measurable functions g_1 and g_2 such that $\xi_1 \overset{\mathrm{d}}{\neq} \xi_2$, $g_1 \neq g_2$, but $g_1(\xi_1) \overset{\mathrm{d}}{=} g_2(\xi_2)$ interesting. The multi-dimensional case is also of interest.

(i) Consider the r.v. $\xi_1 \sim \mathcal{N}(0, \frac{1}{2})$, normally distributed with zero mean and variance $\frac{1}{2}$ and the r.v. $\xi_2 \sim \gamma(a)$, $a > 0$, i.e. ξ_2 has a *gamma distribution* with density $(1/\Gamma(a))x^{a-1}\mathrm{e}^{-x}$, if $x > 0$ and 0 otherwise. Take also the functions $g_1(x) = |x|^\rho$ and $g_2(x) = |x|^\beta$, $x \in \mathbb{R}^1$, where $\rho > 0$ and $\beta > 0$ are arbitrary numbers. Let us see how the two r.v.s $\eta_1 = g_1(\xi_1) = |\xi_1|^\rho$ and $\eta_2 = g_2(\xi_2) = |\xi_2|^\beta = \xi_2^\beta$ are connected. If f_1 and f_2 are the densities of η_1 and η_2 respectively, we find that $f_1(x) = f_2(x) = 0$ if $x \leq 0$ and, for $x > 0$,

$$f_1(x) = \frac{2}{\rho\sqrt{\pi}} x^{1/\rho - 1} \exp(-x^{2/\rho}), \quad f_2(x) = \frac{1}{\beta\Gamma(a)} x^{a/\beta - 1} \exp(-x^{1/\beta}).$$

Now let us keep ρ fixed and take $a = \frac{1}{2}$ and $\beta = \rho/2$. Hence $\xi_1 \sim \mathcal{N}(0, \frac{1}{2})$ as before, $\xi_2 \sim \gamma(\frac{1}{2})$ and taking into account that $\Gamma(\frac{1}{2}) = \sqrt{\pi}$ and comparing f_1 and f_2, we conclude that the r.v.s η_1 and η_2 have the same distribution.

Therefore two different r.v.s $\xi_1 \sim \mathcal{N}(0, \frac{1}{2})$ and $\xi_2 \sim \gamma(\frac{1}{2})$ can be transformed by different functions to identically distributed r.v.s:

$$|\xi_1|^\rho \overset{\mathrm{d}}{=} |\xi_2|^{\rho/2}, \ \rho > 0.$$

(ii) Here is another case involving more than two variables. Take three r.v.s ξ, η and θ where $\xi \sim \mathcal{N}(0, 1)$, $\eta \sim \mathcal{E}xp(1)$ (exponential distribution with parameter 1) and θ follow the arcsine law, that is, the density of θ is $1/(\pi\sqrt{x(1-x)})$ on (0,1) and 0 otherwise. Consider now three new r.v.s, namely

$$\log(\tfrac{1}{2}\xi^2), \ \log(\eta), \ \log(\theta)$$

and denote by ψ_1, ψ_2, ψ_3 their ch.f.s, respectively. Then

$$\psi_1(t) = \Gamma(\tfrac{1}{2} + it)/\sqrt{\pi}, \ \psi_2(t) = \Gamma(1 + it), \ \psi_3(t) = \Gamma(1 + it)\Gamma(\tfrac{1}{2} + it)/\sqrt{\pi}.$$

By using the obvious identity $\psi_1(t)\psi_2(t) = \psi_3(t)$ for all t and assuming that η and θ are independent, we easily arrive at the relation

$$\xi^2 \overset{\mathrm{d}}{=} 2\eta\theta.$$

Note that the same conclusion can be derived directly by writing

$$\xi^2 = (\xi^2 + \tilde{\xi}^2)\frac{\xi^2}{\xi^2 + \tilde{\xi}^2}$$

where $\tilde{\xi} \sim \mathcal{N}(0,1)$ is independent of ξ and showing that the two factors $(\xi^2 + \tilde{\xi}^2)$ and $\xi^2/(\xi^2 + \tilde{\xi}^2)$ are independent and follow exponential and arcsine laws respectively.

The final remark is that all r.v.s considered in cases (i) and (ii) are absolutely continuous. Similar examples for discrete variables can also be constructed. This is left to the reader (try to avoid some trivial cases).

5.4. A function which is a metric on the space of distributions but not on the space of random variables

Let us define the distance $r(X, Y)$ between any two r.v.s X and Y by

$$r(X,Y) = \sup_x |\mathbf{P}\{X \le x\} - \mathbf{P}\{Y \le x\}| = \sup_x |F_X(x) - F_Y(x)|, \quad x \in \mathbb{R}^1.$$

(r is called the *uniform distance* or *Kolmogorov distance*).

Another suitable notation for $r(X, Y)$ is $r(F_X, F_Y)$, where F_X and F_Y are d.f.s of X and Y respectively. The function r, considered on the space of all distribution functions on $\mathbb{R}^1$, is a metric. Indeed, it is easy to see that: (i) $r(X,Y) > 0$ and $r(X,Y) = 0$ iff $F_X = F_Y$; (ii) $r(X,Y) = r(Y,X)$; (iii) $r(X,Y) \le r(X,Z) + r(Z,Y)$. In (i)–(iii), X, Y and Z are arbitrary r.v.s.

Suppose now that the function r is considered on the space of the r.v.s on the underlying probability space. Then, referring to Example 5.1 above, we conclude that r is not a metric because it violates the condition that $r(X,Y) = 0$ implies $X \overset{\text{a.s.}}{=} Y$.

5.5. On the n-dimensional distribution functions

Comparing the definitions of one-dimensional and multi-dimensional d.f.s, we see that in the one-dimensional case condition (d) is implied by (a_1). However, if $n > 1$ then (d) is no longer a consequence of (a_1) and even for $n = 2$ it is easy to construct a function $F(x, y)$ satisfying (a_1), (b_1) and (c_1) but not (d). For example, take the function

$$F(x,y) = \begin{cases} 0, & \text{if } x \le 0 \text{ or } x + y \le 1 \text{ or } y \le 0 \\ 1, & \text{otherwise.} \end{cases}$$

Obviously (a_1), (b_1) and (c_1) are satisfied. Suppose F is a d.f. of some random vector, say (X, Y). Then for every parallelepiped $Q = [a_1, b_1] \times [a_2, b_2]$ (here a rectangle) we might have $\mathbf{P}\{(X,Y) \in Q\} \ge 0$. However, if $Q = [\frac{1}{3}, 1] \times [\frac{1}{3}, 1]$ then

$$\mathbf{P}\{(X,Y) \in Q\} = F(1,1) - F(1,\tfrac{1}{3}) - F(\tfrac{1}{3},1) + F(\tfrac{1}{3},\tfrac{1}{3}) = -1$$

which is impossible. Therefore conditions (a_1), (b_1) and (c_1) are not sufficient for F to be a d.f. in the n-dimensional case when $n \geq 2$.

Let us suggest one additional example. Define

$$G(x, y) = \begin{cases} 0, & \text{if } x \leq 0 \text{ or } y \leq 0 \\ \min\{1, \max\{x, y\}\}, & \text{otherwise.} \end{cases}$$

It can be checked that G satisfies conditions (a_1), (b_1) and (c_1) but not (d). It is sufficient here to take the rectangle $R = [x_1, x_2] \times [y_1, y_2]$, where $0 \leq x_1 \leq y_1 < x_2 \leq y_2 \leq 1$ and calculate the probability $\mathbf{P}[(X, Y) \in R]$.

5.6. The continuity property of one-dimensional distributions may fail in the multi-dimensional case

Let X be a r.v. on the probability space $(\Omega, \mathcal{F}, \mathbf{P})$ and F its d.f. Suppose the values of X fill some interval (finite or infinite) in $\mathbb{R}^1$ and for each x of this interval, $\mathbf{P}[\omega{:}X(\omega) = x] = 0$, that is the probability measure $\mathbf{P}$ has no atoms. Then $F(x)$ is continuous in x everywhere in this interval. Thus we come naturally to the following question: does an analogous property hold in the multi-dimensional case? By an example for $n = 2$ we show that in general the answer is negative. Indeed, consider the following function in the plane:

$$F(x, y) = \begin{cases} xy, & \text{if } 0 \leq x \leq 1, 0 \leq y \leq \frac{1}{2} \\ \frac{x}{2} & \text{if } 0 \leq x \leq 1, \frac{1}{2} < y < \infty \\ y, & \text{if } 1 < x < \infty, 0 \leq y \leq 1 \\ 1, & \text{if } x > 1, y > 1 \\ 0, & \text{otherwise.} \end{cases}$$

It is easy to check that F is a two-dimensional d.f. Denote by (X, Y) the random vector whose d.f. is F. We shall also use the notation in figure 1, where we have indicated the domains in $\mathbb{R}^2$ and the corresponding values of F. Note that the vector (X, Y) takes values in the quadrant $Q = \{(x, y) : 0 \leq x < \infty, 0 \leq y < \infty\}$, and moreover each point $(x, y) \in Q$ has zero probability. Following the one-dimensional case we could expect that $F(x, y)$ is continuous everywhere in Q. But this conjecture is false. Indeed, it is easily seen that every point with coordinates $(1, y)$ where $\frac{1}{2} < y < \infty$ is a discontinuity point of F. If $\Delta(1, y) := F(1, y) - F(1 - 0, y - 0)$ is the size of the jump in F at the point $(1, y)$, we find that $\Delta(1, y) = y - \frac{1}{2}$ for $\frac{1}{2} < y < 1$ and $\Delta(1, y) = \frac{1}{2}$ for $1 \leq y \leq \infty$. The reason for the existence of this discontinuity of the d.f. F is that there is a 'hyperplane' with strongly positive probability, namely $\mathbf{P}[X = 1, \frac{1}{2} \leq Y \leq 1] = \frac{1}{2}$ (see the bold vertical segment in Figure 1).

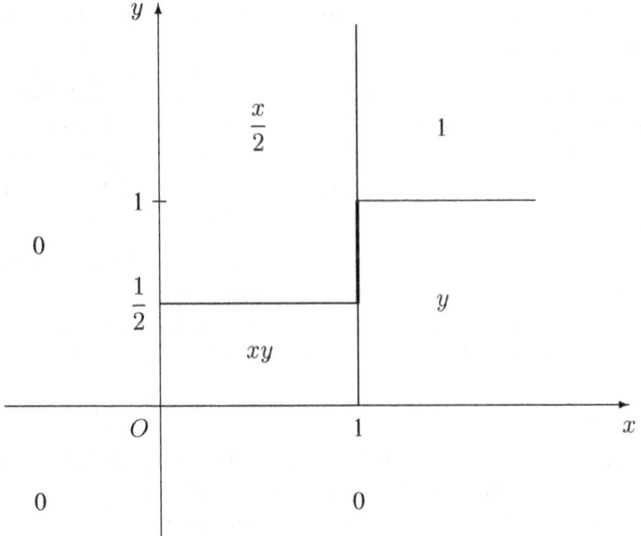

Figure 1

5.7. On the absolute continuity of the distribution of a random vector and of its components

Consider for simplicity the two-dimensional case. Suppose (X_1, X_2) has an absolutely continuous distribution. Then it is easy to see that each of X_1 and X_2 also has an absolutely continuous distribution. The question now is whether the converse is true. To see this, take X_1 to be absolutely continuous and let $X_2 \equiv X_1$, that is $X_2(\omega) = X_1(\omega)$ for each $\omega \in \Omega$. Evidently X_2 is absolutely continuous. Suppose the vector (X_1, X_2) has an absolutely continuous distribution with some density, say f. Then the following relation would hold:

$$\mathbf{P}\{(X_1, X_2) \in B\} = \iint_B f(x_1, x_2)\,\mathrm{d}x_1\mathrm{d}x_2 \quad \text{for any set } B \in \mathbb{R}^2. \tag{1}$$

However, all values of the vector (X_1, X_2) belong to the line $l : x_2 = x_1$. If we take $B = l = \{(x_1, x_2) : x_2 = x_1\}$ then the left-hand side of (1) is 1, but the right-hand side is 0 since the line l has a plane measure 0. Hence (X_1, X_2) is not absolutely continuously distributed, but each of its components is.

Note that if X_1 and X_2 are independent and absolutely continuous, then (X_1, X_2) is also absolutely continuous.

5.8. There are infinitely many multi-dimensional probability distributions with given marginals

If the random vector $(X_1, \ldots, X_n)$ has a d.f. $F(x_1, \ldots, x_n)$ then the *marginal distributions* $F_k(x_k) = \mathbf{P}[X_k \leq x_k]$, $k = 1, \ldots, n$ are uniquely determined. By a few examples we show that the converse is not true. It is sufficient here just to consider the two-dimensional case. The examples treat the discrete, absolutely continuous and the general cases.

(i) Let $p = \{p_{ij}, i, j = 1, 2, \ldots\}$ be a two-dimensional discrete distribution. Select two points, say (x_1, y_1) and (x_1, x_2), each with positive probability and such that $x_1 \neq x_2, y_1 \neq y_2$. We can choose a small ε satisfying the relations $0 < \varepsilon \leq p_{11}$, and $0 < \varepsilon \leq p_{22}$. Consider the set $q = \{q_{ij}, i, j = 1, 2, \ldots\}$ defined as follows:

$$q_{11} = p_{11} - \varepsilon,\ q_{12} = p_{12} + \varepsilon,\ q_{21} = p_{21} + \varepsilon,\ q_{22} = p_{22} - \varepsilon$$

and for all $i, j \neq 1, 2$, we put $q_{ij} = p_{ij}$. Then it is easy to check that q is a two-dimensional distribution. Moreover, the marginal distributions of q are the same as those of p for each ε as chosen above, even though $p \neq q$.

(ii) Consider the following two functions:

$$f(x_1, x_2) = \begin{cases} \frac{1}{4}(1 + x_1 x_2), & \text{if } -1 \leq x_1 \leq 1, -1 \leq x_2 \leq 1 \\ 0, & \text{otherwise,} \end{cases}$$

$$g(x_1, x_2) = \begin{cases} \frac{1}{4}, & \text{if } -1 \leq x_1 \leq 1, -1 \leq x_2 \leq 1 \\ 0, & \text{otherwise.} \end{cases}$$

Then f and g are both two-dimensional probability density functions. For the marginal densities we find

$$f_1(x_1) = \begin{cases} \frac{1}{2}, & \text{if } -1 \leq x_1 \leq 1 \\ 0, & \text{otherwise,} \end{cases} \qquad f_2(x_2) = \begin{cases} \frac{1}{2}, & \text{if } -1 \leq x_2 \leq 1 \\ 0, & \text{otherwise,} \end{cases}$$

$$g_1(x_1) = \begin{cases} \frac{1}{2}, & \text{if } -1 \leq x_1 \leq 1 \\ 0, & \text{otherwise,} \end{cases} \qquad g_2(x_2) = \begin{cases} \frac{1}{2}, & \text{if } -1 \leq x_2 \leq 1 \\ 0, & \text{otherwise} \end{cases}$$

thus $f_1 = g_1$ and $f_2 = g_2$, but obviously $f \neq g$.

(iii) Here is another specific but interesting example. For arbitrary positive constants a, b, c consider the functions

$$f(x, y) = \begin{cases} \frac{\Gamma(a+b+c)}{\Gamma(a)\Gamma(b)\Gamma(c)} x^{a-1} y^{b-1} (1 - x - y)^{c-1}, & \text{if } 0 \leq x, y,\ x + y \leq 1 \\ 0, & \text{otherwise} \end{cases}$$

and

$$\tilde{f}(x, y) = \begin{cases} \frac{1}{\mathrm{B}(a,b+c)\mathrm{B}(b,a+c)} x^{a-1} (1 - x)^{b+c-1} y^{b-1} (1 - y)^{a+c-1}, & \text{if } 0 \leq x,\ y \leq 1 \\ 0 & \text{otherwise.} \end{cases}$$

Here $\Gamma(\cdot)$ and $\mathrm{B}(\cdot,\cdot)$ are the well known gamma and beta functions of Euler. Both f and $\tilde{f}$ are two-dimensional probability density functions. Note that f is the density of the so-called *Dirichlet distribution*. Denote by (X, Y) and $(\tilde{X}, \tilde{Y})$ the random vectors whose densities are f and $\tilde{f}$ respectively. Direct computations show that X and $\tilde{X}$ have beta distribution with parameters $(a, b+c)$ and Y and $\tilde{Y}$ have beta distribution with parameters $(b, a+c)$. Thus, again, the marginal densities of the vectors (X, Y) and $(\tilde{X}, \tilde{Y})$ are identical, but obviously $f \neq \tilde{f}$.

(iv) Suppose F_1 and F_2 are d.f.s obeying densities f_1 and f_2 respectively. Consider the function

$$f(x_1, x_2) = f_1(x_1) f_2(x_2)[1 + \varepsilon(2F_1(x_1) - 1)(2F_2(x_2) - 1)], \quad (x_1, x_2) \in \mathbb{R}^2$$

where ε is an arbitrary number, $|\varepsilon| \leq 1$. Then f is a density function and for each ε the marginal densities are f_1 and f_2 respectively.

The answer to the question formulated at the beginning of this example can be also given in terms of the d.f.s only. Indeed, let F_1, F_2 be any d.f.s and ε any real number, $|\varepsilon| \leq 1$. Then by direct computation we see that

$$F(x_1, x_2) = F_1(x_1) F_2(x_2)[1 + \varepsilon(1 - F_1(x_1))(1 - F_2(x_2))], \quad (x_1, x_2) \in \mathbb{R}^2$$

is a two-dimensional d.f. whose marginals are just F_1 and F_2.

(v) Let F and G be arbitrary d.f.s. in $\mathbb{R}^1$. Define

$$H_1(x, y) = \max\{0, F(x) + G(y) - 1\}, \quad H_2(x, y) = \min\{F(x), G(y)\}.$$

For any $c_1, c_2 \geq 0$ with $c_1 + c_2 = 1$ let

$$H(x, y) = c_1 H_1(x, y) + c_2 H_2(x, y).$$

Then it is not difficult to check that $H(x, y)$, $(x, y) \in \mathbb{R}^2$ are two-dimensional d.f.s (*Fréchet distributions*). Moreover, any d.f. of this class has F and G as its marginals.

Hence there are infinitely many multi-dimensional distributions with the same marginal distributions. In other words, the marginal distributions do not uniquely determine the corresponding joint distribution. The only exception is the case when the random vector consists of independent components. In this case, the joint distribution equals the product of the marginals.

5.9. The continuity of a two-dimensional probability density does not imply that the marginal densities are continuous

Let $f(x, y)$, $(x, y) \in \mathbb{R}^2$ be a probability density function which is continuous. Denote by $f_1(x)$, $x \in \mathbb{R}^1$ and $f_2(y)$, $y \in \mathbb{R}^1$ the corresponding marginal densities. There are problems which require the use of the marginal densities f_1 and f_2 and their continuity properties. Intuitively we might expect that f_1 and f_2 are continuous if f is. However, such a conjecture is not generally true. Indeed, consider the following function:

$$(1) \qquad f(x,y) = \frac{1}{2\sqrt{2\pi}}|x|\exp(-|x| - \tfrac{1}{2}x^2y^2), \quad (x,y) \in \mathbb{R}^2.$$

It is easy to check that f is a probability density function. For the first marginal density f_1 we find

$$(2) \qquad f_1(x) = \begin{cases} 0, & \text{if } x = 0 \\ \frac{1}{2}\exp(-|x|), & \text{if } x \neq 0. \end{cases}$$

Clearly f_1 is discontinuous at $x = 0$ despite the fact that f is continuous.

Notice that the marginal density f_1 is discontinuous at one point only. Now, using (1), we construct a new probability density function which will be continuous, but one of whose marginal densities will have infinitely many points of discontinuity.

Let $r_1, r_2, \ldots$ be rational numbers in some order and let

$$(3) \qquad g(x,y) = \sum_{n=1}^{\infty} \frac{1}{2^n} f(x - r_n, y).$$

Since f given by (1) is bounded on $\mathbb{R}^2$, the series on the right-hand side of (3) is uniformly convergent on $\mathbb{R}^2$. Moreover, g is a probability density function which is everywhere continuous. The marginal density g_1 of g is

$$(4) \qquad g_1(x) = \sum_{n=1}^{\infty} \frac{1}{2^n} f_1(x - r_n)$$

with f_1 given by (2). The boundedness of f_1 implies the uniform convergence of (4). It follows from (4) that g_1 is discontinuous at all rational points $r_1, r_2, \ldots$, though it is continuous at every irrational point of $\mathbb{R}^1$.

5.10. The convolution of a unimodal probability density function with itself is not always unimodal

We present two examples and then discuss them briefly.

(i) Consider the following function:

$$f(x) = \begin{cases} 0, & \text{if } x < -\frac{1}{30} \text{ and } x > \frac{5}{6} \\ 5, & \text{if } -\frac{1}{30} \leq x \leq 0 \\ 1, & \text{if } 0 < x \leq \frac{5}{6}. \end{cases}$$

It is easy to see that f is a probability density function which is unimodal. Direct calculation shows that the convolution $f^{*2}(x) := (f * f)(x)$ is:

$$f^{*2}(x) = \begin{cases} 0, & \text{if } x \le -\frac{1}{15} \text{ and } x \ge \frac{5}{3} \\ 25x + \frac{5}{3}, & \text{if } -\frac{1}{15} < x \le -\frac{1}{30} \\ -15x + \frac{1}{3}, & \text{if } -\frac{1}{30} < x \le 0 \\ x + \frac{1}{3}, & \text{if } 0 < x \le \frac{4}{5} \\ -9x + \frac{25}{3}, & \text{if } \frac{4}{5} < x \le \frac{5}{6} \\ \frac{5}{3} - x, & \text{if } \frac{5}{6} < x \le \frac{5}{3}. \end{cases}$$

Obviously f^{*2} has two local maxima at $x = -\frac{1}{30}$ and $x = \frac{4}{5}$, $f^{*2}\left(-\frac{1}{30}\right) = \frac{5}{6}$, $f^{*2}\left(\frac{4}{5}\right) = \frac{17}{15}$ and one minimum equal to $\frac{1}{3}$ at the point $x = 0$.

Hence the convolution operation does not preserve unimodality.

(ii) Suppose a and b are positive numbers. Denote by u_a and v_a the densities of uniform distributions on $(0, a)$ and $(-\frac{1}{2}a, \frac{1}{2}a)$ respectively. Let $f = \frac{1}{2}(u_a + u_b)$, $g = \frac{1}{2}(v_a + v_b)$. Then each of f and g is a unimodal density and, moreover, g is symmetric. We want to know whether the convolution $f * g$ is unimodal. To see this we use the equality

$$f * g = \tfrac{1}{4}[(u_a * v_a) + (u_b * v_b) + (u_a * v_b) + (u_b * v_a)].$$

Considering separately each of the terms on the right-hand side of this representation we arrive at the following conclusions:

(1) $u_a * v_a$ linearly decreases on $(\frac{1}{2}a, \frac{3}{2}a)$ with slope $(-a^{-2})$ and vanishes on $(\frac{3}{2}a, \infty)$;
(2) $u_b * v_b$ linearly increases on $(-\frac{1}{2}b, \frac{1}{2}b)$ with slope b^{-2};
(3) $u_a * v_b$ is constant on $(a - \frac{1}{2}b, \frac{1}{2}b)$;
(4) $u_b * v_a$ is constant on $(\frac{1}{2}a, b - \frac{1}{2}a)$ and then decreases linearly.

Now choose the parameters a, b such that $b > 3a$. From (1)–(4) it follows that $f * g$ is decreasing in the interval $(\frac{1}{2}a, \frac{3}{2}a)$ and is increasing in $(\frac{3}{2}a, \frac{1}{2}b)$, but this means that $f * g$ is not unimodal.

Let us note that in case (i) the density f is unimodal but not symmetric, while in case (ii) both densities f and g are unimodal, g is symmetric and f is not symmetric. We have seen that the convolutions $f * f$ and $f * g$ are not unimodal. Thus in general the convolution operation does not preserve the unimodality property. Note that if f and g are unimodal densities and both are symmetric then their convolution $f * q$ is unimodal (Lukacs 1970; Dharmadhikari and Joag-Dev 1988).

5.11. The convolution of unimodal discrete distributions is not always unimodal

Recall first the definition of unimodality. Let $\mathcal{P} = \{p_n, n \in \mathbb{N}_0\}$ be a probability distribution on the set of the non-negative integer numbers $\mathbb{N}_0$ or on some subset (or even on a countable subset of $\mathbb{R}^1$). We say that $\mathcal{P}$ is *unimodal* if there is an integer k_0 such that p_k is non-decreasing for $k \leq k_0$ and non-increasing for $k \geq k_0$. The value k_0 is called a mode. We wish to know if the unimodal property is preserved under the convolution operation. Example 5.9 shows that the answer is negative in the absolutely continuous case. Let us find the answer in the discrete case.

Consider two independent r.v.s., say ξ and η with values in the sets $\{0, 1, \ldots, m\}$ and $\{0, 1, \ldots, n\}$ respectively. For the probabilities $p_i = \mathbf{P}[\xi = i]$ and $q_j = \mathbf{P}[\eta = j]$ we suppose that

$$p_0 = \frac{m+2}{2m+2}, \; p_1 = \ldots = p_m = \frac{1}{2m+2}; \; q_0 = \frac{n+2}{2n+2}, \; q_1 = \ldots = q_n = \frac{1}{2n+2}.$$

Then each of the distributions $\mathcal{P}_\xi = \{p_i, i = 0, 1, \ldots, m\}$ and $\mathcal{P}_\eta = \{q_j, j = 0, 1, \ldots, n\}$ is unimodal. The sum $\theta = \xi + \eta$ is a r.v. with values in the set $\{0, 1, \ldots, m+n\}$ and its distribution $\mathcal{P}_\theta = \{r_k, k = 0, 1, \ldots, m+n\}$, in view of the independence of ξ and η, is equal to the convolution of $\mathcal{P}_\xi$ and $\mathcal{P}_\eta$: $\mathcal{P}_\theta = \mathcal{P}_\xi * \mathcal{P}_\eta$. This means that

$$r_k = \mathbf{P}[\theta = k] = \mathbf{P}[\xi + \eta = k] = \sum p_i q_j, \quad k = 0, 1, \ldots, m+n$$

where the summation is over all $i \in \{0, 1, \ldots, m\}$ and all $j \in \{0, 1, \ldots, n\}$ with $i + j = k$. In particular we can easily find that

$$r_0 = p_0 + q_0 = \frac{(m+2)(n+2)}{(2m+2)(2n+2)}, \; r_1 = p_0 q_1 + p_1 q_0 = \frac{m+n+4}{(2m+2)(2n+2)},$$
$$r_2 = p_0 q_2 + p_1 q_1 + p_2 q_0 = \frac{m+n+5}{(2m+2)(2n+2)}, \text{ etc.}$$

Comparing r_0, r_1 and r_2 we see that $r_0 > r_1$ but $r_1 < r_2$. Even without additional calculations this is enough to conclude that the distribution $\mathcal{P}_\theta$, that is the convolution $\mathcal{P}_\xi * \mathcal{P}_\eta$ is not unimodal even though both $\mathcal{P}_\xi$ and $\mathcal{P}_\eta$ are unimodal.

5.12. Strong unimodality is a stronger property than the usual unimodality

The d.f. G is called *strongly unimodal* if the convolution $G * F$ is unimodal for every unimodal F. (This notion was introduced by I. Ibragimov in 1956.)

Note that several useful distributions are indeed strongly unimodal: the normal distribution $\mathcal{N}(a, \sigma^2)$; the uniform distribution on the interval $[a, b]$; the gamma distribution with a shape parameter $a \geq 1$; the beta distribution with parameters (a, b), $a \geq 1$, $b \geq 1$, etc.

However we have seen (see Example 5.9) that the convolution of two unimodal distributions is, in general, not unimodal. This implies that strong unimodality is a stronger property than (usual) unimodality. Obviously, Example 5.9 deals with absolutely continuous distributions. Hence it is of interest to consider such a case involving discrete distributions.

Let F_k denote the uniform distribution on the finite set $\{0, 1, \dots, k\}$ and let $F = \frac{1}{2}(F_0 + F_{m+1})$ for a fixed $m \geq 3$. Then F is unimodal and our goal is to look at the convolution $G = F * F$. The distribution G is concentrated on the set $\{0, 1, 2, \dots, 2m-2\}$ and if g_i, $i = 0, 1, 2, \dots, 2m-2$, are the masses of its 'atoms', then we easily find that

$$4g_0 = 1 + 2m^{-1} + m^{-2},\ 4g_1 = 2m^{-1} + 2m^{-2},\ 4g_2 = 2m^{-1} + 3m^{-2}.$$

It follows immediately that

$$g_0 - g_1 = \tfrac{1}{4}\left(1 - m^{-2}\right) > 0, \quad g_1 - g_2 = \tfrac{1}{4}\left(-m^{-2}\right) < 0.$$

Thus $g_1 < \min\{g_0, g_2\}$ and therefore the distribution $G = F * F$ is not unimodal. In other words, F is unimodal but not strongly unimodal.

5.13. Every unimodal distribution has a unimodal concentration function, but the converse does not hold

Let X be a r.v. with a d.f. F and μ_F be the measure on $(\mathbb{R}^1, \mathcal{B}^1)$ induced by F. Recall that the function

$$(1) \qquad Q_F(L) = \begin{cases} 0, & \text{if } l < 0 \\ \sup\limits_{x \in \mathbb{R}} \mu_F([-\frac{1}{2}l, \frac{1}{2}l] + x), & \text{if } l \geq 0 \end{cases}$$

is said to be a *concentration function* (of P. Lévy) corresponding to F and also to μ_F. (Here the sum of sets is defined in the usual sense: $A + B = \{a + b : a \in A, b \in B\}$.) Important results concerning concentration functions and their applications have been summarized by Hengartner and Theodorescu (1973). From (1) we can easily derive that $Q_F(l), l \in \mathbb{R}^1$ is a d.f.

Let us mention the following result (Hengartner and Theodorescu 1973). If $F(x)$, $x \in \mathbb{R}^1$ is a unimodal d.f. with mode $x^* = 0$, then the concentration function $Q_F(l)$, $l \in \mathbb{R}^1$ is unimodal with mode $l^* = 0$.

By a concrete example we can show that the converse is not always true. We give below the d.f. F and its concentration function Q_F calculated by (1), namely:

$$F(x) = \begin{cases} 0, & \text{if } x < 0 \\ \frac{1}{4}x, & \text{if } 0 \leq x < 1 \\ \frac{1}{4}, & \text{if } 1 \leq x < 2 \\ \frac{1}{4}(x-1), & \text{if } 2 \leq x < 4 \\ \frac{1}{8}(x+2), & \text{if } 4 \leq x < 6 \\ 1, & \text{if } x \geq 6, \end{cases} \qquad Q_F(l) = \begin{cases} 0, & \text{if } l < 0 \\ \frac{1}{4}l, & \text{if } 0 \leq l < 2 \\ \frac{1}{8}(l+2), & \text{if } 2 \leq l < 6 \\ 1, & \text{if } l \geq 6. \end{cases}$$

Clearly Q_F is unimodal but F is not unimodal.

SECTION 6. EXPECTATIONS AND CONDITIONAL EXPECTATIONS

For any r.v. X on a given probability space $(\Omega, \mathcal{F}, \mathbf{P})$ we can define an important characteristic which is called an *expectation*, or an expected value, and is denoted by $\mathbf{E}X$. If $X > 0$ and $\mathbf{P}[X = \infty] > 0$ we put $\mathbf{E}X = \infty$, while if $\mathbf{P}[X = \infty] = 0$ we define

$$\mathbf{E}X = \lim_{n\to\infty} \sum_{k=1}^{\infty} \frac{k}{2^n} \mathbf{P}\left[\frac{k}{2^n} < X \le \frac{k+1}{2^n}\right]. \tag{1}$$

For an arbitrary r.v. X let $X^+ = \max\{X, 0\}$ and $X^- = \max\{-X, 0\}$. Since X^+ and X^- are non-negative, their expectations $\mathbf{E}[X^+]$ and $\mathbf{E}[X^-]$ can be obtained by (1), and if either $\mathbf{E}[X^+] < \infty$ or $\mathbf{E}[X^-] < \infty$ then

$$\mathbf{E}X = \mathbf{E}[X^+] - \mathbf{E}[X^-].$$

The expectation $\mathbf{E}X$ is also called the Lebesgue integral of the $\mathcal{F}$-measurable function X with respect to the probability measure $\mathbf{P}$. We say that the expectation of X is finite if both $\mathbf{E}[X^+]$ and $E[X^-]$ are finite. Since $|X| = X^+ + X^-$, the finiteness of $\mathbf{E}X$ is equivalent to $\mathbf{E}[|X|] < \infty$. In this case the r.v. X is said to be *integrable*. If X is absolutely continuous with a density f, then X is integrable iff $\int_{-\infty}^{\infty} |x| f(x)\, \mathrm{d}x < \infty$ and $\mathbf{E}X = \int_{-\infty}^{\infty} x f(x)\, \mathrm{d}x$. If X is discrete, $\mathbf{P}[X = x_n] = p_n$, $p_n > 0$, $n = 1, 2, \ldots, \sum_n p_n = 1$, then X is integrable iff $\sum_n |x_n| p_n < \infty$ and $\mathbf{E}X = \sum_n x_n p_n$.

For some purposes it is necessary to consider the integral of X over the set $A \in \mathcal{F}$. In such a case $\int_A X\, \mathrm{d}\mathbf{P} = \int_\Omega X(\omega) 1_A(\omega)\, \mathrm{d}\mathbf{P}(\omega)$.

It is convenient to introduce here the space $\mathrm{L}^r(\Omega, \mathcal{F}, \mathbf{P})$, or simply L^r, of all r-integrable r.v.s where $r > 0$ and $X \in \mathrm{L}^r$ iff $\mathbf{E}[|X|^r] < \infty$.

In addition to the expectation $\mathbf{E}X$, important characteristics of the r.v. X are the numbers (if defined)

$$\mathbf{E}[(X-c)^k],\ \mathbf{E}[|X-c|^k],\ k = 1, 2, \ldots, c \in \mathbb{R}^1$$

which are known as the *kth non-central moment* and *kth non-central absolute moment* of X about c respectively. If $c = \mathbf{E}X$ these moments are called *central*. In this section and later we use the notation $m_k = \mathbf{E}[X^k]$ for the kth moment of X. In the particular case when $k = 2$ and $c = \mathbf{E}X$ we get the quantity $\mathbf{E}[(X - \mathbf{E}X)^2]$ which is said to be the *variance* of X and is denoted by $\mathbf{V}X : \mathbf{V}X = \mathbf{E}[(X - \mathbf{E}X)^2]$.

The expectation possesses several properties. We mention here only a few of them. If X_1 and X_2 are integrable r.v.s and $c_1, c_2 \in \mathbb{R}^1$ then $X_1 + X_2$ and $c_i X_i$ are also integrable and

$$\mathbf{E}[c_1 X_1 + c_2 X_2] = c_1 \mathbf{E}X_1 + c_2 \mathbf{E}X_2 \quad \text{(linearity)},$$
$$\mathbf{E}X_1 \le \mathbf{E}X_2 \quad \text{if} \quad X_1 \le X_2 \quad \text{(monotonicity)}.$$

Other properties such as additivity over disjoint sets and different kinds of convergence theorems can be found in the literature (Chung 1974; Chow and Teicher 1978; Laha and Rohatgi 1979; Shiryaev 1995).

The family $\{X_n : n \geq 1\}$ of r.v.s is said to be *uniformly integrable* if

$$\sup_n \int_{[|X|>a]} |X_n| \, \mathrm{d}\mathbf{P}(\omega) \to 0 \text{ as } a \to \infty$$

or, in another equivalent form, if

$$\sup_n \mathbf{E}[|X_n| I_{[|X_n|>a]}] \to 0 \text{ as } a \to \infty.$$

Suppose now that X is a r.v. on the given probability space and $\mathcal{D}$ is a sub-σ-field of $\mathcal{F}$. Following the same steps as for the definition of the expectation $\mathbf{E}X$, we can define the *conditional expectation* $\mathbf{E}[X|\mathcal{D}]$ of the r.v. X with respect to the σ-field $\mathcal{D}$. So, if X is an integrable r.v., then $\mathbf{E}[X|\mathcal{D}]$ is a $\mathcal{D}$-measurable r.v. such that for every $A \in \mathcal{D}$ we have

$$\int_A \mathbf{E}[X|\mathcal{D}] \, \mathrm{d}\mathbf{P} = \int_A X \, \mathrm{d}\mathbf{P} \quad \text{a.s.}$$

The existence of $\mathbf{E}[X|\mathcal{D}]$, up to equivalence, is a consequence of the Radon–Nikodym theorem (Chow and Teicher 1978; Shiryaev 1995). Here are some properties of conditional expectations:

(i) if $X = c$ a.s. where $c =$ constant, then $\mathbf{E}[X|\mathcal{D}] = c$ a.s.;
(ii) $X_1 \leq X_2 \Rightarrow \mathbf{E}[X_1|\mathcal{D}] \leq \mathbf{E}[X_2|\mathcal{D}]$ a.s.;
(iii) if X_1 and X_2 are integrable r.v.s and $c_1, c_2 \in \mathbb{R}^1$, then

$$\mathbf{E}[c_1 X_1 + c_2 X_2|\mathcal{D}] = c_1 \mathbf{E}[X_1|\mathcal{D}] + c_2 \mathbf{E}[X_2|\mathcal{D}] \quad \text{a.s.};$$

(iv) $\mathbf{E}\{\mathbf{E}[X|\mathcal{D}]\} = \mathbf{E}X$;
(v) if $\mathcal{D}_1 \subset \mathcal{D}_2 \subset \mathcal{F}$, then $\mathbf{E}\{\mathbf{E}[X|\mathcal{D}_2]|\mathcal{D}_1\} = \mathbf{E}[X|\mathcal{D}_1]$ a.s.;
(vi) if X is independent of the σ-field $\mathcal{D}$ (that is, X is independent of I_A, $A \in \mathcal{D}$), then $\mathbf{E}[X|\mathcal{D}] = \mathbf{E}X$ a.s.;
(vii) if X is $\mathcal{D}$-measurable and $\mathbf{E}[|XY|] < \infty$, then

$$\mathbf{E}[XY|\mathcal{D}] = X\mathbf{E}[Y|\mathcal{D}] \quad \text{a.s.}$$

Finally, let us mention an important particular case of the conditional expectation. For any event $A \in \mathcal{F}$ the conditional expectation $\mathbf{E}[I_A|\mathcal{D}]$ is denoted by $\mathbf{P}(A|\mathcal{D})$ and is called the *conditional probability* of the event A with respect to the σ-field $\mathcal{D}$ (also see Example 2.4).

This section includes examples devoted to various properties of expectations, conditional expectations and moments (in both one-dimensional and multi-dimensional cases). The Fubini theorem is introduced and analysed in Example 6.6, and conditional medians are considered in Example 6.10.

6.1. On the linearity property of expectations

If one operates with expectations such as $\mathbf{E}[X+Y]$ and $\mathbf{E}[X+Y+Z]$ it is generally accepted that $\mathbf{E}[X+Y] = \mathbf{E}X + \mathbf{E}Y$ and $\mathbf{E}[X+Y+Z] = \mathbf{E}X + \mathbf{E}Y + \mathbf{E}Z$. (Analogous relations can be written for more than three terms.) This is just the so-called *linearity property of expectations*. Its meaning is that the value of $\mathbf{E}[\cdot]$ depends on the variables in $[\cdot]$ only through their marginal distributions.

Recall that in the case of two r.v.s the linearity holds if $\mathbf{E}[X+Y]$ is defined (in the sense that $\mathbf{E}[(X+Y)^+]$ and/or $\mathbf{E}[(X+Y)^-]$ are finite). Of course, if $\mathbf{E}X$ and $\mathbf{E}Y$ both exist then $\mathbf{E}[X+Y]$ exists and equals their sum. Moreover, the linearity holds even when $\mathbf{E}X$ and $\mathbf{E}Y$ are not defined, or if one of them equals $+\infty$ and the other equals $-\infty$ (Simons 1977).

Now the question is: what happens if we consider three variables? Does the linearity property of expectations still remain valid? The answer will follow from the example below.

Let ξ denote a r.v. distributed uniformly on [0,1]. Then $1-\xi$ and $\eta = |2\xi - 1|$ have the same distribution as ξ. Define three new r.v.s, say X, Y and Z, in two different ways.

Case I. $X = Y = \tan(\pi\xi/2)$, $Z = -2X$.

Case II. $X' = \tan(\pi\xi/2)$, $Y' = \tan(\pi(1-\xi)/2)$, $Z' = -2\tan(\pi\eta/2)$.

It is evident that $X \stackrel{\mathrm{d}}{=} X'$, $Y \stackrel{\mathrm{d}}{=} Y'$, $Z \stackrel{\mathrm{d}}{=} Z'$. Our purpose now is to find the expectations $\mathbf{E}[X+Y+Z]$ and $\mathbf{E}[X'+Y'+Z']$. In Case I, $X+Y+Z=0$ and hence $\mathbf{E}[X+Y+Z] = 0$. In Case II we have: $Y' = \cot(\pi\xi/2)$, $Z' = \tan(\pi\xi/2) - \cot(\pi\xi/2)$ if $0 < \xi < \frac{1}{2}$ and $Z' = \cot(\pi\xi/2) - \tan(\pi\xi/2)$ if $\frac{1}{2} < \xi < 1$. Thus $X'+Y'+Z' = 2\tan(\pi\xi/2) = 2X$ if $0 < \xi < 1$ and $X'+Y'+Z' = 2\cot(\pi\xi/2) = 2Y$ if $\frac{1}{2} < \xi < 1$. Hence $\mathbf{P}[X'+Y'+Z' > 0] = 1$. Moreover, it is easy to calculate that $\mathbf{E}[X'+Y'+Z'] = (4/\pi)\log 2$.

Comparing the results from Cases I and II we see that the linearity property described above for two r.v.s can fail for three variables. Note that if one considers $\tilde{X} = X+Y$ and $\tilde{Y} = Z$ then $\mathbf{E}[X+Y+Z] = \mathbf{E}[\tilde{X}+\tilde{Y}]$ and $\mathbf{E}[\tilde{X}+\tilde{Y}]$, when defined, depends on $\tilde{X}$ and $\tilde{Y}$ only through the distribution of $\tilde{X}$ and the distribution of $\tilde{Y}$. But $\tilde{X} = X+Y$ and thus the value of the expectation $\mathbf{E}[X+Y+Z]$, when defined, depends on X, Y, Z through the bivariate distribution of X and Y, and the distribution of Z.

The reader could try to clarify how the linearity property of expectations is expressed when considering more than three variables. In general we have to be careful when taking expectations of even such simple expressions like sums of r.v.s.

6.2. An integrable sequence of non-negative random variables need not have a bounded supremum

Let $\{X_n, n \geq 1\}$ be a sequence of non-negative r.v.s such that for some $p > 0$, X_n^p is integrable for each n, and, moreover, let $\sup_n \mathbf{E}[X_n^p] < \infty$. Then intuitively we could expect that the variables X_n as well as $\sup_n X_n$ are bounded. Let us show that such a conjecture need not be true.

Consider the sequence $X_1, X_2, \dots$ of i.i.d. r.v.s whose common d.f. is $F(x) = 0$ if $x \leq 0$ and $F(x) = 1 - \mathrm{e}^{-x}$ if $x > 0$ (exponential distribution with parameter 1). Then for any $p > 0$ we have $\mathbf{E}[X_n^p] = \Gamma(p+1) < \infty$ and thus $\sup_n \mathbf{E}[X_n^p] < \infty$. Further, for $x > 0$ and $m = 1, 2, \dots$ we find

$$\mathbf{P}[\max_{1 \leq j \leq m} X_j \leq x] = (\mathbf{P}[X_j \leq x])^m = (1 - \mathrm{e}^{-x})^m.$$

Passing to the limit in both parameters m and x we get

$$\lim_{m \to \infty} \mathbf{P}[\max_{1 \leq j \leq m} X_j \leq x] = \mathbf{P}[\sup_j X_j \leq x] = 0 \quad \text{for all } x > 0$$

and

$$\lim_{x \to \infty} \mathbf{P}[\sup_j X_j \leq x] = \mathbf{P}[\sup_j X_j < \infty] = 0.$$

Therefore we have shown that in general the integrability of any order $p > 0$ of members of the sequence $\{X_n, n \geq 1\}$ does not imply boundedness of the supremum of this sequence.

6.3. A necessary condition which is not sufficient for the existence of the first moment

Let X be a r.v. with d.f. F. It is well known and easy to check that the condition $\lim_{x \to \infty} x(1 - F(x)) = 0$ is necessary for the existence of the expectation $\mathbf{E}X$. Thus we arrive at the inverse question: if F is such that $x(1 - F(x)) \to 0$ as $x \to 0$, does this imply that $\mathbf{E}X$ exists? The example below shows that in general the answer is negative. To see this take the following d.f.:

$$F(x) = \begin{cases} 0, & \text{if } x \leq 1 \\ 1 - \frac{1}{kx}, & \text{if } \mathrm{e}^{k-1} < x \leq \mathrm{e}^k, \ k = 1, 2, \dots. \end{cases}$$

Direct reasoning shows that $x(1 - F(x)) \to 0$ as $x \to 0$ while $\int_0^\infty (1 - F(x))\,\mathrm{d}x = \infty$ and since $\mathbf{E}X = \int_0^\infty (1 - F(x))\,\mathrm{d}x$, then $\mathbf{E}X$ does not exist.

We can say even more: if $\mathbf{E}[|X|^\alpha] < \infty$ for some $\alpha > 0$, then necessarily $n^\alpha \mathbf{P}[|X| > n] \to 0$ as $n \to \infty$ (e.g. see Rohatgi 1976).

Let us take $\alpha = 1$ and illustrate once again that a condition like $n\mathbf{P}[X > n] \to 0$ as $n \to \infty$ is not significant for the existence of $\mathbf{E}X$. Indeed, let us consider the following discrete r.v. X defined by $\mathbf{P}[X = n] = c/(n^2 \log n)$, $n = 2, 3, \dots$, c is a norming constant. We can then show that for large n, $\mathbf{P}(X > n) \sim c/(n \log n)$ implying that $n\mathbf{P}[X > n] \to 0$ as $n \to \infty$. However $\mathbf{E}X$ should be equal $\sum_{n=2}^{\infty} c/(n \log n)$ and the divergence of this series shows that the expectation $\mathbf{E}X$ does not exist.

Finally, note that if $n^{\alpha+\delta}\mathbf{P}[|X| > n] \to 0$ as $n \to \infty$ for some $\delta > 0$, then $\mathbf{E}[|X|^{\alpha}]$ does exist.

6.4. A condition which is sufficient but not necessary for the existence of moment of order (-1) of a random variable

The moments of negative orders of r.v.s are used in some probabilistic problems and it is of interest to know the conditions which ensure their existence.

If X is a r.v. with a discrete distribution having positive mass at 0, then $\mathbf{E}[X^{-1}]$ is infinite. The same holds if X is absolutely continuous and its density f satisfies the condition $f(0) > 0$. The following useful result is proved by Piegorsch and Casella (1985): let X be a r.v. with density $f(x)$, $x \in (0, \infty)$ which is continuous and satisfies the condition

$$(1) \qquad \lim_{x \to 0} x^{\alpha} f(x) = 0 \ \text{ for some } \alpha > 0$$

then $\mathbf{E}[X^{-1}] < \infty$.

By an example we aim to show that $\mathbf{E}[X^{-1}]$ can be finite even if (1) fails: that is, in general, condition (1) is sufficient but not necessary for the moment of order minus one to be finite. Indeed, define the family of functions $\{f_n, n \geq 1\}$ by

$$(2) \qquad f_n(x) = |\log^n x|^{-1} \Big/ \int_0^c |\log^n u|^{-1}\, du\,, \quad 0 < x < c$$

where $c =$ constant, $c \in (0, 1)$. It is easy to check that for each n, f_n is a probability density function of some r.v., say X_n. Since

$$\int_0^c |\log^n u|^{-1}\, du < \infty, \quad \lim_{x \to 0} f_n(x) = 0, \quad \lim_{x \downarrow 0} x^{-\alpha} f_n(x) = \infty$$

for every $\alpha > 0$, it follows that (1) is not satisfied. It then remains for us to determine whether $\mathbf{E}[X_n^{-1}]$ exists. By (2) we find that $\mathbf{E}[X_n^{-1}]$ is finite iff

$$(3) \qquad \int_0^c (x |\log^n x|)^{-1}\, dx < \infty.$$

For $n = 1$ the integral in (3) diverges for all $c \in (0, 1)$, but if $n = 2, 3, \ldots$, this integral is finite for any $c \in (0, 1)$. Consequently $\mathbf{E}[X_n^{-1}] < \infty$ iff $n \geq 2$. So, for $n \geq 2$, we have $\mathbf{E}[X_n^{-1}] < \infty$ but condition (1) does not hold.

6.5. An absolutely continuous distribution need not be symmetric even though all its central odd-order moments vanish

Let $F(x)$, $x \in \mathbb{R}^1$ be an absolutely continuous d.f. with a density f. Suppose F is *symmetric*, that is $F(-x) = 1 - F(x)$, or, equivalently, $f(-x) = f(x)$ for all $x \in \mathbb{R}^1$. Suppose F has moments of all orders. Then the central odd-order moments of X

$$m_{2n+1} = \mathbf{E}[X^{2n+1}] = \int_{-\infty}^{\infty} x^{2n+1} f(x)\,\mathrm{d}x$$

are zero for all $n = 0, 1, \ldots$ since the integrand $x^{2n+1} f(x)$ is an odd function and the integral is taken over the interval $(-\infty, \infty)$, which is symmetric with respect to the origin 0.

Suppose now that the distribution $G(x)$, $x \in \mathbb{R}^1$ has all its central odd-order moments vanishing. The question is: does it follow from this condition that G is symmetric? The answer is negative as illustrated by the following example.

Let the function $g(x)$, $x \in \mathbb{R}^1$ be defined by

$$(1)\qquad g(x) = \begin{cases} \frac{1}{48}\exp\left(-|x|^{1/4}(1 + \sin|x|^{1/4}\right), & \text{if } x < 0 \\ \frac{1}{48}\exp\left(-x^{1/4}(1 - \sin x^{1/4}\right), & \text{if } x \geq 0. \end{cases}$$

It is easy to verify that g is a probability density function. Denote by Y a r.v. with this density. Then we can calculate explicitly the moments $m_n = \mathbf{E}[Y^n]$ for each n, $n = 0, 1, \ldots$. The result is

$$m_{2n+1} = 0, \quad m_{2n} = \tfrac{1}{6}(8n+3)!.$$

Thus all central odd-order moments of Y are zero, but obviously the distribution of Y defined by the density (1) is not symmetric. (Also see Example 11.12.)

6.6. A property of the moments of random variables which does not have an analogue for random vectors

Let $(X_1, \ldots, X_n)$ be a random vector on a given probability space $(\Omega, \mathcal{F}, \mathbf{P})$. Let $k_1, \ldots, k_n$ be non-negative integers. If $\mathbf{E}[|X_1|^{k_1} \cdots |X_n|^{k_n}]$ exists, then the number

$$m_{k_1 \ldots k_n} = \mathbf{E}[X_1^{k_1} \cdots X_n^{k_n}]$$

is called a $(k_1, \ldots, k_n)$th *mixed central moment* of the random vector $(X_1, \ldots, X_n)$ and $k = k_1 + \cdots + k_n$ is its order.

If $n = 1$ we have one r.v. X_1 only, and it is well known that the existence of the kth moment m_k implies the existence of all moments m_j for $0 \leq j \leq k$. It suffices to recall the Lyapunov inequality $(\mathbf{E}[|X_1|^j])^{1/j} \leq (\mathbf{E}[|X_1|^k])^{1/k}$, $0 \leq j \leq k$, or to use the elementary inequality $|x|^j \leq 1 + |x|^k$, $x \in \mathbb{R}^1$, $0 \leq j \leq k$. This observation in the one-dimensional case leads to the following question: does a similar statement hold in the multi-dimensional case? The answer is negative and can be expressed as follows: in the case $n > 1$ the existence of a moment $m_{k_1,\dots,k_n}$ does not imply the existence of all moments $m_{j_1,\dots,j_n}$ for $0 \le j_i \le k_i$, $i = 1, \dots, n$. To see this, take $\Omega = (0,1)$, $\mathcal{F} = \mathcal{B}_{(0,1)}$ and $\mathbf{P}$ the Lebesgue measure. For fixed numbers c_1 and c_2, $0 < c_1 \le c_2 < 1$, define the following r.v.s:

$$X_1 = \begin{cases} \omega^{-1}, & \text{if } 0 < \omega < c_2 \\ 0, & \text{if } c_2 \le \omega < 1, \end{cases} \qquad X_2 = \begin{cases} 0, & \text{if } 0 < \omega \le c_1 \\ (1-\omega)^{-1}, & \text{if } c_1 < \omega < 1. \end{cases}$$

It is easy to check that the product $X_1 \cdot X_2$ is integrable, but neither X_1 nor X_2 is. Thus the moment $m_{1,1}$ of the vector (X_1, X_2) exists, but $m_{0,1}$ and $m_{1,0}$ do not exist. Obviously, if $c_1 < c_2$, then $m_{1,1} > 0$ and if $c_1 = c_2$, then $m_{1,1} = 0$.

6.7. On the validity of the Fubini theorem

Let $(\Omega_1, \mathcal{F}_1, \mathbf{P}_1)$ and $(\Omega_2, \mathcal{F}_2, \mathbf{P}_2)$ be two probability spaces. Then there exists only one probability $\mathbf{P}$ on the product $(\Omega_1 \times \Omega_2, \mathcal{F}_1 \times \mathcal{F}_2)$ such that

$$\mathbf{P}(A_1 \times A_2) = \mathbf{P}_1(A_1)\mathbf{P}_2(A_2), \quad A_1 \in \mathcal{F}_1,\ A_2 \in \mathcal{F}_2.$$

Further, for every non-negative (or quasi-integrable) r.v. X defined on the product space $(\Omega_1 \times \Omega_2, \mathcal{F}_1 \times \mathcal{F}_2, \mathbf{P})$, the following formula is both meaningful and valid:

$$\begin{aligned} \int_{\Omega_1 \times \Omega_2} &= \int_{\Omega_1} \mathbf{P}_1(\mathrm{d}\omega_1) \int_{\Omega_2} X_{\omega_1}(\omega_2)\mathbf{P}_2(\mathrm{d}\omega_2) \\ &= \int_{\Omega_2} \mathbf{P}_2(\mathrm{d}\omega_2) \int_{\Omega_1} X_{\omega_2}(\omega_1)\mathbf{P}_1(\mathrm{d}\omega_1) \end{aligned} \tag{1}$$

(for the proof see the books of Gihman and Skorohod (1974/1979) and Neveu (1965)).

Our purpose now is to show that the assumption that $\int X\,\mathrm{d}\mathbf{P}$ exists is essential for the validity of (1). Let $Z \ge 0$ be a non-integrable r.v. on $(\Omega_1, \mathcal{F}_1, \mathbf{P}_1)$ and define the variable X on the product of this space with the discrete space $\{0, 1\}$, both points having equal probabilities, by

$$X(\omega, 0) = Z(\omega), \quad X(\omega, 1) = -Z(\omega).$$

Then it is elementary to check that the second equality in (1) is violated.

6.8. A non-uniformly integrable family of random variables

Consider the sequence of r.v.s $\{X_n, n \ge 1\}$ where

$$\mathbf{P}[X_n = 2^n] = 2^{-n}, \quad \mathbf{P}[X_n = 0] = 1 - 2^{-n}.$$

($\{X_n\}$ arises in the so-called St. Petersburg paradox, see e.g. Székely 1986.) Then the following relation clearly holds:

$$\int_{|X_n| \ge a} |X_n|\,\mathrm{d}\mathbf{P} = \begin{cases} 0, & \text{if } a > 2^n \\ 1, & \text{if } a \le 2^n \end{cases}.$$

This means that $\int_{|X_n \geq a} |X_n| \, \mathrm{d}\mathbf{P}$ does not tend to zero uniformly in n as $a \to \infty$. However, for each n, X_n is integrable since $\mathbf{E}X_n = 1$.

Hence $\{X_n\}$ is an integrable but not uniformly integrable family of r.v.s.

6.9. On the relation $\mathbf{E}[\mathbf{E}(X|Y)] = \mathbf{E}X$

The definition of the conditional expectation of the r.v. X given another r.v. Y or some σ-field, requires X to be integrable: $\mathbf{E}[|X|] < \infty$. In this case the equality $\mathbf{E}[\mathbf{E}(X|Y)] = \mathbf{E}X$ holds. However, the following 'reasoning' appears to contradict this result.

Let Y be a positive r.v. whose density $g_\nu(y)$ is given by

$$(1) \qquad g_\nu(y) = (\tfrac{1}{2}\nu)^{\frac{1}{2}\nu}(\Gamma(\tfrac{1}{2}\nu))^{-1} y^{\frac{1}{2}\nu - 1} \mathrm{e}^{-\frac{1}{2}\nu y}, \quad y > 0, \ \nu > 0$$

(compare this with a gamma distribution). Suppose the conditional distribution of X given $Y = y$ is specified for $y > 0$ by the following probability density function:

$$(2) \qquad f(x|y) = (2\pi)^{-\frac{1}{2}} y^{\frac{1}{2}} \mathrm{e}^{-\frac{1}{2}yx^2}, \quad x \in \mathbb{R}^1.$$

Therefore

$$\mathbf{E}[X|Y = y] = \int_{-\infty}^{\infty} x f(x|y) \, \mathrm{d}x = 0 \Rightarrow \mathbf{E}[X|Y] = 0 \Rightarrow \mathbf{E}[\mathbf{E}(X|Y)] = 0.$$

On the other hand, (1) and (2) imply that the marginal density of X is

$$(3) \quad h_\nu(x) = \Gamma(\tfrac{1}{2}(\nu + 1))(\Gamma(\tfrac{1}{2}\nu)(\pi\nu)^{\frac{1}{2}})^{-1}(1 + x^2/\nu)^{-\frac{1}{2}(\nu+1)}, \quad x \in \mathbb{R}^1$$

that is, X has a *Student distribution* with ν degrees of freedom. In particular, for $\nu = 1$, X has a Cauchy distribution. In this case $\mathbf{E}X$ does not exist and hence $\mathbf{E}[\mathbf{E}(X|Y)] \neq \mathbf{E}X$.

The reason for this 'contradiction' is in the approach used above: we started from (2), which yields $\mathbf{E}(X|Y) = 0$, then from (1) and (2) derived (3) which is a density of a r.v. without expectation.

6.10. Is it possible to extend one of the properties of the conditional expectation?

Consider three r.v.s, say X, Y, Z. Suppose X is integrable and, moreover, X and Z are independent. Then $\mathbf{E}[X|Z] = \mathbf{E}X$ a.s. Having this property we can assume that

$$(1) \qquad \mathbf{E}[X|Y, Z] = \mathbf{E}[X|Y] \ \text{a.s.}$$

Our purpose now is to show that in general such an 'extension' is impossible. To see this, take $\Omega = [0, 1]$, $\mathcal{F} = \mathcal{B}_{[0,1]}$ and $\mathbf{P}$ the Lebesgue measure. Define the following r.v.s:

$$X(\omega) = \begin{cases} 1, & \text{if } \omega \in [0, \frac{1}{2}) \\ 0, & \text{if } \omega \in [\frac{1}{2}, 1], \end{cases}$$
$$Y(\omega) = \begin{cases} 1, & \text{if } \omega \in [0, \frac{3}{4}) \\ 0, & \text{if } \omega \in [\frac{3}{4}, 1], \end{cases}$$
$$Z(\omega) = \begin{cases} 1, & \text{if } \omega \in [\frac{1}{4}, \frac{3}{4}) \\ 0, & \text{if } \omega \notin [\frac{1}{4}, \frac{3}{4}]. \end{cases}$$

Then we can check that X and Z are independent. Furthermore,

$$\mathbf{E}[X|Y] = \begin{cases} \frac{2}{3}, & \text{if } \omega \in [0, \frac{3}{4}) \\ 0, & \text{otherwise}, \end{cases} \quad \mathbf{E}[X|Y,Z] = \begin{cases} 0, & \text{if } \omega \in [\frac{3}{4}, 1] \\ \frac{1}{2}, & \text{if } \omega \in [\frac{1}{4}, \frac{3}{4}) \\ 1, & \text{if } \omega \in [0, \frac{1}{4}). \end{cases}$$

Therefore $\mathbf{E}[X|Y,Z] \neq \mathbf{E}[X|Y]$ and in general (1) does not hold.

6.11. The mean–median–mode inequality may fail to hold

Suppose X is a r.v. with mean μ. A number m is called a *median* of X if $\mathbf{P}(X \geq m) \geq \frac{1}{2}$ and $\mathbf{P}(X \leq m) \geq \frac{1}{2}$. It is easy to see that such m always exists, but in general X may have several medians. If X is unimodal and M is its mode, then the median m is unique and for M, m and μ we have either $M \leq m \leq \mu$ or $M \geq m \geq \mu$—the median falls between the mean and the mode. A result of this kind is referred to as a *mean–median–mode inequality*.

Recall that the symbol $>_s$ is used to denote a *stochastic domination*: for two r.v.s ξ and η, $\xi >_s \eta \iff \mathbf{P}[\xi > x] \geq \mathbf{P}[\eta > x]$ for all x.

Let us cite the following statement (Dharmadhikari and Joag-Dev 1988): if X is a unimodal r.v. with mode M, median m and mean μ and $(X-m)^+ >_s (X-m)^-$, then $M \leq m \leq \mu$.

Our goal now is to describe a case when the mean–median–mode inequality does not hold. Consider a r.v. X with density

$$f(x) = \begin{cases} 0, & \text{if } x \leq 0 \\ x, & \text{if } 0 < x \leq c \\ ce^{-\lambda(x-c)}, & \text{if } x > c. \end{cases}$$

Here c and λ are positive constants and f is density iff $c^2/2 + c/\lambda = 1$. We can easily find the mean, the median and the mode of X:

$$\mu = \frac{c^3}{3} + \frac{c^2}{\lambda} + \frac{c}{\lambda^2}, \quad m = 1, \quad M = c.$$

Now let $c \to 1$. Then $\lambda \to 2$, $\mu \to \frac{13}{12} > 1$ and if c is sufficiently close to 1 but $c > 1$, then $\mu > c$ and $M = c > 1$. Here the median $m\,(=1)$ does not fall between the mean $\mu\,(>1)$ and the mode $M\,(>1)$, i.e. the mean–median–mode inequality does not hold despite the fact that the density f is unimodal.

6.12. Not all properties of conditional expectations have analogues for conditional medians

Recall that the *conditional median* of the r.v. X with respect to the σ-field $\mathcal{D}$ is defined as a $\mathcal{D}$-measurable r.v., say $\tilde{m}$, such that

$$\mathbf{P}[X \geq \tilde{m}|\mathcal{D}] \geq \tfrac{1}{2} \leq \mathbf{P}[X \leq \tilde{m}|\mathcal{D}].$$

By using the notation $\mu(X|\mathcal{D})$ for the conditional median we want to see if the properties of conditional expectations can be extended to conditional medians.

In the examples below X and Y are r.v.s, $\mathcal{D}$ is a σ-field, $\mathcal{F}_0$ is the trivial σ-field ($\mathcal{F}_0 = \{\emptyset, \Omega\}$) and $I(\cdot)$ is the indicator function.

(i) It is not always possible to find conditional medians satisfying

$$\mu(X + Y|\mathcal{D}) = \mu(X|\mathcal{D}) + \mu(Y|\mathcal{D}).$$

Indeed, let X_1 and X_2 be i.i.d. r.v.s with $\mathbf{P}[X_1 = 0] = \frac{1}{3} = 1 - \mathbf{P}[X_1 = 1]$ and put $X = X_1X_2$, $Y = X_1 - X_1X_2$. Then $\mu(X|\mathcal{F}_0) = 0$, $\mu(Y|\mathcal{F}_0) = 0$ and even $XY = 0$ while $\mu(X + Y|\mathcal{F}_0) = \mu(X_1|\mathcal{F}_0) = 1$. Thus the linear property of the conditional expectation ($\mathbf{E}[X + Y|\mathcal{D}] = \mathbf{E}[X|\mathcal{D}] + \mathbf{E}[Y|\mathcal{D}]$) does not in general hold for conditional medians.

(ii) It is not always possible to find conditional medians satisfying

$$\mu(\mu(X|\mathcal{D})|\mathcal{D}_1) = \mu(X|\mathcal{D}), \quad \mathcal{D}_1 \subset \mathcal{D}.$$

Consider the r.v.s X and Y where $\mathbf{P}[Y = k] = \frac{1}{3}$, $k = 0, 1, 2$; $\mathbf{P}[X = 1|Y = k] = \frac{3}{8} = 1 - \mathbf{P}[X = 0|Y = k]$, $k = 0, 1$, and $\mathbf{P}[X = 1|Y = 2] = \frac{7}{8} = 1 - \mathbf{P}[X = 0|Y = 2]$. Let $\mathcal{D}$ be the σ-field generated by Y. Since $\mathbf{P}[X = 1] = \frac{13}{24}$ then $\mu(X|\mathcal{F}_0) = 1$. However, $\mu(X|\mathcal{D}) = \mu(X|Y) = I(Y = 2)$ so $\mu(\mu(X|\mathcal{D})|\mathcal{F}_0) = 0$. Therefore the smoothing property ($\mathbf{E}[\mathbf{E}(X|\mathcal{D})|\mathcal{D}_1] = \mathbf{E}[X|\mathcal{D}_1]$) also does not in general hold for conditional medians.

(iii) If the r.v. X is independent of the σ-field $\mathcal{D}$, it does not necessarily follow that every conditional median $\mu(X|\mathcal{D})$ is constant. To see this we need the following result (Tomkins 1975a): if X is independent of $\mathcal{D}$, then every median $\mu(X|\mathcal{F}_0)$ of X is a conditional median of X with respect to $\mathcal{D}$.

Now consider two independent r.v.s X and Y, each taking the values 1 and 0 with probability $\frac{1}{2}$. Let $\mathcal{D} = \mathcal{D}^Y$ be the σ-field generated by Y. Then X is independent of $\mathcal{D}^Y$ but the conditional median of X with respect to $\mathcal{D}^Y$ is equal to Y, that is it is not constant.

SECTION 7. INDEPENDENCE OF RANDOM VARIABLES

Two r.v.s X_1 and X_2 on a given probability space $(\Omega, \mathcal{F}, \mathbf{P})$ are called *independent* if

$$\mathbf{P}[X_1 \in B_1, X_2 \in B_2] = \mathbf{P}[X_1 \in B_1]\mathbf{P}[X_2 \in B_2] \tag{1}$$

for any $B_1, B_2 \in \mathcal{B}^1$. If $F(x_1, x_2)$, $(x_1, x_2) \in \mathbb{R}^2$ is the joint d.f. of X_1 and X_2 and $F_1(x_1)$, $x \in \mathbb{R}^1$ and $F_2(x_2)$, $x_2 \in \mathbb{R}^1$ are their respective marginal d.f.s then (1) is expressed as

(2) $$F(x_1, x_2) = F_1(x_1)F_2(x_2) \quad \text{for all } x_1, x_2 \in \mathbb{R}^1.$$

In the absolutely continuous case the independence of X_1 and X_2 can be written in terms of the corresponding densities by

(3) $$f(x_1, x_2) = f_1(x_1)f_2(x_2) \quad \text{for all } x_1, x_2 \in \mathbb{R}^1.$$

If X_1 and X_2 are discrete r.v.s with $\mathbf{P}[X_1 = x_{1i}] = p_{1i}$, $p_{1i} > 0$, $i \geq 1$, $\sum_i p_{1i} = 1$ and $\mathbf{P}[X_2 = x_{2j}] = p_{2j}$, $p_{2j} > 0$, $j \geq 1$, $\sum_j p_{2j} = 1$, then X_1 and X_2 are independent iff

(4) $$\mathbf{P}[X_1 = x_{1i}, X_2 = x_{2j}] = \mathbf{P}[X_1 = x_{1i}]\mathbf{P}[X_2 = x_{2j}]$$

or, equivalently, $\tilde{p}_{ij} = p_{1i}p_{2j}$ for all possible i, j, where $\tilde{p}_{ij} := \mathbf{P}[X_1 = x_{1i}, X_2 = x_{2j}]$.

We say that $X_1, \dots, X_n$ is a family of *mutually independent r.v.s* if for every k, $2 \leq k \leq n$ and $1 \leq i_1 < i_2 < \dots < i_k \leq n$ the following relation holds:

(5) $$\mathbf{P}[X_{i_1} \in B_{i_1}, \dots, X_{i_k} \in B_{i_k}] = \mathbf{P}[X_{i_1} \in B_{i_1}] \dots \mathbf{P}[X_{i_k} \in B_{i_k}]$$

for arbitrary Borel sets $B_{i_1}, \dots, B_{i_k}$. If (5) is valid only for $k = 2$, the variables $X_1, \dots, X_n$ are called *pairwise independent*. It is clear how mutual independence and pairwise independence of r.v.s can be expressed through the corresponding d.f.s (see (2)), and how to do this in the absolutely continuous case (see (3)) and in the discrete case (see (4)).

Parallel to the notion of independence we can introduce the closely related notion of conditional independence. Let $\mathcal{D}$ be a σ-field, $\mathcal{D} \subset \mathcal{F}$ and $\mathcal{D}_1$, $\mathcal{D}_2$ be classes of events. Then $\mathcal{D}_1$ and $\mathcal{D}_2$ are said to be *conditionally independent* given $\mathcal{D}$ if, for all $D_1 \in \mathcal{D}_1$ and $D_2 \in \mathcal{D}_2$, the following relation holds:

$$\mathbf{P}(D_1 D_2 | \mathcal{D}) = \mathbf{P}(D_1 | \mathcal{D})\mathbf{P}(D_2 | \mathcal{D}) \quad \text{a.s.}$$

Obviously this definition includes the conditional independence of random events and of random variables.

Let X and Y be r.v.s with $0 < \mathbf{V}X < \infty$, $0 < \mathbf{V}Y < \infty$. The quantity

$$\rho(X, Y) = \frac{\mathbf{E}[(X - \mathbf{E}X)(Y - \mathbf{E}Y)]}{(\mathbf{V}X\mathbf{V}Y)^{1/2}}$$

is said to be a *correlation coefficient* between X and Y (simply, a correlation of X and Y). If $\rho(X, Y) = 0$, the variables X and Y are called *uncorrelated*.

We refer the reader to the books by Feller (1968, 1971), Chung (1974), Chow and Teicher (1978), Laha and Rohatgi (1979), Shiryaev (1995) for a detailed treatment of the notion of independence and several related topics.

The examples in this section examine the relationship between independence, dependence and related properties of r.v.s.

7.1. Discrete random variables which are pairwise but not mutually independent

Using some of the examples in Section 3 we can easily construct sets of r.v.s with different independence/dependence properties.

(i) Let (X, Y, Z) take each of the values (1,0,0), (0,1,0), (0,0,1), (1,1,1) with probability $\frac{1}{4}$. Then X, Y and Z are pairwise independent. For example, it is easy to see that $\mathbf{P}[X = 1, Z = 0] = \frac{1}{4} = \frac{1}{2} \cdot \frac{1}{2} = \mathbf{P}[X = 1]\mathbf{P}[Z = 0]$. However,

$$\mathbf{P}[X = 1, Y = 1, Z = 1] = \tfrac{1}{4} \neq \tfrac{1}{8} = \mathbf{P}[X = 1]\mathbf{P}[Y = 1]\mathbf{P}[Z = 1],$$

and hence the three variables are not mutually independent.

(ii) Let Ω consist of nine points: the permutations of 1, 2, 3 and the triplets (1,1,1), (2,2,2), (3,3,3). Each has probability $\frac{1}{9}$. Introduce three r.v.s, say X_1, X_2, X_3, where X_k equals the number appearing at the kth place. The possible values of these variables are 1, 2, 3 and we can easily show that

$$\text{(1)} \quad \mathbf{P}[X_k = i] = \tfrac{1}{3}, \ \mathbf{P}[X_k = i, X_l = j] = \tfrac{1}{9}, \ k, l = 1, 2, 3, \ i, j = 1, 2, 3.$$

It follows immediately from (1) that X_1, X_2, X_3 are pairwise independent. Since X_1 and X_2 uniquely determine X_3, the three variables are not mutually independent.

(iii) Let us continue the construction in case (ii). Consider new triplets $(X_4, X_5, X_6), (X_7, X_8, X_9), \ldots$, similar in structure to (X_1, X_2, X_3) and each independent of (X_1, X_2, X_3). Thus we obtain an infinite sequence of r.v.s $X_1, X_2, \ldots, X_n, \ldots$. Clearly, any two members X_k, X_l of this sequence satisfy relations (1). However the product rule does not hold for any k, $k \geq 3$, of these variables. Thus the r.v.s $\{X_n, n \geq 1\}$ are only pairwise independent.

7.2. Absolutely continuous random variables which are pairwise but not mutually independent

Let ξ and η be two independent r.v.s uniformly distributed in the interval $(0, \pi)$. Define the variables $X_1 = \tan \xi$, $X_2 = \tan \eta$, $X_3 = -\tan(\xi + \eta)$. The variables X_1 and X_2 are independent, as a consequence of the independence of ξ and η. By finding the distribution of X_3 we can establish that X_3 and X_1 are independent, as are X_3 and X_2. However, these variables are functionally dependent by the relation $X_1 + X_2 + X_3 = X_1 X_2 X_3$ and thus they cannot be mutually independent.

Thus we have constructed a triplet of r.v.s which are pairwise but not mutually independent (equivalently, independent at level 2 and dependent at level 3).

7.3. A set of dependent random variables such that any of its subsets consists of mutually independent variables

If $X_1, \ldots, X_n$ are r.v.s, $n \geq 3$, and we know that they are mutually independent, then any proper subset of them consists of mutually independent variables. However, in general the converse statement is not true (see Examples 7.1 and 7.2, or construct analogues to some of the examples in Section 3). Here we shall consider two examples covering the discrete and the absolutely continuous cases.

(i) Let $n \geq 3$ and $A \subset \mathbb{R}^{n-1}$ be the set of all $(n-1)$-dimensional vectors of the type $a = (\alpha_1, \ldots, \alpha_{n-1})$ where $\alpha_i = 1$ or 0, $i = 1, \ldots, n-1$. Obviously A contains 2^{n-1} elements (vectors): $|A| = 2^{n-1}$. Let $I(a) = \alpha_1 + \cdots + \alpha_{n-1}$, so $I(a)$ takes values $0, 1, \ldots, n-1$. Let $B \subset \mathbb{R}^n$ be the set of all vectors b where

$$b = \begin{cases} (\alpha_1, \ldots, \alpha_{n-1}, 1), & \text{if } I(a) \text{ is even} \\ (\alpha_1, \ldots, \alpha_{n-1}, 0), & \text{if } I(a) \text{ is odd.} \end{cases}$$

Then $T : a \mapsto b$ is a one–one mapping of A onto B, so $|B| = 2^{n-1}$ and, moreover, B is permutation invariant.

Let $a^{(j)}$ be an $(n-1)$-dimensional vector obtained from b by eliminating the jth component of b. Denote by $A^{(j)}$ the set of all such vectors $a^{(j)}$. Thus we have defined the mapping $T_j^{-1} : B \mapsto A^{(j)}$. Clearly, $A^{(n)} = A$ and since B is permutation invariant, we have $A^{(j)} = A^{(n)}$ for all $j = 1, \ldots, n-1$ and hence $A^{(j)} = A$ for all $j = 1, \ldots, n$.

Now define the n-dimensional random vector $X = (X_1, \ldots, X_n)$ taking values in the set B and with a distribution given by

$$\text{(1)} \qquad \mathbf{P}[X = x] = \begin{cases} 2^{-(n-1)}, & \text{if } x \in B \\ 0, & \text{otherwise.} \end{cases}$$

Let $X^{(j)} = T_j^{-1}(X) = (X_1, \ldots, X_{j-1}, X_{j+1}, \ldots, X_n)$. Since T_j^{-1} are one–one mappings of B onto A, we find easily that the distribution of $X^{(j)}$ is given by

$$\text{(2)} \qquad \mathbf{P}[X^{(j)} = x^{(j)}] = \begin{cases} 2^{-(n-1)}, & \text{if } x^{(j)} \in A \\ 0, & \text{otherwise.} \end{cases}$$

The next step is to use relation (2) in order to find the marginal distribution of each of the components X_i of the vector X. We have

$$\text{(3)} \qquad \mathbf{P}[X_i = x_i] = \begin{cases} \frac{1}{2}, & \text{if } x_i = 0 \text{ or } 1 \\ 0, & \text{otherwise.} \end{cases}$$

Now, comparing (1), (2) and (3) we arrive at the following conclusion: we have constructed n dependent discrete r.v.s $X_1, \ldots, X_n$ which are $(n-1)$-wise independent, that is any proper subset of which consists of mutually independent variables being in this case even identically distributed.

(ii) Let X be a r.v. with density function f and mean $\mu = \mathbf{E}X$. Let $X_1, \ldots, X_n$, $n \geq 3$, be r.v.s and take a function of the following type as their joint density:

$$(4) \qquad g_n(x_1, \ldots, x_n) = \Big[\prod_{i=1}^{n} f(x_i)\Big]\Big[1 + \prod_{j=1}^{n}(x_j - \mu)f(x_j)\Big], \text{ each } x_j \in \mathbb{R}^1.$$

We consider g_n only for those $x_j \in \mathbb{R}^1$, $j = 1, \ldots, n$, for which $|x_j - \mu| f(x_j) < 1$. Otherwise we put $g_n(\cdot) = 0$. Then g_n is a non-negative function. In order for (1) to be a density function, the integral of g_n over the range of $(x_1, \ldots, x_n)$ described above must be equal to 1. This leads to the condition

$$(5) \qquad \int_{-\infty}^{\infty} (x - \mu) f^2(x) \mathrm{d}x = 0.$$

Notice that (5) is satisfied if, for example, the density f is symmetric about its mean value μ.

Let the density f satisfy (5), g_n be defined by (4), and $X_1, \ldots, X_n$ be r.v.s with density g_n. Our purpose now is to establish what dependence there is between these n variables.

By direct integration of (4) we find that each of the r.v.s $X_1, \ldots, X_n$ has as its density the given function f. Suppose we have chosen k of the Xs, without restriction we can choose $X_1, \ldots, X_k$, $2 \leq k < n$. Denote by $h_k(x_1, \ldots, x_k)$, $(x_1, \ldots, x_k) \in \mathbb{R}^k$ the joint density of $X_1, \ldots, X_k$. Then from (4) and (5) we can easily show that

$$h_k(x_1, \ldots, x_k) = f(x_1) \ldots f(x_k).$$

Obviously this relation implies that $X_1, \ldots, X_k$ are mutually independent. Of course, the same holds for any k-subset of $X_1, \ldots, X_n$ where $2 \leq k < n$. Nevertheless all n r.v.s $X_1, \ldots, X_n$ are not mutually independent because (4) implies that $g_n(x_1, \ldots, x_n) \neq f(x_1) \ldots f(x_n)$.

It is useful to consider the following case. Let X be distributed uniformly on the interval $(0, c)$, $0 < c < \infty$. Its density is $f(x) = 1/c$ for $0 < x < c$ and 0 otherwise. Then $\mu = \mathbf{E}X = \frac{1}{2}c$ and (5) is satisfied. Take the random vector $X_1, \ldots, X_n$ with density

$$g_n(x_1, \ldots, x_n) = \begin{cases} c^{-n}[1 + \prod_{i=1}^{n}(x_i - \frac{1}{2}c)c^{-1}], & \text{if } 0 < x_i < c, i = 1, \ldots, n \\ 0, & \text{otherwise.} \end{cases}$$

Clearly g_n is not the uniform density on the n-dimensional cube $(0, c)^n$ in $\mathbb{R}^n$ and $X_1, \ldots, X_n$ cannot be mutually independent. However, any k of them, $2 \leq k < n$, will be distributed uniformly in the cube $(0, c)^k$ in $\mathbb{R}^k$ and these k variables are mutually independent.

Hence we have described collections of n dependent absolutely continuous r.v.s which are $(n - 1)$-wise independent.

(iii) Consider the following function

$$f(x_1, \dots, x_n) = \begin{cases} (2\pi)^{-n}[1 - \cos x_1 \dots \cos x_n], & \text{if } (x_1, \dots, x_n) \in Q_n \\ 0, & \text{otherwise} \end{cases}$$

where Q_n is the n-dimensional cube $[0, 2\pi]^n$ in $\mathbb{R}^n$. It is easy to check that f is non-negative and the integral of f over $\mathbb{R}^n$ equals 1. Hence f is a probability density function of a random vector in $\mathbb{R}^n$, say of $(X_1, \dots, X_n)$.

Denoting by $f_k(x_k)$ the marginal density of the component X_k, we find that

$$f_k(x_k) = \begin{cases} 1/(2\pi), & \text{if } 0 \le x_k \le 2\pi \\ 0, & \text{otherwise} \end{cases}$$

implying that X_k is uniformly distributed on the interval $[0, 2\pi]$ and this holds for any (single) r.v. $X_1, X_2, \dots, X_n$. The form of their joint density f shows that these variables are not independent. If, however, we take any k of them, we conclude that for $2 \le k \le n-1$ they are independent (their joint density is equal to $1/(2\pi)^k$ on the cube $Q_k = [0, 2\pi]^k$ in $\mathbb{R}^k$).

Therefore $X_1, \dots, X_k$ is another collection of n dependent r.v.s which are $(n-1)$-wise independent. (Compare with case (ii) above.)

7.4. Collection of n dependent random variables which are k-wise independent

In Example 7.3 we have described collections of n dependent r.v.s which are $(n-1)$-wise independent. Thus it is of a general interest to see collections of n dependent r.v.s which are k-wise independent with $k < n - 1$.

We present two examples: in the first we have $n = 4$, $k = 2$, while in the second $n = 5$, $k = 3$.

(i) Let F_1, F_2, F_3, F_4 be d.f.s on $\mathbb{R}^1$ (or on its subsets). Denote $G_j = 1 - F_j$ and define the function $H_{1234}(x_1, x_2, x_3, x_4)$, $(x_1, x_2, x_3, x_4) \in \mathbb{R}^4$ as follows (for simplicity we omit the arguments but we know they are real):

$$H_{1234} = F_1F_2F_3F_4\{1 + \varepsilon_1 G_2G_3G_4 + \varepsilon_2 G_1G_3G_4 + \varepsilon_3 G_1G_2G_4 + \varepsilon_4 G_1G_2G_3\}.$$

Our first claim is that if ε_1, ε_2, ε_3, ε_4, are non-zero numbers in the interval $(-1, 1)$ and $|\varepsilon_1| + |\varepsilon_2| + |\varepsilon_3| + |\varepsilon_4| < 1$, then H_{1234} is a four-dimensional d.f. Let $(\xi_1, \xi_2, \xi_3, \xi_4)$ be a random vector whose d.f. is just H_{1234}. We are interested in the independence/dependence properties of the components of this vector, so we need to know its k-dimensional marginal distributions for $k = 3$, 2 and 1. For example, if H_{123}, H_{12} and H_1 are the d.f.s of (ξ_1, ξ_2, ξ_3), (ξ_1, ξ_2) and ξ_1 respectively, we easily find that

$$H_{123} = F_1F_2F_3\{1 + \varepsilon_4 G_1G_2G_3\}, \; H_{12} = F_1F_2 \text{ and } H_1 = F_1.$$

It is quite clear how to write down the d.f. of any possible subset of components of the vector $(\xi_1, \xi_2, \xi_3, \xi_4)$.

Thus we arrive at the following conclusions:

(a) ξ_j has a d.f. equal to F_j, $j = 1, 2, 3, 4$;
(b) any two of the r.v.s $\xi_1, \xi_2, \xi_3, \xi_4$ are independent;
(c) any three of them as well as all four are dependent.

Therefore $\{\xi_1, \xi_2, \xi_3, \xi_4\}$ is a collection of dependent r.v.s which are twice-wise (= pairwise) independent.

(ii) Suppose that we have five d.f.s F_1, F_2, F_3, F_4, F_5 and as above we use the notation $G_j = 1 - F_j$, $j = 1, \dots, 5$. Define the function $H_{12345}(x_1, x_2, x_3, x_4, x_5)$, $(x_1, x_2, x_3, x_4, x_5) \in \mathbb{R}^5$ as follows:

$$\begin{aligned} H_{12345} &= F_1F_2F_3F_4F_5\{1 + \varepsilon_1 G_2G_3G_4G_5 + \varepsilon_2 G_1G_3G_4G_5 \\ &\quad + \varepsilon_3 G_1G_2G_4G_5 + \varepsilon_4 G_1G_2G_3G_5 + \varepsilon_5 G_1G_2G_3G_4\}. \end{aligned}$$

If $\varepsilon_1, \varepsilon_2, \varepsilon_3, \varepsilon_4, \varepsilon_5$ are non-zero numbers in the interval $(-1, 1)$ and $|\varepsilon_1| + |\varepsilon_2| + |\varepsilon_3| + |\varepsilon_4| + |\varepsilon_5| < 1$, then H_{12345} is a five-dimensional d.f. of a random vector in $\mathbb{R}^5$, say $(\eta_1, \eta_2, \eta_3, \eta_4, \eta_5)$. In order to clarify what kind of independence/dependence there exists between the components of this vector, we have first to find all k-dimensional marginal distributions for k = 4, 3, 2, 1. In particular, if H_{1234}, H_{123}, H_{12} and H_1 are the d.f.s of $(\eta_1, \eta_2, \eta_3, \eta_4)$, (η_1, η_2, η_3), (η_1, η_2) and η_1 respectively, we find that

$$H_{1234} = F_1F_2F_3F_4\{1 + \varepsilon_5 G_1G_2G_3G_4\}, \ H_{123} = F_1F_2F_3, \ H_{12} = F_1F_2, \ H_1 = F_1.$$

Similarly we can write the d.f.s in all the remaining cases, thus arriving at the following conclusions:

(a) η_j has a d.f. equal to F_j, $j = 1, 2, 3, 4, 5$;
(b) any two of the r.v.s $\eta_1, \eta_2, \eta_3, \eta_4, \eta_5$ are independent;
(c) any three of them are independent;
(d) any four, as well as all five, variables are dependent.

Hence $\{\eta_1, \eta_2, \eta_3, \eta_4, \eta_5\}$ is a collection of dependent r.v.s which are three-wise independent.

Note finally that a similar idea can be used when describing n dependent r.v.s which are m-wise independent. In cases (i) and (ii) above, as well as in the general case, the description can be done in terms of probability density functions.

7.5. An independence-type property for random variables

Let $X_1, X_2, \dots$ be positive integer-valued r.v.s and $S_k = X_1 + \cdots + X_k$. Suppose that $Y_1, Y_2, \dots$ is another sequence of i.i.d. positive integer-valued r.v.s with $\mathbf{P}[Y_1 = i] = p_i$, $p_i > 0$, $\sum_{i=1}^{\infty} p_i = 1$, and for all $k \geq 1$ and $i \geq 1$ the following relation holds:

$$\text{(1)} \qquad \mathbf{P}[S_k = i] = \mathbf{P}[Y_1 + \cdots + Y_k = i].$$

For various purposes one needs to find $\mathbf{P}[S_1 = i_1, S_2 = i_2, \ldots, S_k = i_k]$. Taking into account (1), the equalities $S_2 = i_1 + X_2$, $S_3 = i_2 + X_3, \ldots, S_k = i_{k-1} + X_k$ and the independence of Ys, we can suppose that

$$(2) \qquad \mathbf{P}[S_1 = i_1, S_2 = i_2, \ldots, S_k = i_k] = p_{i_1} p_{i_2 - i_1} \cdots p_{i_k - i_{k-1}}.$$

Obviously (2) is satisfied if the variables $X_1, X_2, \ldots$ are independent. Thus we want to know whether or not relation (2) holds for any choice of the sequence $\{X_k\}$.

Let p_1, p_2, p_3 be positive numbers with $p_1 + p_2 + p_3 = 1$. Denote by Y a r.v. taking the values 1, 2, 3 with probabilities p_1, p_2, p_3 respectively, and let $\{Y_k, k \geq 1\}$ be a sequence of independent copies of Y.

Now define the pair of r.v.s (X_1, X_2) as follows:

$$\mathbf{P}[X_1 = i, X_2 = j] = p_i p_j + \varepsilon_{ij}, \quad i, j = 1, 2, 3$$

with $\varepsilon_{11} = \varepsilon_{22} = \varepsilon_{33} = 0$, $\varepsilon_{21} = \varepsilon_{32} = \varepsilon_{13} = \varepsilon$ and $\varepsilon_{12} = \varepsilon_{23} = \varepsilon_{31} = -\varepsilon$ where the real number ε is chosen so that $|\varepsilon| \leq \min\{p_1p_2, p_2p_3, p_1p_3\}$. Let$(X_3, X_4), (X_5, X_6), \ldots$ be independent copies of the pair (X_1, X_2). Thus we obtain the sequence $X_1, X_2, \ldots, X_n, \ldots$.

We want to determine whether the sequences $\{X_k\}$ and $\{Y_k\}$ just defined satisfy conditions (1) and (2). Evidently, for all i, j we have

$$\mathbf{P}[X_1 = i] = p_i \quad \text{and} \quad \mathbf{P}[X_1 + X_2 = j] = \mathbf{P}[Y_1 + Y_2 = j]$$

and (1) holds. Furthermore, if $\varepsilon \neq 0$ then

$$\mathbf{P}[S_1 = 2, S_2 = 3] = \mathbf{P}[X_1 = 2, X_2 = 1] = p_2 p_1 + \varepsilon \neq p_2 p_1$$

and hence (2) is not satisfied. Therefore the independence property for the sequence $\{X_k\}$ is essential for the validity of (2).

7.6. Dependent random variables X and Y such that X^2 and Y^2 are independent

It is well known that if X and Y are independent r.v.s, then for any continuous functions g and h, the r.v.s $g(X)$ and $h(Y)$ are also independent (see Gnedenko 1962; Feller 1971). The converse statement is true if the functions g and h are one–one mappings of $\mathbb{R}^1$ to $\mathbb{R}^1$. However, we can choose functions g and h without this condition such that $g(X)$ and $h(Y)$ are independent r.v.s but X and Y themselves are not. We present two examples treating the discrete and the absolutely continuous cases.

(i) Consider the two-dimensional random vector (X, Y) with

$$p_{i,j} := \mathbf{P}[X = i, Y = j], \quad i, j = -1, 0, 1$$

where

$$p_{1,1} = p_{-1,1} = \tfrac{1}{32}, \quad p_{-1,-1} = p_{1,-1} = p_{1,0} = p_{0,1} = \tfrac{3}{32},$$
$$p_{-1,0} = p_{0,-1} = \tfrac{5}{32}, \quad p_{0,0} = \tfrac{8}{32}.$$

It is easy to check that X^2 and Y^2 are independent r.v.s but X and Y are not.

(ii) Let X_1 and X_2 be two independent absolutely continuous r.v.s. Take another r.v. Y which is independent on X_1, X_2 and assumes the values $+1$ and -1 with probability $\frac{1}{2}$ each. Define two new r.v.s, say Z_1 and Z_2, by

$$Z_1 = YX_1, \quad Z_2 = YX_2.$$

The absolute continuity of X_1 and X_2 implies that Z_1 and Z_2 are absolutely continuous. Obviously, Z_1 and Z_2 are functionally connected and thus they cannot be independent. However, $Z_1^2 = X_1^2$, $Z_2^2 = X_2^2$ and, since X_1 and X_2 are independent, Z_1^2 and Z_2^2 are independent.

(iii) Here is another illustration. Let the random vector (X, Y) have the following density (compare with Example 5.8(ii)):

$$f(x,y) = \begin{cases} \frac{1}{4}(1+xy), & \text{if } |x| < 1 \text{ and } |y| < 1 \\ 0, & \text{otherwise.} \end{cases}$$

We easily find the marginal densities $f_1(x)$ of X and $f_2(y)$ of Y:

$$f_1(x) = \begin{cases} \frac{1}{2}, & \text{if } |x| < 1 \\ 0, & \text{otherwise,} \end{cases} \qquad f_2(y) = \begin{cases} \frac{1}{2}, & \text{if } |y| < 1 \\ 0, & \text{otherwise.} \end{cases}$$

Obviously $f(x,y) \neq f_1(x)f(y)$ for all x and y, hence X and Y are dependent.

Each of the variables X^2 and Y^2 takes values in (0,1) and for $x \in (0,1)$ and $y \in (0,1)$ we find

$$\begin{aligned} \mathbf{P}[X^2 < x, Y^2 < y] &= \mathbf{P}[-\sqrt{x} < X < \sqrt{x}, -\sqrt{y} < Y < \sqrt{y}] \\ &= \frac{1}{4}\int_{-\sqrt{x}}^{\sqrt{x}}\int_{-\sqrt{y}}^{\sqrt{y}} (1+uv)\,\mathrm{d}u\,\mathrm{d}v \\ &= \sqrt{x}\sqrt{y} = \mathbf{P}[X^2 < x]\mathbf{P}[Y^2 < y], \quad x, y \in (0,1). \end{aligned}$$

Thus X^2 and Y^2 are independent r.v.s.

7.7. The independence of random variables in terms of characteristic functions

If X is a r.v. defined on a given probability space $(\Omega, \mathcal{F}, \mathbf{P})$, then the function $\phi(t) = \mathbf{E}[\mathrm{e}^{itX}]$, $t \in \mathbb{R}^1$, $i = \sqrt{-1}$ is called a *characteristic function* (ch.f.) of X. An extensive treatment of ch.f.s is given in Section 8. Here we illustrate the independence property of r.v.s in terms of the corresponding ch.f.s.

Let X_1, X_2 be independent r.v.s and ϕ_1, ϕ_2 their characteristic functions (ch.f.s) respectively. Then the ch.f. ϕ of the sum $X_1 + X_2$ is $\phi_1\phi_2$:

$$(1) \qquad \phi(t) = \phi_1(t)\phi_2(t) \quad \text{for all } t \in \mathbb{R}^1.$$

We can pose the converse question: if ϕ_1, ϕ_2 and ϕ are the ch.f.s of X_1, X_2, and $X_1 + X_2$ and (1) holds, does it follow that X_1 and X_2 are independent? Let us show that the answer to this question is negative.

(i) Let the random vector (X_1, X_2) have density

$$(2) \qquad f(x_1, x_2) = \begin{cases} \frac{1}{4}[1 + x_1x_2(x_1^2 - x_2^2)], & \text{if } |x_1| \le 1 \text{ and } |x_2| \le 1 \\ 0, & \text{otherwise.} \end{cases}$$

First, we find from (2) the marginal densities f_1 and f_2 of X_1 and X_2, namely

$$f_1(x_1) = \begin{cases} \frac{1}{2}, & \text{if } |x_1| \le 1 \\ 0, & \text{otherwise,} \end{cases} \qquad f_2(x_2) = \begin{cases} \frac{1}{2}, & \text{if } |x_2| \le 1 \\ 0, & \text{otherwise.} \end{cases}$$

Since $f(x_1, x_2) \ne f_1(x_1)f_2(x_2)$, the r.v.s X_1 and X_2 are not independent.

Second, the variables X_1 and X_2 are identically distributed and for their ch.f.s ϕ_1 and ϕ_2 we can easily show that

$$\phi_1(t) = \phi_2(t) = t^{-1}\sin t, \quad t \in \mathbb{R}^1.$$

Third, denote by g the density of the sum $X_1 + X_2$. Then g is expressed by f from (2) as $g(x) = \int_{-\infty}^{\infty} f(x_1, x - x_1)\,\mathrm{d}x_1$ and a direct integration yields

$$g(x) = \begin{cases} \frac{1}{4}(2 + x), & \text{if } -2 \le x \le 0 \\ \frac{1}{4}(2 - x), & \text{if } 0 < x \le 2 \\ 0, & \text{if } |x| > 2. \end{cases}$$

Having g, we find that the ch.f. ϕ of $X_1 + X_2$ is

$$\phi(t) = t^{-2}\sin^2 t.$$

Therefore $\phi(t) = \phi_1(t)\phi_2(t)$, that is relation (1) is satisfied, but, as we saw above, the variables X_1 and X_2 are dependent.

(ii) Take $X_1 = X_2 = X$ where X has a Cauchy distribution with density $1/[\pi(1 + x^2)]$, $x \in \mathbb{R}^1$. If ϕ_1, ϕ_2 and ϕ are the ch.f.s of X_1, X_2 and $X_1 + X_2$ respectively, we have $\phi_1(t) = \phi_2(t) = \mathrm{e}^{-|t|}$, $\phi(t) = \mathrm{e}^{-2|t|}$. Hence $\phi(t) = \phi_1(t)\phi_2(t)$ for all $t \in \mathbb{R}^1$, but clearly X_1 and X_2 are not independent.

Finally, let us recall that the r.v.s $X_1, \dots, X_n$ with ch.f.s $\phi_1, \dots, \phi_n$ are independent iff for all real $t_1, \dots, t_n$

$$\mathbf{E}[\exp(i(t_1X_1 + \cdots + t_nX_n))] = \phi_1(t_1)\dots\phi_n(t_n).$$

Comparing (1) with this general condition enables us to explain the conclusions obtained in the examples above.

7.8. The independence of random variables in terms of generating functions

If X is a non-negative integer-valued r.v., then the function $p(z) = \mathbf{E}[z^X]$ is called a *probability generating function* (p.g.f.). Recall that $p(z)$ is defined for all complex numbers z with $|z| \le 1$. Further, if X is an arbitrary r.v., then the function $M(z) = \mathbf{E}[e^{zX}]$, z complex, is called a *moment generating function* (m.g.f.) of X. More on p.g.f.s and m.g.f.s is included in Section 8. Here we are interested in expressing the independence property of r.v.s by the corresponding generating functions.

(i) Let X and Y be independent non-negative integer-valued r.v.s. Denote by p_X, p_Y, and p_{X+Y} the probability generating functions of X, Y and $X+Y$ respectively. Then

$$p_{X+Y}(z) = p_X(z)p_Y(z). \tag{1}$$

It is natural to ask the following question: if X and Y are non-negative, integer-valued r.v.s such that (1) is satisfied, does it follow that X and Y are independent? We show by an example that in general the answer is negative.

Let ξ and η be independent r.v.s such that ξ takes the values 0, 1 and 2 with probability $\frac{1}{3}$ each, and η takes the values 0 and 1 with probabilities $\frac{1}{3}$ and $\frac{2}{3}$ respectively. Define $X = \xi$ and $Y = \xi + \eta \pmod 3$. Then Y takes the values 0, 1 and 2 with probability $\frac{1}{3}$ each. Further, the sum $X + Y$ takes the values 0, 1, 2, 3 and 4 with probabilities $\frac{1}{9}$, $\frac{2}{9}$, $\frac{4}{9}$, $\frac{2}{9}$ and $\frac{1}{9}$ respectively. Obviously relation (1) is satisfied for the p.g.f.s of X, Y and $X+Y$. However, the variables X and Y are not independent; they are functionally dependent.

In addition, we can show that X and Y are uncorrelated (for this property see Examples 7.9 and 7.10 below).

(ii) If X and Y are arbitrary r.v.s which are independent and M_X, M_Y and M_{X+Y} are the m.g.f.s of X, Y and $X+Y$ respectively, then

$$M_{X+Y}(z) = M_X(z)M_Y(z). \tag{2}$$

As in case (i) we want to know if (2) implies the independence of X and Y. The answer will follow from the example below.

Let (X, Y) be a two-dimensional random vector defined by the table:

X \ Y	1	2	3
1	$\frac{2}{18}$	$\frac{1}{18}$	$\frac{3}{18}$
2	$\frac{3}{18}$	$\frac{2}{18}$	$\frac{1}{18}$
3	$\frac{1}{18}$	$\frac{3}{18}$	$\frac{2}{18}$

We can easily find that X and Y are identically distributed r.v.s taking each of the values 1, 2, 3 with probability $\frac{1}{3}$. The sum $Z = X + Y$ is a r.v. taking the values 2, 3, 4, 5, 6 with probabilities $\frac{1}{9}, \frac{2}{9}, \frac{3}{9}, \frac{2}{9}, \frac{1}{9}$ respectively. Since X, Y and $X + Y$ are non-negative and integer-valued, we can study their properties in terms of the p.g.f.s. But in all cases we can use m.g.f.s. Thus for the m.g.f.s we get

$$\begin{aligned} M_X(z) &= \mathbf{E}[e^{zX}] = M_Y(z) = \mathbf{E}[e^{zY}] = \tfrac{1}{3}(e^z + e^{2z} + e^{3z}), \\ M_Z(z) &= M_{X+Y}(z) = \tfrac{1}{9}(e^{2z} + 2e^{3z} + 3e^{4z} + 2e^{5z} + e^{6z}). \end{aligned}$$

Clearly $M_{X+Y}(z) = M_X(z)M_Y(z)$, i.e. relation (2) is satisfied. However, the r.v.s X and Y are not independent as can be seen easily from the table above: $\mathbf{P}[X = i, Y = j] \neq \mathbf{P}[X = i]\mathbf{P}[Y = j]$ for all $i \neq j$.

Finally, let us comment on both cases (i) and (ii). The independence of two (or more) r.v.s can be expressed in terms of the p.g.f.s or the m.g.f.s. Let us illustrate this for two variables.

If (X_1, X_2) is a random vector whose components X_1 and X_2 are non-negative integer-valued r.v.s, then its p.g.f., say $p(z_1, z_2)$, is defined as

$$p(z_1, z_2) = \mathbf{E}[z_1^{X_1} z_2^{X_2}], \text{ complex } z_1, z_2, \ |z_1| \leq 1, |z_2| \leq 1.$$

Denote by $p_1(z_1)$ the p.g.f. of X_1 and $p_2(z_2)$ the p.g.f. of X_2. Then X_1 and X_2 are independent iff $p(z_1, z_2) = p_1(z_1)p_2(z_2)$ for all z_1 and z_2. For $z_1 = z_2 = z$, the function $p(z, z) = \mathbf{E}[z^{X_1+X_2}]$ is the p.g.f. of the sum $X_1 + X_2$ in which case $p(z, z) = p_1(z)p_2(z)$. This is exactly case (i) above where we do not have $p(z_1, z_2) = p(z_1)p(z_2)$ for all z_1, z_2, i.e. we do not have independent X_1 and X_2.

For an arbitrary random vector (X_1, X_2) the m.g.f. is defined by

$$M(z_1, z_2) = \mathbf{E}[\exp(z_1 X_1 + z_2 X_2)], \quad z_1, z_2 \text{ complex.}$$

Denote by $M_1(z_1)$ and $M_2(z_2)$ the m.g.f.s of X_1 and X_2 respectively, and $|z_1| \leq r$ and $|z_2| \leq r$, given $r \geq 0$. Then X_1 and X_2 are independent iff $M(z_1, z_2) = M_1(z_1)M_2(z_2)$ for all z_1, z_2. If we take $z_1 = z$, $z_2 = z$ we get the function $M(z, z)$ which is the m.g.f. of the sum $X_1 + X_2$. Obviously in this case $M(z, z) = M_1(z)M_2(z)$. We met this equality in case (ii) above. However, in this case $M(z_1, z_2) = M_1(z_1)M_2(z_2)$ does not hold for all z_1 and z_2. This explains why X_1 and X_2 are not independent.

7.9. The distribution of a sum can be expressed by the convolution even if the variables are dependent

If X_1 and X_2 are r.v.s with d.f.s F_1 and F_2 respectively, and X_1, X_2 are independent, the distribution of the sum $X_1 + X_2$ is $F_1 * F_2$. If X_1 and X_2 are absolutely continuous with densities f_1 and f_2 respectively, then the density of $X_1 + X_2$ is $f_1 * f_2$.

Now we are interested in the converse: what is the connection between the r.v.s X_1 and X_2 if we know that the sum $X_1 + X_2$ has distribution $F_1 * F_2$ or density $f_1 * f_2$? The answer will follow from an example based on the Cauchy distribution.

Let $f_a(x) = a/[\pi(a^2 + x^2)]$, $x \in \mathbb{R}^1$ be the density of a Cauchy distribution, where $a > 0$. It is easy to check, for example by using ch.f.s, that the family of Cauchy densities is closed under convolutions. Consider two independent r.v.s ξ and η each with density f_a. Let $X = \alpha\xi + \beta\eta$, $Y = \gamma\xi + \delta\eta$ where α, β, γ, δ are arbitrary real numbers. Then the sum $X + Y$ has density $f_{(\alpha+\beta+\gamma+\delta)a}$, which is the convolution of the densities $f_{(\alpha+\beta)a}$ of X and $f_{(\gamma+\delta)a}$ of Y. Nevertheless, X and Y are not independent.

7.10. Discrete random variables which are uncorrelated but not independent

It is a well known result that if X and Y are integrable and independent r.v.s, they are uncorrelated. The property of uncorrelatedness is weaker than independence. This will be demonstrated by a few examples. Here we consider discrete r.v.s; the absolutely continuous case is treated in Example 7.11.

(i) Let X and Y be r.v.s such that $p_{i,j} = \mathbf{P}[X = i, Y = j]$ are given by

$$p_{1,1} = p_{-1,1} = p_{1,-1} = p_{-1,-1} = \tfrac{1}{4}\varepsilon,\ p_{0,1} = p_{0,-1} = p_{1,0} = p_{-1,0} = \tfrac{1}{4}(1-\varepsilon)$$

where $0 < \varepsilon < 1$. It is easy to find the marginal distributions of X, Y and compute that $\mathbf{E}X = 0$, $\mathbf{E}Y = 0$. Moreover, we also find that $\mathbf{E}[XY] = 0$ and hence the variables X and Y are uncorrelated. However,

$$\mathbf{P}[X = 0, Y = 0] = 0 \neq \mathbf{P}[X = 0]\mathbf{P}[Y = 0] = \tfrac{1}{4}(1-\varepsilon)^2$$

and thus X and Y are not independent.

(ii) Let $\Omega = \{1, 2, 3\}$ and let each $\omega \in \Omega$ have probability $\frac{1}{3}$. Define two r.v.s X and Y by

$$X(\omega) = \begin{cases} 1, & \text{if } \omega = 1 \\ 0, & \text{if } \omega = 2 \\ -1, & \text{if } \omega = 3, \end{cases} \qquad Y(\omega) = \begin{cases} 0, & \text{if } \omega = 1 \\ 1, & \text{if } \omega = 2 \\ 0, & \text{if } \omega = 3. \end{cases}$$

Then $\mathbf{E}X = 0$, $\mathbf{E}[XY] = 0$, so X and Y are uncorrelated. But

$$\mathbf{P}[X = 1, Y = 1] = 0 \neq \tfrac{1}{3} \cdot \tfrac{1}{3} = \mathbf{P}[X = 1]\mathbf{P}[Y = 1]$$

and therefore X and Y are not independent.

(iii) Let X and Y be r.v.s each taking the values $-1, 0, 1$. The joint probability $p_{i,j} = \mathbf{P}[X = i, Y = j]$ is given by

$$p_{1,0} = p_{-1,0} = p_{0,1} = p_{0,-1} = \tfrac{1}{4}.$$

Then obviously $\mathbf{E}X = 0$, $\mathbf{E}Y = 0$ and $\mathbf{E}[XY] = 0$. Thus X and Y are uncorrelated. Further,

$$\mathbf{P}[X = 1] = \mathbf{P}[X = -1] = \tfrac{1}{4}, \quad \mathbf{P}[X = 0] = \tfrac{1}{2},$$
$$\mathbf{P}[Y = 1] = \mathbf{P}[Y = -1] = \tfrac{1}{4}, \quad \mathbf{P}[Y = 0] = \tfrac{1}{2}$$

and clearly the relation $\mathbf{P}[X = i, Y = j] = \mathbf{P}[X = i]\mathbf{P}[Y = j]$ is not valid for all pairs (i, j). So the variables X and Y are dependent.

(iv) Let ξ be a r.v. taking the values 0, $\frac{1}{2}\pi$ and π with probability $\frac{1}{3}$ each. Then it is easy to see that $X = \sin\xi$ and $Y = \cos\xi$ are uncorrelated. However, they are not independent. Moreover, X and Y are functionally connected: $X^2 + Y^2 = 1$.

7.11. Absolutely continuous random variables which are uncorrelated but not independent

(i) Let X_1 and X_2 have a joint uniform distribution on the unit disk, i.e. the probability density function f is $f(x_1, x_2) = \pi^{-1}$ if $x_1^2 + x_2^2 \le 1$ and $f(x_1, x_2) = 0$ otherwise. Simple computation shows that $\mathbf{E}[X_1 X_2] = 0$. Thus the variables X_1 and X_2 are uncorrelated. It is very easy to find the marginal densities f_1 and f_2 and see that $f(x_1, x_2) \ne f_1(x_1) f_2(x_2)$. This means that X_1 and X_2 are not independent.

(ii) If ξ is uniformly distributed on the interval $(0, 2\pi)$ then $X_1 = \sin\xi$ and $X_2 = \cos\xi$ satisfy the relations:

$$\mathbf{E}X_1 = 0, \quad \mathbf{E}X_2 = 0, \quad \mathbf{E}[X_1 X_2] = 0.$$

Therefore X_1 and X_2 are uncorrelated but not independent. They are functionally dependent: $X_1^2 + X_2^2 = 1$. (See case (iv) of Example 7.10.)

(iii) Recall that the r.v. X is said to be *normally distributed* with parameters a and σ^2, where $a \in \mathbb{R}^1$, $\sigma^2 > 0$, if X is absolutely continuous and has a density $(\sqrt{2\pi}\sigma)^{-1} \exp[-\frac{1}{2}(x - a)^2/\sigma^2]$, $x \in \mathbb{R}^1$. In such a case we use the following standard notation: $X \sim \mathcal{N}(a, \sigma^2)$. Several properties of the normal distribution will be discussed further in Section 10.

Let $X \sim \mathcal{N}(0, 1)$ and $X_2 = X_1^2 - 1$. Then $\mathbf{E}X_2 = 0$, $\mathbf{E}[X_1 X_2] = 0$. Hence X_1 and X_2 are uncorrelated. However they are functionally dependent.

7.12. Independent random variables have zero correlation ratio, but the converse is not true

Let X and Y be r.v.s such that $\mathbf{E}Y$ and $\mathbf{V}Y$ exist. As usual, $\mathbf{E}[Y|X]$ denotes the conditional expectation of Y given X. The quantity

$$K_X(Y) = \mathbf{V}[\mathbf{E}(Y|X)]/\mathbf{V}Y$$

is called a *correlation ratio* of Y with respect to X. Obviously $0 \leq K_X(Y) \leq 1$ and $K_X(Y)$ is defined for Y with $\mathbf{V}Y > 0$ (see Rényi 1970). Note that $K_X(Y)$ gives us information about the mutual dependence of X and Y.

Obviously, if X and Y are independent and $0 < \mathbf{V}Y < \infty$ then $K_X(Y) = 0$, but not conversely. To see this, take (X, Y) to be uniformly distributed on the unit disk $x^2 + y^2 < 1$. Let $g(y|x)$ be the conditional density of Y given $X = x$. We have

$$g(y|x) = 1/[2(1-x^2)^{\frac{1}{2}}] \quad \text{for} \quad |y| < (1-x^2)^{\frac{1}{2}} \quad \text{and} \quad |x| < 1.$$

Hence $\mathbf{E}[Y|X] = 0$ and consequently $K_X(Y) = 0$, though the variables X and Y are not independent.

7.13. The relation $\mathbf{E}[Y|X] = \mathbf{E}Y$ almost surely does not imply that the random variables X and Y are independent

If X and Y are independent r.v.s on a given probability space and Y is integrable, then $\mathbf{E}[Y|X] = \mathbf{E}Y$ a.s. Now the question is whether or not the converse is true.

Let Z be any integrable r.v. which is distributed symmetrically with respect to zero, and let X be a r.v. independent of Z and such that $X \geq 1$ a.s. Let $Y = Z/X$. Then Y is integrable and the conditional expectation $\mathbf{E}[Y|X]$ is well defined. Obviously we have

$$\mathbf{E}Z = 0, \quad \mathbf{E}Y = 0, \quad \mathbf{E}[Y|X] = 0.$$

Therefore the relation $\mathbf{E}[Y|X] = \mathbf{E}Y$ a.s. is satisfied but the variables X and Y are dependent.

7.14. There is no relationship between the notions of independence and conditional independence

Intuitively the notions of independence and conditional independence are close to each other (see the introductory notes to this section). By a few examples we can show that neither of them implies the other one. (Also see Example 3.6.)

(i) Let $X_n, n \geq 1$ be independent Bernoulli r.v.s, that is X_n are i.i.d. and each takes two values, 1 and 0, with probabilities p and $1-p$ respectively. As usual, let $S_n = X_1 + \ldots + X_n$. Then obviously for $S_2 = 0$ or 2, we have $\mathbf{P}[X_1 = 1|S_2] > 0$ and $\mathbf{P}[X_2 = 1|S_2] > 0$, whereas for $S_2 = 0$, $\mathbf{P}[X_1 = 1, X_2 = 1|S_2] = 0$; that is, the equality

$$\mathbf{P}[X_1 = 1, X_2 = 1|S_2] = \mathbf{P}[X_1 = 1|S_2]\mathbf{P}[X_2 = 1|S_2]$$

is not satisfied. Therefore the independence property of r.v.s can be lost under conditioning.

(ii) Let $X_n, n \geq 1$ be independent integer-valued r.v.s and $S_n = X_1 + \ldots + X_n$. Then clearly the r.v.s $S_n, n \geq 1$ are dependent. However, given that the event $[S_2 = k]$ has a positive probability and occurs, we can easily show that

$$\begin{aligned}
\mathbf{P}[S_1 = i, S_3 = j|S_2] &= \frac{\mathbf{P}[S_1 = i, S_2 = k, S_3 = j]}{\mathbf{P}[S_2 = k]} \\
&= \frac{\mathbf{P}[S_1 = i]\mathbf{P}[X_2 = k - i]\mathbf{P}[X_3 = j - k]}{\mathbf{P}[S_2 = k]} \\
&= \mathbf{P}[S_1 = i|S_2]\frac{\mathbf{P}[X_3 = j - k]\mathbf{P}[S_2 = k]}{\mathbf{P}[S_2 = k]} \\
&= \mathbf{P}[S_1 = i|S_2]\mathbf{P}[S_3 = j|S_2].
\end{aligned}$$

Therefore there are dependent r.v.s which are conditionally independent.

(iii) Consider three r.v.s, X, Y and Z, with the following joint distribution:

$$\mathbf{P}[X = k, Y = m, Z = n] = p^3 q^{m-3}$$

where $0 < p < 1$, $q = 1 - p$, $k = 1, \dots, m - 1$, $m = 2, \dots, n - 1$, $n = 3, 4, \dots$.

Firstly, we can easily find the distributions of the pairs (X, Y), (X, Z) and (Y, Z), then the marginal distribution of each of X, Y and Z and see in particular that the r.v.s Z and X are dependent.

Further, we have

$$\begin{aligned}
&\mathbf{P}[X = k, Y = m] = p^2 q^{m-2}, && k = 1, \dots, m - 1, \quad m = 2, 3, \dots, \\
&\mathbf{P}[Z = n|X = k, Y = m] = pq^{n-m-1}, && k = 1, \dots, m - 1, \quad m = 2, \dots, n - 1.
\end{aligned}$$

Hence for $k = 1, \dots, m - 1$ and $m = 2, 3, \dots$ we can obtain that

$$\mathbf{E}[Z|X = k, Y = m] = \sum_{n=m+1}^{\infty} n\, pq^{n-m-1} = m + \frac{1}{p}$$

and write the relation

$$\mathbf{E}[Z|X, Y] = Y + \frac{1}{p} \quad \text{a.s.}$$

Moreover, for any measurable and bounded function g,

$$\mathbf{E}[g(Z)|X = k, Y = m] = \sum_{j=1}^{\infty} g(j + m) pq^{j-1}$$

so that

$$\mathbf{E}[g(Z)|X, Y] = \sum_{j=1}^{\infty} g(Y + j) pq^{j-1} \quad \text{a.s.}$$

Obviously the right-hand side of the last equality does not depend on X, which means that Z is conditionally independent of X given Y despite the fact (mentioned above) that Z and X are dependent r.v.s.

7.15. Mutual independence implies the exchangeability of any set of random variables, but not conversely

Let $X_1, \ldots, X_n$ be i.i.d. r.v.s. Clearly for any permutation $(i_1, \ldots, i_n)$ of $(1, \ldots, n)$, the random vectors $(X_1, \ldots, X_n)$ and $(X_{i_1}, \ldots, X_{i_n})$ have the same distribution. Thus $X_1, \ldots, X_n$ is a set of *exchangeable variables*. However, the converse is not generally true and this is illustrated by the following examples.

(i) Let θ be an arbitrary r.v. with values in the interval (0, 1). Let $Y_1, Y_2, \ldots$ be r.v.s which, conditionally on θ, are independent and take the values 1 and 0 with probabilities θ and $1 - \theta$ respectively. Then for any sequence $u_1, \ldots, u_n$ of 0s and 1s, we have

(1) $$\mathbf{P}[Y_1 = u_1, Y_2 = u_2, \ldots, Y_n = u_n \,|\, \theta\,] = \theta^k (1-\theta)^{n-k}$$

where $k = u_1 + \ldots + u_n$ and n is an arbitrary natural number. We are interested in the properties of the set of r.v.s $Y_1, \ldots, Y_n$. Taking the expectation of both sides of (1) we find that the probability

$$\mathbf{P}[Y_1 = u_1, \ldots, Y_n = u_n] = \mathbf{E}[\mathbf{P}(Y_1 = u_1, \ldots, Y_n = u_n|\theta)] = \mathbf{E}[\theta^k(1-\theta)^{n-k}]$$

depends only on the sum $u_1 + \ldots + u_n$, which is k, and on n of course. Therefore $Y_1, \ldots, Y_n$, for any n, is a set of exchangeable variables. However, $Y_1, \ldots, Y_n$ are not mutually independent. Indeed, $\mathbf{P}[Y_j = 1] = \mathbf{E}\theta$ for each $j \geq 1$. Further, (1) implies that $\mathbf{P}[Y_1 = 1, \ldots, Y_n = 1] = \mathbf{E}[\theta^n]$. On the other hand $\prod_{j=1}^n \mathbf{P}[Y_j = 1] = (\mathbf{E}\theta)^n$. But θ is an arbitrary r.v. with values in (0, 1). If, for example, θ is uniformly distributed on (0, 1), then $(\mathbf{E}\theta)^n = (\frac{1}{2})^n \neq 1/(n+1) = \mathbf{E}[\theta^n]$. This justifies our statement that $Y_1, \ldots, Y_n$ are not mutually independent.

(ii) Suppose that an urn containing balls of two colours, say w white and b black, is used, and after each draw the chosen ball is returned, together with s balls of the same colour. Introduce the r.v.s $Y_1, \ldots, Y_n$ such that

$$Y_i = \begin{cases} 1, & \text{if the } i\text{th draw is black} \\ 0, & \text{if the } i\text{th draw is white.} \end{cases}$$

It can be shown that the variables $Y_1, \ldots, Y_n$ are not independent but they are exchangeable. The last statement follows from the fact that $\mathbf{P}[\bigcap_{i=1}^n (Y_i = y_i)]$ depends only on the sum $\sum_{i=1}^n y_i$ (for details we refer the reader to Johnson and Kotz (1977)).

7.16. Different kinds of monotone dependence between random variables

Recall that the r.v. Y is said to be *completely dependent* on the r.v. X if there exists a function g such that

$$\mathbf{P}[Y = g(X)] = 1.$$

Another measure of dependence between two non-degenerate r.v.s X and Y is that of sup correlation, defined by

$$\tilde{\rho}(X, Y) = \sup \rho(f(X), g(Y))$$

where the supremum is taken over all measurable f and g with $0 < \mathbf{V}[f(X)] < \infty$, $0 < \mathbf{V}[g(Y)] < \infty$ and ρ is the ordinary correlation coefficient.

Let X and Y be absolutely continuous r.v.s. They are called *monotone dependent* if there exists a monotone function g for which $\mathbf{P}[Y = g(X)] = 1$.

The quantity

$$\rho^*(X, Y) = \sup \rho(f(X), g(Y))$$

where the supremum is taken over all monotone functions f and g such that $0 < \mathbf{V}[f(X)] < \infty$ and $0 < \mathbf{V}[g(Y)] < \infty$, is said to be a *monotone correlation.*

Let us try to compare these kinds of monotone dependence. It is clear that if X and Y are monotone dependent, then their monotone correlation is 1. However, the converse statement is false. Indeed, let (X, Y) have a uniform distribution over the region $[(0, 1) \times (0, 1)] \cup [(1, 2) \times (1, 2)]$. Then

$$\rho^*(X, Y) \geq \rho(I_{(0,1)}(X), I_{(0,1)}(Y)) = 1$$

but X and Y are not monotone dependent.

Further, it is obvious that

$$|\rho(X, Y)| \leq \rho^*(X, Y) \leq \tilde{\rho}(X, Y). \tag{1}$$

For a bivariate normally distributed (X, Y), it is well known that $|\rho(X, Y)| = \tilde{\rho}(X, Y)$, and in this case we should have equalities in (1). On the other hand, it can easily be seen that in general ρ^* is not equal to $\tilde{\rho}$. Indeed, take (X, Y) with a uniform distribution on the region

$$[(0, 1) \times (0, 1)] \cup [(0, 1) \times (2, 3)] \cup [(1, 2) \times (1, 2)] \cup [(2, 3) \times (2, 3)].$$

Let $f(x) = I_{(0,1)}(x) + I_{(2,3)}(x)$. Then $\rho^*(X, Y) < 1$, but

$$\tilde{\rho}(X, Y) \geq \rho(f(X), f(Y)) = 1.$$

SECTION 8. CHARACTERISTIC AND GENERATING FUNCTIONS

Let X be a r.v. defined on the probability space $(\Omega, \mathcal{F}, \mathbf{P})$. The function

$$\phi(t) = \mathbf{E}[\mathrm{e}^{itX}], \quad t \in \mathbb{R}^1, \quad i = \sqrt{-1} \tag{1}$$

is said to be a **characteristic function** (ch.f.) of X. If $F(x)$, $x \in \mathbb{R}^1$, is the d.f. of X then $\phi(t) = \int_{-\infty}^{\infty} \mathrm{e}^{itx}\, \mathrm{d}F(x)$. Thus $\phi(t) = \int_{-\infty}^{\infty} \mathrm{e}^{itx} f(x)\, \mathrm{d}x$ if X is absolutely continuous with density f and $\phi(t) = \sum_n e^{itx_n} p_n$ if X is discrete with $\mathbf{P}[X = x_n] = p_n$, $p_n > 0$, $\sum_n p_n = 1$. Recall some of the basic properties of a ch.f.

(i) $\phi(0) = 1$, $\phi(-t) = \bar{\phi}(t)$, $|\phi(t)| \le 1$, $t \in \mathbb{R}^1$.

(ii) If $\mathbf{E}[X^n]$ exists, then $\phi^{(n)}(0)$ exists and $\mathbf{E}[X^n] = i^{-n}\phi^{(n)}(0)$.

(iii) If $\phi^{(n)}(0)$ exists and n is even, then $\mathbf{E}[X^n]$ exists; if n is odd, then $\mathbf{E}[X^{n-1}]$ exists.

(iv) If $\mathbf{E}[X^n]$ exists (and hence $\mathbf{E}[X^k]$ exists for $k < n$) then

$$\phi(t) = \sum_{k=0}^{n} \frac{(it)^k}{k!} \mathbf{E}[X^k] + o(t^n)$$

in the neighbourhood of the origin.

(v) $\phi(t), t \in \mathbb{R}^1$ is a ch.f. iff $\phi(0) = 1$ and ϕ is positive definite.

(vi) If X_1 and X_2 are r.v.s with ch.f.s ϕ_1 and ϕ_2, and X_1 and X_2 are independent, then the ch.f. ϕ of the sum $X_1 + X_2$ is given by

$$\phi(t) = \phi_1(t)\phi_2(t), \quad t \in \mathbb{R}^1.$$

(vii) If we know the ch.f. ϕ of a r.v. X then we can find the d.f. F of X by the so-called inversion formula and, moreover, if ϕ is absolutely integrable over $\mathbb{R}^1$ then X is absolutely continuous and its density is the inverse Fourier transform of ϕ.

Let us introduce two other functions which, like the ch.f. ϕ, are essentially used in probability theory. For an arbitrary r.v. X with a d.f. F denote

$$M(z) = \mathbf{E}[e^{zX}] = \int e^{zx}\, dF(x), \quad z \text{ a complex number.} \tag{2}$$

Suppose for some real $r > 0$ the function $M(z)$ is well defined for all z, $|z| < r$. In such a case M is called a **moment generating function** (m.g.f.) of X and also of F. The relationship between the m.g.f. M and the ch.f. ϕ, see (1), is obvious: $M(it) = \phi(t)$ for real t.

If X is a non-negative integer valued r.v. we can introduce the function

$$p(z) = \mathbf{E}[z^X], \quad z \text{ complex} \tag{3}$$

which is called a **probability generating function** (p.g.f.) of X. (Note that the m.g.f. and p.g.f. were briefly introduced in Example 7.8 and used to analyse the independence property.)

Some of the properties of ϕ listed above can be reformulated for the generating functions M and p. However, note that the ch.f. of a distribution always exists while the m.g.f. need not always exist (excluding the trivial case when $t = 0$).

The ch.f. ϕ is called *analytic* if there is a number $r > 0$ such that ϕ can be represented by a convergent power series in the interval $(-r, r)$, that is if $\phi(t) = \sum_{k=0}^{\infty} a_k t^k/k!$, $t \in (-r, r)$, with some complex coefficients a_k. The following important result is often used (see Lukacs 1970; Chow and Teicher 1978). If F and ϕ are a pair of a d.f. and a ch.f., then the following conditions are equivalent: (a) ϕ is r-analytic; (b) the moments $m_k = \int x^k \, \mathrm{d}F(x)$, $k \geq 1$ are finite and ϕ admits the representation $\phi(t) = \sum_{k=0}^{\infty} m_k (it)^k/k!$, $t \in (-r, r)$; (c) $\int \mathrm{e}^{t|x|} \, \mathrm{d}F(x) < \infty$, $0 \leq t < r$. Usually (c) is called *Cramér condition.*

Clearly, the m.g.f. M does exist iff the corresponding ch.f. ϕ is analytic.

We say that the ch.f. ϕ is *decomposable* (or *factorizable*) if

$$\phi(t) = \phi_1(t)\phi_2(t), \quad t \in \mathbb{R}^1$$

where ϕ_1 and ϕ_2 are both ch.f.s of non-degenerate distributions. If ϕ admits only a trivial product representation (that is, if ϕ_1 or ϕ_2 is of the form e^{iat}, a=constant), it is called *indecomposable.*

We refer the reader to the books by Lukacs (1970), Ramachandran (1967), Feller (1971), Chow and Teicher (1978), Rao (1984), Shiryaev (1995) and Bauer (1996) where the theory of characteristic functions and related topics can be found in detail.

In this section we have included various counterexamples which explain the meaning of some of the properties of ch.f.s and of generating functions.

8.1. Different characteristic functions which coincide on a finite interval but not on the whole real line

Suppose ϕ_1, ϕ_2 are ch.f.s such that $\phi_1(t) = \phi_2(t)$ for $t \in [-l, l]$ where l is an arbitrary positive number. Does it then follow that $\phi_1(t)$ coincides with $\phi_2(t)$ for all $t \in \mathbb{R}^1$? This important problem was considered and solved almost 60 years ago by Gnedenko (1937). Let us present his solution.

Consider the function $h(x) = 0$ if $|x| > \pi/2$ and $h(x) = x$, if $|x| \leq \pi/2$. If $c(t) = \int_{-\infty}^{\infty} h(x)h(x+t) \, \mathrm{d}x$, then the ratio $\phi_1(t) = c(t)/c(0)$ is a ch.f. An easy calculation shows that

$$\phi_1(t) = \begin{cases} 1 + 3\pi^{-1}t - 2\pi^{-3}t^3, & \text{if } -\pi \leq t \leq 0 \\ 1 - 3\pi^{-1}t + 2\pi^{-3}t^3, & \text{if } 0 \leq t \leq \pi \\ 0, & \text{if } |t| > \pi. \end{cases}$$

Now introduce another function, say ϕ_2, as follows:

$$\phi_2(t) = \phi_1(t), \text{ if } |t| \leq \pi$$
$$\phi_2(t + 2\pi) = \phi_2(t), \text{ if } t \in \mathbb{R}^1.$$

Let us show that ϕ_2 is a ch.f. Obviously ϕ_2 is an even function with the Fourier expansion

(1) $$\frac{1}{2}a_0 + \sum_{n=1}^{\infty} a_n \cos nt.$$

A standard calculation shows that

$$a_0 = 0,\ a_n = 6\pi^{-2}[n^{-2}(1+\cos n\pi) + 4\pi^{-2}n^{-4}(1-\cos n\pi)],\ n = 1, 2, \ldots .$$

Thus the series (1) converges uniformly, its coefficients are non-negative and their sum equals $\phi_2(0) = 1$. Hence $\phi_2(t) = \int_{-\infty}^{\infty} e^{itx}\, dF(x)$ for some d.f. F. This means that ϕ_2 is a ch.f.

Therefore we have that $\phi_2(t) = \phi_1(t)$ for $t \in [-\pi, \pi]$ but not for all $t \in \mathbb{R}^1$. In a similar way we can construct two ch.f.s ϕ_1 and ϕ_2 which coincide on the interval $[-l, l]$ for large enough l but not for all $t \in \mathbb{R}^1$.

Note finally that at the end of the Gnedenko's paper we can find a very important remark made by A. Ya. Khintchine concerning the above result. Let F_1 and F_2 be the d.f.s corresponding to ϕ_1 and ϕ_2. The above reasoning implies the equality

$$\phi_1(t)\phi_1(t) = \phi_1(t)\phi_2(t) \quad \text{for all } t \in \mathbb{R}^1$$

which is equivalent to the relation

(2) $$F_1 * F_1 = F_1 * F_2.$$

Equation (2) states that there exists a d.f. whose convolutions with two different d.f.s coincide. In other words, the convolution equality (2) does not in general imply that $F_1 = F_2$.

8.2. Discrete and absolutely continuous distributions can have characteristic functions coinciding on the interval $[-1, 1]$

Let X be a r.v. whose ch.f. ϕ_1 is given by

(1) $$\phi_1(t) = \begin{cases} 1 - |t|, & \text{if } |t| \le 1 \\ 0, & \text{otherwise.} \end{cases}$$

Obviously ϕ_1 is absolutely integrable on $\mathbb{R}^1$ and the density f of X is

$$f(x) = \frac{1}{2\pi}\int_{-\infty}^{\infty} e^{-itx}\phi_1(t)\, dt = \frac{1-\cos x}{\pi x^2}, \qquad x \in \mathbb{R}^1.$$

Consider now the r.v. Y where

$$\mathbf{P}[Y = 0] = \tfrac{1}{2},\ \mathbf{P}[Y = (2k-1)\pi] = \frac{2}{(2k-1)^2\pi^2},\ k = 0, \pm 1, \pm 2, \ldots .$$

If ϕ_2 is the ch.f. of Y, then

$$\phi_2(t) = \frac{1}{2} + \frac{4}{\pi^2}\sum_{k=1}^{\infty} \frac{\cos(2k-1)\pi t}{(2k-1)^2}. \tag{2}$$

Let us show that ϕ_1 given by (1) equals ϕ_2 given by (2) for each $t \in [-1, 1]$. The function $h(t) = |t|$ has the following Fourier expansion: $h(t) = \frac{1}{2}a_0 + \sum_{n=1}^{\infty} a_n \cos n\pi t$, $|t| \le 1$, where $a_0 = 1$, $a_n = 2(\cos n\pi - 1)/(n^2\pi^2)$. For even n, $a_n = 0$ and for odd n, that is for $n = 2k - 1$, we have $a_{2k-1} = -4/((2k-1)^2\pi^2)$. Now comparing $\phi_1(t)$ and $\phi_2(t)$ we conclude that $\phi_1(t) = \phi_2(t)$ for each $t \in [-1, 1]$. Nevertheless ϕ_1 and ϕ_2 correspond to quite different distributions, one of which is absolutely continuous while the other is purely discrete. Note additionally that $\phi_1(t) \neq \phi_2(t)$ for $|t| > 1$.

8.3. The absolute value of a characteristic function is not necessarily a characteristic function

If ϕ is a ch.f., then it is of general interest to know whether $|\phi|$ is also a ch.f. Consider the function

$$\phi(t) = \tfrac{1}{8}(1 + 7e^{it}), \quad t \in \mathbb{R}^1.$$

Obviously ϕ is a ch.f. of a r.v. taking two values. We now want to know whether

$$|\phi(t)| = (|\phi(t)|^2)^{1/2} = (\phi(t)\,\overline{\phi(t)})^{1/2} = \tfrac{1}{8}(50 + 7e^{-it} + 7e^{it})^{1/2}$$

is a ch.f. If the answer were positive then $\psi := |\phi|$ must be of the form

$$\psi(t) = pe^{itx_1} + (1-p)e^{itx_2}$$

where $0 < p < 1$ and x_1, x_2 are different real numbers. Comparing $|\psi|^2$ and $|\phi|^2$ we see that p should satisfy the relations

$$p^2 = (1-p)^2 = \tfrac{7}{64}, \quad 2p(1-p) = \tfrac{50}{64}$$

which are obviously incompatible. Hence $|\phi|$ is not a ch.f. although ϕ is.

8.4. The ratio of two characteristic functions need not be a characteristic function

Let ϕ_1 and ϕ_2 be ch.f.s. Is it true that the ratio ϕ_1/ϕ_2 is also a ch.f.? The answer is based on the following result (see Lukacs 1970). A necessary condition for a function, analytic in some neighbourhood of the origin, to be a ch.f., is that in either half-plane the singularity nearest to the real axis is located on the imaginary axis.

Consider the following two functions

$$\phi_1(t) = \left[\left(1 - \frac{it}{a}\right)\left(1 - \frac{it}{a+ib}\right)\left(1 - \frac{it}{a-ib}\right)\right]^{-1},$$

$$\phi_2(t) = \left(1 - \frac{it}{a}\right)^{-1}, \ t \in \mathbb{R}^1$$

where $a \geq b > 0$. One can check that both ϕ_1 and ϕ_2 are analytic ch.f.s. Furthermore, their quotient $\psi(t) = \phi_1(t)/\phi_2(t)$ satisfies some of the elementary properties of ch.f.s, namely $\psi(-t) = \overline{\psi(t)}$, $|\psi(t)| \leq \psi(0) = 1$ for all $t \in \mathbb{R}^1$. However, the condition in the result cited above is violated since ψ has no singularity on the imaginary axis while it has a pair of conjugate complex poles $\pm b - ia$.

Therefore in general the ratio of two ch.f.s is not a ch.f.

8.5. The factorization of a characteristic function into indecomposable factors may not be unique

We shall give two examples concerning the discrete and the absolutely continuous case respectively.

(i) The function $\phi(t) = \frac{1}{6}\sum_{k=0}^{5} e^{itk}$ is the ch.f. of a discrete uniform distribution on the set $\{0, 1, 2, 3, 4, 5\}$. Take the functions

$$\phi_1(t) = \tfrac{1}{3}(1 + e^{2it} + e^{4it}), \quad \phi_2(t) = \tfrac{1}{2}(1 + e^{it})$$
$$\psi_1(t) = \tfrac{1}{3}(1 + e^{it} + e^{2it}), \quad \psi_2(t) = \tfrac{1}{2}(1 + e^{3it}).$$

Obviously we have

$$\phi(t) = \phi_1(t)\phi_2(t) = \psi_1(t)\psi_2(t), \quad t \in \mathbb{R}^1.$$

It is easy to see that ϕ_1, ϕ_2, ψ_1, ψ_2 are all ch.f.s of some (discrete) distributions. Moreover, ϕ_2 and ψ_2 correspond to two-point distributions and hence they are indecomposable (see Gnedenko and Kolmogorov 1954; Lukacs 1970). Thus it only remains to show that ϕ_1 and ψ_1 are also indecomposable. Suppose that $\psi_1(t) = \psi_{11}(t)\psi_{12}(t)$, where ψ_{11} and ψ_{12} are non-trivial factors. Clearly ψ_1 corresponds to a distribution, say G_1, concentrated at three points, 0, 1, 2 each with probability $\frac{1}{3}$. However, the discontinuity points of G_1 are of the type $x_j + y_k$ where x_j and y_k are discontinuity points of the distributions corresponding to the ch.f.s ψ_{11} and ψ_{12} respectively (see Lukacs 1970).

Since G_1 has three discontinuity points and ψ_{11}, ψ_{12} are non-trivial, we conclude that

$$\psi_{11}(t) = pe^{itx_1} + (1-p)e^{itx_2}, \quad \psi_{12}(t) = qe^{ity_1} + (1-q)e^{ity_2}$$

where $0 < p < 1$, $0 < q < 1$. But $\psi_1(t) = \psi_{11}(t)\psi_{12}(t)$ implies that p, q must satisfy the relations

$$pq = (1-p)(1-q) = p(1-q) + q(1-p) = \tfrac{1}{3}.$$

Clearly this is not possible.

We have therefore shown that ψ_1 is indecomposable and, since $\phi_1(t) = \psi_1(2t)$, we conclude that ϕ_1 is also indecomposable.

(ii) Consider now a uniform distribution over the interval $(-1, 1)$. The ch.f. ϕ of this distribution is

$$\phi(t) = t^{-1} \sin t, \quad t \in \mathbb{R}^1.$$

Using the elementary formula $t^{-1} \sin t = \cos(t/2)(t/2)^{-1} \sin(t/2)$ we obtain

$$\phi(t) = t^{-1} \sin t = \left[\prod_{k=1}^{n} \cos(t/2^k) \right] (t/2^n)^{-1} \sin(t/2^n).$$

Passing to the limit in n, as $n \to \infty$, we get the following well known representation:

$$(1) \qquad \phi(t) = t^{-1} \sin t = \prod_{k=1}^{\infty} \cos(t/2^k).$$

Now it only remains for us to show that $\cos(t/2^k)$ is an indecomposable ch.f. This is a consequence of the equality $\cos(t/2^k) = \frac{1}{2}(\mathrm{e}^{it/2^k} + \mathrm{e}^{-it/2^k})$ which implies that $\cos(t/2^k)$ is a ch.f. of a distribution concentrated at two points and hence it is indecomposable.

Another factorization can be obtained by using the formula

$$t^{-1} \sin t = (t/3)^{-1} \sin(t/3)[2 \cos(2t/3) - 1]/3.$$

In this case we have

$$(2) \qquad \phi(t) = t^{-1} \sin t = \tfrac{1}{3}[2 \cos(2t/3) - 1] \prod_{k=1}^{\infty} \cos(t/3 \cdot 2^k).$$

It follows from (2) that ϕ is a product of indecomposable factors and obviously (1) and (2) are different factorizations of the ch.f. ϕ.

8.6. An absolutely continuous distribution can have a characteristic function which is not absolutely integrable

Let ϕ be a ch.f. and F be its d.f. Recall that if ϕ is absolutely integrable on $\mathbb{R}^1$, then F is absolutely continuous and the density $f = F'$ is the inverse Fourier transform of ϕ (see Feller 1971; Lukacs 1970; Loève 1977/1978). Let us now clarify if the converse statement holds. For this purpose we shall use the following theorem of G. Pólya (see Lukacs 1970; Feller 1971).

Let $\psi(t), t \in \mathbb{R}^1$ be a real-valued continuous function such that: (i) $\psi(0) = 1$; (ii) $\psi(-t) = \psi(t)$; (iii) $\psi(t)$ is convex for $t > 0$; (iv) $\lim_{t \to \infty} \psi(t) = 0$. Then ψ is a ch.f. of a distribution which is absolutely continuous.

Take for example the following two functions:

$$\psi_1(t) = \frac{1}{1+|t|}, \quad \text{if } t \in \mathbb{R}^1 \text{ and } \psi_2(t) = \begin{cases} 1 - |t|, & \text{if } 0 \le |t| \le \frac{1}{2} \\ 1/(4|t|), & \text{if } |t| > \frac{1}{2}. \end{cases}$$

According to the result cited above we conclude that ψ_1 and ψ_2 are ch.f.s which correspond to absolutely continuous distributions. However, it is easy to check that ψ_1 and ψ_2 are not absolutely integrable.

Finally, suppose X is a r.v. exponentially distributed, $X \sim \mathcal{E}xp(\lambda)$. By definition X is absolutely continuous (its density is $\lambda e^{-\lambda x}$, $x > 0$). However its ch.f. is equal to $\lambda/(\lambda - it)$, and obviously this function is not absolutely integrable on $\mathbb{R}^1$.

Therefore the absolute integrability condition for the ch.f. is sufficient but not necessary for the corresponding d.f. to be absolutely continuous.

8.7. A discrete distribution without a first-order moment but with a differentiable characteristic function

This and the next example are given to show that the existence of the derivative $\phi^{(n)}(0)$ for odd n does not necessarily imply that the moment $m_n = \mathbf{E}[X^n]$ exists. To see this, consider the r.v. X with

$$\mathbf{P}[X = \pm n] = \frac{c}{n^2 \log n}, \quad n = 2, 3, \ldots, \quad c = \left[2 \sum_{n=2}^{\infty} \frac{1}{n^2 \log n} \right]^{-1}.$$

Then the ch.f. ϕ of X is

$$\phi(t) = 2c \sum_{n=2}^{\infty} \frac{\cos nt}{n^2 \log n}, \quad t \in \mathbb{R}^1. \tag{1}$$

Since the partial sums of the series $\sum_{n=2}^{\infty} (\sin nt)/n$ are uniformly bounded, the series $\sum_{n=2}^{\infty} (\sin nt)/(n \log n)$ obtained from (1) by differentiation is uniformly convergent (see Zygmund 1968). This implies the uniform differentiability of the ch.f. $\phi(t)$ for all $t \in \mathbb{R}^1$. In particular, if $t = 0$, $\phi'(0) = 0$ but the expectation $\mathbf{E}X$ does not exist because the series $\sum_{n=2}^{\infty} 1/(n \log n)$ is divergent.

8.8. An absolutely continuous distribution without expectation but with a differentiable characteristic function

(i) Let X be a r.v. with the following density:

$$f(x) = \begin{cases} 0, & \text{if } |x| \le 2 \\ c/(x^2 \log |x|), & \text{if } |x| > 2 \end{cases}$$

where c is a norming constant, $0 < c < \infty$ (the exact value is not essential). Since $\int_2^\infty (x \log x)^{-1}\,\mathrm{d}x = \infty$, the expectation $\mathbf{E}X$ does not exist. Nevertheless we can ask whether the ch.f. $\phi(t)$ of X is differentiable at $t = 0$. Since

$$\phi(t) = 2c \int_2^\infty \frac{\cos tx}{x^2 \log x}\,\mathrm{d}x$$

is even, we can write the difference $[1 - \phi(t)]/(2c)$ for $t > 0$ in the following way:

$$\frac{1 - \phi(t)}{2c} = \int_2^{1/t} \frac{1 - \cos tx}{x^2 \log x}\,\mathrm{d}x + \int_{1/t}^\infty \frac{1 - \cos tx}{x^2 \log x}\,\mathrm{d}x.$$

Obviously $1 - \phi(t)$ is a real-valued and non-negative function. For an arbitrary $u \in \mathbb{R}^1$ we have $0 \le 1 - \cos u \le \min\{2, u^2\}$. This implies that $1 - \phi(t)$ is not greater than some constant multiplied by the function $h(t)$ where

$$h(t) = t^2 \int_2^{1/t} \frac{\mathrm{d}x}{\log x} + 2 \int_{1/t}^\infty \frac{\mathrm{d}x}{x^2 \log x}.$$

However, since $h(t) = O(-t/\log t) = o(t)$ as $t \to 0$, we find that

$$\phi(t) = 1 + o(t) \quad \text{as} \quad t \to 0.$$

Therefore the ch.f. $\phi(t)$ is differentiable at $t = 0$ and $\phi'(0) = 0$.

(ii) Let us extend case (i). Suppose now that X is a r.v. with the following density (K is a norming constant):

$$f(x) = \begin{cases} 0, & \text{if } |x| \le 2 \\ K/(x^4 \log |x|), & \text{if } |x| > 2. \end{cases}$$

It can be shown that the ch.f. $\phi(t) = \int_{-\infty}^\infty e^{itx} f(x)\,\mathrm{d}x$, $t \in \mathbb{R}^1$ is differentiable at $t = 0$ three times and e.g. $\phi^{(3)}(0) = 0$ ($\phi^{(4)}(t)$ does not exist at $t = 0$). However $\mathbf{E}[|X|^3] = \int_{-\infty}^\infty |x|^3 f(x)\,\mathrm{d}x = \infty$, i.e. m_3, the third-order moment of X, does not exist. (For details see Rao 1984.)

8.9. The convolution of two indecomposable distributions can even have a normal component

Let F_1, F_2 be d.f.s and ϕ_1, ϕ_2 their ch.f.s respectively. If at least one of ϕ_1, ϕ_2 is decomposable, then the convolution $F_1 * F_2$ has a ch.f. $\phi_1\phi_2$ which is also decomposable. If F_1 and F_2 are both indecomposable, is it true that $F_1 * F_2$ is indecomposable? Regardless of our intuition we shall show that $F_1 * F_2$ can contain a decomposable component which in particular can be chosen to be normal. To see this, let us consider the d.f. F with the following ch.f.: $\phi(t) = (1-t^2)e^{-t^2/2}$, $t \in \mathbb{R}^1$.

According to Linnik and Ostrovskii (1977) any ch.f. of the form $(1-b^2t^2)\exp[ict - b^2t^2/2]$ where $b, c \in \mathbb{R}^1$, $b \neq 0$, is indecomposable. So, ϕ is indecomposable. Denote by ψ the ch.f. of the d.f. $G := F * F$. Then $\psi(t) = \phi^2(t) = (1-t^2)^2 e^{-t^2}$. Write $\psi(t)$ in the form $\psi(t) = \psi_1(t)\psi_2(t)$ where

$$\psi_1(t) = (1-t^2)^2 \exp(-3t^2/4), \quad \psi_2(t) = \exp(-t^2/4).$$

It is then not difficult to check that the integral $\int_{-\infty}^{\infty} \phi_1(t)\exp(-itx)\,dt$ is real-valued and non-negative for all $x \in \mathbb{R}^1$. This implies that ψ_1 is a ch.f. of some distribution (it is not important which one). On the other hand, ψ_2 can be identified as the ch.f. of the normal distribution $\mathcal{N}(0, \frac{1}{2})$ since in general the normal distribution $\mathcal{N}(a, \sigma^2)$ has a ch.f. equal to $\exp[iat - \frac{1}{2}\sigma^2 t^2]$, $t \in \mathbb{R}^1$.

Hence the indecomposability property is not preserved under convolution.

The same example considered above can be interpreted as follows. Let X_1, X_2 be independent r.v.s with a common d.f. F. Then the sum $X_1 + X_2$ has a d.f. G and, moreover, the following relation holds:

$$X_1 + X_2 \stackrel{\mathrm{d}}{=} Y_1 + Y_2$$

where Y_1, Y_2 are independent r.v.s such that Y_1 has a ch.f. ψ_1, $Y_2 \sim \mathcal{N}(0, \frac{1}{2})$.

8.10. Does the existence of all moments of a distribution guarantee the analyticity of its characteristic and moment generating functions?

Let X be a r.v. with ch.f. ϕ and m.g.f. M. Then if $\phi(t)$ and $M(z)$ are analytic functions for $t \leq t_0$ or $|z| \leq r_0$ with $t_0 > 0$, $r_0 > 0$, the r.v. X possesses moments of all orders. Thus we come to the question of whether the converse of the statement is true.

Suppose Z is a r.v. with density

$$f(x) = \begin{cases} 0, & \text{if } x < 0 \\ \frac{1}{2}\exp(-\sqrt{x}), & \text{if } x \geq 0. \end{cases}$$

Then we have $m_k = \mathbf{E}[Z^k] = (2k+1)!$, $k = 0, 1, 2, \ldots$, hence Z possesses moments of any order. For clarifying the properties of the ch.f. of Z we need the following result (see Laha and Rohatgi 1979). The ch.f. ϕ of the r.v. X is analytic iff: (a) X has moments m_k of any order k, $k \geq 1$; (b) there exists a constant $c > 0$ such that $|m_k| \leq k!c^k$ for all $k \geq 1$.

Since in our example $m_k = (2k+1)!$ we can easily find that

$$(|m_k|/k!)^{1/k} = k^{(k+1)/k}\left[\left(1+\frac{1}{k}\right)\cdots\left(1+\frac{k+1}{k}\right)\right]^{1/k} > k$$

and clearly condition (b) in the above result is not satisfied. Therefore the ch.f. ϕ of Z cannot be analytic. It follows that the m.g.f. M does not exist. Note that the last statement can be derived directly. Indeed, $M(z)$ can be written in the form

$$M(z) = \frac{1}{2}\int_0^\infty \exp(zx - \sqrt{x})\, dx.$$

If $\varepsilon > 0$ is small enough then for every z with $0 < z < \varepsilon$ we have $zx - \sqrt{x} \to \infty$ as $x \to \infty$. This implies that $\int_0^\infty \exp(zx - \sqrt{x})\, dx = \infty$. Therefore $M(z)$ does not exist in spite of the fact that all moments of Z do exist.

Finally, let us show a case which is an extension of the above example. Suppose U is a r.v. with density

$$g(x) = c \exp(-|x|^\gamma), \quad x \in \mathbb{R}^1$$

where $0 < \gamma < 1$ and c is a norming constant, $c^{-1} := \int_{-\infty}^\infty \exp(-|x|^\gamma)\, dx$. Then $\mathbf{E}[|U|^k] < \infty$ for every $k \geq 1$, so U possesses moments of any order. Nevertheless, the ch.f. of U is not analytic and consequently the m.g.f. of U does not exist.

SECTION 9. INFINITELY DIVISIBLE AND STABLE DISTRIBUTIONS

Let X be a r.v. with d.f. F and ch.f. ϕ. We say that X, as well as F and ϕ, are **infinitely divisible** if for each $n \geq 1$ there exist i.i.d. r.v.s $X_{n1}, \ldots, X_{nn}$ such that

$$X \stackrel{\mathrm{d}}{=} X_{n1} + \ldots + X_{nn}$$

or equivalently, if for a d.f. F_n and a ch.f. ϕ_n,

$$F = F_n * \cdots * F_n = (F_n)^{*n} \quad \text{and} \quad \phi = (\phi_n)^n.$$

Let us note the following properties.

(i) A distribution F with bounded support is infinitely divisible iff it is degenerate.
(ii) The infinitely divisible ch.f. does not vanish.
(iii) The product of a finite number of infinitely divisible ch.f.s is a ch.f. which is again infinitely divisible.
(iv) The r.v. X can be a limit of sums $S_n = \sum_{k=1}^n X_{nk}$ iff X is infinitely divisible.

Fundamental in this field is the following result (see Feller 1971; Chow and Teicher 1978; Shiryaev 1995). The r.v. X with ch.f. ϕ is infinitely divisible iff ϕ admits the following canonical representation known as the Lévy–Khintchine representation:

(1) $$\phi(t) = \exp\left\{ i\gamma t + \int_{-\infty}^\infty \left(e^{itu} - 1 - \frac{itu}{1+u^2} \right) \frac{1+u^2}{u^2}\, dG(u) \right\}$$

where $\gamma \in \mathbb{R}^1$, and $G(x)$, $x \in \mathbb{R}^1$, is non-decreasing left-continuous function of bounded variation and $G(-\infty) = 0$.

Now let us introduce another notion. The r.v. X, its d.f. F and its ch.f. ϕ are called **stable** if for every $n \geq 1$ there exist constants a_n and $b_n > 0$ and independent r.v.s $X_1, \ldots, X_n$ distributed like X such that

$$b_n X + a_n \stackrel{\mathrm{d}}{=} X_1 + \ldots + X_n$$

or, equivalently, $F\left(\frac{x-a_n}{b_n}\right) = [F(x)]^{*n}$, or $[\phi(t)]^n = \phi(b_n t)\mathrm{e}^{ia_n t}$.

The basic result concerning stable distributions is as follows (see Chow and Teicher 1978; Zolotarev 1986). The r.v. X with ch.f. ϕ is stable iff ϕ admits the following canonical representation:

$$\phi(t) = \exp\left\{i\gamma t - c|t|^\alpha \left[1 + i\beta \frac{t}{|t|} w(t, \alpha)\right]\right\} \tag{2}$$

where $\gamma \in \mathbb{R}^1$, $0 < \alpha \leq 2$, $|\beta| \leq 1$, $c \geq 0$ and

$$w(t, \alpha) = \begin{cases} \tan \frac{1}{2}\pi\alpha, & \text{if } \alpha \neq 1 \\ (2/\pi) \log |t|, & \text{if } \alpha = 1. \end{cases}$$

Recall that (2) is also known as the Lévy–Khintchine representation. In particular, if $\gamma = 0$, $\beta = 0$, we obtain the symmetric stable distributions. They have ch.f.s of the type $\exp(-c|t|^\alpha)$ where $c \geq 0$, $0 < \alpha \leq 2$.

A detailed investigation of the infinitely divisible distributions and the stable distributions can be found in the books by Gnedenko and Kolmogorov (1954), Lukacs (1970), Feller (1971), Linnik and Ostrovskii (1977), Loéve (1978), Chow and Teicher (1978) and Zolotarev (1986).

The next examples illustrate different properties of infinitely divisible and stable distributions. Two examples deal with random vectors.

9.1. A non-vanishing characteristic function which is not infinitely divisible

Let the r.v. X with ch.f. $\phi(t)$, $t \in \mathbb{R}^1$, be infinitely divisible. Then ϕ does not vanish. The example below shows that in general the converse is not true.

Consider the discrete r.v. X which takes the values -1, 0, 1 with probabilities $\frac{1}{8}$, $\frac{3}{4}$, $\frac{1}{8}$ respectively. The ch.f. ϕ of X is

$$\phi(t) = \tfrac{1}{8}\mathrm{e}^{-it} + \tfrac{3}{4}\mathrm{e}^{it0} + \tfrac{1}{8}\mathrm{e}^{it} = \tfrac{1}{8}(3 + \cos t).$$

Obviously $\phi(t) > 0$ for all $t \in \mathbb{R}^1$, so ϕ does not vanish. Nevertheless, X is not infinitely divisible. To see this, let us assume that X can be written as

$$X \stackrel{\mathrm{d}}{=} X_1 + X_2 \tag{1}$$

where X_1 and X_2 are i.i.d. r.v.s. Since X has three possible values, it is clear that each of X_1 and X_2 can take only two values, say a and b, $a < b$. Let $\mathbf{P}[X_1 = a] = p$, $\mathbf{P}[X_1 = b] = 1 - p$ for some p, $0 < p < 1$. Then $X_1 + X_2$ takes the values $2a$, $a + b$ and $2b$ with probabilities p^2, $2p(1-p)$ and $(1-p)^2$ respectively. So it should be

$$2a = -1, \quad a + b = 0, \quad 2b = 1, \quad p^2 = \tfrac{1}{8}, \quad 2p(p+1) = \tfrac{3}{4}, \quad (1-p)^2 = \tfrac{1}{8}$$

which are clearly incompatible. Hence the representation (1) is not possible, implying that X is not infinitely divisible.

9.2. If $|\phi|$ is an infinitely divisible characteristic function, this does not always imply that ϕ is also infinitely divisible

Recall that if ϕ is an infinitely divisible ch.f. then its absolute value $|\phi|$ is so. It is not so trivial that in general the converse statement is false. This was discovered by Gnedenko and Kolmogorov (1954) and we present here their example of a ch.f. ϕ such that $|\phi|$ is infinitely divisible, but ϕ is not. Consider the function

$$\phi(t) = \frac{1-b}{1-a}\,\frac{1+a\mathrm{e}^{-it}}{1-b\mathrm{e}^{it}}, \quad t \in \mathbb{R}^1 \tag{1}$$

where $0 < a \le b < 1$. Obviously ϕ is continuous, $\phi(0) = 1$ and

$$\phi(t) = \frac{1-b}{1-a}\left[a\mathrm{e}^{-it} + (1+ab)\sum_{k=0}^{\infty} b^k \mathrm{e}^{itk}\right].$$

It follows that ϕ is the ch.f. of a r.v. X with

$$\mathbf{P}[X = -1] = \frac{(1-b)a}{1-a}, \quad \mathbf{P}[X = k] = \frac{(1-b)(1+ab)b^k}{1-a}, \quad k = 0, 1, 2, \dots .$$

Let us show that ϕ is not infinitely divisible. Indeed, we find that

$$\log \phi(t) = \sum_{k=1}^{\infty}\left[\frac{(-1)^{k-1}a^k}{k}(\mathrm{e}^{-itk} - 1) + \frac{b^k}{k}(\mathrm{e}^{itk} - 1)\right]. \tag{2}$$

We can also write $\log \phi(t)$ in its canonical form (see the introductory notes to this section; the Lévy–Khintchine formula) by taking $\gamma = \sum_{k=1}^{\infty}(b^k + (-1)^k a^k)/(k^2+1)$ and $G(x)$ to be a function of bounded variation with jumps of size $kb^k/(k^2+1)$ at $x = k$ and $(-1)^{k-1}ka^k/(k^2+1)$ at $x = -k$ for $k = 1, 2, \dots$. However, G is not monotone, which automatically implies that ϕ cannot be infinitely divisible.

Furthermore, the function

$$\bar{\phi}(t) = \frac{1-b}{1-a}\,\frac{1+a\mathrm{e}^{it}}{1-b\mathrm{e}^{-it}}$$

is also a ch.f. but not infinitely divisible. Our next step is to show that the function

$$\psi(t) = |\phi(t)|^2 = \phi(t)\bar{\phi}(t)$$

is infinitely divisible. Note that ψ is a ch.f. as a product of two ch.f.s. It is easy to write firstly $\log \bar{\phi}(t)$ in the form (2) and then obtain $\log \psi(t)$, namely

$$\log \psi(t) = \sum_{k=1}^{\infty} \frac{1}{k}\left[b^k+(-1)^{k-1}a^k\right](\mathrm{e}^{-itk}-1)+\sum_{k=1}^{\infty} \frac{1}{k}\left[b^k+(-1)^{k-1}a^k\right](\mathrm{e}^{itk}-1).$$

Thus in the Lévy–Khintchine formula for $\log \psi(t)$ we can take $\gamma = 0$ and $G(x)$ to be a non-decreasing function with jumps of size $k(k^2+1)^{-1}\left[b^k + (-1)^{k-1}a^k\right]$ at the points $x = \pm k$, $k = 1, 2, \ldots$. Since this representation of $\log \psi(t)$ is unique, we conclude that the ch.f. ψ is infinitely divisible. Moreover $|\phi| = (|\phi|^2)^{\frac{1}{2}}$ is also infinitely divisible despite the fact that ϕ given by (1) is not. Another interesting observation is that the infinitely divisible ch.f. ψ is the product of the two non-infinitely divisible ch.f.s ϕ and $\bar{\phi}$.

9.3. The product of two independent non-negative and infinitely divisible random variables is not always infinitely divisible

(i) Define two independent r.v.s X and Y having values in the sets $\{0, 1, 2, 3, \ldots\}$ and $\{1, 1+c, 1+2c, 1+3c, \ldots\}$ respectively where $1 < c < \frac{3}{2}$. The corresponding probabilities for X and Y are $\{p_0, p_1, p_2, \ldots\}$ and $\{q_0, q_1, q_2, \ldots\}$ where $p_j > 0$, $\Sigma p_j = 1$, $q_j > 0$, $\Sigma q_j = 1$.

Consider the product $Z = XY$ and suppose it is infinitely divisible. Then

$$Z \stackrel{\mathrm{d}}{=} Z_1 + Z_2 \tag{1}$$

where Z_1 and Z_2 are i.i.d. r.v.s. Evidently, the 'first' six possible values of Z are 0, $1, 2, 1+c, 3, 1+2c$. It follows that 0, 1 and $1+c$ are among the values of Z_1 (and hence of Z_2). But this implies that $2+c$ is a possible value of Z. Since $2+c < 1+2c$ we get a contradiction. Consequently a relation similar to (1) is not possible. Thus Z cannot be infinitely divisible.

Notice that X and Y take their values from different sets. The same answer concerning the non-infinite divisibility of the product XY can be obtained in the case of X and Y taking values in the same space.

(ii) Let us exhibit now an example in which the reasoning is based on the following (see Katti 1967). Suppose $\{p_n, n \in \mathbb{N}_0\}$ is a distribution with $p_0 > 0$ and $p_1 > 0$. Then $\{p_n\}$ is infinitely divisible iff the numbers r_k, $k = 0, 1, \ldots$, defined by

$$(n+1)p_{n+1} = \sum_{k=0}^{n} r_k p_{n-k}, \quad n = 0, 1, 2, \ldots \tag{2}$$

are all non-negative.

Let us use this result to prove a new and not too well known statement: let ξ and η be independent r.v.s each having a Poisson distribution $\mathcal{P}_0(\lambda)$. Then both ξ and η are infinitely divisible, but the product $X = \xi\eta$ is not.

Indeed, take $n > 1$ such that $n+1$ is a prime number. Then

$$\begin{aligned} p_{n+1} &= \mathbf{P}[X = n+1] = \mathbf{P}[\xi\eta = n+1] \\ &= \mathbf{P}[\xi = 1, \eta = n+1] + \mathbf{P}[\xi = n+1, \eta = 1] = 2\lambda^{n+1}\mathrm{e}^{-2\lambda}/(n+1)!. \end{aligned}$$

The number n itself is even and hence n has at least two (integer) factorizations: $n = 1 \cdot n = 2 \cdot (n/2)$. Therefore

$$p_n = \mathbf{P}[X = n] > 2 \cdot (\lambda^2\mathrm{e}^{-\lambda}/2!) \cdot (\lambda^{n/2}\mathrm{e}^{-\lambda}/(n/2)!) = \lambda^{n/2+2}\mathrm{e}^{-2\lambda}/(n/2)!.$$

Obviously $p_0 = \mathbf{P}[X = 0] = 1 - (1 - \mathrm{e}^{-\lambda})^2 > 0$, $p_1 = \lambda^2\mathrm{e}^{-2\lambda} > 0$, and so $r_0 = p_1/p_0 > 0$. Further, suppose that $r_1, r_2, \ldots, r_{n-1}$ in (2) are all non-negative. Let us check the sign of r_n. We have

$$\begin{aligned} r_n p_0 &\le (n+1)p_{n+1} - r_0 p_n < 2\lambda^{n+2}\mathrm{e}^{-2\lambda}/n! - r_0\lambda^{n/2+2}\mathrm{e}^{-2\lambda}/(n/2)! \\ &= \mathrm{e}^{-2\lambda}\lambda^{n/2+2}\left[2\lambda^{n/2}/n! - r_0/(n/2)!\right]. \end{aligned}$$

Since $\lambda > 0$ is fixed and $1/n!$ goes to zero as $n \to \infty$ faster than $1/(n/2)!$, we conclude that for sufficiently large n the number r_n becomes negative. This does not agree with the property in (2) that all r_n are non-negative. Hence the product $\eta\xi$ of two independent Poisson r.v.s is not infinitely divisible.

9.4. Infinitely divisible products of non-infinitely divisible random variables

There are many examples of the following kind: if X is a r.v. which is absolutely continuous and infinitely divisible and X_1, X_2 are independent copies of X, then the product X_1X_2 is again infinitely divisible.

As a first example take $X \sim \mathcal{N}(0,1)$. Then X_1X_2 has a ch.f. equal to $1/(1+t^2)^{1/2}$ and hence X_1X_2 is infinitely divisible.

As a second example, take $X \sim \mathcal{C}(0,1)$, i.e. X has a Cauchy density $f(x) = 1/(\pi(1+x^2))$, $x \in \mathbb{R}^1$. If X_1 and X_2 are independent copies of X, it can be checked that the ch.f. of the product X_1X_2 is infinitely divisible.

These and other examples (discussed by Rohatgi *et al* (1990)) lead to the following question. Suppose X_1 and X_2 are independent copies of the absolutely continuous r.v. X. Suppose further that the product $Y = X_1X_2$ is infinitely divisible. Does this imply that X itself is infinitely divisible?

Let Y be a r.v. distributed $\mathcal{N}(0,1)$. Then there exists a r.v. X such that by taking two independent copies, X_1 and X_2, we obtain $X_1X_2 \stackrel{\mathrm{d}}{=} Y$ (for details see Groeneboom and Klaassen (1982)). Thus $\mathbf{P}[|Y| > x^2] \ge (\mathbf{P}[|X| > x])^2$ which implies that

$$\mathbf{P}[|X| > x] \le (\mathbf{P}[|Y| > x^2])^{1/2} = O(\mathrm{e}^{-x^4/4}) \quad \text{as} \quad x \to \infty.$$

Referring to the paper of Steutel (1973) for details we conclude that X cannot be infinitely divisible. Hence the answer to the above question is negative.

9.5. Every distribution without indecomposable components is infinitely divisible, but the converse is not true

Following tradition, denote by I_0 the class of distributions which have no indecomposable components. Recall that $F \in I_0$ means that the ch.f. ϕ of F cannot be represented in the form $\phi = \phi_1\phi_2$ where ϕ_1 and ϕ_2 are ch.f.s of non-degenerate distributions. Detailed study of the class I_0 is due to A. Ya. Khintchine. In this connection see Linnik and Ostrovskii (1977) where among a variety of results, the following theorem is proved: the class I_0 is a subclass of the class of infinitely divisible distributions.

Let us show that this inclusion is strong. Indeed, take the following ch.f.:

$$\phi(t) = \frac{1-a}{1-ae^{it}}, \quad 0 < a < 1, \quad t \in \mathbb{R}^1. \tag{1}$$

The representation

$$\phi(t) = \exp[\log(1-a) - \log(1-ae^{it})] = \exp\left[\sum_{n=1}^{\infty} \frac{a^n}{n}(e^{itn} - 1)\right]$$

shows that ϕ is a limit of products of ch.f.s corresponding to Poisson distributions. Then (see Gnedenko and Kolmogorov 1954; Loève 1977/1978) the ch.f. ϕ is infinitely divisible.

Further, the identity $1/(1-x) = \prod_{k=0}^{\infty}(1 + x^{2^k})$, $|x| < 1$ implies that

$$\phi(t) = \prod_{k=0}^{\infty}(1 + a^{2^k}e^{it2^k})/(1 + a^{2^k}).$$

Recall that $(1 + a^{2^k}e^{it2^k})/(1 + a^{2^k})$ is the ch.f. of a distribution concentrated at two points, namely 0 and 2^k. However, such a distribution is indecomposable (see Example 9.1). Hence the ch.f. ϕ defined by (1) is infinitely divisible but ϕ does not belong to the class I_0.

9.6. A non-infinitely divisible random vector with infinitely divisible subsets of its coordinates

Let (X_1, X_2, X_3) be a random vector and $\psi(t_1, t_2, t_3)$, $t_1, t_2, t_3 \in \mathbb{R}^1$ its ch.f.:

$$\psi(t_1, t_2, t_3) = \mathbf{E}\left[\exp[i(t_1X_1 + t_2X_2 + t_3X_3)]\right].$$

The vector (X_1, X_2, X_3) is said to be *infinitely divisible* if for each $\alpha > 0$, $\psi^\alpha(t_1, t_2, t_3)$ is again a ch.f. Obviously this notion can be introduced for random vectors in $\mathbb{R}^n$ with $n > 3$. We confine ourselves to the three-dimensional case for simplicity.

Let us note that if (X_1, X_2, X_3) is infinitely divisible, then each subset of its coordinates X_1, X_2, X_3 is infinitely divisible. This follows easily from the properties of the usual one-dimensional infinitely divisible distributions. Thus it is natural to ask whether the converse statement is true.

Consider two independent r.v.s X and Y each $\mathcal{N}(0, 1)$. Let

$$Z_1 = X^2, \quad Z_2 = XY, \quad Z_3 = Y^2.$$

It is easy to check that each of Z_1, Z_2, Z_3 is infinitely divisible. Moreover, any of the two-dimensional random vectors (Z_1, Z_2), (Z_1, Z_3) and (Z_2, Z_3) is also infinitely divisible. However, the vector (Z_1, Z_2, Z_3) is indecomposable, it has trivariate gamma distribution which is not infinitely divisible. For details we refer the reader to works by Lévy (1948), Griffiths (1970) and Rao (1984).

9.7. A non-infinitely divisible random vector with infinitely divisible linear combinations of its components

If (X, Y) is an infinitely divisible random vector, then any linear combination $Z = \alpha_1 X + \alpha_2 Y$, $\alpha_1, \alpha_2 \in \mathbb{R}^1$ is an infinitely divisible r.v. The question to be considered is whether the converse is true. This problem, posed by C.R. Rao, has been solved by Ibragimov (1972). The following example shows that the answer is negative.

For $x = (x_1, x_2) \in \mathbb{R}^2$, $|x| = (x_1^2 + x_2^2)^{1/2}$ and $0 < \varepsilon < \frac{1}{4}$, define the function

$$a_\varepsilon(x) = \begin{cases} 1, & \text{if } |x| \le \frac{1}{2} - \varepsilon \text{ or } \frac{1}{2} + \varepsilon < |x| \le 1 \\ 0, & \text{if } |x| > 1 \\ -\varepsilon, & \text{if } \frac{1}{2} - \varepsilon < |x| \le \frac{1}{2} + \varepsilon. \end{cases}$$

Let $A_\varepsilon(U)$, $U \in \mathbb{R}^2$, be the signed measure with density a_ε, that is

$$A_\varepsilon(U) = \int_U a_\varepsilon(x)\,\mathrm{d}x$$

and also introduce the function

$$(1) \quad \psi_\varepsilon(t) = \exp\left[\int_{\mathbb{R}^2} (\mathrm{e}^{i\langle t,x\rangle} - 1)\,\mathrm{d}A_\varepsilon(x)\right], \quad t, x \in \mathbb{R}^2, \quad \langle t, x\rangle = t_1 x_1 + t_2 x_2.$$

For all sufficiently small $\varepsilon > 0$, ψ_ε is positive definite and hence ψ_ε is the ch.f. of some d.f. F_ε in $\mathbb{R}^2$. Indeed, from (1)

$$\psi_\varepsilon(t) = c\left[1 + \sum_{n=1}^{\infty} \frac{1}{n!}\left(\int_{\mathbb{R}^2} \mathrm{e}^{i\langle t,x\rangle}\,\mathrm{d}A_\varepsilon(x)\right)^n\right]$$

where $c = \exp[-A_\varepsilon(\mathbb{R}^2)]$. Thus F_ε can be written in the form $F_\varepsilon = c(G_0 + G_\varepsilon)$ where G_0 is a probability measure with $G_0(\{0\}) = 1$ and G_ε is a measure with density $\gamma_\varepsilon(x) = \sum_{n=1}^{\infty} \tilde{a}_\varepsilon^{(n)}(x)/n!$. Furthermore we can check that for all small ε,

$$\tilde{a}_\varepsilon^{(2)}(x) = \int_{\mathbb{R}^2} a_\varepsilon(x-u)a_\varepsilon(u)\,\mathrm{d}u \geq 0 \quad \text{and} \quad \tilde{a}_\varepsilon^{(3)}(x) = \tilde{a}_\varepsilon^{(2)} * a_\varepsilon(x) \geq 0.$$

Hence for $n \geq 4$ we have $\tilde{a}_\varepsilon^{(n)}(x) = \tilde{a}_\varepsilon^{(n-2)} * \tilde{a}_\varepsilon^{(2)}(x) \geq 0$. For small ε, $a_\varepsilon(x)$ is close to the function $\tilde{a}_0^{(2)}(x)$. It is easy to see that for $\frac{1}{4} \leq x < \frac{3}{4}$ we have $\inf_x \tilde{a}_0^{(2)}(x) = c_1 > 0$. Thus for small ε, $\tilde{a}_\varepsilon^{(2)}(x) > 2\varepsilon$ if $\frac{1}{2} - \varepsilon < |x| \leq \frac{1}{2} + \varepsilon$. Evidently this implies that $\gamma_\varepsilon(x) \geq 0$ for all $x \in \mathbb{R}^2$. Therefore F_ε described above is a probability measure in $\mathbb{R}^2$.

Denote by (X_1, X_2) a random vector with d.f. F_ε. Since A_ε is a signed measure (its values are not only positive), F_ε cannot be infinitely divisible.

It remains to be shown that any linear combination $\alpha_1 X_1 + \alpha_2 X_2$, $\alpha_1, \alpha_2 \in \mathbb{R}^1$, has a distribution which is infinitely divisible. Indeed, for $s \in \mathbb{R}^1$

$$\begin{aligned}\phi_\alpha(s) &:= \mathbf{E}\{\exp[is(\alpha_1 X_1 + \alpha_2 X_2)]\} = \psi_\varepsilon(\alpha_1 s, \alpha_2 s) \\ &= \exp\left[\int_{\mathbb{R}^2} (\mathrm{e}^{is\langle\alpha, x\rangle} - 1)\,\mathrm{d}A_\varepsilon(x)\right].\end{aligned}$$

Denoting $\langle\alpha, x\rangle = u$ where $u \in \mathbb{R}^1$ we can write $\phi_\alpha(s)$ in the form

$$\phi_\alpha(s) = \exp\left[\int_{-\infty}^{\infty} (\mathrm{e}^{isu} - 1)\,\mathrm{d}H_\alpha(u)\right], \quad \mathrm{d}H_\alpha(u) = \int_{\langle\alpha,x\rangle \leq u} \mathrm{d}A_\varepsilon(x)\,\mathrm{d}u.$$

Since for sufficiently small ε every strip $\{x : u_1 \leq \langle\alpha, x\rangle \leq u_2\}$ has positive A_ε-measure, we conclude (again see Ibragimov 1972) that the function $H_\alpha(u)$, $u \in \mathbb{R}^1$, is a d.f. and moreover, $\phi_\alpha(s) = \psi_\varepsilon(\alpha_1 s, \alpha_2 s)$, $s \in \mathbb{R}^1$, is a ch.f. of a distribution which is infinitely divisible.

Thus we have established that any linear combination $\alpha_1 X_1 + \alpha_2 X_2$ is an infinitely divisible r.v. but (X_1, X_2) is not an infinitely divisible vector.

9.8. Distributions which are infinitely divisible but not stable

Usually we introduce and study the class of infinitely divisible distributions and then the class of stable distributions. One of the first observed properties is that every stable distribution is infinitely divisible. Let us show that the converse is not always true.

(i) Let X be a r.v. with Poisson distribution, $X \sim \mathcal{P}o(\lambda)$, that is

$$\mathbf{P}[X = n] = \frac{\lambda^n \mathrm{e}^{-\lambda}}{n!}, \quad n = 0, 1, 2, \ldots$$

where the parameter $\lambda > 0$ is given. If ϕ is the ch.f. of X then

(1) $$\phi(t) = \exp[\lambda(e^{it} - 1)], \quad t \in \mathbb{R}^1.$$

Since $\phi(t) = [\phi_n(t)]^n$ for $\phi_n(t) = \exp[\lambda n^{-1}(e^{it} - 1)]$ and ϕ_n is again a ch.f. (of $\mathcal{P}_0(\lambda/n)$) then X is infinitely divisible. However, ϕ from (1) does not satisfy any relation of the type $\phi(b_1 t)\phi(b_2 t) = \phi(bt)e^{i\gamma t}$ (see the introductory notes). Hence the Poisson distribution is not stable despite the fact that it is infinitely divisible.

(ii) Let Y be a r.v. with *Laplace distribution*, that is, its density g is

$$g(x) = \frac{1}{2\lambda} \exp\left(-\frac{|x - \mu|}{\lambda}\right), \quad x \in \mathbb{R}^1$$

where $\mu \in \mathbb{R}^1$, $\lambda > 0$. For the ch.f. ψ of Y we have

(2) $$\psi(t) = e^{it\mu}/(1 + t^2\lambda^2), \quad t \in \mathbb{R}^1.$$

It is not difficult to verify that ψ is infinitely divisible. But ψ from (2) does not satisfy any relation of the type $\psi(b_1 t)\psi(b_2 t) = \psi(bt)e^{i\gamma t}$ and hence ψ is not stable.

Therefore the Laplace distribution is an example of an absolutely continuous distribution which is infinitely divisible without being stable.

(iii) Suppose the gamma distributed r.v. Z has a density

$$g(x) = \frac{1}{\sqrt{2\pi}} x^{-1/2} e^{-x}, \quad x > 0$$

(and $g(x) = 0$ for $x \leq 0$). Then by using the explicit form of the ch.f. of Z we can show that Z is infinitely divisible but not stable.

An additional example of a distribution which is infinitely divisible but not stable is given in Example 21.8.

9.9. A stable distribution which can be decomposed into two infinitely divisible but not stable distributions

Let X be a r.v. with Cauchy distribution $\mathcal{C}(1, 0)$, that is, its density is

$$f(x) = \frac{1}{\pi(1 + x^2)}, \quad x \in \mathbb{R}^1.$$

If ϕ denotes the ch.f. of X, then $\phi(t) = e^{-|t|}$, $t \in \mathbb{R}^1$. It is well known that this distribution is stable. Let us show that X can be written as

(1) $$X \stackrel{d}{=} X_1 + X_2$$

where X_1 and X_2 are independent r.v.s whose distributions are infinitely divisible but not stable. For introduce the following two functions:

$$\phi_1(t) = \exp[-|t| + 1 - e^{-|t|}], \quad \phi_2(t) = \exp[e^{-|t|} - 1].$$

We claim that ϕ_1 and ϕ_2 are ch.f.s of distributions which are infinitely divisible. This follows from the fact that each of ϕ_1, ϕ_2 can be expressed in the form $\exp[-\int_0^t \int_0^u \psi(v)\,dv du]$ with a suitable integrand ψ and the only assumption is that ψ is a ch.f. Then our conclusion concerning ϕ_1 and ϕ_2 is a consequence of a result of Lukacs (1970, Th. 12.2.8).

It is easy to verify that ϕ_1 and ϕ_2 are not stable ch.f.s. Thus we have

$$\phi(t) = \mathrm{e}^{-|t|} = \phi_1(t)\phi_2(t). \tag{2}$$

Now take two independent r.v.s, say X_1 and X_2, whose ch.f.s are ϕ_1 and ϕ_2 respectively. It only remains to see that (2) implies (1).

Therefore we have constructed two r.v.s X_1 and X_2 which are independent, both are infinitely divisible but not stable and they are such that the sum $X_1 + X_2$ has a stable distribution.

SECTION 10. NORMAL DISTRIBUTION

We say that the r.v. X has a **normal distribution** with parameters a and σ^2, $a \in \mathbb{R}^1$, $\sigma > 0$, if X is absolutely continuous and has a density

$$f(x) = \frac{1}{\sigma\sqrt{2\pi}} \exp\left[-\frac{(x-a)^2}{2\sigma^2}\right], \quad x \in \mathbb{R}^1. \tag{1}$$

In such a case we use the notation $X \sim \mathcal{N}(a, \sigma^2)$. It is easy to write explicitly the d.f. corresponding to (1).

Consider the particular case when $a = 0$, $\sigma = 1$. We obtain the functions

$$\varphi(x) = \frac{1}{\sqrt{2\pi}} \exp(-\tfrac{1}{2}x^2), \quad x \in \mathbb{R}^1 \tag{2}$$

and

$$\Phi(x) = \frac{1}{\sqrt{2\pi}} \int_{-\infty}^{x} \exp(-\tfrac{1}{2}u^2)\,du, \quad x \in \mathbb{R}^1. \tag{3}$$

These two functions, φ and Φ, are called a *standard normal density function*, and a *standard normal d.f.* respectively. They correspond to a r.v. $\mathcal{N}(0, 1)$.

Recall that the r.v. $X \sim \mathcal{N}(a, \sigma^2)$ has $\mathbf{E}X = a$, $\mathbf{V}X = \sigma^2$ and a ch.f $\phi(t) = \exp(iat - \frac{1}{2}\sigma^2 t^2)$. If $a = 0$, then all odd-order moments are zero, that is, $m_{2n+1} = \mathbf{E}[X^{2n+1}] = 0$, while the even-order moments are $m_{2n} = \mathbf{E}[X^{2n}] = \sigma^{2n}(2n-1)!!$.

Consider now the random vector $X = (X_1, \dots, X_n)$. If $\mathbf{E}X_i = a_i$, $i = 1, \dots, n$, then $a = (a_1, \dots, a_n)$ is called a *mean value vector* (or vector of the expectation) of X. The matrix $C = (c_{ij})$ where $c_{ij} = \mathbf{E}[(X_i - a_i)(X_j - a_j)]$, $i, j = 1, \dots, n$ is called a *covariance matrix* of X. We say that X has an n-**dimensional normal distribution** if X possesses a density function

(4) $f(x_1, \ldots, x_n) =$

$$(2\pi)^{-n/2}|D|^{1/2} \exp\left\{ -\frac{1}{2} \sum_{i,j=1}^{n} d_{ij}(x_i - a_i)(x_j - a_j) \right\}, \quad (x_1, \ldots, x_n) \in \mathbb{R}^n.$$

Here the matrix $D = (d_{ij})$ is the inverse matrix to C. Clearly, D exists if C is positive definite and $|D| := \det D$.

Note that we could start with the vector $a = (a_1, \ldots, a_n) \in \mathbb{R}^n$ and the symmetric positive definite matrix $C = (c_{ij})$, then invert C to yield matrix D, and finally use the vector a and the matrix D to write the function f as in (4). This function f is an n-dimensional density and thus there is a random vector, say $(X_1, \ldots, X_n)$, whose density is f. By definition this vector is called normally distributed, and (4) defines an n-dimensional normal density.

For some of the examples below we need the explicit form of (4) when $n = 2$. The two-dimensional (or bivariate) normal density can be written as

(5) $f(x_1, x_2) = \dfrac{1}{2\pi\sigma_1\sigma_2\sqrt{1-\rho^2}} \times$

$$\times \exp\left\{ -\frac{1}{2(1-\rho^2)} \left[\frac{(x_1 - a_1)^2}{\sigma_1^2} - 2\rho \frac{(x_1 - a_1)(x_2 - a_2)}{\sigma_1\sigma_2} + \frac{(x_2 - a_2)^2}{\sigma_2^2} \right] \right\}$$

where $\sigma_1, \sigma_2 > 0$ and $|\rho| < 1$. If (X_1, X_2) is a random vector with density (5) then $\mathbf{E}X_1 = a_1$, $\mathbf{E}X_2 = a_2$, $\mathbf{V}X_1 = \sigma_1^2$, $\mathbf{V}X_2 = \sigma_2^2$ and ρ equals the correlation coefficient $\rho(X_1, X_2)$.

The normal distribution over $\mathbb{R}^1$ and $\mathbb{R}^n$ is considered in almost all textbooks and lecture notes. We refer the reader to the books by Anderson (1958), Parzen (1960), Gnedenko (1962), Papoulis (1965), Thomasian (1969), Feller (1971), Laha and Rohatgi (1979), Rao (1984), Shiryaev (1995) and Bauer (1996).

In this section we have given various examples which clarify the properties of the normal distribution.

10.1. Non-normal bivariate distributions with normal marginals

(i) Take two independent r.v.s ξ_1 and ξ_2, each distributed $\mathcal{N}(0, 1)$. Consider the following two-dimensional random vector:

$$(X_1, X_2) = \begin{cases} (\xi_1, |\xi_2|), & \text{if } \xi_1 \geq 0 \\ (\xi_1, -|\xi_2|), & \text{if } \xi_1 < 0. \end{cases}$$

Obviously the distribution of (X_1, X_2) is not bivariate normal, but each of the components X_1 and X_2 is normally distributed.

(ii) Suppose $h(x)$, $x \in \mathbb{R}^1$, is any odd continuous function vanishing outside the interval $[-1, 1]$ and satisfying the condition $|h(x)| \leq (2\pi\mathrm{e})^{-1/2}$. Using the standard normal density φ we define the function

$$f(x,y) = \varphi(x)\varphi(y) + h(x)h(y). \tag{1}$$

It is easy to check that $f(x,y)$, $(x,y) \in \mathbb{R}^2$, is a two-dimensional density function and $f(x,y)$ is not bivariate normal, but the marginal densities $f_1(x)$ and $f_2(y)$ are both normal.

The function h in (1) can be chosen as follows:

$$h(x) = (2\pi \mathrm{e})^{-1/2} x^3 I_{[-1,1]}(x), \quad x \in \mathbb{R}^1$$

where $I_{[-1,1]}(\cdot)$ is the indicator function of the interval $[-1,1]$.

(iii) For any number ε, $|\varepsilon| \le 1$, define the function

$$H(x,y) = \Phi(x)\Phi(y)[1 + \varepsilon(1-\Phi(x))(1-\Phi(y))], \quad (x,y) \in \mathbb{R}^2$$

(Φ is the standard normal d.f.). It is easy to check that H is a two-dimensional d.f. with marginal distributions $\Phi(x)$ and $\Phi(y)$ respectively. Obviously, if $\varepsilon \ne 0$, H is non-normal.

Another possibility is to take the function

$$h(x,y) = \varphi(x)\varphi(y)[1 + \varepsilon(2\Phi(x) - 1)(2\Phi(y) - 1)].$$

Then $h(x,y)$ is a two-dimensional density function with marginals $\varphi(x)$ and $\varphi(y)$ respectively, and $h(x,y)$ is non-normal if $\varepsilon \ne 0$.

(iv) Consider the following function:

$$f(x,y) = \begin{cases} [1/(\pi(1-\rho^2)^{1/2})] \exp[-\frac{1}{2}\rho^{-2}(x^2 - 2\rho xy + y^2)], & \text{if } xy \ge 0 \\ 0, & \text{if } xy < 0 \end{cases}$$

where $\rho \in (-1,1)$. It is easy to verify that f is a two-dimensional density function. Denote by (X,Y) the random vector whose density is $f(x,y)$. Obviously the distribution of (X,Y) is not normal, but each of the components X and Y is distributed $\mathcal{N}(0,1)$.

10.2. If (X_1, X_2) has a bivariate normal distribution then X_1, X_2 and $X_1 + X_2$ are normally distributed, but not conversely

Let (X_1, X_2) have a bivariate normal distribution and $f(x_1, x_2)$, $(x_1, x_2) \in \mathbb{R}^2$ be its density. Then each of the r.v.s X_1, X_2 and $X_1 + X_2$ has a one-dimensional normal distribution. We are interested in whether the converse is true.

Suppose X_1, X_2 are independent r.v.s each distributed $\mathcal{N}(0,1)$. Then their joint density is $f(x_1,x_2) = \varphi(x_1)\varphi(x_2)$ (φ is the standard normal density). The function $f(x_1,x_2)$ which is symmetric in both arguments x_1, x_2 will be changed as follows.

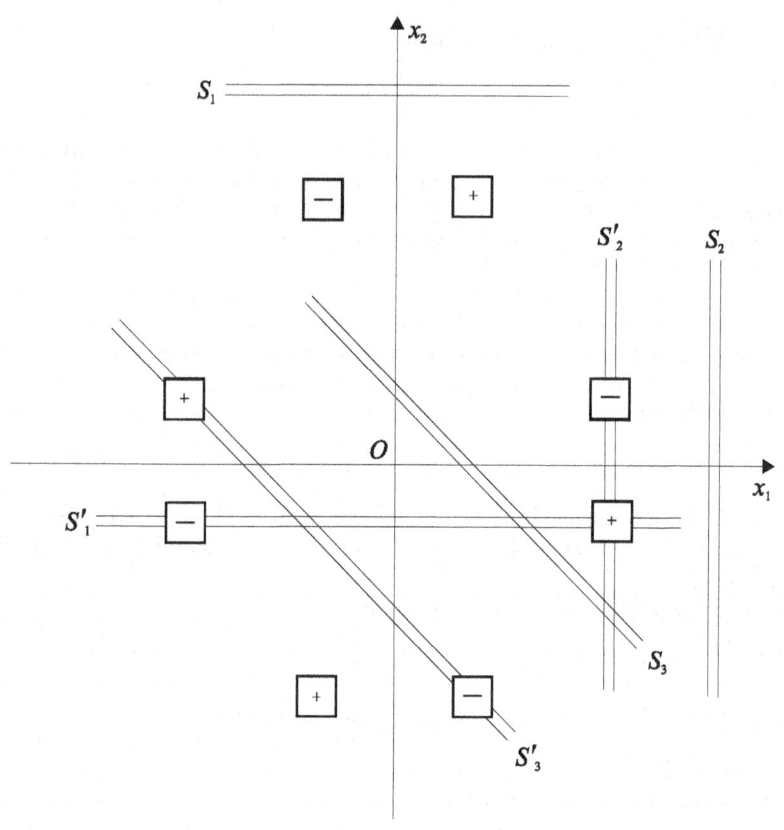

Figure 2

Firstly, let us draw eight equal squares at a fixed distance from the origin O and located symmetrically about the axes Ox_1 and Ox_2 as shown in figure 2. Put alternately the signs $(+)$ and $(-)$ in the squares. Let the small positive number ε denote the amount of 'mass' which we transfer from a square with $(-)$ to a square with $(+)$. Now define the function

$$(1) \qquad g(x_1, x_2) = \begin{cases} f(x_1, x_2) + \varepsilon, & \text{if } (x_1, x_2) \in Q^+ \\ f(x_1, x_2) - \varepsilon, & \text{if } (x_1, x_2) \in Q^- \\ f(x_1, x_2), & \text{if } (x_1, x_2) \notin Q^+ \cup Q^- \end{cases}$$

where Q^+ is the union of the squares with $(+)$ and Q^- the union of those with $(-)$.

For such squares we can choose $\varepsilon > 0$ sufficiently small such that $g(x_1, x_2) \geq 0$ for all $(x_1, x_2) \in \mathbb{R}^2$. From (1) we find immediately that $\int\int_{\mathbb{R}^2} g(x_1, x_2) \mathrm{d}x_1 \, \mathrm{d}x_2 = 1$. Hence g is a density function of a two-dimensional random vector, say (Y_1, Y_2). Next we want to find the distributions of Y_1, Y_2 and $Y_1 + Y_2$. The strips drawn in figure 2 will help us to do this. These strips can be arbitrarily wide and arbitrarily located but parallel to Ox_1 or Ox_2 or to the bisector of quadrants II and IV. Evidently the strips either do not intersect any of the squares, or each intersects just two of them, one signed by $(+)$ and another by $(-)$. Since the total mass in any strip remains unchanged and we know the distribution of (X_1, X_2) (recall it is normal), then we easily conclude that $Y_1 \sim \mathcal{N}(0,1)$, $Y_2 \sim \mathcal{N}(0,1)$, $Y_1 + Y_2 \sim \mathcal{N}(0,1)$. For example, look at the pairs of strips (S_1, S_1'), (S_2, S_2'), (S_3, S_3'). However it is clear that the distribution of (Y_1, Y_2) given by the density (1) is not bivariate normal.

Therefore the normality of Y_1, Y_2 and $Y_1 + Y_2$ is not enough to ensure that (Y_1, Y_2) is normally distributed.

10.3. A non-normally distributed random vector such that any proper subset of its components consists of jointly normally distributed and mutually independent random variables

We present here two examples based on different ideas. The first one is related to Example 7.3.

(i) Let the r.v. X have a distribution $\mathcal{N}(a, \sigma^2)$ and let f be its density. Take n r.v.s, say $X_1, \ldots, X_n$, $n \geq 3$, and define their joint density g_n as follows:

$$(1) \qquad g_n(x_1, \ldots, x_n) = \left[\prod_{i=1}^n f(x_i)\right]\left[1 + \prod_{j=1}^n (x_j - a) f(x_j)\right], \quad (x_1, \ldots, x_n) \in \mathbb{R}^n.$$

Firstly we have to check that g_n is a probability density. Since in this case we know f explicitly, $f(x) = (2\pi\sigma^2)^{-1/2} \exp[-(x-a)^2/2\sigma^2]$, we can easily find that $\int_{-\infty}^{\infty} (x - a) f^2(x) \, \mathrm{d}x = 0$. Then we can derive that g_n is non-negative and its integral over $\mathbb{R}^n$ is 1. Thus g_n, given by (1), is a density and, as we accepted, g_n is the density of the vector $(X_1, \ldots, X_n)$.

Let us choose k of the variables $X_1, \ldots, X_n$, $2 \leq k \leq n - 1$. Without loss of generality assume that $X_1, \ldots, X_k$ is our choice. Denote by g_k the density function of $(X_1, \ldots, X_k)$. From (1) we obtain

$$(2) \qquad g_k(x_1, \ldots, x_k) = f(x_1) \ldots f(x_k)$$

(recall that f is the density of X and $X \sim \mathcal{N}(a, \sigma^2)$). Therefore the variables $X_1, \ldots, X_k$ are jointly normally distributed and, moreover, they are independent. This conclusion holds for all choices of k variables among $X_1, \ldots, X_n$ and, let us repeat, $2 \leq k \leq n - 1$. It is also clear that each X_j, $j = 1, \ldots, n$ has a normal distribution $\mathcal{N}(a, \sigma^2)$.

Therefore we have described a set of n r.v.s which, according to (1), are dependent but, as it follows from (2), are $(n-1)$-wise independent.

(ii) Let $(X_1, \ldots, X_n)$ be an n-dimensional normally distributed random vector. Then its distribution is determined uniquely if we know the distributions of all pairs (X_i, X_j), $i, j = 1, \ldots, n$. This observation leads to the question of whether $(X_1, \ldots, X_n)$ is necessarily normal if all the pairs (X_i, X_j) are two-dimensional normal vectors. We shall show that the joint normality of all pairs and even of all $(n-1)$-tuples does not imply that $(X_1, \ldots, X_n)$ is normally distributed. (Look at case (i) above.)

Firstly, let $n \geq 3$ and $(\xi_1, \ldots, \xi_n)$ be such that $\xi_j = \pm 1$ and any particular sign vector $(y_1, \ldots, y_n)$ is taken with probability p if $\prod_{j=1}^n y_j = +1$ and with probability $q = 2^{-(n-1)} - p$ if $\prod_{j=1}^n y_j = -1$. Here $0 \leq p \leq 2^{-(n-1)}$. It is not difficult to see that all subsets of $n-1$ of the r.v.s $\xi_1, \ldots, \xi_n$ are independent (that is, any $n-1$ of them are mutually independent). Moreover, if $p \neq 2^{-n}$, all n variables are not mutually independent. Indeed, if $1 \leq k \leq n$ the vector $(y_{i_1}, \ldots, y_{i_k})$ can be extended in 2^{n-k-1} ways to a vector $(y_1, \ldots, y_n)$ with $\prod_{j=1}^n y_j = 1$ and in as many ways to one for which $\prod_{j=1}^n y_j = -1$. Thus

$$\mathbf{P}[\xi_{i_1} = y_{i_1}, \ldots, \xi_{i_k} = y_{i_k}] = 2^{n-k-1}p + 2^{n-k-1}q = 2^{-k}$$

and this equality holds for any $k \leq n-1$. Hence $\xi_1, \ldots, \xi_n$ are $(n-1)$-wise independent. Since $\mathbf{P}[\xi_1 = 1, \ldots, \xi_n = 1] = p$ it is obvious that $\xi_1, \ldots, \xi_n$ are not independent when $p \neq 2^{-n}$.

Now take $Z_1, \ldots, Z_n$ to be n mutually independent standard normal r.v.s which are independent of the vector $(\xi_1, \ldots, \xi_n)$. Define a new vector $(X_1, \ldots, X_n)$ where $X_j = \xi_j |Z_j|$, $j = 1, \ldots, n$. Then clearly the X_j are again standard normal. The independence of the Z_j together with the above reasoning concerning the properties of the vector $(\xi_1, \ldots, \xi_n)$ imply that all subsets of $n-1$ of the variables $X_1, \ldots, X_n$ are independent. Thus any $(n-1)$-tuple out of $X_1, \ldots, X_n$ has an $(n-1)$-dimensional normal distribution. It remains for us to clarify whether all n variables $X_1, \ldots, X_n$ are independent. It is easy to see that

$$\mathbf{P}[X_1 > 0, \ldots, X_n > 0] = \mathbf{P}[\xi_1 = 1, \ldots, \xi_n = 1] = p.$$

We conclude from this that if $p \neq 2^{-n}$ then the variables $X_1, \ldots, X_n$ are not independent and not normally distributed.

Let us note finally that in both cases, (i) and (ii), the joint normality and the mutual independence of any $n-1$ of the variables $X_1, \ldots, X_n$ do not imply that the vector $(X_1, \ldots, X_n)$ is normally distributed.

10.4. The relationship between two notions: normality and uncorrelatedness

Let (X, Y) be a random vector with normal distribution. Recall that both X and Y are also normally distributed, and if X and Y are uncorrelated, they are independent.

The examples below will show how important the normality of (X, Y) is.

(i) Let $X \sim \mathcal{N}(0,1)$. For a fixed number $c \geq 0$ define the r.v. Y by

$$Y = \begin{cases} X, & \text{if } |X| \leq c \\ -X, & \text{if } |X| \geq c. \end{cases}$$

It is easy to see that $Y \sim \mathcal{N}(0,1)$ for each c. Further,

$$\mathbf{E}[XY] = \mathbf{E}[X^2 I(|X| \leq c)] - \mathbf{E}[X^2 I(|X| > c)].$$

This implies that $\mathbf{E}[XY] = -1$ if $c = 0$ and $\mathbf{E}[XY] \to 1$ as $c \to \infty$. Since $\mathbf{E}[XY]$ depends continuously on c, there exists c_0 for which $\rho(X,Y) = \mathbf{E}[XY] = 0$. In fact, $c_0 \approx 1.54$ is the only solution of the equation $\mathbf{E}[XY] = 4\int_0^{c_0} x^2 \varphi(x)\,dx - 1 = 0$ (φ is the standard normal density). For this c_0 the r.v.s X and Y are uncorrelated. However, $\mathbf{P}[X > c, Y > c] = 0 \neq \mathbf{P}[X > c]\mathbf{P}[Y > c]$ and hence X and Y are not independent.

(ii) Let $\varphi_1(x,y)$ and $\varphi_2(x,y)$, $(x,y) \in \mathbb{R}^2$ be standard bivariate normal densities with correlation coefficients ρ_1 and ρ_2 respectively. Define

$$f(x,y) = c_1\varphi_1(x,y) + c_2\varphi_2(x,y), \quad (x,y) \in \mathbb{R}^2$$

where c_1, c_2 are arbitrary numbers, $c_1, c_2 \geq 0$, $c_1 + c_2 = 1$.

One can see that f is non-normal if $\rho_1 \neq \rho_2$. If we denote by (X,Y) a random vector with density f then we can easily find that $X \sim \mathcal{N}(0,1)$, $Y \sim \mathcal{N}(0,1)$. Moreover, the correlation coefficient between X and Y is $\rho = c_1\rho_1 + c_2\rho_2$. Choosing c_1, c_2, ρ_1, ρ_2 such that $c_1\rho_1 + c_2\rho_2 = 0$, we obtain two normally distributed and uncorrelated r.v.s X and Y. However, they are not independent.

(iii) Let (X,Y) be a two-dimensional random vector with density

$$f(x,y) = \frac{1}{2\pi\sqrt{3}} \left\{ \exp\left[-\tfrac{2}{3}(x^2 + xy + y^2)\right] + \exp\left[-\tfrac{2}{3}(x^2 - xy + y^2)\right] \right\}, \quad (x,y) \in \mathbb{R}^2.$$

Obviously the distribution of (X,Y) is not bivariate normal. Direct calculation shows that $X \sim \mathcal{N}(0,1)$, $Y \sim \mathcal{N}(0,1)$ and $\mathbf{E}[XY] = 0$. Thus X and Y are uncorrelated but dependent.

(iv) Let $X = \xi_1 + i\xi_2$ and $Y = \xi_3 + i\xi_4$ where $i = \sqrt{-1}$ and $(\xi_1, \xi_2, \xi_3, \xi_4)$ is a normally distributed random vector with zero mean and covariance matrix

$$C = \begin{bmatrix} 1 & 0 & 0 & -1 \\ 0 & 1 & -1 & 0 \\ 0 & -1 & 1 & 0 \\ -1 & 0 & 0 & 1 \end{bmatrix}.$$

The reader can check that C is a covariance matrix. Since X and Y are complex-valued, their covariance is

$$\mathbf{E}[X\bar{Y}] = \mathbf{E}[\xi_1\xi_3 + \xi_2\xi_4] + i\,\mathbf{E}[\xi_2\xi_3 - \xi_1\xi_4] = 0 + i(-1+1) = 0.$$

Hence X and Y are uncorrelated. Let us see if they are independent. If so, then ξ_1 and ξ_4 would be independent, and thus uncorrelated. But $\mathbf{E}[\xi_1\xi_4] = -1$. This contradiction shows that X and Y are dependent.

10.5. It is possible that $X, Y, X+Y, X-Y$ are each normally distributed, X and Y are uncorrelated, but (X, Y) is not bivariate normal

Consider the following function:

$$f(x,y) = \frac{1}{2\pi}\exp\left[-\tfrac{1}{2}(x^2+y^2)\right]\left\{1 + xy(x^2-y^2)\exp\left[-\tfrac{1}{2}(x^2+y^2+2\varepsilon)\right]\right\}$$

where $(x,y) \in \mathbb{R}^2$ and the constant $\varepsilon > 0$ is chosen in such a way that

$$\left|xy(x^2-y^2)\exp\left[-\tfrac{1}{2}(x^2+y^2+2\varepsilon)\right]\right| \le 1.$$

In order to establish that f is a two-dimensional probability density function and then derive some other properties, it is best to find first the Fourier transform ϕ of f. We have

$$\begin{aligned}\phi(s,t) &= \iint_{\mathbb{R}^2} \exp(isx + ity) f(x,y)\,\mathrm{d}x\,\mathrm{d}y \\ &= \exp\left[-\tfrac{1}{2}(s^2+t^2)\right] + \tfrac{1}{32}st(s^2-t^2)\exp\left[-\varepsilon - \tfrac{1}{4}(s^2+t^2)\right],\ (s,t)\in\mathbb{R}^2\end{aligned}$$

From this we deduce the following conclusions.

(1) Since $\phi(0,0) = 1$, $f(x,y)$ is the density of a two-dimensional vector (X,Y).
(2) $\phi(t,0) = \phi(0,t) = \exp(-\frac{1}{2}t^2) \Rightarrow X \sim \mathcal{N}(0,1)$, $Y \sim \mathcal{N}(0,1)$.
(3) $\phi(t,t) = \exp(-t^2)$, that is $X+Y \sim \mathcal{N}(0,2)$.
(4) $X - Y$ is also normally distributed.
(5) X and Y are uncorrelated.

However, the random vector (X,Y) as defined by the density $f(x,y)$ is not bivariate normal despite the fact that properties (2) – (5) are satisfied.

10.6. If the distribution of $(X_1, \ldots, X_n)$ is normal, then any linear combination and any subset of $X_1, \ldots, X_n$ is normally distributed, but there is a converse statement which is not true

This example can be considered as a natural continuation of Examples 10.2 and 10.5. Let us introduce the function

(1) $f_\varepsilon(x_1,\dots,x_n) =$

$$(2\pi)^{-n/2}\prod_{k=1}^{n}\varphi_0(x_k)\left[1+\varepsilon(x_1^2-x_2^2)\prod_{k=1}^{n}x_kI_{(-1,1)}(x_k)\exp\left(\frac{1}{2}\sum_{j=1}^{n}x^2\right)\right]$$

where $(x_1,\dots,x_n)\in\mathbb{R}^n$, $\varphi_0(x)=\exp(-\frac{1}{2}x^2)$, $I_{(-1,1)}$ is the indicator function of the interval $(-1,1)$ and the constant ε is chosen such that

$$|\varepsilon(x_1^2-x_2^2)\prod_{k=1}^{n}x_kI_{(-1,1)}(x_k)\exp\left(\frac{1}{2}\sum_{j=1}^{n}x_j^2\right)|\le 1.$$

Under this condition we can check that f_ε is a density of some n-dimensional random vector, say $(X_1,\dots,X_n)$. Evidently the density f_ε defined by (1) is not normal.

Now let us derive some statements for the distributions of the components of $(X_1,\dots,X_n)$. For this purpose we find the ch.f. ϕ of f_ε explicitly, namely

(2) $$\phi(t_1,\dots,t_n)=\prod_{k=1}^{n}\varphi_0(t_k)+\varepsilon\left[(\psi(t_1)\tilde\psi(t_2)-\tilde\psi(t_1)\psi(t_2))\prod_{k=3}^{n}\psi(t_k)\right]$$

where

$$\psi(t)=\begin{cases}(2i/t^2)(\sin t-t\cos t), & \text{if } t\ne 0\\ 0, & \text{if } t=0,\end{cases}$$

$$\tilde\psi(t)=\begin{cases}(2i/t)+(6/t^2)\psi(t), & \text{if } t\ne 0\\ 0, & \text{if } t=0.\end{cases}$$

From (2) one can draw the following conclusions.

(a) Each of the components $X_1,\dots,X_n$ is distributed $\mathcal{N}(0,1)$.
(b) For each k, $k<n$, the vector $(X_{i_1},\dots,X_{i_k})$ is normally distributed.
(c) If $U=X_1\pm X_2$ and V is any linear combination of the variables $X_3,\dots,X_n$, then $U+V$ is normally distributed.
(d) If $a_1,\dots,a_n$ are real numbers such that $a_k\ne 0$ for $k=1,\dots,n$ and $|a_1|\ne|a_2|$, then $\sum_{k=1}^n a_kX_k$ is not normally distributed.
(e) $\mathbf{E}[\prod_{k=1}^n X_k]=0$.

For the particular case $n=2$ (which can be compared with Example 10.5) we obtain that: (a) $\Rightarrow$ X_1 and X_2 have standard normal distribution; (c) $\Rightarrow$ X_1+X_2 and X_1-X_2 are normally distributed; (e) $\Rightarrow$ X_1 and X_2 are uncorrelated. However (X_1,X_2) is not normal, which follows from (d).

Return again to the general case. Let $U=X_1\pm X_2$, U_1 be a linear combination of any k of the variables $X_3,X_4,\dots,X_n$, $0\le k\le n-2$, and Y be a linear combination of the remaining $n-k-2$ of these variables. Then the r.v.s $X=U+U_1$ and Y are independent and normally distributed. Indeed, X and Y are uncorrelated and normal r.v.s and (c) implies that a countably infinite number of distinct linear combinations of them are distributed normally.

10.7. The condition characterizing the normal distribution by normality of linear combinations cannot be weakened

Let us start with the formulation of the following result (Hamedani and Tata 1975). Suppose $\{(a_k, b_k), k = 1, 2, \ldots\}$ is a countable 'distinct' sequence in $\mathbb{R}^2$ such that for each k, $a_k X + b_k Y$ is a normal r.v. Then (X, Y) has a bivariate normal distribution. (Here 'distinct' means that the parametric equations $t_1 = a_k t, t_2 = b_k t$ represent an infinite number of lines in $\mathbb{R}^2$.)

We are now interested in whether the condition of this theorem can be weakened. More precisely, let X and Y be r.v.s satisfying the following condition:

(C_N) for given N pairs $(a_k, b_k), k = 1, \ldots, N$, N a fixed natural number, the linear combinations $a_k X + b_k Y, k = 1, \ldots, N$ are normally distributed.

The question is of whether (C_N) implies that (X, Y) has a bivariate normal distribution. To see this, consider the following function:

$$\phi(s,t) = \exp\left[-\tfrac{1}{2}(s^2+t^2)\right] + \exp\left[-\varepsilon - \tfrac{1}{2}c(s^2+t^2)\right]\left[\prod_{k=1}^{N}(b_k^2 s^2 - a_k^2 t^2)\right]$$

where $s, t \in \mathbb{R}^1, \varepsilon, c \in \mathbb{R}^+$.

Firstly, we shall show that for a suitable choice of ε and c, $\phi(s,t)$ is the ch.f. of some two-dimensional distribution. Indeed, denoting by $f(x,y)$ the inverse Fourier transform of $\phi(s,t)$, we obtain:

$$\begin{aligned} f(x,y) &= (2\pi)^{-2}\iint_{\mathbb{R}^2} \exp(-isx - ity)\phi(s,t)\,\mathrm{d}s\,\mathrm{d}t \\ &= (2\pi)^{-1}\exp\left[-\tfrac{1}{2}(x^2+y^2)\right] + h_{\varepsilon,c}(x,y), \end{aligned}$$

$$\begin{aligned} h_{\varepsilon,c}(x,y) := (2\pi)^{-2}\mathrm{e}^{-\varepsilon}\iint_{\mathbb{R}^2} &\exp(-isx - ity)\exp\left[-\tfrac{1}{2}c(s^2+t^2)\right] \\ \times &\textstyle\prod_{k=1}^{N}(b_k^2 s^2 - a_k^2 t^2)\,\mathrm{d}s\,\mathrm{d}t. \end{aligned}$$

Further, we need an estimate for the function $h_{\varepsilon,c}$ which has just been introduced. It can be shown (Hamedani and Tata 1975) that for suitably chosen constants ε and c

$$|h_{\varepsilon,c}(x,y)| \le (2\pi)^{-1}\exp\left[-\tfrac{1}{2}(x^2+y^2)\right] \quad \text{for all } (x,y) \in \mathbb{R}^2. \tag{1}$$

Since $\phi(s,t)$ is a continuous function, $\phi(0,0) = 1$ and $f(x,y)$ is real-valued, we conclude that f and ϕ is a pair of functions where f is a two-dimensional density and ϕ its corresponding ch.f. Denote by (ξ, η) the random vector whose density is f. Further, the definition of ϕ immediately implies that

$$\phi(a_k t, b_k t) = \exp\left[-\tfrac{1}{2}(a_k^2 + b_k^2)t^2\right], \quad k = 1, \ldots, N.$$

This means that the r.v. $a_k\xi + b_k\eta$ is normally distributed, $\mathcal{N}(0, a_k^2 + b_k^2)$, for $k = 1, \ldots N$. However, ϕ itself is not the ch.f. of a bivariate normal distribution.

Therefore we have constructed a pair of r.v.s, ξ and η, for which condition (C_N) is satisfied, but (ξ, η) is not normal. Thus (C_N) is not enough for normality of (ξ, η). It should be noted that condition (1) holds only if N is finite.

10.8. Non-normal distributions such that all or some of the conditional distributions are normal

(i) Let $f(x, y)$, $(x, y) \in \mathbb{R}^2$ be a bivariate normal density. Then it is easy to check that each of the conditional densities $f_1(x|y)$ and $f_2(y|x)$ is normal. This observation leads naturally to the question of whether the converse statement is true. We shall show that the answer is negative.

Consider the following function:

$$g(x, y) = C \exp[-(1 + x^2)(1 + y^2)], \quad (x, y) \in \mathbb{R}^2. \tag{1}$$

Here $C > 0$ is a norming constant such that $\iint_{\mathbb{R}^2} g(x, y)\, dx\, dy = 1$.

A standard calculation shows that the conditional densities $g_1(x|y)$ and $g_2(y|x)$ of $g(x, y)$ are expressed as follows:

$$g_1(x|y) = (2\pi\sigma_y^2)^{-1/2} \exp(-x^2/2\sigma_y^2),$$
$$g_2(y|x) = (2\pi\sigma_x^2)^{-1/2} \exp(-y^2/2\sigma_x^2)$$

where $\sigma_y^2 = 1/(2(1 + y^2))$, $\sigma_x^2 = 1/(2(1 + x^2))$, $x \in \mathbb{R}^1$, $y \in \mathbb{R}^1$.

Obviously $g_1(x|y)$ and $g_2(y|x)$ are normal densities of $\mathcal{N}(0, \sigma_y^2)$ and $\mathcal{N}(0, \sigma_x^2)$ respectively. However, $g(x, y)$ given by (1) is not a two-dimensional normal density.

Therefore the normality of the conditional densities does not imply that the two-dimensional density is normal. Let us note that similar properties hold for any density (non-normal) of the type

$$\tilde{g}(x, y) = \tilde{C} \exp\left[-\sum_{i,j=0}^{2} b_{ij} x^i y^j\right]$$

(for details see Castillo and Galambos (1987, 1989)).

One particular case of $\tilde{g}$ is the function g given by (1).

(ii) Consider now another interesting situation. Let ξ be a r.v. distributed uniformly on the interval [0, 1] and $\eta_1, \eta_2, \eta_3, \eta_4, \eta_5$ be independent r.v.s each with distribution $\mathcal{N}(0, 1)$. Suppose additionally that ξ and η_k, $k = 1, \ldots 5$ are independent. Define the r.v.s

$$X_1 = \sqrt{\xi}\eta_1 + \sqrt{1 - \xi}\eta_2, \; X_2 = \sqrt{\xi}\eta_3 + \sqrt{1 - \xi}\eta_2, \; X_3 = \sqrt{\xi}\eta_4 + \sqrt{1 - \xi}\eta_5.$$

It is then not difficult to check that each of X_1, X_2, X_3 has a standard normal distribution $\mathcal{N}(0, 1)$. Further, if $\phi_{3|1,2}(t)$ denotes the conditional ch.f. of X_3 given X_1, X_2, then we find

$$\phi_{3|1,2}(t) = \mathbf{E}[e^{itX_3}|X_1, X_2] = \mathbf{E}\{\mathbf{E}[e^{itX_3}|X_1, X_2, \xi]\} = e^{-t^2/2}$$

and hence X_3, conditionally on X_1 and X_2, has a normal distribution $\mathcal{N}(0,1)$. So, given these properties we conjecture that the vector (X_1, X_2, X_3) has a trivariate normal distribution. Let us check whether or not this is correct. For this purpose we compute the joint ch.f. $\psi(t_1, t_2)$ of X_1 and X_2 as follows

$$\begin{aligned}\psi(t_1,t_2) &= \mathbf{E}\{\exp(it_1X_1+it_2X_2)\} = \mathbf{E}\{\mathbf{E}[\exp(it_1X_1+it_2X_2)|\xi]\} \\ &= \mathbf{E}\{\exp[-\tfrac{1}{2}t_1^2\xi - \tfrac{1}{2}t_2^2\xi - \tfrac{1}{2}(t_1+t_2)^2(1-\xi)]\} \\ &= \exp[-\tfrac{1}{2}(t_1+t_2)^2]\mathbf{E}[\exp(t_1t_2\xi)] = (t_1t_2)^{-1}(\mathrm{e}^{t_1t_2}-1)\mathrm{e}^{-\frac{1}{2}(t_1+t_2)^2}.\end{aligned}$$

This form of the ch.f. $\psi(t_1, t_2)$ shows that the distribution of (X_1, X_2) is not bivariate normal. Therefore the vector (X_1, X_2, X_3) cannot have a trivariate normal distribution despite the normality of each of the components X_1, X_2, X_3 and the conditional normality of X_3 given X_1, X_2.

Note that under some additional assumptions the conditional normality of the components of a random vector will imply the normality of the vector itself (see Ahsanullah 1985; Ahsanullah and Sinha 1986; Bischoff and Fieger 1991; Hamedani 1992).

10.9. Two random vectors with the same normal distribution can be obtained in different ways from independent standard normal random variables

Recall that there are a few equivalent definitions of a multi-variate normal distribution. According to one of them a set of r.v.s $X_1, \ldots, X_n$ with zero means is said to have a multi-variate normal distribution if these variables are linear combinations of independent r.v.s, say $\xi_1, \ldots, \xi_M$, where $\xi_j \sim \mathcal{N}(0,1)$, for $j = 1, \ldots, M$. That is, we have

$$X_i = \sum_{j=1}^{M} c_{ij}\xi_j, \; i = 1, \ldots, N. \tag{1}$$

Note that there is no restriction on M. It may be possible that $M < N$, $M = N$ or $M > N$.

Suppose we are given the r.v.s $\xi_1, \ldots, \xi_M$ which are independent and distributed $\mathcal{N}(0,1)$. Any fixed matrix (c_{ij}) generates by (1) a random vector with a multi-variate normal distribution. Then the natural question which arises is whether different matrices generate random vectors with different (multi-variate normal) distributions. To find the answer we need the following result (see Breiman 1969): if both random vectors $(X_1, \ldots, X_N)$ and $(Y_1, \ldots, Y_N)$ have a multi-variate normal distribution with the same mean and the same covariance matrix, then they have the same distribution.

Now we shall use this result to answer the above question. According to (1) each of the vectors $(X_1, \ldots, X_N)$ and $(Y_1, \ldots, Y_N)$ is a transformation of independent

$\mathcal{N}(0,1)$ r.v.s obtained by using a matrix. Thus the question to be considered is whether this matrix is unique for both vectors. By a simple example we can show that the answer is negative.

Take ξ_1 and ξ_2 to be independent $\mathcal{N}(0,1)$ r.v.s and let

$$X_1 = \xi_1 + \xi_2, \quad X_2 = 2\xi_1 + \xi_2.$$

Define also

$$Y_1 = \sqrt{2}\xi_1, \quad Y_2 = \frac{3}{\sqrt{2}}\xi_1 + \frac{1}{\sqrt{2}}\xi_2.$$

Thus we obtain two random vectors, (X_1, X_2) and (Y_1, Y_2). It is easy to see that (X_1, X_2) has zero mean and covariance matrix $\begin{pmatrix} 2 & 3 \\ 3 & 5 \end{pmatrix}$. Further, (Y_1, Y_2) has zero mean and the same covariance matrix.

Moreover, both vectors, (X_1, X_2) and (Y_1, Y_2) are multi-variate normal. By the above result, (X_1, X_2) and (Y_1, Y_2) have the same distribution. However, as we have seen, these identically distributed vectors are obtained in quite different ways from independent $\mathcal{N}(0,1)$ r.v.s.

10.10. A property of a Gaussian system may hold even for discrete random variables

A set of r.v.s $\{\xi_1, \dots, \xi_n\}$ is said to be a *Gaussian system* if any of its subsets has a Gaussian (normal) distribution. Suppose for convenience that each ξ_j has zero mean and denote $c_{ij} = \mathrm{Cov}(\xi_i, \xi_j) = \mathbf{E}[\xi_i\xi_j]$. Then for an arbitrary choice of four indices i, j, k, l (including any possible number of coincidences) the following relation holds:

$$\mathbf{E}[\xi_i\xi_j\xi_k\xi_l] = c_{ij}c_{kl} + c_{ik}c_{jl} + c_{il}c_{jk}. \tag{1}$$

Note that a similar property is satisfied also for a larger even number of variables chosen from the given Gaussian system (including coincidences of indices). To prove such a property it is enough to use the ch.f. of the random vector whose components are involved in the product.

The above property has some useful applications but it is also of independent interest. It is natural to ask if this property holds for Gaussian systems only. If the answer were positive, then (1) would be a property characterizing the given system of r.v.s as Gaussian. It turns out, however, this is not the case. Here is a simple illustration.

Consider the sequence $\eta_1, \eta_2, \dots$ of i.i.d. r.v.s such that

$$\mathbf{P}[\eta_1 = -\sqrt{3}] = \mathbf{P}[\eta_1 = \sqrt{3}] = \tfrac{1}{6}, \quad \mathbf{P}[\eta_1 = 0] = \tfrac{2}{3}.$$

It is easy to see that for all choices of indices (including possible coincidences)

$$\mathbf{E}\eta_i = 0, \quad \mathbf{E}[\eta_i\eta_j] = c_{ij} \quad \text{and} \quad \mathbf{E}[\eta_i\eta_j\eta_k] = 0,$$

where $c_{ij} = 1$, if $j = i$ and $c_{ij} = 0$, if $j \neq i$. Direct calculations show that

$$\mathbf{E}[\eta_i\eta_i\eta_i\eta_i] = 3, \quad \mathbf{E}[\eta_i\eta_i\eta_j\eta_j] = 1 \quad \text{for} \quad j \neq i \quad \text{and} \quad \mathbf{E}[\eta_i\eta_j\eta_k\eta_l] = 0$$

for all other choices of indices. All these facts taken together justify the following relation (compare with (1)):

$$\mathbf{E}[\eta_i\eta_j\eta_k\eta_l] = c_{ij}c_{kl} + c_{ik}c_{jl} + c_{il}c_{jk}$$

which is valid for arbitrary indices i, j, k, l.

Hence (1) is satisfied for a collection of r.v.s which are far from being Gaussian.

SECTION 11. THE MOMENT PROBLEM

Let $\{m_0 = 1, m_1, m_2, \ldots\}$ be a sequence of real numbers and I be a fixed interval, $I \subset \mathbb{R}^1$. Suppose that $\{m_n\}$ are the moments of some d.f. $F(x)$, $x \in I$, i.e.

$$m_n = \int_I x^n \, \mathrm{d}F(x), \quad n = 0, 1, 2 \ldots .$$

If F is uniquely determined by the moment sequence $\{m_n\}$ we say that the **moment problem** has a unique solution or that it is *determinate*. Otherwise the moment problem has more than one solution or that it is *indeterminate*. We also say that the r.v. $X \sim F$ is determinate or indeterminate. Note that the moment problem in the case $I = [0, \infty)$ is called the *Stieltjes moment problem*, while in the case $I = (-\infty, \infty)$ we speak of the *Hamburger moment problem*.

Below are some criteria for the moment problem to be determinate or indeterminate.

Criterion (C$_1$). Let $F(x)$, $x \in \mathbb{R}^1$ be a d.f. whose ch.f. $\phi(t)$, $t \in \mathbb{R}^1$ is r-analytic for some $r > 0$. Then F is uniquely determined by its moment sequence $\{m_n\}$ where $m_n = \int_{-\infty}^{\infty} x^n \, \mathrm{d}F(x)$. Further, the ch.f. ϕ is r-analytic for some $r > 0$ iff

$$\overline{\lim_{n\to\infty}} \frac{(m_{2n})^{1/(2n)}}{2n} < \infty.$$

This is equivalent to the existence of the m.g.f. $M(t) = \mathbf{E}[\mathrm{e}^{tX}]$, $|t| < t_0$, $t_0 > 0$ (*Cramér condition*).

Criterion (C$_2$). Let $\{m_0 = 1, m_1, m_2, \ldots\}$ be the moments of a d.f. $F(x)$, $x \in \mathbb{R}^1$ and let

$$(1) \qquad \sum_{n=1}^{\infty} (m_{2n})^{-1/(2n)} = \infty \quad \textit{(Carleman condition)}.$$

Then F is uniquely determined by $\{m_n\}$. If the d.f. F has as support the interval $[0, \infty)$ (instead of $(-\infty, \infty)$) then a sufficient condition for uniqueness is $\sum_{n=1}^{\infty} (m_n)^{-1/(2n)} = \infty$.

Criterion ($\mathbf{C}_3$). (a) Suppose the d.f. $F(x)$, $x \in \mathbb{R}^1$ is absolutely continuous with density $f(x) > 0$, $x \in \mathbb{R}^1$ and let F have moments of all orders. If

$$\int_{-\infty}^{\infty} \frac{-\log f(x)}{1+x^2}\, \mathrm{d}x < \infty \quad \textit{(Krein condition)} \tag{2a}$$

then the distribution F is indeterminate.

(b) Let the d.f. $F(x)$, $x \in \mathbb{R}^+$ ($F(0) = 0$) be absolutely continuous with density $f(x) > 0$, $x > 0$ ($f(x) = 0$, $x \le 0$) and let F have moments of all orders. If

$$\int_{a}^{\infty} \frac{-\log f(x^2)}{1+x^2}\, \mathrm{d}x < \infty \quad \textit{(Krein condition)} \tag{2b}$$

for some $a \ge 0$, then the distribution F is indeterminate.

The proof of criteria (C_1) and (C_2) can be found in Shohat and Tamarkin (1943). Criterion (C_3) was suggested by Krein (1944) and discussed intensively by Akhiezer (1965) and Berg (1995). In these sources, as well as in Kendall and Stuart (1958), Feller (1971), Chow and Teicher (1978) and Shiryaev (1995), the reader will find discussions of these and other related topics.

The examples in this section clarify the role of the conditions which guarantee the uniqueness of the moment problem and reveal the relationships between different sufficient conditions.

11.1. The moment problem for powers of the normal distribution

If ξ is a r.v., $\xi \sim \mathcal{N}(a, \sigma^2)$, then the distribution of ξ (the normal distribution) as well as that of ξ^2 (χ^2-distribution) are uniquely determined by the corresponding moment sequences. These facts are well known but also they can be easily checked by, e.g. the Carleman criterion. Thus a reasonable question is: what can we say about higher powers of ξ? It turns out 'the picture' changes even for ξ^3. The first observation is that all moments $\mathbf{E}[(\xi^3)^k]$, $k = 1, 2, \ldots$, exist, however $\mathbf{E}[\exp(t\xi^3)]$ exists only if $t = 0$. Hence the m.g.f. does not exist, however we cannot conclude (see Criterion (C_1)) that the distribution of ξ^3 is indeterminate.

The case ξ^3 allows us to make a more detailed analysis. For let us take a r.v. $\eta \sim \mathcal{N}(0, \frac{1}{2})$ whose density is $\pi^{-1/2}\exp(-x^2)$, $x \in \mathbb{R}^1$. Then the new r.v. $X = \eta^3$ has the following density:

$$f(x) = (1/3\sqrt{\pi})|x|^{-2/3}\exp(-|x|^{2/3}), \quad x \in \mathbb{R}^1.$$

By using some standard integrals ($\int_0^\infty (1+x^2)^{-1}\, \mathrm{d}x = \pi/2$, $\int_0^\infty [(\log x)/(1+x^2)]\, \mathrm{d}x = 0$, $\int_0^\infty [x^\delta/(1+x^2)]\, \mathrm{d}x = \pi/2\cos(\delta\pi/2)$, $-1 < \delta < 1$) we can easily conclude that

$$\int_{-\infty}^{\infty} [-\log f(x)/(1+x^2)]\, \mathrm{d}x < \infty.$$

Hence, according to the Krein criterion (C_3), the distribution of the r.v. $X = \eta^3$ is not determined uniquely by the moment sequence $\{m_k = \mathbf{E}[X^k], k = 1, 2, \ldots\}$.

In a similar way we can show that the moment problem is indeterminate for the distribution of any r.v. η^{2n+1}, $n = 1, 2, \ldots$.

Let us return to the r.v. $X = \eta^3$. Knowing that the distribution of η^3 is indeterminate, we should like to describe explicitly another r.v. with the same moments as those of η^3. One possible way to do this is to consider the following function:

$$f_\varepsilon(x) = f(x)\left\{1 + \varepsilon\left[\cos(\sqrt{3}|x|^{2/3}) - \sqrt{3}\sin(\sqrt{3}|x|^{2/3})\right]\right\}, \quad x \in \mathbb{R}^1.$$

It can be shown that for $\varepsilon \in [-\frac{1}{2}, \frac{1}{2}]$, $f_\varepsilon(x), x \in \mathbb{R}^1$, is a probability density function. Denote by X_ε a r.v. having f_ε as its density. Obviously $f_\varepsilon \neq f$, except for the trivial case $\varepsilon = 0$. Our further reasoning is based on the equality

$$\int_{-\infty}^{\infty} x^k f(x)[\cos(\sqrt{3}|x|^{2/3}) - \sqrt{3}\sin(\sqrt{3}|x|^{2/3})]\,dx = 0, \quad k = 1, 2, \ldots.$$

This immediately implies that

$$\mathbf{E}[X_\varepsilon^k] = \mathbf{E}[X^k], \quad k = 1, 2, \ldots$$

despite the fact that X_ε and X are r.v.s with different distributions since for their densities one holds: $f_\varepsilon \neq f$, $\varepsilon \in [-\frac{1}{2}, \frac{1}{2}]$ (except $\varepsilon = 0$).

It is interesting (and even curious) to note that the distribution of the absolute value $|X|$ is determinate! Indeed, the r.v. $|X| = |\eta|^3$, where $\eta \sim \mathcal{N}(0, \frac{1}{2})$, has a density $(2/3\sqrt{\pi})x^{-2/3}\exp(-x^{2/3})$ for $x > 0$ (and 0 for $x \leq 0$). Then for the moment $m_k = \mathbf{E}[|X|^k]$ of order k, $k = 1, 2, \ldots$, we have $m_k = (1/\sqrt{\pi})\Gamma((3k+1)/2)$. For large k, $m_k \sim ck^{3/2}$, $c =$ constant, implying that $\sum_{k=1}^{\infty}(m_k)^{-1/(2k)} = \infty$, i.e. the Carleman condition is satisfied. Therefore in this case the moment problem is determinate.

11.2. The lognormal distribution and the moment problem

Let X be a r.v. such that $\log X \sim \mathcal{N}(0, 1)$. In this case we say that X has a (standard) *lognormal distribution*. The density f of X is given by

$$f(x) = \begin{cases} (2\pi)^{-1/2}x^{-1}\exp[-\frac{1}{2}(\log x)^2], & \text{if } x > 0 \\ 0, & \text{if } x \leq 0. \end{cases} \tag{1}$$

The moments $m_n = \mathbf{E}[X^n]$, can be calculated explicitly, namely $m_n = e^{n^2/2}$, $n \geq 1$. It is easy to check that the moments $\{m_n\}$ do not satisfy the Carleman condition $\sum_{n=1}^{\infty}(m_n)^{-1/(2n)} = \infty$. Since this condition is only sufficient, we cannot say whether the sequence $\{m_n\}$ determines uniquely the d.f. F of X. Further, we have the following relations:

$$\int_0^\infty (1+x^2)^{-1}|\log x|^k \, dx = \int_{-\infty}^\infty (1+e^{2y})^{-1}|y|^k e^y \, dy$$
$$\leq \int_{-\infty}^0 |y|^k e^y \, dy + \int_0^\infty |y|^k e^y \, dy < \infty, \quad k = 0, 1, 2, \ldots .$$

From this we conclude that the density (1) does satisfy the Krein condition (2*b*). According to Criterion (C_3) the lognormal distribution is not determined uniquely by its moments. Alternatively, the same conclusion is derived by referring to Criterion (C_1) after showing that $\mathbf{E}[e^{tX}] = \infty$ for each $t > 0$ meaning that the m.g.f. of X does not exist.

Thus we come to the following interesting question: is it possible to find explicitly other distributions with the same moments as those of the lognormal distribution? Two possible answers are given below.

(i) Let $\{f_\varepsilon(x), x \in \mathbb{R}^1, \varepsilon \in [-1, 1]\}$ be a family of functions defined as:

$$f_\varepsilon(x) = \begin{cases} f(x)[1 + \varepsilon \sin(2\pi \log x)], & \text{if } x > 0 \\ 0, & \text{if } x \leq 0 \end{cases} \tag{2}$$

where f is given by (1). Obviously, $f_\varepsilon(x) \geq 0$ for all $x \in \mathbb{R}^1$ and any $\varepsilon \in [-1, 1]$. In order to establish other properties of f_ε, we have to prove that

$$J_k := \int_0^x x^k f(x) \sin(2\pi \log x) \, dx = 0, \quad k = 0, 1, 2, \ldots . \tag{3}$$

Indeed, by the substitution $\log x = u = y + k$ we reduce J_k to the integrals

$$\frac{1}{\sqrt{2\pi}} \int_{-\infty}^\infty \exp\left(-\frac{1}{2}u^2 + ku\right) \sin(2\pi u) \, du$$
$$= \frac{1}{\sqrt{2\pi}} \exp\left(\frac{1}{2}k^2\right) \int_{-\infty}^\infty \exp\left(-\frac{1}{2}y^2\right) \sin(2\pi y) \, dy.$$

The last integral is zero since the integrand is an odd function and the interval $(-\infty, \infty)$ is symmetric with respect to 0.

So, based on (3) we draw the following conclusions concerning the family (2). If $k = 0$ then for any $\varepsilon \in [-1, 1]$, $f_\varepsilon(x)$, $x \in \mathbb{R}^1$ is a probability density function of some r.v., say X_ε. Obviously, if $\varepsilon = 0$, f_ε and X_ε are respectively f and X defined at the beginning of this section. Moreover, we have

$$\mathbf{E}[X_\varepsilon^k] = \mathbf{E}[X^k] \quad \text{for any } k, \quad k = 1, 2, \ldots$$

despite the fact that $f_\varepsilon \neq f$ for $\varepsilon \neq 0$.

Therefore we have described explicitly the family $\{X_\varepsilon\}$ containing infinitely many absolutely continuous r.v.s having the same moments as those of the r.v. X with lognormal distribution. This example, after the paper by Heyde (1963a) appeared, became one of the most popular examples illustrating the classical moment problem.

(ii) Now we shall exhibit another family of d.f.s $\{H_a, a > 0\}$ having the same moments as the lognormal distribution (1). Let us announce *a priori* that H_a, for each $a > 0$, will correspond to some discrete r.v. Y_a.

For $a > 0$ consider the function

$$h_a(t) = \sum_{n=-\infty}^{\infty} a^{-n} \mathrm{e}^{-n^2/2} \exp(ia\mathrm{e}^n t), \quad t \in \mathbb{R}^1. \tag{4}$$

It is easy to see that the series in (4) is convergent for all $t \in \mathbb{R}^1$ and all $a > 0$. Moreover, the functions $h_a(t)$, $t \in \mathbb{R}^1$ are continuous and positive definite in the standard sense: $\sum_{j,k} z_j \bar{z}_k h_a(t_j - t_k) \geq 0$, $t_j, t_k \in \mathbb{R}^1$, z_j, z_k are complex numbers. By the Bochner theorem (see Feller 1971) the function

$$\psi_a(t) = h_a(t)/h_a(0), \quad t \in \mathbb{R}^1$$

is a ch.f. Denote respectively by H_a and Y_a the d.f. and the r.v. whose ch.f. is ψ_a. The explicit form (4) of the function h_a allows us to describe explicitly the r.v. Y_a. We have

$$\mathbf{P}[Y_a = a\mathrm{e}^k] := p_a(k) = a^{-k} \mathrm{e}^{-k^2/2}/h_a(0), \quad k = 0, \pm 1, \pm 2, \ldots.$$

The next step is to find the moments $\widetilde{m}_n = \mathbf{E}[Y_a^n] = \sum_{k=-\infty}^{\infty} (a\mathrm{e}^k)^n p_a(k)$, $n = 1, 2, \ldots$. Since

$$\left(\frac{\mathrm{d}}{\mathrm{d}t}\right)^n h_a(t)\Big|_{t=0} = \sum_{k=-\infty}^{\infty} a^{-k} (ia\mathrm{e}^k)^n \mathrm{e}^{-k^2/2}$$

we can easily obtain

$$\begin{aligned} \mathrm{e}^{-n^2/2} \widetilde{m}_n &= i^{-n} \sum_k a^{-k} (ia\mathrm{e}^k)^n \mathrm{e}^{-(n^2+k^2)/2}/h_a(0) \\ &= \sum_k a^{-(k-n)} \mathrm{e}^{-(k-n)^2/2}/h_a(0) = 1. \end{aligned}$$

It follows from this that

$$\mathbf{E}[Y_a^n] = \widetilde{m}_n = \mathrm{e}^{n^2/2} = \mathbf{E}[X^n], \quad n = 1, 2, \ldots.$$

Therefore there is an infinite family $\{Y_a\}$ of discrete r.v.s such that Y_a has the same moments as the r.v. X with lognormal distribution. We refer the reader to papers by Leipnik (1981) and Pakes and Khattree (1992) for more comments about this example and other related topics.

11.3. The moment problem for powers of an exponential distribution

Let ξ be a r.v., $\xi \sim \mathcal{E}xp(1)$. The density of ξ is e^{-x} for $x > 0$ and 0 for $x \leq 0$, so the moment of order k is $m_k = \mathbf{E}[\xi^k] = k!$, $k = 1, 2, \ldots$. The distribution of ξ, i.e. the exponential distribution is uniquely determined by the moment sequence $\{m_k\}$. This follows from the Carleman criterion as well as from the existence of the m.g.f. of ξ and referring to Criterion (C_1).

Now we want to clarify whether powers of ξ have determinate distributions. For let $\delta > 0$ and let $X = \xi^\delta$. If f is the density of X, we easily find that

$$f(x) = (1/\delta)x^{1/\delta - 1}\exp(-x^{1/\delta}), \text{ if } x > 0; \quad f(x) = 0 \text{ for } x \leq 0.$$

The moments of X exist and $\mathbf{E}[X^k] = \Gamma(\delta k + 1)$, $k = 1, 2, \ldots$. We use the density f and find that $\int_0^\infty [-(\log f(x^2))/(1 + x^2)]\,\mathrm{d}x < \infty$ iff $\delta > 2$. Hence for $\delta > 2$ the distribution of the r.v. $X = \xi^\delta$ is indeterminate.

Let us show that in this case, ξ^δ with $\delta > 2$, there is a family of r.v.s all having the same moments as those of ξ^δ. Indeed, consider the function:

$$f_\varepsilon(x) = f(x)\{1 + \varepsilon[\cos(c_\delta x^{1/\delta}) - (1/c_\delta)\sin(c_\delta x^{1/\delta})]\}, \quad x > 0$$

where $c_\delta = \tan(\pi/\delta)$ and $|\varepsilon| \leq \sin(\pi/\delta)$. The equality

$$\int_0^\infty x^k f(x)[\cos(c_\delta x^{1/\delta}) - (1/c_\delta)\sin(c_\delta x^{1/\delta})]\,\mathrm{d}x = 0, \quad k = 0, 1, 2, \ldots$$

shows that f_ε is a probability density function of a r.v. X_ε and

$$\mathbf{E}[X_\varepsilon^k] = \mathbf{E}[X^k], \quad k = 1, 2, \ldots$$

even though $f_\varepsilon \neq f$ (except the trivial case $\varepsilon = 0$).

For completeness we have to consider the case $\delta \in (0, 2]$. Since $m_k = \mathbf{E}[(\xi^\delta)^k] = \Gamma(\delta k + 1)$, we can use the properties of the gamma function $\Gamma(\cdot)$ and show that the Carleman condition is satisfied. Thus we conclude that if $\xi \sim \mathcal{E}xp(1)$ and $0 < \delta \leq 2$, then the distribution of ξ^δ is uniquely determined by its moment sequence (also see Example 11.4).

11.4. A class of hyper-exponential distributions with an indeterminate moment problem

Recall first that the one-sided *hyper-exponential distribution* $\mathcal{H}^+(a, b, c)$, where a, b, c are positive numbers, is given by the density function

$$(1) \qquad f(x) = \begin{cases} \dfrac{cb^{-a/c}}{\Gamma(a/c)} x^{a-1}\exp(-x^c/b), & \text{if } x > 0 \\ 0, & \text{if } x \leq 0. \end{cases}$$

(Notice that the gamma distribution $\gamma(a, b)$, and the exponential distribution $\mathcal{E}xp(\lambda)$ are special cases of the hyper-exponential distributions.)

It can be shown (for details see Hoffmann-Jorgensen 1994) that if X is a r.v., $X \sim \mathcal{H}^+(a, b, c)$, then the quantity $\mathbf{E}[X^k]$ does exist for any $k \geq 0$ (not just for integer k) and, moreover,

$$\mathbf{E}[X^k] = b^{k/c}\Gamma\left(\frac{k+a}{c}\right)/\Gamma\left(\frac{a}{c}\right).$$

Hence the r.v. X has finite moments $m_k = \mathbf{E}[X^k]$, $k = 1, 2, \ldots$, and the question is whether or not the moment sequence $\{m_k\}$ determines uniquely the distribution of X.

Let us take some $a > 0$, $b > 0$ and $0 < c < \frac{1}{2}$. Then we can choose $\rho > 0$ such that $j := a + \rho$ is an integer number, set $r = \rho/c$, $\lambda = r + 1/\beta$, $s = \tan(c\pi)$ and introduce the function

$$\psi(u) = u^\rho \exp(-ru^c)\sin(\lambda s u^c), \quad u > 0.$$

Since $e^{-x} \leq r^r x^{-r}$ for all $x > 0$, we easily see that $|\psi(u)| \leq 1$, $u > 0$. Let k be a fixed non-negative integer. Then $n = k + j$ is an integer, $v = c\pi \in (0, \frac{\pi}{2})$ and substituting $x = u^c$ yields

$$c\int_0^\infty u^{k+a-1}\psi(u)\exp(-u^c/b)\,\mathrm{d}u = \int_0^\infty x^{(n\pi/\gamma)-1}e^{-\lambda x}\sin(\lambda s x)\,\mathrm{d}x = 0.$$

This implies that for any non-negative integer k and any real ε the following relation holds:

$$\int_0^\infty u^k f_\varepsilon(u)\,\mathrm{d}u = \int_0^\infty u^k f(u)\,\mathrm{d}u$$

where f is the hyper-exponential density (1) and

$$f_\varepsilon(x) = \begin{cases} f(x)[1 + \varepsilon\psi(x)], & \text{if } x > 0 \\ 0, & \text{if } x \leq 0. \end{cases}$$

Since $|\psi(x)| \leq 1$, $x > 0$, it is easy to see that for any $\varepsilon \in [-1, 1]$, f_ε is a probability density function of a r.v. X_ε and

$$\mathbf{E}[X_\varepsilon^k] = \mathbf{E}[X^k], \quad k = 1, 2, \ldots$$

despite the fact that $f_\varepsilon \neq f$ (except the trivial case $\varepsilon = 0$).

Therefore for $a > 0$, $b > 0$ and $c \in (0, \frac{1}{2})$ the hyper-exponential distribution $\mathcal{H}^+(a, b, c)$ is not determined uniquely by its moment sequence. It can be shown, however, that for $a > 0$, $b > 0$ and $c \geq \frac{1}{2}$, the moment problem for $\mathcal{H}^+(a, b, c)$ has a unique solution (Berg 1988).

Since for the exponential distribution $\mathcal{E}xp(\lambda)$, the gamma distribution $\gamma(a,b)$ and the hyper-exponential distribution $\mathcal{H}^+(a,b,c)$ we have

$$\mathcal{E}xp(\lambda) = \gamma(1, 1/\lambda), \quad \gamma(a,b) = \mathcal{H}^+(a,b,1)$$

it follows that if $X \sim \mathcal{E}xp(\lambda)$ or $X \sim \gamma(a,b)$, where $\lambda > 0$, $a > 0$ and $b > 0$ are arbitrary, then the distribution of X^δ for $\delta > 2$ is not determined uniquely by the corresponding moment sequence. (Also see Example 11.3.)

11.5. Different distributions with equal absolute values of the characteristic functions and the same moments of all orders

Let us start with the number sequence $\{a_k, k = 1,2,\ldots\}$ where $a_k > 0$ and $a := \sum_{k=1}^{\infty} a_k < \infty$. We shall consider a special sequence of distributions and study the corresponding sequence of the ch.f.s.

The uniform distribution on $[-a_k, a_k]$ has a ch.f. $\sin(a_k t)/(a_k t)$. Denote by $f_k(x)$, $x \in [-2a_k, 2a_k]$ the convolution of this distribution with itself. Then the ch.f. ϕ_k of f_k is $\phi_k(t) = [\sin(a_k t)/(a_k t)]^2$, $k = 1,2,\ldots$, $t \in \mathbb{R}^1$. The product $\prod_{j=1}^{m} \phi_j(t)$ converges, as $m \to \infty$, to the function $\phi(t)$ where

$$\phi(t) := \prod_{j=1}^{\infty} \phi_j(t), \quad t \in \mathbb{R}^1. \tag{1}$$

Using the Taylor expansion of $\sin x$ for small x, we find that

$$\phi(t) \approx \exp\left(-\tfrac{1}{3}\tilde{a}t^2\right) \quad \text{as } t \to 0$$

where $\tilde{a} = \sum_{j=1}^{\infty} a_j^2$. Therefore according to Lukacs (1970, Th. 3.6.1), $\phi(t)$, $t \in \mathbb{R}^1$ is a ch.f. The d.f. corresponding to ϕ is absolutely continuous. Let its density be denoted by f. Clearly f is an infinite convolution: that is, $f = f_1 * f_2 * \cdots$. By the inversion formula (Feller 1971; Lukacs 1970; Shiryaev 1995) we find that $f(0) = (2\pi)^{-1}\int_{-\infty}^{\infty} \phi(t)\,\mathrm{d}t$. Since $\phi(t) \geq 0$, for all $t \in \mathbb{R}^1$, we can construct the density f_0 by setting $f_0(x) = (2\pi f(0))^{-1}\phi(x)$, $x \in \mathbb{R}^1$. If ϕ_0 denotes the ch.f. of f_0 then

$$\begin{aligned}\phi_0(t) &= \textstyle\int_{-\infty}^{\infty} \exp(itx) f_0(x)\,\mathrm{d}x \\ &= (2\pi f(0))^{-1} \textstyle\int_{-\infty}^{\infty} \exp(-itx)\phi(x)\,\mathrm{d}x = f(t)/f(0).\end{aligned}$$

Note especially that the support of ϕ_0 is contained in the interval $[-2a, 2a]$. Using the function ϕ_0 and the function ϕ from (1) we define the following four functions:

$$\begin{aligned}\psi_1(t) &= \phi_0(t) + \tfrac{1}{2}[\phi_0(t+4a) + \phi_0(t-4a)], \\ \psi_2(t) &= \phi_0(t) - \tfrac{1}{2}[\phi_0(t+4a) + \phi_0(t-4a)], \\ g_1(x) &= (2\pi f(0))^{-1}\phi(x)(1 + \cos 4ax), \\ g_2(x) &= (2\pi f(0))^{-1}\phi(x)(1 - \cos 4ax).\end{aligned}$$

The above reasoning shows that $g_1(x)$, $x \in \mathbb{R}^1$ and $g_2(x)$, $x \in \mathbb{R}^1$ are probability density functions. Moreover, ψ_1 and ψ_2 given by (2) and (3) are just their ch.f.s. Also $|\psi_1(t)| = |\psi_2(t)|$ for each $t \in \mathbb{R}^1$.

Denote by X_1 and X_2 r.v.s with densities g_1 and g_2 respectively. If $c_n^{-1} := \pi f(0) \prod_{k=1}^n a_k^2$ then it is easy to derive from (4) that $|g_1(x)| \le c_n |x|^{-2n}$ for each $n \in \mathbb{N}$. The same estimation holds for g_2. Hence both variables X_1 and X_2 possess moments of all orders. As a consequence we obtain (see Feller 1971) that the ch.f.s ψ_1 and ψ_2 have derivatives of all orders, and, since $|\psi_1(t)| = |\psi_2(t)|$, $\psi_1(t) = \psi_2(t)$ for t in a small neighbourhood of 0. This implies that the moments of X_1 and X_2 of all orders coincide. Looking again at the pairs (g_1, g_2), (ψ_1, ψ_2) and (X_1, X_2) we conclude that $|\psi_1(t)| = |\psi_2(t)|$ for all $t \in \mathbb{R}^1$, $\mathbf{E}[X_1^k] = \mathbf{E}[X_2^k]$ for each $k \in \mathbb{N}$ but nevertheless $g_1 \ne g_2$.

11.6. Another class of absolutely continuous distributions which are not determined uniquely by their moments

Consider the r.v. X with the following density:

$$f(x) = \begin{cases} c\,\exp(-\alpha x^\lambda), & \text{if } x > 0 \\ 0, & \text{if } x \le 0. \end{cases}$$

Here $\alpha > 0$, $0 < \lambda < \frac{1}{2}$ and c is a norming constant.

For $\varepsilon \in (-1, 1)$ and $\beta = \alpha \tan \lambda\pi$ define the function f_ε by

$$f_\varepsilon(x) = \begin{cases} c\,\exp(-\alpha x^\lambda)(1 + \varepsilon\,\sin(\beta x^\lambda)), & \text{if } x > 0 \\ 0, & \text{if } x \le 0. \end{cases}$$

Obviously $f_\varepsilon(x) \ge 0$, $x \in \mathbb{R}^1$. Next we shall use the relation

$$\text{(1)} \qquad \int_0^\infty x^n \exp(-\alpha x^\lambda) \sin(\beta x^\lambda)\,\mathrm{d}x = 0.$$

Let us establish the validity of (1). If $p > 0$ and q is a complex number with $\Re q > 0$, then we use the well known identity

$$\int_0^\infty t^{p-1} \mathrm{e}^{-qt}\,\mathrm{d}t = \Gamma(p)/q^p.$$

Denoting $p = (n+1)/\lambda$, $q = a + ib$, $t = x^\lambda$, we find

$$\int_0^\infty x^{\lambda[(n+1)/\lambda - 1]} \exp[-(a+ib)x^\lambda] \lambda x^{\lambda-1}\,\mathrm{d}x = \lambda \int_0^\infty x^n \exp[-(a+ib)x^\lambda]\,\mathrm{d}x$$

$$= \lambda \int_0^\infty x^n \exp(-ax^n) \cos(bx^\lambda)\,\mathrm{d}x - i\lambda \int_0^\infty x^n \exp(-ax^\lambda) \sin(bx^\lambda)\,\mathrm{d}x$$

$$= \Gamma((n+1)/\lambda) \left[a^{(n+1)/\lambda}(1 + i\,\tan\lambda\pi)^{(n+1)/\lambda}\right].$$

The last ratio is real-valued because $\sin[\pi(n+1)] = 0$ and

$$\begin{aligned}(1 + i\tan\lambda\pi)^{(n+1)/\lambda} &= (\cos\lambda\pi)^{-(n+1)/\lambda}(\cos\lambda\pi + i\sin\lambda\pi)^{(n+1)/\lambda}\\ &= (\cos\lambda\pi)^{-(n+1)/\lambda}e^{i\pi(n+1)}\\ &= (\cos\lambda\pi)^{-(n+1)/\lambda}\cos\pi(n+1).\end{aligned}$$

Thus (1) is proved. Taking $n = 0$ we see that f_ε is a probability density function. Denote by X_ε a r.v. whose density is f_ε. The relationship between f_ε and f, together with (1), imply that

$$\mathbf{E}[X_\varepsilon^n] = \mathbf{E}[X^n] \quad \text{for each } n = 1, 2 \dots .$$

Therefore we have constructed infinitely many r.v.s X_ε with the same moments as those of X though their densities f_ε and f are different ($f_\varepsilon = f$ only if $\varepsilon = 0$). So in this case the moment problem is indeterminate. However, this fact is not surprising because the density f does satisfy criterion (C_3).

11.7. Two different discrete distributions on a subset of natural numbers both having the same moments of all orders

Let $q \geq 2$ be a fixed natural number and $M_q = \{q^j : j = 0, 1, 2, \dots\}$. Clearly $M_q \subset \mathbb{N}$ for $q = 2, 3, \dots$. If $n \in M_q$ then n has the form q^j and we can define p_n by $p_n = e^{-q}q^j/j!$. It is easy to see that $\{p_n\}$ is a discrete probability distribution over the set M_q. Denote by X a r.v. with values in M_q and a distribution $\{p_n\}$. In this case we say that X has a *log-Poisson distribution*. Then the kth-order moment m_k of X is

$$m_k = \mathbf{E}[X^k] = \sum_{j=0}^{\infty} e^{-q}q^{kj}q^j/j! = \exp[q(q^k - 1)] < \infty.$$

Our purpose now is to construct many other r.v.s with the same moments. Consider the function $h(z) := \prod_{k=1}^{\infty}(1 - zq^{-k})$. Since $\sum_{k=1}^{\infty} q^{-k} < \infty$ for $q > 1$ then $h(z)$ is an analytic function in the whole complex plane. Let $h(z) = \sum_{j=0}^{\infty} c_j z^j$ be its Taylor expansion around 0. Taking into account the equality $h(qz) = (1 - z)h(z)$ we have the relation $c_j/c_{j-1} = -(q^j - 1)^{-1}$ where $c_0 = 1$ and for $j \geq 1$ we find

$$c_j = (-1)^j[(q-1)(q^2-1)\dots(q^j-1)]^{-1}.$$

Setting $a_j = j!c_j$ we see that $|a_j| \leq 1$ for all j. This implies that

$$(1) \qquad e^{-q}\sum_{j=1}^{\infty} q^{kj}a_j q^j/j! = e^{-q}h(q^{k+1}) = 0 \quad \text{for all} \quad k = 0, 1, 2, \dots .$$

Now introduce the number set $\{p_n^{(\varepsilon)}, n \in M_q\}$ where

$$\begin{aligned} p_n^{(\varepsilon)} &:= p_n(1+\varepsilon a_j) \\ &= \mathrm{e}^{-q} q^j [(j!)^{-1} + \varepsilon(-1)^j ((q-1)(q^2-1)\dots(q^j-1))^{-1}], \quad n = q^j. \end{aligned}$$

Here ε is any number in the interval $[-1,1]$. Obviously $p_n^{(\varepsilon)} \geq 0$ and (1) implies that $\sum_{n \in M_q} p_n^{(\varepsilon)} = 1$ for any $\varepsilon \in [-1,1]$. Therefore $\{p_n^{(\varepsilon)}\}$ defines a discrete probability distribution over the set M_q. Let X_ε be a r.v. with values in M_q and a distribution $\{p_n^{(\varepsilon)}\}$. Using (1) again we conclude that $\mathbf{E}[X_\varepsilon^k] = \mathbf{E}[X^k]$ for each $k = 1, 2, \dots$ and $\varepsilon \in [-1,1]$.

So, excluding the trivial case $\varepsilon = 0$, we have constructed discrete r.v.s X and X_ε whose distributions are different but whose moments of all orders coincide.

11.8. Another family of discrete distributions with the same moments of all orders

Let $\overline{\mathbb{N}} = \{0, \pm 1, \pm 2, \dots\}$ and X be a r.v. with the following distribution:

$$\mathbf{P}[X = \mathrm{e}^{8n}] = c\mathrm{e}^{-n^2}, \quad n \in \overline{\mathbb{N}}, \quad c^{-1} = \sum{}^* \mathrm{e}^{-n^2}.$$

Here and below $\sum{}^* \mathrm{e}^{-n^2}$ is the sum over all $n \in \overline{\mathbb{N}}$. For any positive integer k we can calculate explicitly the moment $m_k = \mathbf{E}[X^k]$ of order k, namely

$$m_k = \sum{}^* \mathrm{e}^{8kn} c\mathrm{e}^{-n^2} = \mathrm{e}^{16k^2} \sum{}^* c \exp[-(n+4k)^2] = \mathrm{e}^{16k^2}.$$

Now we shall construct a family consisting of infinitely many r.v.s with the same moments.

For any $\varepsilon \in (0,1)$ define the function

$$h_\varepsilon(n) = \begin{cases} 0, & \text{if } n \equiv 0 \pmod 4 \\ \varepsilon, & \text{if } n \equiv 1 \pmod 4 \\ 0, & \text{if } n \equiv 2 \pmod 4 \\ -\varepsilon, & \text{if } n \equiv 3 \pmod 4). \end{cases}$$

In the sequel we use the evident properties: for any fixed ε, $h_\varepsilon(n)$ is an odd function of n, that is $h_\varepsilon(-n) = -h_\varepsilon(n)$; $h_\varepsilon(n)$ is a periodic function of period equal to 4; $h_\varepsilon(n+4k)$ is an odd function in n for each integer k.

The next crucial step is to evaluate the sum S_k where

$$S_k = \sum{}^* \mathrm{e}^{8kn} h_\varepsilon(n) c\mathrm{e}^{-n^2}, \quad k = 0, 1, 2, \dots.$$

We have

$$\begin{aligned} S_k &= c \sum{}^* h_\varepsilon(n) \exp(8kn - n^2) \\ &= c \exp(16k^2) \sum{}^* h_\varepsilon(n) \exp[-(n-4k)^2] \\ &= c \exp(16k^2) \sum{}^* h_\varepsilon(u+4k) \exp(-u^2). \end{aligned}$$

The last sum is zero because $h_\varepsilon(u+4k)$ is an odd function of u for all $k = 0, 1, 2, \ldots$. Thus we have established that

$$(1) \qquad S_k = 0 \quad \text{for all } k = 0, 1, 2, \ldots \quad \text{and all } \varepsilon \in (0, 1).$$

As a consequence of (1) we derive that

$$q_\varepsilon(n) = c\mathrm{e}^{-n^2}(1 - h_\varepsilon(n)) \geq 0$$

for each $n \in \overline{\mathbb{N}}$ and any $\varepsilon \in (0, 1)$ and moreover $\sum{}^* q_\varepsilon(n) = 1$. This means that the set of numbers $\{q_\varepsilon(n), n \in \overline{\mathbb{N}}\}$ can be regarded as a discrete probability distribution of some r.v. which will be denoted by X_ε.

Thus we have constructed a r.v. X_ε whose values are the same as those of X but whose distribution is

$$\mathbf{P}[X_\varepsilon = \mathrm{e}^{8n}] = c\mathrm{e}^{-n^2}(1 - h_\varepsilon(n)).$$

Since (1) is satisfied for any $k = 1, 2, \ldots$ we find that

$$\mathbf{E}[X_\varepsilon^k] = \mathbf{E}[X^k] = \exp(16k^2).$$

Therefore for any $\varepsilon \in (0, 1)$ we have $X_\varepsilon \overset{\mathrm{d}}{\neq} X$ but nevertheless X_ε and X have the same moments of all orders.

11.9. On the relationship between two sufficient conditions for the determination of the moment problem

(i) Let $Z = X \log(1 + Y)$ where X and Y are independent r.v.s each distributed exponentially with parameter 1. Obviously Z is absolutely continuous and takes non-negative values. The nth moment m_n of Z is

$$m_n = n!\nu_n \quad \text{with} \quad \nu_n = \int_0^\infty [\log(1 + x)]^n \mathrm{e}^{-x}\, \mathrm{d}x.$$

It can be shown (the details are left to the reader) that

$$(1) \qquad e^{-1} \log(1 + n) < (\nu_n)^{1/n} < c \log(1 + n), \quad c = \text{constant}$$

Thus

$$(m_n/n!)^{1/n} = (\nu_n)^{1/n} > \mathrm{e}^{-1} \log(1 + n) \to \infty \quad \text{as } n \to \infty$$

which implies that the series $\sum_{n=1}^\infty m_n t^n/n!$ does not converge for any $t \neq 0$. Therefore we cannot apply Criterion (C_1) (see the introductory notes to this section) to decide whether the d.f. of Z is determined by its moments.

From (1) we obtain that for large n,

$$(m_n)^{1/n} \approx \mathrm{e}^{-1} n (\nu_n)^{1/n} < c\mathrm{e}^{-1} n \log(1+n)$$

and hence $\sum_n (m_n)^{-1/(2n)} = \infty$ because the series $\sum_{j=1}^{\infty} 1/(j \log(1+j))$ is divergent.

Therefore, according to Criterion (C_2), the d.f. F of the r.v. Z is determined uniquely by its moments. Note, however, that we used the Carleman criterion (C_2) whereas Criterion (C_1), based on the existence of the m.g.f. does not help in this case.

(ii) In case (i) we considered an absolutely continuous r.v. Here we take a discrete r.v. and tackle the same questions as raised in case (i).

So, let Z be a r.v. whose set of values is $\{3, 4, 5, \ldots\}$ and whose distribution is given by

$$\mathbf{P}[Z = j] = c \exp(-j/\log j), \quad j = 3, 4, \ldots$$

where c is a norming constant obtained from the condition $\mathbf{P}[Z \geq 3] = 1$.

Our purpose is to verify whether Criteria (C_1) and (C_2) apply. Firstly we derive suitable upper and lower bounds for the moment m_{2n} of order $2n$.

Introducing the function

$$h(x) = x^n \exp(-x/\log x)$$

where $x \geq 3$, $n \geq 4$, we can easily show that

$$\frac{h'(x)}{h(x)} = \frac{n}{x} - \frac{1}{\log x} + \frac{1}{\log^2 x} = 0$$

iff $x = x_n$ where

$$n = \frac{n}{\log x_n}\left(1 - \frac{1}{\log x_n}\right) \leq \frac{x_n}{\log x_n}.$$

Since n and x_n tend to infinity simultaneously, we have $(n \log x_n)/x_n \to 1$ whence $\frac{1}{2} n \log n \leq x_n \leq 2n \log n$ for $n \geq n_0$. If we define M_n by

$$M_n = \max_{x \geq 3} [x^n \exp(-x/\log x)] = x_n^n \exp(-x_n/\log x_n)$$

then for $n \geq n_0$ we obtain the following estimate:

$$\begin{aligned} m_{2n} &= c \sum_{j=3}^{\infty} j^{2n} \mathrm{e}^{-j/\log j} \leq c M_{2n+2} \sum_{j=3}^{\infty} \frac{1}{j^2} \\ &\leq c M_{2n+2} \leq c[(4n+4)\log(2n+2)]^{2n+2} \mathrm{e}^{-2n-2}. \end{aligned}$$

Therefore for all sufficiently large n

$$\begin{aligned} (m_{2n})^{1/(2n)} &\leq 4c^{1/(2n)} [(2n+2)\log(2n+2)]^{1+(1/n)} \\ &\leq 8(2n+2)\log(2n+2). \end{aligned} \tag{2}$$

On the other hand,

(3) $\qquad (m_{2n})^{1/(2n)} > (cM_{2n})^{1/(2n)} > \mathrm{e}^{-1}(\tfrac{1}{2}n\log n) \quad$ for all large n.

Now using (2) and (3) we easily find that

$$\overline{\lim_{n\to\infty}}\,[(m_{2n})^{1/(2n)}/(2n)] = \infty \quad \text{and} \quad \sum_{n=1}^{\infty}(m_{2n})^{-1/(2n)} = \infty.$$

Therefore the ch.f. of the r.v. Z is not analytic and we cannot apply Criterion (C_1) to say whether the moment sequence $\{m_k\}$ determines uniquely the distribution of Z. However, the Carleman criterion (C_2) guarantees that the distribution of Z is uniquely determined by its moments.

11.10. The Carleman condition is sufficient but not necessary for the determination of the moment problem

In two different ways we now illustrate that the Carleman condition is not necessary for the moment problem to be determined.

(i) Let F_{H} be a symmetric distribution on $(-\infty, \infty)$ and F_{S} a distribution on $[0, \infty)$. (The subscripts H and S correspond to the Hamburger and the Stieltjes cases.) By the relations

$$F_{\mathrm{H}}(x) = \begin{cases} \frac{1}{2}[1 + F_{\mathrm{S}}(x^2)], & \text{if } x \geq 0 \\ \frac{1}{2}[1 - F_{\mathrm{S}}(x^2)], & \text{if } x < 0 \end{cases}$$

we can define a one–one correspondence between the set of symmetric distributions on $(-\infty, \infty)$ and the set of distributions on $[0, \infty)$. It is clear that F_{H} possesses moments $\{\tilde{m}_n\}$ of all orders iff F_{S} possesses moments $\{m_n\}$ of all orders. In this case

$$\tilde{m}_{2n} = m_n, \quad \tilde{m}_{2n+1} = 0, \quad n = 0, 1, 2, \ldots.$$

Thus we conclude that the Hamburger problem for $\{\tilde{m}_n\}$ is determinate iff the corresponding Stieltjes problem is determinate. Moreover the Carleman condition $\sum(m_{2n})^{-1/(2n)} = \infty$ for the determination of the Hamburger case becomes $\sum(m_n)^{-1/(2n)} = \infty$ in the Stieltjes case. We shall use this result later but let us now formulate the following result (see Heyde 1963b): if a set $\{m_n\}$ of moments corresponds to a determinate Stieltjes problem, the solution of which has no point of discontinuity at the origin, then the set $\{m_n\}$ also corresponds to a determinate Hamburger problem.

Consider the r.v. X with density f given by

$$f(x) = \begin{cases} [1/\Gamma(1/\beta)]\exp(-x^{\beta}), & \text{if } x > 0 \\ 0, & \text{if } x \leq 0 \end{cases}$$

where $0 < \beta < 1$. One can show that $m_n = \mathbf{E}[X^n] = \Gamma((n+1)/\beta)/\Gamma(1/\beta)$, $n = 0, 1, 2, \ldots$, so $(m_n)^{1/n} \sim Kn^{1/\beta}$ for some constant K. Then $\sum(m_n)^{-1/(2n)} = \infty$

for $\frac{1}{2} \le \beta < 1$, and by the Carleman criterion (for the Stieltjes case) the Stieltjes problem for these moments is determinate for $\frac{1}{2} \le \beta < 1$. Since the distribution with density f has no discontinuity at the origin, from the above result we conclude that the Hamburger problem corresponding to the moments $m_n = \Gamma((n+1)/\beta)/\Gamma(1/\beta)$, $n = 0, 1, 2, \dots$ with $\frac{1}{2} \le \beta < 1$ is also determinate. However, it is easy to check that $\sum (m_{2n})^{-1/(2n)} < \infty$ for $0 < \beta < 1$. Hence the Carleman condition is not necessary for the determination of the moment problem (on $(-\infty, \infty)$).

(ii) Now we shall use the following interesting and intuitively unexpected result (see Heyde 1963b): let the moments $\{1, m_1, m_2, \dots\}$ correspond to a determinate Stieltjes problem. After a mass ε, $0 < \varepsilon < 1$, has been added at the origin and the distribution has been renormalized, it is possible for the new set of moments $\{1, m_1(1+\varepsilon)^{-1}, m_2(1+\varepsilon)^{-1}, \dots\}$ to correspond to an indeterminate Stieltjes problem.

So, let $\{1, m_1, m_2, \dots\}$ and $\{1, m_1(1+\varepsilon)^{-1}, m_2(1+\varepsilon)^{-1}, \dots\}$, $0 < \varepsilon < 1$, be sets of moments corresponding respectively to a determinate and an indeterminate Stieltjes problem. Suppose the Carleman condition is necessary for the determination of the moment problem. Then we should have $\sum (m_n)^{-1/(2n)} = \infty$, which is impossible because $\sum (m_n)^{-1/(2n)}(1+\varepsilon)^{1/(2n)} < \infty$.

11.11. The Krein condition is sufficient but not necessary for the moment problem to be indeterminate

As mentioned before the Krein condition is sufficient for the moment problem to be indeterminate. Let us consider examples showing the role of this condition.

(i) Let X be a r.v., $X \sim \mathcal{N}(0, \frac{1}{2})$, and $\delta > 0$. Then the density of the r.v. $|X|^\delta$ is

$$f_\delta(x) = \frac{2}{\delta\sqrt{\pi}} x^{1/\delta - 1} \exp(-x^{2/\delta}), \text{ if } x > 0; \quad f_\delta(x) = 0, \text{ if } x \le 0.$$

All moments $\mathbf{E}[(|X|^\delta)^k]$, $k = 1, 2, \dots$, exist. Berg (1988) has shown that the distribution of $|X|^\delta$ is determinate for $\delta \le 4$ and indeterminate for $\delta > 4$. Berg did not use the Krein condition. For the density f_δ, $\delta > 4$, we find that $\int_0^\infty \{-\log f_\delta(x^2)/(1+x^2)\}\, dx < \infty$, i.e. the Krein condition is satisfied, and this is the easiest way to show that the moment problem is indeterminate.

(ii) Take the function $h(x) = \exp(-x^\gamma)$, if $x > 0$ and $h(x) = \exp(x)$, if $x \le 0$. Here $\gamma \in (0, \frac{1}{2})$ and let c_γ be a constant such that $g_\gamma(x) = c_\gamma h(x)$, $x \in \mathbb{R}^1$ is a probability density. If Y is a r.v. with density g_γ, then all moments $\mathbf{E}[Y^k]$, $k = 1, 2, \dots$, exist. Moreover, $\int_{-\infty}^\infty \{-\log g_\gamma(x)/(1+x^2)\}\, dx = \infty$. Hence the Krein condition is not satisfied but the distribution of Y is indeterminate as follows from Example 11.6.

11.12. An indeterminate moment problem and non-symmetric distributions whose odd-order moments all vanish

In Example 6.5 we described a r.v. Y such that $m_{2n+1} = \mathbf{E}[Y^{2n+1}] = 0$ for all $n = 0, 1, 2, \ldots$ but the distribution of Y is non-symmetric. However, we did not discuss the reason for this fact. Let us note that the distribution of the r.v. Y in Example 6.5 is not determined uniquely by its moments. Now we shall show that the vanishing of the odd-order moments of a non-symmetric distribution is closely related to indeterminate Stieltjes problems.

From Example 11.7 we know that there are indeterminate Stieltjes problems. Let the d.f.s F_1 and F_2 be two distinct solutions of such a problem for a given set of moments $\{1, m_1, m_2, \ldots\}$. Then

$$F(x) = \tfrac{1}{2}F_1(x) + \tfrac{1}{2}[1 - F_2(-x-0)], \quad x \in \mathbb{R}^1$$

is a d.f. which evidently is non-symmetric. Moreover, F has the following moments: 1, 0, m_2, 0, m_4, 0, ... Therefore any odd-order moment of F is zero despite its non-symmetry.

Finally, let us present one additional example based on the lognormal distribution considered in Example 11.2. Once again, let

$$f(x) = (2\pi)^{-1/2}x^{-1}\exp[-\tfrac{1}{2}(\log x)^2],\ f_1(x) = f(x)[1 - \sin(2\pi \log x)], \quad x > 0.$$

Denote by Z a r.v. whose density g is defined as follows:

$$g(x) = \begin{cases} \tfrac{1}{2}f(x), & \text{if } x > 0 \\ \tfrac{1}{2}f_1(-x), & \text{if } x \le 0. \end{cases}$$

Then one can check that all the moments of Z are finite, all odd-order moments $\mathbf{E}[Z^{2n+1}]$ are zero but Z is non-symmetric.

11.13. A non-symmetric distribution with vanishing odd-order moments can coincide with the normal distribution only partially

Let us recall that in general no probability distribution is determined by a finite number of moments. The previous examples show that the distribution cannot be determined uniquely even if we know all (and hence an infinite number of) its moments. However, if we specify the class of distributions, then a member of this class could be determined by a finite number of moments. For example, a member of the so-called class of Pearson distributions is specified by a knowledge of at most four moments (Feller 1971; Heyde 1975). Certainly we have to indicate the normal distribution $\mathcal{N}(a, \sigma^2)$ which is determined uniquely by its first two moments only. Thus we come to the following question: does there exist a r.v. X such that for infinitely many k, but not for all $k \ge 1$, we have $\mathbf{E}[X^k] = \mathbf{E}[Z^k]$ where $Z \sim \mathcal{N}(a, \sigma^2)$ but nevertheless $X \overset{\mathrm{d}}{\neq} Z$?

Let Z be a r.v. distributed $\mathcal{N}(0,1)$. We shall construct a r.v. X such that

$$(1) \qquad \mathbf{E}[X^{2k+1}] = \mathbf{E}[Z^{2k+1}], \quad k = 0, 1, 2, \ldots,$$

$$\mathbf{E}[X^2] = \mathbf{E}[Z^2], \quad \mathbf{E}[X^4] = \mathbf{E}[Z^4] \quad \text{but} \quad X \overset{\mathrm{d}}{\neq} Z.$$

If (1) holds we can speak about a partial, but not full, coincidence of the distributions of X and Z.

So, let Y_1 be a r.v. with density

$$g(x) = \tfrac{1}{48} c^4 \exp(-c|x|^{1/4})[1 - \varepsilon \operatorname{sign} x \sin(c|x|^{1/4})], \quad x \in \mathbb{R}^1$$

where $c > 0$, $\varepsilon \neq 0$, $|\varepsilon| < 1$. Obviously g is non-symmetric. The moments $m_k = \mathbf{E}[Y_1^k]$ can be calculated explicitly, namely

$$m_{2k+1} = 0, \quad m_{2k} = \tfrac{1}{6} c^{-8k}(8k+3)!, \quad k = 0, 1, 2, \ldots$$

(see also Example 6.5). By choosing $c = (11!/6)^{1/8}$ we get $m_2 = \mathbf{E}[Y_1^2] = 1$.

Take now a r.v. Y_2 which is independent of Y_1 and takes the values 1 and -1 with probability $\frac{1}{2}$ each. For some constant β, $0 < \beta < 1$ which will be specified later, put

$$X = (1-\beta)^{1/2} Y_1 + \beta^{1/2} Y_2.$$

Clearly the distribution of X is non-normal and non-symmetric,

$$\mathbf{E}[X^{2k+1}] = 0 = \mathbf{E}[Z^{2k+1}] \quad \text{for each } k = 0, 1, 2, \ldots, \quad \mathbf{E}[X^2] = 1 = \mathbf{E}[Z^2].$$

Finally we find

$$\mathbf{E}[X^4] = \beta^2 + 6\beta(1-\beta) + (1-\beta)^2 \mathbf{E}[Y_1^4].$$

It remains to choose β such that $\mathbf{E}[X^4] = 3 = \mathbf{E}[Z^4]$. Indeed, if the kurtosis coefficient of the r.v. Y_1 is $\gamma_2 = \mathbf{E}[Y_1^4] - 3$, then take $\beta = \left(\gamma_2 - \sqrt{2\gamma_2}\right)/(\gamma_2 - 2)$ Since c was already fixed ($c = (11!/16)^{1/8}$) then γ_2 and hence β have definite values.

Thus we have constructed a r.v. X which coincides partially but not fully with the standard normal r.v. Z in the sense of (1).

SECTION 12. CHARACTERIZATION PROPERTIES OF SOME PROBABILITY DISTRIBUTIONS

There are probability distributions which can be characterized uniquely by some properties. In such cases it is natural to use the term 'characterization properties'.

Let us formulate two important results connected with the most popular distributions, the normal distribution and the Poisson distribution.

Cramér theorem. *If the sum $X_1 + X_2$ of the r.v.s X_1 and X_2 is normally distributed and these variables are independent, then each of X_1 and X_2 is normally distributed* (Cramér 1936).

Raikov theorem. *If X_1 and X_2 are non-negative integer-valued r.v.s such that $X_1 + X_2$ has a Poisson distribution and X_1 and X_2 are independent, then each of X_1 and X_2 has a Poisson distribution* (Raikov 1938).

These important theorems, several useful corollaries and other characterization theorems can be found in the books by Fisz (1963), Moran (1968), Feller (1971), Kagan *et al* (1973), Chow and Teicher (1978) and Galambos and Kotz (1978).

Let us note that some of the examples dealt with in Section 10 can be compared with the Cramér theorem. In particular this comparison shows that the assumption of the independence of X_1 and X_2 is essential.

We present below various examples of discrete and absolutely continuous distributions and clarify whether or not some properties are characterization properties.

12.1. A binomial sum of non-binomial random variables

Let the r.v.s. X and Y be non-negative integer-valued and let their sum $Z = X + Y$ have a binomial distribution with parameters (n, p), $Z \sim \mathcal{B}i(n, p)$. Then the probability generating function of Z is $\mathbf{E}[s^Z] = (ps+q)^n$. If additionally we suppose that X and Y are independent, then $(ps + q)^n = \mathbf{E}[s^X]\mathbf{E}[s^Y]$. Since all factors of the polynomial $(ps + q)^n$ have the form $(ps + q)^k$, $k = 0, 1, \dots, n$, it follows that each of the variables X and Y is also binomially distributed. This observation leads to the following question: does this conclusion hold without the hypothesis of independence between X and Y? Let us show that the answer is negative.

Let ζ be any non-negative integer-valued r.v. Suppose ζ takes more than two different values. Define the r.v.s ξ and η by

$$\xi = \left[\tfrac{1}{2}\zeta\right], \quad \eta = \left[\tfrac{1}{2}(\zeta + 1)\right]$$

where $[x]$ denotes the 'integer part' of x. Obviously

$$\zeta = \xi + \eta.$$

Moreover, knowing the distribution of ζ we can easily compute $\mathbf{P}[\xi = k]$ and $\mathbf{P}[\eta = m]$ for all possible values of k, m. Since $\mathbf{P}[\xi = k, \eta = m] = 0$ for those k, m satisfying the relation $|k - m| > 1$, we see that the r.v.s ζ and η are not independent. Note that this property holds irrespective of the distribution of ζ. In particular, suppose ζ is binomially distributed. Then neither ξ nor η is binomial, but their sum $\xi + \eta$, which is equal to ζ, has a binomial distribution. Recall that ζ and η are dependent.

12.2. A property of the geometric distribution which is not its characterization property

Recall that the r.v. X has a *geometric distribution* with parameter p, $0 < p < 1$, if $\mathbf{P}[X = n] = pq^n$, $q = 1 - p$, $n = 0, 1, \ldots$.

Let X_1 and X_2 be independent r.v.s each distributed as X. From the definition of a conditional probability we can easily derive that

$$(1) \qquad \mathbf{P}[X_1 = k | X_1 + X_2 = n] = \frac{1}{n+1}, \quad k = 0, 1, \ldots, n.$$

That is, $X_1 | X_1 + X_2 = n$ is discrete uniform on $\{0, 1, \ldots, n\}$.

We are interested now in whether (1) is a characterization property of the geometric distribution. More precisely: suppose X_1, X_2 are integer-valued independent r.v.s which satisfy relation (1), does it follow that X_1, X_2 are geometrically distributed?

To find the answer let us consider the set $\Omega = \{\omega_{kn} : k = 0, 1, \ldots, n, n = 1, 2, \ldots\}$ and let p_n, $n = 0, 1, \ldots$ be positive numbers with $\sum_{n=0}^{\infty} p_n = 1$. Define a probability $\mathbf{P}$ on Ω as follows:

$$\mathbf{P}(\omega_{kn}) = p_n/(n+1).$$

This means that $\Omega = \bigcup_{n=0}^{\infty} \Omega_n$ where $\Omega_n = \{\omega_{kn}, k = 0, 1, \ldots n\}$, $\mathbf{P}(\Omega_n) = p_n$ and each of the outcomes ω_{kn} has probability $p_n/(n+1)$. Introduce two r.v.s, Y_1 and Y_2, such that $Y_1(\omega_{kn}) = k$, $Y_2(\omega_{kn}) = n - k$. Then for $k = 0, 1, \ldots, n$,

$$\begin{aligned} \mathbf{P}[Y_1 = k | Y_1 + Y_2 = n] &= \frac{\mathbf{P}[Y_1 = k, Y_1 + Y_2 = n]}{\mathbf{P}[Y_1 + Y_2 = n]} \\ &= \frac{\mathbf{P}(\omega_{kn})}{\mathbf{P}(\Omega_n)} = \frac{p_n/(n+1)}{p_n} = \frac{1}{n+1}. \end{aligned}$$

Thus relation (1) is true for the r.v.s Y_1 and Y_2. However, the distribution of Y_1 is

$$\mathbf{P}[Y_1 = k] = \mathbf{P}[\{\omega_{kn} : n = k, k+1, \ldots\}] = \sum_{n=k}^{\infty} \frac{p_n}{n+1}$$

and since the p_n are arbitrary (with $\sum_n p_n = 1$), $\mathbf{P}[Y_1 = k]$, $k = 0, 1, \ldots$ can be very different from the geometric distribution.

Therefore (1) is not a characterization property of the geometric distribution. If additionally we suppose that X_1 and X_2 are independent and identically distributed, it can be proved that each of these variables has a geometric distribution.

12.3. If the random variables X, Y and their sum $X + Y$ each have a Poisson distribution, this does not imply that X and Y are independent

(i) Let X and Y be independent r.v.s each with a Poisson distribution. Then their sum $X + Y$ also has a Poisson distribution. We want to know whether the converse of the above statement is true: if X and Y are integer-valued and each of X, Y and $X+Y$ has a Poisson distribution, are the variables X and Y independent? It turns out that the answer to this question is negative.

Take two r.v.s, ξ and η each with a Poisson distribution of a given rate. Denote their individual distributions by $\{q_i, i = 0,1,\dots\}$ and $\{r_j, j = 0,1,\dots\}$ where $q_i = \mathbf{P}[\xi = i]$ and $r_j = \mathbf{P}[\eta = j]$. Introduce the sets $\Lambda_1 = \{(0,1),(1,2),(2,0)\}$ and $\Lambda_2 = \{(0,2),(2,1),(1,0)\}$.The joint distribution of ξ and η, $p_{ij} := \mathbf{P}[\xi = i, \eta = j]$ will be defined in the following way:

$$(1) \qquad p_{ij} = \begin{cases} q_i r_j + \varepsilon, & \text{if } (i,j) \in \Lambda_1 \\ q_i r_j - \varepsilon, & \text{if } (i,j) \in \Lambda_2 \\ q_i r_j, & \text{otherwise.} \end{cases}$$

Here ε is a real number such that $|\varepsilon| < \min_{ij} q_i r_j$, $(i,j) \in \Lambda_1 \cup \Lambda_2$.

It is easy to check that $\{p_{ij}, i = 0,1,\dots, j = 0,1,\dots\}$ is a two-dimensional discrete probability distribution. Moreover, using (1) we find that the sum $\xi + \eta$ has a Poisson distribution. By definition ξ and η also have Poisson distributions. However, (1) implies that the r.v.s ξ and η are not independent.

(ii) Here is a case slightly similar to (i). For fixed $\lambda > 0$ let ε be an arbitrary number in the interval $\left(0, \frac{1}{6}\lambda^4 e^{-2\lambda}\right)$. Defin8e p_{ij}, $i = 0,1,\dots$, $j = 0,1,\dots$, as follows:

$$p_{11} = \lambda^2 e^{-2\lambda} + \varepsilon, \quad p_{13} = \tfrac{1}{6}\lambda^4 e^{-2\lambda} - \varepsilon, \quad p_{31} = \tfrac{1}{6}\lambda^4 e^{-2\lambda} - \varepsilon,$$
$$p_{33} = \tfrac{1}{36}\lambda^6 e^{-2\lambda} + \varepsilon, \quad p_{ij} = \tfrac{1}{i!j!}\lambda^{i+j} e^{-2\lambda} \quad \text{for all other } i \text{ and } j.$$

Direct calculations lead to the following conclusions:

1) $\{p_{ij}, i,j = 0,1,\dots\}$ is a two-dimensional discrete distribution of a random vector, say (X,Y);
2) $X \sim \mathcal{P}o(\lambda)$ and $Y \sim \mathcal{P}o(\lambda)$;
3) $X + Y \sim \mathcal{P}o(2\lambda)$.

However the two components X and Y are not independent.

12.4. The Raikov theorem does not hold without the independence condition

Recall that the independence of the variables X and Y is one of the hypotheses in the Raikov theorem (see the introductory notes). We are now interested in what happens if we do not assume that X and Y are independent. Our reasoning is similar to that used in Example 12.1.

Let ζ be a r.v. with a Poisson distribution. Define the r.v.s ξ and η by

$$\xi = \left[\tfrac{1}{2}\zeta\right], \quad \eta = \left[\tfrac{1}{2}(\zeta + 1)\right]$$

(here $[x]$ denotes the integer part of x). It is easy to verify that each of ξ and η is an integer-valued r.v., neither ξ nor η has a Poisson distribution, ξ and η are not independent, but the sum $\xi + \eta = \zeta$ has a Poisson distribution.

Therefore, as expected, the independence condition in the Raikov theorem cannot be dropped.

12.5. The Raikov theorem does not hold for a generalized Poisson distribution of order k, $k \geq 2$

We say that the integer-valued non-negative r.v. X has a *generalized Poisson distribution* of order k and parameter λ, $\lambda > 0$, if

$$\text{(1)} \qquad \mathbf{P}[X = n] = \sum_{j_1,\dots,j_k} \frac{\lambda^{j_1+\cdots+j_k} e^{-k\lambda}}{j_1!\dots j_k!}$$

where the summation is taken over all non-negative integers $j_1, \dots, j_k$ such that $j_1 + 2j_2 + \cdots + kj_k = n$. If $k = 1$, then (1) defines the usual Poisson distribution. By using (1) we find explicitly the p.g.f. $g(s) = \mathbf{E}[s^X]$ (see Philippou 1983):

$$\text{(2)} \qquad g(s) = \exp\left[-\lambda\left(k - \sum_{i=1}^{k} s^i\right)\right], \quad |s| \leq 1.$$

Suppose now that Y_1, Y_2 are independent r.v.s taking values in the set $\{0, 1, 2 \dots\}$ and such that the sum $Y_1 + Y_2$ has a generalized Poisson distribution of order k. The question is: does it follow from this that each of the variables Y_1, Y_2 has a generalized Poisson distribution of order k?

Note that in the particular case $k = 1$ the usual Poisson distribution is obtained and it follows from the Raikov theorem that the answer to this question is positive (see Example 12.4). We have to find an answer for $k \geq 2$.

Consider two independent r.v.s, Z_1 and Z_2 where Z_1 has a generalized Poisson distribution of order $(k-1)$ and a parameter λ, and Z_2 has the following distribution:

$$\mathbf{P}[Z_2 = j] = \frac{\lambda^{j/k} e^{-\lambda}}{(j/k)!}, \quad j = 0, k, 2k, 3k, \dots .$$

We shall use the explicit form of the p.g.f.s $g_1(s)$ and $g_2(s)$ of Z_1 and Z_2 respectively. Taking (2) into account we find that

$$g_1(s) = \exp\left[-\lambda\left(k - 1 - \sum_{i=1}^{k-1} s^i\right)\right], \quad |s| \leq 1.$$

On the other hand, direct computation shows that

$$g_2(s) = \exp[-\lambda(1 - s^k)], \quad |s| \leq 1.$$

Since Z_1 and Z_2 are independent, the p.g.f. g_3 of the sum $Z_1 + Z_2$ is the product of g_1 and g_2. Thus

$$g_3(s) = \exp\left[-\lambda\left(k - \sum_{i=1}^{k} s^i\right)\right], \quad |s| \leq 1.$$

But, looking at (2), we see that g_3 is the p.g.f. of a generalized Poisson distribution of order k. Therefore the r.v. $Z = Z_1 + Z_2$ has just this distribution. Moreover, Z is decomposed into a sum of two independent r.v.s Z_1 and Z_2, neither of which has a generalized Poisson distribution of order k. The Raikov theorem is therefore not valid for generalized Poisson distributions of order $k \geq 2$.

12.6. A case when the Cramér theorem is not applicable

Recall first that the Cramér theorem can be reformulated in the following equivalent form. Let $F_1(x)$, $x \in \mathbb{R}^1$ and $F_2(x)$, $x \in \mathbb{R}^1$ be non-degenerate d.f.s satisfying the relation

$$(1) \qquad (F_1 * F_2)(x) = \Phi_{a,\sigma}(x) \quad \text{for all } x \in \mathbb{R}^1$$

where $\Phi_{a,\sigma}$ is a d.f. corresponding to $\mathcal{N}(a, \sigma^2)$. Then each of F_1 and F_2 is a normal d.f.

Suppose now that the condition (1) is satisfied only for $x \leq x_0$, where x_0 is a fixed number, $x_0 < \infty$ (i.e. not for all $x \in \mathbb{R}^1$). Is it true in this case that F_1 and F_2 are normal d.f.s? The answer follows from the next example.

Denote by $\Phi = \Phi_{0,1}$ the standard normal d.f. and define the function:

$$F_1(x) = \begin{cases} 2\sum_{n=0}^{\infty}(-1)^n \Phi(x-n), & \text{if } x \leq 0 \\ v(x), & \text{if } x > 0 \end{cases}$$

where $v(x)$ is an arbitrary non-decreasing function defined for $x \in (0, \infty)$ and such that $v(0+) = F_1(0)$ and $v(\infty) = 1$.

It is easy to check that F_1 is a d.f. and let ξ_1 be a r.v. with this d.f. Further, let F_2 be the d.f. of the r.v. ξ_2 taking two values, 0 and 1, each with probability $\frac{1}{2}$. Then we find that

$$(F_1 * F_2)(x) = \tfrac{1}{2}[F_1(x) + F_1(x-1)] = \Phi(x) \quad \text{for all } x \leq 0$$

i.e. condition (1) is satisfied for $x \leq x_0$ with $x_0 = 0$. However if $x > 0$, then $(F_1 * F_2)(x) \neq \Phi(x)$. Obviously F_1 and F_2 are not normal d.f.s.

Hence condition (1) in the Cramér theorem cannot be relaxed.

12.7. A pair of unfair dice may behave like a pair of fair dice

Recall first that a standard and symmetric die (a fair die) is a term used for a 'real' material cube whose six faces are numbered by 1, 2, 3, 4, 5, 6 and such that when rolling this die each of the outcomes has probability $\frac{1}{6}$.

Suppose now we have at our disposal four dice: white, black, blue and red. The available information is that the white and the black dice are standard and symmetric.

Then the sum $X + Y$ of the numbers of these two dice is a r.v. which is easy to describe. Clearly, $X + Y$ takes the values 2, 3, 4, 5, 6, 7, 8, 9, 10, 11 and 12 with probabilities $\frac{1}{36}, \frac{2}{36}, \frac{3}{36}, \frac{4}{36}, \frac{5}{36}, \frac{6}{36}, \frac{5}{36}, \frac{4}{36}, \frac{3}{36}, \frac{2}{36}$ and $\frac{1}{36}$, respectively.

Suppose we additionally know that the blue and the red dice are such that the sum $\xi + \eta$ of the numbers on these two dice is exactly as the sum $X + Y$ obtained when rolling the white and the black dice (i.e. $\xi + \eta$ takes the same values as $X + Y$ with the same probabilities shown above). Does this information imply that the blue die and the red die are fair, i.e. that each is standard and symmetric?

It turns out the answer to this question is negative as can be seen by the following physically realizable situation. Take a pair of ordinary dice changing, however, the numbers on the faces. Namely, the faces of the blue die are numbered $1, 2, 2, 3, 3$ and 4 while those of the red die are numbered $1, 3, 4, 5, 6$ and 8. If ξ and η are the numbers appearing after rolling these two dice, we easily find that indeed $\xi + \eta \stackrel{d}{=} X + Y$.

Hence, despite the facts that the sum $X + Y$ comes from a pair of fair dice and that $X + Y$ has the same distribution as $\xi + \eta$, this does not imply that the blue and the red dice are fair.

The practical advice is: do not rush to pay the same for a pair of dice with fair sums as for a pair of fair dice!

12.8. On two properties of the normal distribution which are not characterizing properties

Let X and Y be independent r.v.s distributed $\mathcal{N}(0, 1)$. Then the ratio X/Y has a distribution

$$\mathbf{P}[X/Y \le z] = \frac{1}{2\pi} \iint\limits_{x/y \le z} \exp\left(-\frac{1}{2}x^2\right) \exp\left(-\frac{1}{2}y^2\right) \mathrm{d}x\mathrm{d}y, \quad z \in \mathbb{R}^1.$$

It is easy to check that

$$(\mathbf{P}[X/Y \le z])'_z = \frac{1}{\pi(1 + z^2)}, \quad z \in \mathbb{R}^1.$$

Hence X/Y has a Cauchy distribution. Let us call this property (N/N→C).

The presence of the property (N/N→C) leads to the following question. Suppose X and Y are i.i.d. r.v.s with zero means such that X/Y has a Cauchy distribution. Is it true that X and Y are normally distributed? By examples we show that the answer to this question is negative.

(i) Consider two i.i.d. r.v.s ξ and η having the density

$$f(x) = \frac{\sqrt{2}}{\pi} \frac{1}{1 + x^4}, \quad x \in \mathbb{R}^1.$$

If $g(z)$, $z \in \mathbb{R}^1$ denotes the density of the ratio ξ/η then we easily find

$$\begin{aligned} g(z) &= \left(\frac{2}{\pi^2}\iint_{x/y\le z}(1+x^4)^{-1}(1+y^4)^{-1}\mathrm{d}x\,\mathrm{d}y\right)'_z \\ &= \left(\frac{2}{\pi^2}\int_{-\infty}^{\infty}(1+y^4)^{-1}(1+z^4y^4)^{-1}|y|\,\mathrm{d}y\right)'_z = \frac{1}{\pi(1+z^2)}. \end{aligned}$$

Therefore the ratio ξ/η has a Cauchy distribution but obviously the variables ξ and η are non-normally distributed.

Thus we have established that the property (N/N→C) is not a characterization property of the normal distribution.

(ii) Let us consider another case on the same topic. It can be checked that

$$f_1(x) = \frac{2x^4}{\pi(1+x^2)(1+x^4)}, \qquad x \in \mathbb{R}^1$$

is a density function. Take two independent r.v.s X_1 and Y_1 each with density f_1. Then the ratio $Z_1 = X_1/Y_1$ has a Cauchy distribution (for details see Steck 1959). Clearly X_1 and Y_1 are not normally distributed.

(iii) The variables X and Y above are independent and this condition is essential for X/Y to be Cauchy distributed. It turns out the ratio X/Y has a Cauchy distribution for some cases when X and Y are dependent. (We can guess that now the reasoning is more complicated.) Indeed, consider the function

$$\psi(t,u) = (1-2iu+t^2)^{-1/2}, \quad t,u \in \mathbb{R}^1.$$

It can be shown that ψ is the ch.f. of a two-dimensional random vector, say (ξ, η) such that: 1) each of ξ and η is not normally distributed (look at the marginal ch.f.s of ξ and η); 2) ξ and η are dependent; 3) the ratio ξ/η has Cauchy distribution. (For details see Rao 1984.)

(iv) Let $X_1,\dots,X_n$ be n independent r.v.s each distributed normally $\mathcal{N}(0,\sigma^2)$, $n \ge 2$. Define

$$\bar{X} = \frac{1}{n}\sum_{i=1}^{n} X_i, \quad s^2 = \frac{1}{n-1}\sum_{i=1}^{n}(X_i-\bar{X})^2, \quad T = \frac{\sqrt{n}\bar{X}}{s}.$$

It is well known (see Feller 1971) that T has a *Student distribution* with $n-1$ degrees of freedom. (Recall that T is often used in mathematical statistics.)

Let us consider the converse. Suppose $X_1,\dots,X_n$ are i.i.d. r.v.s with density $f(x)$, $x \in \mathbb{R}^1$, and we are given that the variable T has a Student distribution. Does it follow from this that f is a normal density? The example below shows that for $n = 2$ the answer is negative.

Let X_1, X_2 be independent r.v.s each with density f and let

$$\bar{X} = \tfrac{1}{2}(X_1 + X_2), \quad s = |X_1 - X_2|/\sqrt{2}, \quad T = \sqrt{2}\bar{X}/s.$$

Our assumption is that T has a Student distribution. Thus the problem is to find the unknown density f.

Let us introduce the functions $h_1(x_1, x_2)$, $h_2(x, y)$ and $h_3(z, y)$ which are the densities of the random vectors (X_1, X_2), $(\bar{X}, s)$ and (T, s) respectively. We find

$$\begin{aligned} h_1(x_1, x_2) &= f(x_1)f(x_2), & x_1 \in \mathbb{R}^1,\ x_2 \in \mathbb{R}^1, \\ h_2(x, y) &= 2^{3/2} f(x + y/\sqrt{2}) f(x - y/\sqrt{2}), & x \in \mathbb{R}^1,\ y \in \mathbb{R}^+, \\ h_3(z, y) &= 2f\left(y(z+1)/\sqrt{2}\right) f\left(y(z-1)/\sqrt{2}\right), & z \in \mathbb{R}^1,\ y \in \mathbb{R}^+. \end{aligned}$$

By the assumption above T has a Student distribution and clearly in this case (of two variables X_1 and X_2) T has a Cauchy distribution, that is the density of T is $1/[\pi(1+z^2)]$, $z \in \mathbb{R}^1$. But the density of T can be obtained from $h_3(z, y)$ by integration. Thus we come to the following relation:

$$(1) \qquad 2\pi \int_0^\infty f\left(y(z+1)/\sqrt{2}\right) f\left(y(z-1)/\sqrt{2}\right) \mathrm{d}y = \frac{1}{1+z^2}.$$

It can be shown (see Mauldon 1956) that f is an even function and that the general solution of (1) is of the form $f(x) = \pi^{-1/2} g(x^2)$, $x \in \mathbb{R}^1$, where

$$(2) \qquad \int_0^\infty g(u)g(au)\,\mathrm{d}u = \frac{1}{1+a}, \qquad a > 0.$$

Furthermore, the integral equation (2) has an infinitely many solutions. However, for our purpose it is enough to take e.g. only two solutions, namely:

$$(\text{a}) \qquad g(u) = \mathrm{e}^{-u} \quad \Rightarrow \quad f(x) = \frac{1}{\sqrt{\pi}}\mathrm{e}^{-x^2}, \quad x \in \mathbb{R}^1$$

$$(\text{b}) \qquad g(u) = \sqrt{2/\pi}(1+u^2) \quad \Rightarrow \quad f(x) = \frac{\sqrt{2}}{\pi}\frac{1}{1+x^4}, \quad x \in \mathbb{R}^1.$$

Obviously in case (a) the variables X_1 and X_2 are distributed normally $\mathcal{N}(0, \frac{1}{2})$, while in case (b) the distributions of X_1 and X_2 are both non-normal. Therefore we have constructed two i.i.d. r.v.s X_1 and X_2 whose distribution is non-normal but the variable T has a Student distribution.

Finally it is interesting to note that the same probability density function, namely $f(x) = (\sqrt{2}/\pi)/(1 + x^4)$, has appeared in both cases (i) and (ii).

(v) Recall the definition of the so-called *beta distribution of the second kind* denoted by $\beta^{(2)}(a, b)$. We say that the r.v. $X \sim \beta^{(2)}(a, b)$ if its density equals $x^{a-1}(1+x)^{-a-b}/B(a, b)$, if $x > 0$ and 0, if $x \le 0$. Here $a > 0$, $b > 0$.

The following result being of independent interest (see Letac 1995) is used for the reasoning below: Z has a Cauchy distribution $\mathcal{C}(0,1) \iff Z$ is symmetric and $|Z|^2 \sim \beta^{(2)}(\frac{1}{2},\frac{1}{2})$.

Consider now two independent and symmetric r.v.s X_1 and X_2 such that

$$|X_1|^2 \sim \beta^{(2)}(\tfrac{1}{2},b) \quad \text{and} \quad |X_2|^2 \sim \beta^{(2)}(\tfrac{1}{2},\tfrac{1}{2}+b) \quad \text{for some} \quad b>0.$$

Then, referring again to the book by Letac (1995) for details, we can show that the quotient X_2/X_1 has a Cauchy distribution $\mathcal{C}(0,1)$.

Hence we have described another case of independent r.v.s X_1 and X_2 such that their quotient X_2/X_1 follows a Cauchy distribution. Obviously, neither X_1 nor X_2 is normally distributed. Note, however, that here we did not make the advance requirement that X_1 and X_2 have the same distribution.

12.9. Another interesting property which does not characterize the normal distribution

Let us start with the following result (for the proof see Baringhaus *et al* 1988). If X and Y are independent r.v.s, $Z = XY/(X^2+Y^2)^{1/2}$ and $X \sim \mathcal{N}(0,\sigma_1^2)$, $Y \sim \mathcal{N}(0,\sigma_2^2)$, then $Z \sim \mathcal{N}(0,\sigma^2)$ with $\sigma^2 = \sigma_1^2\sigma_2^2/(\sigma_1+\sigma_2)^2$.

It is interesting to poit out that Z is a non-linear function of two normally distributed r.v.s and, as stated, Z itself has also a normal distribution. This leads to the inverse question: if X and Y are independent r.v.s and Z is normally distributed, does it follow that $X \sim \mathcal{N}$ and $Y \sim \mathcal{N}$?

Assume the answer is positive. Thus we can suppose e.g. that $Z \sim \mathcal{N}(0,1)$. The definition of Z implies that

$$1/X^2 + 1/Y^2 \stackrel{\mathrm{d}}{=} 1/Z^2.$$

It is easy to find that the distribution of $1/Z^2$ has a Laplace transform $\psi(t) = \exp(-\sqrt{2t})$, $t \geq 0$, meaning that $1/Z^2$ has a stable distribution with parameter $\frac{1}{2}$. Let us show that $1/Z^2$ admits the representation

$$1/Z^2 \stackrel{\mathrm{d}}{=} U_1 + U_2$$

where U_1 and U_2 are independent non-negative r.v.s such that the distribution of each of them does not have an 'atom' at 0 and does not belong to the class of stable distributions with parameter $\frac{1}{2}$. For this we write ψ in the form

$$\psi(t) = \exp\left[-\frac{1}{\sqrt{2\pi}}\int_0^\infty (1-\mathrm{e}^{-tx})x^{-3/2}\,\mathrm{d}x\right], \quad t \geq 0$$

and introduce the following two functions of $x \in \mathbb{R}^1$:

$$h_1(x) = \frac{1}{\sqrt{8\pi}}x^{-1/2}I_{(0,1)}(x), \quad h_2(x) = h_1(x) + \frac{1}{\sqrt{2\pi}}x^{-1/2}I_{(1,\infty)}(x)$$

(as usual $I_A(\cdot)$ is the indicator function of the set A). Denoting

$$\varphi_j(t) = \int_0^\infty (1 - \mathrm{e}^{-tx})x^{-1}h_j(x)\,\mathrm{d}x, \quad j = 1, 2$$

we see that the integrals $\int_1^\infty x^{-1}h_j(x)\,\mathrm{d}x$, $j = 1, 2$, are convergent and both

$$\psi_1(t) = \exp[-\varphi_1(t)],\ t \in \mathbb{R}^1 \quad \text{and} \quad \psi_2(t) = \exp[-\varphi_2(t)],\ t \in \mathbb{R}^1$$

are Laplace transforms of an infinitely divisible distribution with support $[0, \infty)$ (see Feller 1971). Since $\psi_j(t) \to \infty$ as $t \to \infty$, $j = 1, 2$, these distributions do not have 'atoms' at 0.

Suppose now that U_j is a r.v. having ψ_j as its Laplace transform, $j = 1, 2$. We can take U_1 and U_2 to be independent. Then the Laplace transform of the distribution of U_1+U_2 equals $\psi_1(t)\psi_2(t)$. However $\psi_1(t)\psi_2(t) = \psi(t)$. This fact and the reasoning above imply that $1/Z^2$ is the sum of U_1 and U_2 which are independent but, obviously, the distributions of $1/U_1$ and $1/U_2$ are not normal as they might be if the answer to the above question were positive.

The interesting property described at the beginning is therefore not a characterizing property for a normal distribution.

12.10. Can we weaken some conditions under which two distribution functions coincide?

Let us formulate the following result (see Riedel 1975). Suppose $F_1(x)$, $x \in \mathbb{R}^1$, is an infinitely divisible d.f. and $F_2(x) = \Phi(x)$, $x \in \mathbb{R}^1$, where Φ is the standard normal d.f. Then the condition

$$\lim_{x \to -\infty} F_1(x)/F_2(x) = 1 \tag{1}$$

implies that

$$F_1 = F_2. \tag{2}$$

It is interesting to show the importance of the conditions in this result. In particular, the following question is discussed by Blank (1981): does (1) imply that $F_1 = F_2$ if we suppose that F_1 and F_2 are arbitrary infinitely divisible d.f.s, $F_2(x) > 0$, $x \in \mathbb{R}^1$? By an example we show that the answer is negative.

Introduce the functions

$$G_1(x) := \frac{1}{\mathrm{e}} \sum_{k \le x} \frac{1}{k!}, \quad G_2(x) := \frac{1}{\mathrm{e}} \sum_{2k \le x} \frac{1}{k!}$$

and define F_1 and F_2 as convolutions by

$$F_1(x) = (\Phi * G_1)(x), \quad F_2(x) = (\Phi * G_2)(x), \quad x \in \mathbb{R}^1.$$

Then both F_1 and F_2 are infinitely divisible and $F_2(x) > 0$, $x \in \mathbb{R}^1$. Let us now estimate the quantity $[F_1(x)/F_2(x)] - 1$ in two ways. We have

$$\frac{F_1(x)}{F_2(x)} - 1 = \frac{\sum_{k=0}^{\infty}(1/k!)\Phi(x-k)}{\sum_{k=0}^{\infty}(1/k!)\Phi(x-2k)} - 1 \leq \frac{\sum_{k=1}^{\infty}(1/k!)\Phi(x-k)}{\sum_{k=1}^{\infty}(1/k!)\Phi(x-2k)}$$

$$\leq \frac{\sum_{k=1}^{\infty}(1/k!)\Phi(x-k)}{\Phi(x)} \leq \frac{\Phi(x-1)}{\Phi(x)} \sum_{k=1}^{\infty}(1/k!) \to 0 \quad \text{as } x \to -\infty.$$

On the other hand,

$$1 - \frac{F_1(x)}{F_2(x)} \leq \frac{\Phi(x-2)}{\Phi(x)} \sum_{k=1}^{\infty}(1/k!) \to 0 \quad \text{as } x \to -\infty.$$

Thus $\lim_{x\to-\infty}[F_1(x)/F_2(x)] = 1$, that is relation (1) is satisfied, but $F_1 \neq F_2$.

12.11. Does the renewal equation determine uniquely the probability density?

Let us start with a sequence $\{X_i, i = 1, 2, \ldots\}$ of non-negative i.i.d. r.v.s with a common d.f. F and density f. It is accepted to interpret X_i as a lifetime, or renewal time and it is important to know the probability distribution, say H_t, of the variable N_t defined as the number of renewals on the time interval $[0, t]$. In some cases it is even more important to find $U(t) = \mathbf{E}N_t$, the average number of renewals up to time t without asking explicitly for H_t. We have

$$U(t) = F(t) + \int_0^t F(t-s)\,\mathrm{d}U(s) \tag{1}$$

and hence the function $u(t) = \mathrm{d}U(t)/\mathrm{d}t$ (which exists since $f = F'$ exists), called a renewal density, satisfies

$$u(t) = f(t) + \int_0^t f(t-s)u(s)\,\mathrm{d}s \quad \text{for } t > 0. \tag{2}$$

The term *renewal equation* is used for both (1) and (2) and we are interested in how to find U (or u) in terms of F (or f), and conversely. If f^* and u^* are the Laplace transforms of f and u ($f^*(\alpha) = \int_0^\infty \mathrm{e}^{-\alpha t} f(t)\,\mathrm{d}t$, $u^*(\alpha) = \int_0^\infty \mathrm{e}^{-\alpha t} u(t)\,\mathrm{d}t$), we easily find from (2) that

$$u^*(\alpha) = \frac{f^*(\alpha)}{1 - f^*(\alpha)}, \quad \alpha \geq 0. \tag{3}$$

Obviously, (3) and (2) imply that f determines u uniquely. Consequently F determines U uniquely. E.g. if $F \sim \mathcal{E}xp(\lambda)$, then $f^*(\alpha) = \lambda/(\lambda + \alpha)$, $u^*(\alpha) = \lambda/\alpha \Rightarrow u(t) = \lambda$ for all $t \Rightarrow U(t) = \lambda t$, a well known result.

Let us now answer the inverse question: does u determine f uniquely?

Recall first that the classical renewal theorem (Feller 1971) states that in general $\lim_{t\to\infty} u(t) = 1/\mu$, where $\mu = \mathbf{E}[X_1]$ is the average lifetime.

Let us show that there is a renewal density $u(t)$ with $u(t) \to 1/\mu$ as $t \to \infty$ for same $\mu > 0$ and such that $f^*(\alpha) = u^*(\alpha)/[1 + u^*(\alpha)]$ found from (3) leads to a function $f(t)$ which may not be a probability density.

Indeed, take $u(t) = (1 - \mathrm{e}^{-\mu t})/\mu$ for $t > 0$ and fixed μ. Obviously, $u(t) \to 1/\mu$ as $t \to \infty$. The Laplace transform $u^*(\alpha)$ of this $u(t)$ is

$$u^*(\alpha) = \frac{1}{\alpha(\alpha+\mu)} \quad \Rightarrow \quad f^*(\alpha) = \frac{1}{\alpha^2 + \alpha\mu + 1}.$$

Suppose now that $\mu < 2$ (by assumption $\mu > 0$). Inverting f^* we find that $f(t) = \mathrm{e}^{-\mu t/2}(c/2)^{-1}\sin(ct/2)$, $0 < t < \infty$, where $c = \sqrt{4-\mu^2}$ and that $\int_0^\infty f(t)\,\mathrm{d}t = 1$. However the function f is not a probability density.

12.12. A property not characterizing the Cauchy distribution

Suppose the r.v. X has a Cauchy distribution with density $f(x) = 1/[\pi(1+x^2)]$, $x \in \mathbb{R}^1$. Then it is easy to check that the r.v. $1/X$ has the same density. This property leads naturally to the following question. Let X be a r.v. which is absolutely continuous and let its d.f. be denoted by $F(x)$, $x \in \mathbb{R}^1$. Suppose the r.v. $1/X$ has the same d.f. F. Does it follow that F is the Cauchy distribution?

Clearly, if the answer to this question is positive, then the property $X \stackrel{\mathrm{d}}{=} 1/X$ would be a characterizing property of the Cauchy distribution. It turns out, however, that in general the answer is negative. Let us illustrate this by the following example.

Suppose X is a r.v. with density

$$g(x) = \begin{cases} \frac{1}{4}, & \text{if } |x| \le 1 \\ \frac{1}{4}x^{-2}, & \text{if } |x| > 1. \end{cases}$$

Thus X is absolutely continuous and it is easy to check that $1/X$ has the same density g. Hence $X \stackrel{\mathrm{d}}{=} 1/X$. However, that X does not enjoy the Cauchy distribution.

12.13. A property not characterizing the gamma distribution

Let X and Y be independent r.v.s each with a gamma distribution $\gamma(p, a)$, $p > 0$, $a > 0$; that is, the common density is

$$f(x) = \begin{cases} 0, & \text{if } x \le 0 \\ \dfrac{a^p}{\Gamma(p)} x^{p-1}\mathrm{e}^{-ax}, & \text{if } x > 0. \end{cases} \tag{1}$$

Then the ratio $Z = X/Y$ has the following density:

$$g(z) = \begin{cases} 0, & \text{if } z \le 0 \\ \dfrac{1}{\mathrm{B}(p,p)} z^{p-1}(1+z)^{-2p}, & \text{if } z > 0 \end{cases} \tag{2}$$

(beta distribution of the second kind; see Example 12.8).

This connection between gamma and beta distributions leads to the next question. Let X and Y be positive independent r.v.s such that the ratio $Z = X/Y$ has a density given by (2). Does this imply that each of X and Y has gamma distribution?

Let us show that the answer to this question is negative. To see this, introduce the following two functions, where $a > 0$, $p > 0$:

$$f_1(x) = \begin{cases} 0, & \text{if } x \le 0 \\ c_1 x^{-p-1} \mathrm{e}^{-a/x}, & \text{if } x > 0, \end{cases}$$

$$f_2(x) = \begin{cases} 0, & \text{if } x \le 0 \\ c_2 x^p / [(1+x^2)^{p+1/2}], & \text{if } x > 0. \end{cases}$$

It is easy to check that with

$$c_1 = a^p/\Gamma(p) \quad \text{and} \quad c_2 = 2\Gamma(p+\tfrac{1}{2})/[\Gamma(\tfrac{1}{2}p)\Gamma(\tfrac{1}{2}p+\tfrac{1}{2})]$$

f_1 and f_2 are density functions. Take two independent r.v.s, say ξ_1 and η_1, each with density f_1. Then we can establish that the density g_1 of the ratio $\zeta_1 = \xi_1/\eta_1$ coincides with the density g given by (2). Clearly, f_1 does not have the form (1).

The same conclusion can be derived if we start with two independent r.v.s, ξ_2 and η_2, each having the density f_2. In this case again the density of $\zeta_2 = \xi_2/\eta_2$ coincides with (2) while f_2 is not of the form (1).

12.14. An interesting property which does not characterize uniquely the inverse Gaussian distribution

We say that the r.v. X has an *inverse Gaussian distribution* with parameters $\mu > 0$ and $\lambda > 0$ if the density f of X is given by

$$f(x) = \begin{cases} \left(\dfrac{\lambda}{2\pi x^3}\right)^{1/2} \exp\left[-\dfrac{\lambda}{2\mu^2}\dfrac{(x-\mu)^2}{x}\right], & \text{if } x > 0 \\ 0, & \text{if } x \le 0. \end{cases} \tag{1}$$

It is easy to see that all moments $m_n = \mathbf{E}[X^n]$, $n = 1, 2, \ldots$, exist. Moreover, X has an analytic ch.f., hence this distribution is determined uniquely by its moment sequence $\{m_n, n = 1, 2, \ldots\}$.

It is interesting to note that all negative-order moments of X are also finite, that is $\mathbf{E}[X^{-n}]$ exists for each positive integer n. Further, a standard transformation leads to the following interesting relation:

$$\mathbf{E}[X^{-n}] = \mathbf{E}[X^{n+1}]/(\mathbf{E}X)^{2n+1}, \quad n = 1, 2, \ldots. \tag{2}$$

This relation and the uniqueness of the moment problem mentioned above motivate the conjecture: if X is a positive r.v. such that all moments $\mathbf{E}[X^n]$ and $\mathbf{E}[X^{-n}]$, $n = 1, 2, \ldots$, exist and satisfy (2), then X has an inverse Gaussian distribution. It turns out, however, that this conjecture is not correct.

Note firstly that $\mathbf{E}X = \mu$ and let for simplicity $\mu = 1$. Then (2) has the form $\mathbf{E}[X^{-n}] = \mathbf{E}[X^{n+1}]$. Further, the density (1) satisfies the relation

$$xf(x) = \frac{1}{x^2} f\left(\frac{1}{x}\right), \quad x > 0. \tag{3}$$

Thus the density f of the inverse Gaussian distribution can be considered as a solution of the functional equation (3).

Let Y be a r.v. whose density g satisfies (3). Then it is easy to check that the relation (2) is fulfilled for Y. So it is clear that if (3) has a unique solution, namely f by (1), then our conjecture will be true; otherwise it will be false. To clarify this consider the function g given by

$$g(x) = \begin{cases} \dfrac{2}{\pi} \dfrac{1}{\sqrt{x}} \dfrac{1}{(1+x)^2}, & \text{if } x > 0 \\ 0, & \text{if } x \leq 0. \end{cases} \tag{4}$$

It can be verified directly that g is a probability density function which satisfies (3). As a consequence, Y satisfies (2).

Therefore we have found two r.v.s, X and Y, whose densities (1) and (4) are different, and nevertheless both satisfy relation (2). Thus the relation (2) is not a characterizing property of the inverse Gaussian distribution.

Finally we suggest that the reader considers equation (3) and tries to find other solutions to it which will provide new r.v.s satisfying relation (2).

SECTION 13. DIVERSE PROPERTIES OF RANDOM VARIABLES

In this section we consider examples devoted to different properties of r.v.s and their numerical characteristics. Some notions are defined in the examples themselves.

13.1. On the symmetry property of the sum or the difference of two symmetric random variables

Recall first that the r.v. X is called *symmetric* about 0 if $X \stackrel{d}{=} (-X)$. In terms of the d.f. F, the density f and the ch.f. φ this property is expressed as follows: $F(-x) = 1 - F(x)$ for all $x \geq 0$; $f(-x) = f(x)$ for all $x \in \mathbb{R}^1$; $\varphi(t)$, $t \in \mathbb{R}^1$ takes only real values. By the examples below we analyse the symmetry and the independence properties under summation or subtraction.

(i) If X and Y are identically distributed and independent r.v.s, then their difference $X - Y$ is symmetric about 0. Suppose we know that $X \stackrel{d}{=} Y$ and that the difference

$X - Y$ is symmetric. Does it follow that X and Y are independent? To see this consider the random vector (X, Y) defined as follows:

X \ Y	1	2	3
1	$\frac{1}{12}$	$\frac{2}{12}$	$\frac{1}{12}$
2	$\frac{1}{12}$	0	$\frac{1}{12}$
3	$\frac{2}{12}$	0	$\frac{4}{12}$.

It is easy to check that X and Y have the same distribution, each taking the values $1, 2$ and 3 with probability equal to $\frac{2}{6}$, $\frac{1}{6}$ and $\frac{3}{6}$ respectively. Obviously, X and Y are not independent. Further, the difference $Z = X - Y$ takes the values $-2, -1, 0, 1, 2$ with probabilities $\frac{6}{36}, \frac{5}{36}, \frac{14}{36}, \frac{5}{36}, \frac{6}{36}$. Clearly Z and $(-Z)$ have the same distribution. In other words, $Z = X - Y$ is a symmetric r.v. despite the fact that the variables X and Y are not independent.

(ii) If X and Y are symmetric and independent r.v.s, then the sum $Z = X + Y$ is again symmetric. Thus it is of interest to discuss the following question. Suppose X and Y are independent r.v.s and we know that X is symmetric and that the sum $Z = X + Y$ is also symmetric. Is it true that Y is symmetric? Intuitively we could expect a positive answer. It turns out, however, in general the answer is negative. This is illustrated by the following example.

Let ξ be a r.v. with the following ch.f. indicating that ξ is symmetric:

(1)
$$\psi_\xi(t) = \begin{cases} 1 - 2|t|, & \text{if } |t| \le \frac{1}{2} \\ 0, & \text{if } |t| > \frac{1}{2}. \end{cases}$$

Consider two other ch.f.s:

$$h_1(t) = \begin{cases} 1 - |t|, & \text{if } |t| \le \frac{1}{2} \\ 1/(4|t|), & \text{if } |t| > \frac{1}{2}, \end{cases} \quad h_2(t) = \begin{cases} 1 - |t|, & \text{if } |t| \le 1 \\ 0, & \text{if } |t| > 1. \end{cases}$$

Introduce now a r.v. η with ch.f. ψ_η which is the mixture of h_1 and h_2:

$$\psi_\eta(t) = \tfrac{1}{2}\mathrm{e}^{it}h_1(t) + \tfrac{1}{2}\mathrm{e}^{-it}h_2(t), \quad t \in \mathbb{R}^1.$$

Elementary transformations show that

(2)
$$\psi_\eta(t) = \begin{cases} (1 - |t|)\cos t, & \text{if } |t| \le \frac{1}{2} \\ \mathrm{e}^{it}/(8|t|) + \frac{1}{2}\varepsilon(t)\mathrm{e}^{-it}(1 - |t|), & \text{if } |t| > \frac{1}{2} \end{cases}$$

where $\varepsilon(t) = 1$, if $|t| \le 1$ and $\varepsilon(t) = 0$, if $|t| > 1$.

The explicit form (2) of the ch.f. ψ_η shows that the r.v. η is not symmetric.

Thus we have described two r.v.s, ξ and η, the first being symmetric while the second is not. Assuming that ξ and η are independent we look for the properties of the sum $\zeta = \xi + \eta$. Since for the ch.f.s ψ_ξ, ψ_η and ψ_ζ we have $\psi_\zeta = \psi_\xi\psi_\eta$, in view of (1) and (2), it is not difficult to find that

$$\psi_\zeta(t) = \begin{cases} (1-2|t|)(1-|t|)\cos t, & \text{if } |t| \le \frac{1}{2} \\ 0, & \text{if } |t| > \frac{1}{2}. \end{cases}$$

Obviously ψ_ζ takes only real values which means that the r.v. ζ is symmetric.

Therefore the symmetric property of two variables, ξ and $\zeta = \xi + \eta$, together with the independence of ξ and η, do not imply that η is symmetric.

Here is another equivalent interpretation—the difference, and hence the sum, of two dependent r.v.s both symmetric, need not be symmetric.

13.2. When is a mixture of normal distributions infinitely divisible?

Let $G(u)$, $u \in \mathbb{R}^+$ be a d.f. Then the function $\psi(t)$, $t \in \mathbb{R}^1$ where

$$(1) \qquad \psi(t) = \int_0^\infty \exp(-\tfrac{1}{2}t^2 u)\, \mathrm{d}G(u)$$

is a ch.f. The d.f. F with ch.f. ψ is called a *mixture of normal distributions* and G a mixing distribution. Note that the density f of F corresponding to (1) has the form (see Kelker 1971):

$$f(x) = \int_0^\infty (2\pi u)^{-1/2} \exp(-x^2/(2u))\, \mathrm{d}G(u).$$

Since the normal distribution is infinitely divisible it is natural to ask whether such a mixture preserves the infinite divisibility. It is easy to check that if G is an infinitely divisible d.f. then ψ is an infinitely divisible ch.f. Now we want to answer the converse question: if ψ is an infinitely divisible ch.f., does it follow that the mixing distribution G is infinitely divisible?

Consider the function $H(x)$, $x \in \mathbb{R}^1$ where

$$H(x) = 0,\ 0.26,\ 0.52,\ 0.48,\ 0.74 \text{ and } 1$$

respectively in the intervals

$$(-\infty, 1],\ (1, 2],\ (2, 3],\ (3, 4],\ (4, 5] \text{ and } (5, \infty).$$

Clearly H is not a d.f. However, we obtain the following interesting and unexpected fact that the convolution $H * H$ is a d.f. Moreover, the function $\int_0^\infty (2\pi u)^{-1/2} \exp(-x^2/(2u))\, \mathrm{d}H(u)$ is a density. Define G as follows:

$$(2) \qquad G(x) = \mathrm{e}^{-1} \sum_{k=0}^{\infty} (k!)^{-1} H^{*k}(x).$$

We can verify that G given by (2) is a d.f. and find that

$$\int_0^\infty \exp\left(-\frac{1}{2}t^2u\right) \mathrm{d}G(u) = \mathrm{e}^{-1}\sum_{k=0}^\infty \frac{1}{k!}\left[\int_0^\infty \exp\left(-\frac{1}{2}t^2u\right) \mathrm{d}H(u)\right]^k$$
$$= \exp\left[\int_0^\infty \exp\left(-\frac{1}{2}t^2u\right) \mathrm{d}H(u) - 1\right]$$
$$= \exp\left(\int_0^\infty [\cos(tx) - 1]\frac{1}{\sqrt{2\pi}}\int_0^\infty u^{-1/2}\exp[-x^2/(2u)]\,\mathrm{d}H(u)\,\mathrm{d}x\right).$$

It is easy to see that the last expression in this chain of equalities coincides with the Kolmogorov canonical representation for an infinitely divisible ch.f. provided that $\int_0^\infty u^{-1/2}\exp(-x^2/(2u))\,\mathrm{d}H(u) \geq 0$ for all $x > 0$ (see Gnedenko and Kolmogorov 1954). But H satisfies this condition by construction.

Therefore ψ defined by (1), with G given by (2), is an infinitely divisible ch.f.

It remains for us to show that G in (2) is not infinitely divisible. This follows from the Lévy–Khintchine representation for the ch.f. of G and from the fact that H is not non-decreasing.

13.3. A distribution function can belong to the class IFRA but not to IFR

Let $F(x)$, $x \geq 0$ be a d.f. with density f. We say that F is an *increasing failure rate* distribution and write $F \in$ IFR, if its failure rate $r(x) := f(x)/(1 - F(x))$ is increasing in x, $x > 0$. In this case $-\log[1 - F(x)]$ is a convex function in the domain where $-\log[1 - F(x)]$ is finite. This observation motivates the more general definition: $F \in$ IFR if $-\log[1 - F(x)]$ is convex where finite. However, for some problems it is necessary to introduce a considerably weaker restriction on F. For example, if F has density f and failure rate r such that $(1/x)\int_0^x r(u)\,\mathrm{d}u$ is increasing in x, we say that F has an *increasing failure rate average*. In this case we write $F \in$ IFRA. More generally, $F \in$ IFRA if $(-1/x)\log[1 - F(x)]$ is increasing where finite.

Thus we have introduced two classes of distributions, IFR and IFRA, and it is natural to ask what the relationship between them is.

According to Barlow and Proshan (1966), if $F \in$ IFR and $F(0) = 0$ then $F \in$ IFRA. We are interested now in whether the converse is true. To see this, consider the function

$$F(x) = \begin{cases} 0, & \text{if } x \leq 0 \\ (1 - \mathrm{e}^{-x})(1 - \mathrm{e}^{-kx}), & \text{if } x > 0,\ k > 1. \end{cases}$$

It is easy to check that $F \in$ IFRA but $F \notin$ IFR.

13.4. A continuous distribution function of the class NBU which is not of the class IFR

A d.f. F (of a non-negative r.v.) is said to belong to the class NBU (*new and better than used*) if for any $x, y \geq 0$ we have

$$(1) \qquad \bar{F}(x+y) \leq \bar{F}(x)\bar{F}(y) \quad \text{where} \quad \bar{F} = 1 - F.$$

If for any $y > 0$ the function $[F(x+y) - F(x)]/\bar{F}(x)$ is increasing in x, we say that F is of the class IFR (compare this definition with that given in Example 13.3).

It is well known that $F \in$ IFR $\Rightarrow$ $F \in$ NBU, but in general the converse implication is not true (see Barlow and Proshan 1966).

The d.f. $F \in$ IFR has the property that it is continuous o'n the set $\{x : F(x) < 1\}$ and, moreover, $h(x) = -\log F(x)$ is a convex function. However, the elements of the class NBU need not be continuous. This follows from a simple example. Indeed, Consider the function

$$F(x) = 1 - 2^{-k} \quad \text{for} \quad x \in (k, k+1], \quad k = 0, 1, 2, \ldots.$$

It is easy to check that (1) is satisfied and hence $F \in$ NBU. Obviously F is discontinuous and hence $F \notin$ IFR.

Suppose now that $F \in$ NBU and F is continuous. Does it follow from these conditions that $F \in$ IFR? It turns out that the answer is negative. To see this consider the function

$$h(x) = \begin{cases} \sin^2 x, & \text{if } x \in [0, \frac{1}{2}\pi] \\ \frac{1}{2}\pi(x - \frac{1}{2}\pi) + 1, & \text{if } x \in (\frac{1}{2}\pi, \infty). \end{cases}$$

It is easy to check that $F(x) = 1 - \mathrm{e}^{-h(x)}$, $x \geq 0$ is a d.f. and, moreover, that

$$h(x+y) \geq h(x) + h(y), \quad x, y \geq 0.$$

Therefore $F \in$ NBU and clearly F is continuous. Nevertheless $F \notin$ IFR since $h(x) = -\log \bar{F}(x)$ is not a convex function.

13.5. Exchangeable and tail events related to sequences of random variables

Let $\{X_n, n \geq 1\}$ be an infinite sequence of r.v.s defined on the probability space $(\Omega, \mathcal{F}, \mathbf{P})$. Denote by $\sigma\{X_1, \ldots, X_n\}$ the σ-field generated by $X_1, \ldots, X_n$. Then clearly $\bigcup_{k=1}^{\infty} \sigma\{X_n, X_{n+1}, \ldots, X_{n+k}\}$ is a field and let $\sigma\{X_n, X_{n+1}, \ldots\}$ be the σ-field generated by this field. The sequence of σ-fields $\sigma\{X_n, X_{n+1}, \ldots\}$, $\sigma\{X_{n+1}, X_{n+2}, \ldots\}, \ldots$ is non-increasing, its limit exists and is a σ-field. This limit is denoted by

$$\mathcal{T} = \bigcap_{n=1}^{\infty} \sigma\{X_n, X_{n+1}, \ldots\}.$$

$\mathcal{T}$ is called the *tail σ-field* of the sequence $\{X_n, n \geq 1\}$. Any event $A \in \mathcal{T}$ is called a *tail event*, and any function on Ω which is measurable with respect to $\mathcal{T}$ is said to be a *tail function*.

Let us formulate the basic result concerning the tail events and functions.

Kolmogorov 0–1 law. *Let $\{X_n\}$ be a sequence of independent r.v.s and $\mathcal{T}$ be its tail σ-field. Then for any tail event A, $A \in \mathcal{T}$, either $\mathbf{P}(A) = 0$ or $\mathbf{P}(A) = 1$. Moreover, any tail function is a.s. constant, that is, if Y is a r.v. such that $\sigma\{Y\} \subset \mathcal{T}$ then $\mathbf{P}[Y = c] = 1$ with $c =$* constant.

We now introduce another notion. (Also see Example 7.14). We say that the r.v.s $X_1, \ldots, X_n$ are *exchangeable* (another term is *symmetrically dependent*) if for each of the $n!$ permutations $\{i_1, i_2, \ldots, i_n\}$ of $\{1, 2, \ldots, n\}$, the random vectors $(X_{i_1}, X_{i_2}, \ldots, X_{i_n})$ and $(X_1, X_2, \ldots, X_n)$ have the same distribution. Further, an infinite sequence $\{X_n, n \geq 1\}$ is said to be exchangeable if for each n the r.v.s $X_1, \ldots, X_n$ are exchangeable. The $\mathcal{B}^\infty$-measurable function $g(X_1, X_2, \ldots)$ is called exchangeable if it is invariant under all finite permutations of its arguments: $g(X_1, \ldots, X_n, X_{n+1}, \ldots) = g(X_{i_1}, \ldots, X_{i_n}, X_{n+1}, \ldots)$. In particular, $A \in \sigma\{X_1, X_2, \ldots\}$ is called an *exchangeable event* if its indicator function $I(A)$ is an exchangeable function.

Let us formulate a result concerning the exchangeability.

Hewitt–Savage 0–1 law. *Let $\{X_n, n \geq 1\}$ be a sequence of r.v.s which are independent and identically distributed. Then for any exchangeable event $A \in \sigma\{X_1, X_2, \ldots\}$ either $\mathbf{P}(A) = 0$ or $\mathbf{P}(A) = 1$.*

Note that a detailed presentation of the notions and results given briefly above can be found in the books by Feller (1971), Laha and Rohatgi (1979), Chow and Teicher (1978), Aldous (1985) and Galambos (1988).

Obviously tailness and exchangeability are close notions and it would be useful to illustrate by a few examples the relationships between them.

(i) The first question concerns the tail and exchangeable events. Let $\{X_n, n \geq 1\}$ be a sequence of (real-valued) r.v.s and $\mathcal{T}$ its tail σ-field. If $A \in \mathcal{T}$, then for any permutation $\{i_1, \ldots, i_n\}$ of $\{1, \ldots, n\}$, $n \geq 1$, we can write A in the form $\{(X_{n+1}, X_{n+2}, \ldots) \in B_{n+1}\}$ where B_{n+1} is a Borel set in $\mathbb{R}^\infty$, that is, $B_{n+1} \in \mathcal{B}^\infty$. Thus for each n,

$$A = \{(X_1, X_2, \ldots) \in \mathbb{R}^n \times B_{n+1}\} = \{(X_{i_1}, \ldots, X_{i_n}, X_{n+1}, \ldots) \in \mathbb{R}^n \times B_{n+1}\}$$

and since $B_\infty := \mathbb{R}^n \times B_{n+1}$ is a Borel set in $\mathbb{R}^\infty$, this implies that the tail event A is also an exchangeable event.

However, there are exchangeable events which are not tail events. The simplest example is to take $A = \{X_n = 0 \text{ for all } n \geq 1\}$. Obviously A is an exchangeable event. But $A \notin \sigma\{X_n, X_{n+1}, \ldots\}$ for every $n \geq 1$. So A is not a tail event.

(ii) Now let us clarify the possibility of changing some of the conditions in the Hewitt–Savage 0–1 law. Consider the sequence $\{X_n, n \geq 1\}$ of independent r.v.s where $\mathbf{P}[X_1 = 1] = \mathbf{P}[X_1 = -1] = \frac{1}{2}$ and $\mathbf{P}[X_n = 0] = 1$ for $n \geq 2$. The event $A = \{\sum_{j=1}^{n} X_j > 0$ for infinitely many $n\}$ is clearly an exchangeable (but not tail!) event with respect to the infinite sequence of r.v.s $\{X_n, n \geq 1\}$. Moreover, $\mathbf{P}(A) = \mathbf{P}[X_1 > 0] = \frac{1}{2}$ and hence $\mathbf{P}(A)$ is neither 0 nor 1 as we could expect. Here the Hewitt–Savage 0–1 law is not applicable since the independent r.v.s $X_n, n \geq 1$ are not identically distributed.

(iii) Let $X_n, n \geq 1$ be independent r.v.s such that

$$\mathbf{P}[X_n = 1] = 2^{-n}, \quad \mathbf{P}[X_n = 0] = 1 - 2^{-n}, \ n = 1, 2, \ldots$$

and let $A = \{X_n = 0 \text{ for all } n \geq 1\}$. Then A is an exchangeable but not a tail event. Further, we have

$$\mathbf{P}(A) = \prod_{n=1}^{\infty} \mathbf{P}[X_n = 0] = \prod_{n=1}^{\infty} (1 - 2^{-n}).$$

Since $\sum_{n=1}^{\infty} 2^{-n} < \infty$, the infinite product $\prod_{n=1}^{\infty}(1 - 2^{-n})$ converges to a positive limit which is strictly less than 1 and hence $\mathbf{P}(A)$ is neither 0 nor 1.

Therefore again, as in case (ii), the Hewitt–Savage 0–1 law does not hold. Note that here the variables $X_n, n \geq 1$, are independent and take the same values but with different probabilities.

13.6. The de Finetti theorem for an infinite sequence of exchangeable random variables does not always hold for a finite number of such variables

Let $\{X_n, n \geq 1\}$ be an infinite sequence of exchangeable r.v.s each taking two values, 0 and 1. Then according to the de Finetti theorem (see Feller 1971) there is a unique probability measure μ on the Borel σ-field $\mathcal{B}_{[0,1]}$ of the interval [0, 1] such that for each n we have

$$(1) \qquad \mathbf{P}[X_1 = \varepsilon_1, X_2 = \varepsilon_2, \ldots, X_n = \varepsilon_n] = \int_{[0,1]} p^k (1-p)^{n-k} \mu\,(\mathrm{d}p)$$

where $\varepsilon_j = 0$ or 1 and $k = \varepsilon_1 + \ldots + \varepsilon_n$.

In other words, the distribution of the number of occurrences of the outcomes 0 and 1 in a finite segment of length n of the infinite sequence $X_1, X_2, \ldots$ of exchangeable variables is always a mixture of a binomial distribution with some proper distribution over the interval [0, 1]. Thus we come to the question of whether this result holds for a fixed finite exchangeable sequence. The answer can be found from the following two examples.

(i) Consider the case $n = 2$ and the r.v.s X_1 and X_2:

$$\mathbf{P}[X_1 = 0, \ X_2 = 1] = \mathbf{P}[X_1 = 1, \ X_2 = 0] = \tfrac{1}{2},$$

$$\mathbf{P}[X_1 = 0,\ X_2 = 0] = \mathbf{P}[X_1 = 1,\ X_2 = 1] = 0.$$

It is easy to see that X_1 and X_2 are exchangeable. Suppose a representation like (1) holds. Then it would follow that

$$\int_0^1 p^2 \mu\,(\mathrm{d}p) = 0 \quad \text{and} \quad \int_0^1 (1-p)^2 \mu\,(\mathrm{d}p) = 0.$$

This means that μ puts mass one both at 0 and at 1, which is not possible.

(ii) Let $Y_1, \ldots, Y_n$ be n independent r.v.s with some common distribution. Let $S_n = Y_1 + \ldots + Y_n$ and $Z_k = Y_k - n^{-1}S_n$ for $k = 1, \ldots, n-1$. Then it is easy to check that the r.v.s $Z_1, \ldots, Z_{n-1}$ are exchangeable but their joint distribution is not of the form (1). Therefore the de Finetti theorem does not always hold for a finite exchangeable sequence.

13.7. Can we always extend a finite set of exchangeable random variables?

If $\{X_n\}$ is a finite or an infinite sequence of exchangeable r.v.s then any subset consists of r.v.s which are exchangeable.

Suppose now we are given the set $X_1, \ldots, X_m$ of exchangeable r.v.s. We say that $X_1, \ldots, X_m$ can be *extended exchangeable* if there is a finite set $X_1, \ldots, X_m, X_{m+1}, \ldots, X_{m+k}$, $k \geq 1$, or an infinite set $X_1, \ldots, X_m, X_{m+1}, X_{m+2}, \ldots$ of r.v.s which are exchangeable. Thus the question to consider is: can we extend any fixed set of exchangeable r.v.s to an infinite exchangeable sequence? Let us show that in general the answer is negative.

Consider the particular case of three r.v.s X_1, X_2, X_3 each taking the values 0 or 1 with $\mathbf{P}[X_j = 1] = \mathbf{P}[X_j = 0] = \frac{1}{2}$, $j = 1, 2, 3$. Let

$$\mathbf{P}[X_1 = 1, X_2 = 1] = \mathbf{P}[X_1 = 1, X_3 = 1] = \mathbf{P}[X_2 = 1, X_3 = 1] = 0.2.$$

It is easy to see that (X_1, X_2, X_3) is an exchangeable set. Assume this set can be extended to an infinite exchangeable sequence $X_1, X_2, X_3, X_4, X_5, \ldots$. This would mean that for each $n \geq 4$ the set $X_1, X_2, X_3, X_4, \ldots, X_n$ consists of exchangeable variables. Then we can easily show that

$$\begin{aligned} 0 &\leq \mathbf{E}\left[\left(\textstyle\sum_{j=1}^n I(X_j = 1)\right)^2\right] - \left(\mathbf{E}\left[\textstyle\sum_{j=1}^n I(X_j = 1)\right]\right)^2 \\ &= \textstyle\sum_{j=1}^n \sum_{k=1}^n \mathbf{P}[X_j = 1, X_k = 1] - \frac{1}{4}n^2 \\ &= \tfrac{1}{2}n + \textstyle\sum_{j \neq k} \mathbf{P}[X_j = 0, X_k = 1] - \frac{1}{4}n^2 \\ &= \tfrac{1}{2}n + (0.2)n(n-1) - \tfrac{1}{4}n^2 = (0.3)n - (0.05)n^2. \end{aligned}$$

Obviously it follows from this that n must satisfy the restriction $n \leq 6$. However, this contradicts the definition of an infinite exchangeability and therefore the desired extension of a finite to an infinite exchangeable sequence is not always possible.

Interesting results on exchangeability of finite or infinite sequences of random events or r.v.s can be found in the works of Kendall (1967) and Galambos (1988).

Finally let us mention that the variables in an infinite exchangeable sequence are necessarily non-negatively correlated. This follows directly from an examination of the terms of the variance $\mathbf{V}[X_1 + \ldots + X_n]$. However, in the above specific example we have $\rho(X_1, X_2) < 0$.

13.8. Collections of random variables which are or are not independent and are or are not exchangeable

Let $\mathcal{X} := \{X_n, n \geq 2\}$ be a finite or infinite sequence of r.v.s which are independent and identically distributed. Then $\mathcal{X}$ is exchangeable, that is both properties independence and exchangeability hold for $\mathcal{X}$ in this case. If, however, X_n (at least two of them) have different distributions then $\mathcal{X}$ is not exchangeable regardless of whether $\mathcal{X}$ is independent or dependent.

Our goal now is to describe a sequence of r.v.s which are totally dependent and exchangeable. For consider a sequence $\mathcal{X} = \{X_n, n \geq 2\}$ of i.i.d. r.v.s each with zero mean and finite variance. Let ξ be another r.v. independent of $\mathcal{X}$. Assume that ξ is non-degenerate with finite variance, that is, $0 < \mathbf{V}\xi < \infty$. Let us define a new sequence

$$\mathcal{Y} := \{Y_n, n \geq 2\} \quad \text{where} \quad Y_n = X_n + \xi.$$

It is easily seen that $\mathcal{Y}$ is exchangeable. Let us clarify whether or not $\mathcal{Y}$ is independent.

The distribution of $Y_{i_1}, \ldots, Y_{i_k}$ is the same for any possible choice of k variables from $\mathcal{Y}$, $k = 1, 2, \ldots$. Taking $k = 2$ we conclude that $\mathcal{Y}$ is characterized by a common correlation coefficient, say ρ_0 where

$$\rho_0 = \rho(Y_i, Y_j) = (\mathbf{E}[Y_i Y_j] - \mathbf{E}Y_i \mathbf{E}Y_j)/(\mathbf{V}Y_i \mathbf{V}Y_j)^{1/2}$$

for any two representatives Y_i and Y_j of $\mathcal{Y}$. A simple reasoning shows that

$$\rho_0 = (\mathbf{V}\xi)/(\mathbf{V}X_1 + \mathbf{V}\xi)$$

where $\mathbf{V}X_1$ $(= \mathbf{E}[X_1^2])$ is the common variance of the sequence $\mathcal{X}$. The assumption $0 < \mathbf{V}\xi < \infty$ implies that $\rho_0 \neq 0$ (in fact $\rho_0 > 0$) and hence Y_i and Y_j cannot be independent because they are not even uncorrelated. In other words, $\mathcal{Y}$ is totally dependent in the sense that there is no pair of variables in $\mathcal{Y}$ which is independent. Hence the sequence $\mathcal{Y}$, finite or infinite, is dependent and exchangeable.

The final conclusion is that these two properties, independence and exchangeability, are incompatible.

13.9. Integrable randomly stopped sums with non-integrable stopping times

Let X and $X_1, X_2, \ldots$ be i.i.d. r.v.s defined on the probability space $(\Omega, \mathcal{F}, \mathbf{P})$ and $\{\mathcal{F}_n, n \geq 0\}$ where $\mathcal{F}_0 = \{\emptyset, \Omega\}$ is an increasing sequence of sub-σ-fields of $\mathcal{F}$.

Recall that the r.v. τ with possible values in the set $\{1, 2, \ldots, \infty\}$ is said to be a *stopping time* with respect to $\{\mathcal{F}_n\}$ if the set $[\omega : \tau(\omega) = n]$ denoted further simply by $[\tau = n]$ belongs to $\mathcal{F}_n$ for each n. If $S_0 = 0$, $S_n = X_1 + \ldots + X_n$ then S_τ is the sum of the first τ of the variables $X_1, X_2, \ldots$, that is $S_\tau = X_1 + \ldots + X_\tau$. For many problems it is important to have conditions under which the r.v.s X, τ and S_τ are integrable. Let us formulate the following result (see Gut and Janson 1986). Let $r \geq 1$ and $\mathbf{E}X \neq 0$. Then $\mathbf{E}[|X|^r] < \infty$ and $\mathbf{E}[|S_\tau|^r] < \infty$ imply that $\mathbf{E}[\tau^r] < \infty$.

Our aim now is to show that the condition $\mathbf{E}X \neq 0$ is essential for the validity of this result. So, let the r.v. X have $\mathbf{E}X = 0$. In particular, take X such that $\mathbf{P}[X = 1] = \mathbf{P}[X = -1] = \frac{1}{2}$ and $\tau = \min\{n : S_n = 1\}$. Clearly τ is a stopping time with respect to $\{\mathcal{F}_n\}$ where $\mathcal{F}_n = \sigma\{X_1, \ldots, X_n\}$. It is easy to check that the r.v. X and the random sum S_τ have moments of all orders, that is, for any $r > 0$ we have $\mathbf{E}[|X|^r] < \infty$ and $\mathbf{E}[|S_\tau|^r] < \infty$. However, $\mathbf{E}[\tau^{1/2}] = \infty$ and therefore $\mathbf{E}[\tau^r]$ does not exist for any $r \geq \frac{1}{2}$.

Chapter 3

Limit Theorems

Courtesy of Professor A. T. Fomenko of Moscow University.

SECTION 14. VARIOUS KINDS OF CONVERGENCE OF SEQUENCES OF RANDOM VARIABLES

On the probability space $(\Omega, \mathcal{F}, \mathbf{P})$ we have a given r.v. X and a sequence of r.v.s $\{X_n, n \geq 1\}$. Important probabilistic problems require us to find the limit of X_n as $n \to \infty$. However, this limit can be understood in a different way. Let us define basic kinds of convergence considered in probability theory.

(a) We say that $\{X_n\}$ converges **almost surely** (a.s.), or with probability 1, to X as $n \to \infty$ and write $X_n \xrightarrow{\text{a.s.}} X$ if

$$\mathbf{P}[\omega : \lim_{n\to\infty} X_n(\omega) = X(\omega)] = 1.$$

(b) The sequence $\{X_n\}$ is said to converge **in probability** to X as $n \to \infty$ if for any $\varepsilon > 0$ we have

$$\lim_{n\to\infty} \mathbf{P}[\omega : |X_n(\omega) - X(\omega)| \geq \varepsilon] = 0.$$

In this case the following notation is used: $X_n \xrightarrow{\text{P}} X$ or $\text{P} - \lim_{n\to\infty} X_n = X$.
(c) Let F and F_n be the d.f.s of X and X_n respectively. The sequence $\{X_n\}$ is called convergent **in distribution** to X if

$$\lim_{n\to\infty} F_n(x) = F(x)$$

for all $x \in \mathbb{R}^1$ for which $F(x)$ is continuous. Notation: $F_n \xrightarrow{\text{d}} F$, and $X_n \xrightarrow{\text{d}} X$.
(d) Suppose X and X_n, $n \geq 1$, belong to the space L^r for some $r > 0$ (that is $\mathbf{E}[|X|^r] < \infty$, $\mathbf{E}[|X_n|^r] < \infty$). We say that the sequence $\{X_n\}$ converges to X **in** L^r**-sense**, and write $X_n \xrightarrow{\text{L}^r} X$, if

$$\lim_{n\to\infty} \mathbf{E}[|X_n - X|^r] = 0.$$

In particular, the L^r-convergence with $r = 2$ is called square mean (or quadratic mean) convergence and is used so often in probability theory and mathematical statistics.

Note that the convergence in distribution defined in (c) is closely related to the weak convergence treated in Section 16 (see also Example 14.1(iii)). Some notions (complete convergence, weak L^1-convergence and convergence of the Cesàro means) are introduced and analysed in Examples 14.14–14.18.

Practically all textbooks and lecture notes in probability theory deal extensively with the topics of convergence of random sequences. The reader is advised to consider the following references: Parzen (1960), Neveu (1965), Lamperti (1966), Moran (1968), Ash (1970), Rényi (1970), Feller (1971), Roussas (1973), Neuts (1973), Chung (1974), Petrov (1975), Lukacs (1975), Chow and Teicher (1978), Billingsley (1995), Laha and Rohatgi (1979), Serfling (1980) and Shiryaev (1995).

It is usual for any course in probability theory to justify the following scheme.

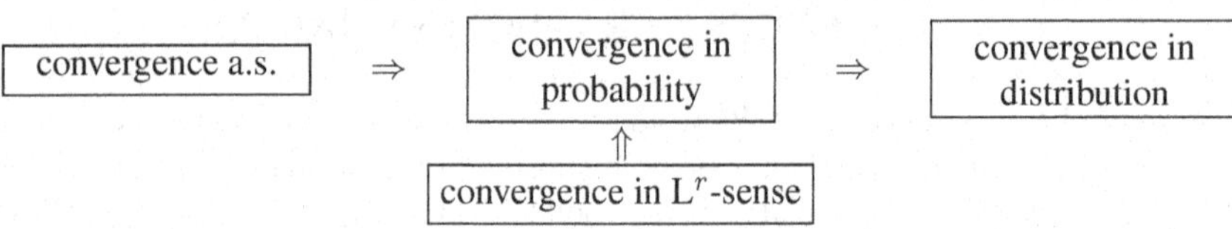

In this section we consider examples which are different in their content and level of difficulty but all illustrate this general scheme clearly. In particular, we show that the inclusions shown above are all strong inclusions. The relationship between these four kinds of convergence and other kinds of convergence of random sequences is also analysed.

14.1. Convergence and divergence of sequences of distribution functions

We now summarize a few elementary statements showing that a sequence of d.f.s $\{F_n, n \geq 1\}$ can have different behaviour as $n \to \infty$. In particular it can be divergent, or convergent, but not to a d.f.

(i) Let $F(x)$, $x \in \mathbb{R}^1$ be a d.f. which is continuous. Consider two sets of d.f.s, $\{F_n, n \geq 1\}$ and $\{G_n, n \geq 1\}$ where

$$F_n(x) = F(x+n), \quad G_n(x) = F(x + (-1)^n n).$$

Obviously $F_n(x) \to 1$ as $n \to \infty$ for each $x \in \mathbb{R}^1$. But a function equal to 1 at all points is not a d.f. Hence $\{F_n\}$ is convergent but the limit $\lim_{n\to\infty} F_n(x)$ is not a d.f.

Further, $G_{2n}(x) \to 1$ whereas $G_{2n+1}(x) \to 0$ as $n \to \infty$ for all $x \in \mathbb{R}^1$. Clearly the family $\{G_n\}$ does not converge.

(ii) Consider the family of d.f.s $\{F_n, n \geq 1\}$ where

$$F_n(x) = \begin{cases} 0, & \text{if } x \leq \frac{1}{n} \\ 1, & \text{if } x > \frac{1}{n}. \end{cases}$$

Then $F_n(x) \to F(x)$ if $x \neq 0$ and $F_n(0) = 0$ for each $n \geq 1$, where F is the d.f. of a degenerate r.v. which is zero with probability 1. Thus $\lim_{n\to\infty} F_n(x)$ exists but is not a d.f. because it is not right-continuous at $x = 0$.

(iii) The following basic result is always used when considering convergence in distribution (see Lukacs 1970; Feller 1971; Chow and Teicher 1978; Billingsley 1995; Shiryaev 1995):

$$(1) \qquad F_n \xrightarrow{\text{d}} F \iff \int_{\mathbb{R}^1} g(x)\,\mathrm{d}F_n(x) \to \int_{\mathbb{R}^1} g(x)\,\mathrm{d}F(x)$$

for all continuous and bounded functions $g(x)$, $x \in \mathbb{R}^1$.

Despite the fact that (1) contains a necessary and sufficient condition it is useful to show that the assumptions for g cannot be weakened. For example take g bounded and measurable (but not continuous), say

$$g(x) = \begin{cases} 0, & \text{if } x \le 0 \\ 1, & \text{if } x > 0. \end{cases}$$

Denote by F and F_n the d.f.s of the r.v.s $X \equiv 0$ and $X_n \equiv 1/n$ respectively. Then $F_n \xrightarrow{\text{d}} F$ and obviously $\int g \, \mathrm{d}F_n = 1$ for each $n \ge 1$ but $\int g \, \mathrm{d}F = 0$. Therefore (1) does not hold, as we of course expected.

Finally, recall that the integral relation (1) can be used as a definition of the weak convergence of d.f.s (see Example 14.9 and the topics discussed in Section 16).

14.2. Convergence in distribution does not imply convergence in probability

We show by a few specific examples that in general as $n \to \infty$,

$$X_n \xrightarrow{\text{d}} X \quad \not\Rightarrow \quad X_n \xrightarrow{\text{P}} X.$$

(i) Let X be a Bernoulli variable, that is X is a r.v. taking the values 1 and 0 with probability $\frac{1}{2}$ each. Let $\{X_n, n \ge 1\}$ be a sequence of r.v.s such that $X_n = X$ for any n. Since $X_n \overset{\text{d}}{=} X$ then $X_n \xrightarrow{\text{d}} X$ as $n \to \infty$. Now let $Y = 1 - X$. Thus $X_n \xrightarrow{\text{d}} Y$ because Y and X have the same distributions. However, X_n cannot converge to Y in any other mode since $|X_n - Y| = 1$ always. In particular,

$$\mathbf{P}[|X_n - Y| > \varepsilon] \not\to 0$$

for an arbitrary $\varepsilon \in (0, 1)$ and therefore $X_n \overset{\text{P}}{\not\to} Y$ as $n \to \infty$.

(ii) Let $\Omega = \{\omega_1, \omega_2, \omega_3, \omega_4\}$, $\mathcal{F}$ be the σ-field of all subsets of Ω and $\mathbf{P}$ the discrete uniform measure. Define the r.v.s X_n, $n \ge 1$, and X where

$$X_n(\omega_1) = X_n(\omega_2) = 1, \quad X_n(\omega_3) = X_n(\omega_4) = 0, \ n \ge 1,$$

$$X(\omega_1) = X(\omega_2) = 0, \quad X(\omega_3) = X(\omega_4) = 1.$$

Then $|X_n(\omega) - X(\omega)| = 1$ for all $\omega \in \Omega$ and $n \ge 1$. Hence as in case (i), X_n cannot converge in probability to X as $n \to \infty$. Further, if F and F_n, $n \ge 1$, are the d.f.s of X and X_n, $n \ge 1$, we have

$$F(x) = \begin{cases} 0, & \text{if } x \le 0 \\ \frac{1}{2}, & \text{if } 0 < x \le 1 \\ 1, & \text{if } x > 1, \end{cases} \qquad F_n(x) = \begin{cases} 0, & \text{if } x \le 0 \\ \frac{1}{2}, & \text{if } 0 < x \le 1 \\ 1, & \text{if } x > 1. \end{cases}$$

Thus $F_n(x) = F(x)$ for all $x \in \mathbb{R}^1$ and trivially $F_n(x) \to F(x)$ at each continuity point of F. Therefore $X_n \xrightarrow{\text{d}} X$ but, as was shown, $X_n \overset{\text{P}}{\not\to} X$.

(iii) Let X be any symmetric r.v., for example $X \sim \mathcal{N}(0,1)$, and put $X_n = -X$ for each $n \geq 1$. Then $X_n \stackrel{\mathrm{d}}{=} X$ and $X_n \xrightarrow{\mathrm{d}} X$. However, $X_n \stackrel{\mathrm{P}}{\not\to} X$ because for an arbitrary $\varepsilon > 0$ we have

$$\mathbf{P}[|X_n - X| > \varepsilon] = \mathbf{P}\left[|X| > \tfrac{1}{2}\varepsilon\right] = 2\left(1 - \Phi\left(\tfrac{1}{2}\varepsilon\right)\right) \not\to 0 \quad \text{as} \quad n \to \infty.$$

14.3. Sequences of random variables converging in probability but not almost surely

(i) Let $\Omega = [0,1]$, $\mathcal{F} = \mathcal{B}_{[0,1]}$ and $\mathbf{P}$ be the Lebesgue measure. For every number $n \in \mathbb{N}$ there is only one pair of integers, m and k, where $m \geq 0, 0 \leq k \leq 2^m - 1$, such that $n = 2^m + k$. Define the sequence of events $A_n = [k2^{-m}, (k+1)2^{-m})$ and put $X_n = X_n(\omega) = 1_{A_n}(\omega)$. Thus we obtain a sequence of r.v.s $\{X_n, n \geq 1\}$. Obviously

$$\mathbf{P}[|X_n| \geq \varepsilon] = \begin{cases} 2^{-m}, & \text{if } 0 \leq \varepsilon < 1 \\ 0, & \text{if } \varepsilon \geq 1. \end{cases}$$

Since $n \to \infty$ iff $m \to \infty$, we can conclude that

(1) $$X_n \xrightarrow{\mathrm{P}} 0 \quad \text{as } n \to \infty.$$

Now we want to see whether in (1) the convergence in probability can be replaced by almost sure convergence.

It is easy to show that for each fixed $\omega \in \Omega$ there are always infinitely many n for which $X_n(\omega) = 1$ and infinitely many n such that $X_n(\omega) = 0$. Indeed, $\omega \in [k2^{-m}, (k+1)2^{-m})$ for exactly one k where $k = 0, 1, \ldots, 2^m - 1$, that is $\omega \in A_{2^m+1}$. Obviously, if $k < 2^m - 1$ then also $\omega \in A_{2^m+k+1}$ and if $k = 2^m - 1$ (and $m \geq 1$) then also $\omega \in A_{2^{m+1}}$. In other words, $\omega \in A_n$ i.o., and also $\omega \in \Omega \backslash A_n$ i.o. which means that $\limsup_{n\to\infty} X_n = 1$ and $\liminf_{n\to\infty} X_n = 0$. Therefore

$$X_n \stackrel{\text{a.s.}}{\not\to} 0 \quad \text{as } n \to \infty.$$

(ii) Consider the sequence $\{X_n, n \geq 1\}$ of independent r.v.s where

$$\mathbf{P}[X_n = 1] = \frac{1}{n}, \quad \mathbf{P}[X_n = 0] = 1 - \frac{1}{n}, \quad n \geq 1.$$

Obviously for any ε, $0 < \varepsilon < 1$ we have

$$\mathbf{P}[|X_n - 0| > \varepsilon] = \mathbf{P}[X_n = 1] = \frac{1}{n} \to 0 \quad \text{as} \quad n \to \infty$$

and thus $X_n \xrightarrow{\mathrm{P}} 0$ as $n \to \infty$. It turns out, however, that the convergence $X_n \xrightarrow{\text{a.s.}} 0$ fails to hold. Let us analyse a more general situation. For given r.v.s X, X_n, $n \geq 1$, define the events

$$A_n(\varepsilon) = \{|X_n - X| > \varepsilon\}, \quad B_m(\varepsilon) = \cup_{n=m}^{\infty} A_n(\varepsilon).$$

Then

(2) $\quad X_n \xrightarrow{\text{a.s.}} X$ as $n \to \infty \iff \mathbf{P}(B_m(\varepsilon)) \to 0$ as $m \to \infty$ for all $\varepsilon > 0$.

Indeed, let

$$C = \{\omega \in \Omega : X_n(\omega) \to X(\omega) \text{ as } n \to \infty\}, \quad A(\varepsilon) = \{\omega \in \Omega : \omega \in A_n(\varepsilon) \text{ i.o.}\}.$$

Then $\mathbf{P}(C) = 1$ iff $\mathbf{P}(A(\varepsilon)) = 0$ for all $\varepsilon > 0$. However $\{B_m(\varepsilon)\}$ is a decreasing sequence of events, $B_m(\varepsilon) \downarrow A(\varepsilon)$ as $m \to \infty$ and so $\mathbf{P}(A(\varepsilon)) = 0$ iff $\mathbf{P}(B_m(\varepsilon)) \to 0$ as $m \to \infty$. Thus (2) is proved.

Using statement (2) for our specific sequence $\{X_n\}$ yields

$$\mathbf{P}(B_m(\varepsilon)) = 1 - \lim_{M \to \infty} \mathbf{P}[X_n = 0 \text{ for all } n \text{ such that } m \le n \le M].$$

By the independence of X_n,

$$\mathbf{P}(B_m(\varepsilon)) = 1 - \left(1 - \frac{1}{m}\right)\left(1 - \frac{1}{m+1}\right)\cdots$$

and since the product $\prod_{k=m}^{\infty}(1 - k^{-1})$ is zero for each $m \in \mathbb{N}$ we conclude that $\mathbf{P}(B_m(\varepsilon)) = 1$ for all m, that is $\mathbf{P}(B_m(\varepsilon))$ does not tend to zero as (2) indicates. Therefore the sequence $\{X_n\}$ does not converge almost surely.

14.4. On the Borel–Cantelli lemma and almost sure convergence

Let $\{X_n, n \ge 1\}$ be a sequence of r.v.s such that for each $\varepsilon > 0$,

(1) $$\sum_{n=1}^{\infty} \mathbf{P}[|X_n| > \varepsilon] < \infty.$$

According to the Borel–Cantelli lemma, if $\{A_n, n \ge 1\}$ is an arbitrary sequence of events and $\sum_{n=1}^{\infty} \mathbf{P}(A_n) < \infty$, then $\mathbf{P}[A_n \text{ i.o.}] = 0$ (see also Example 4.2). This lemma and condition (1) immediately imply that $X_n \xrightarrow{\text{a.s.}} 0$ as $n \to \infty$. Moreover, the same conclusion, $X_n \xrightarrow{\text{a.s.}} 0$ as $n \to \infty$, holds if for any sequence of numbers $\{\varepsilon_n\}$ with $\varepsilon_n \downarrow 0$, we have

(2) $$\sum_{n=1}^{\infty} \mathbf{P}[|X_n| > \varepsilon_n] < \infty.$$

We now want to clarify whether the converse of the statement is true. For this purpose consider the probability space $(\Omega, \mathcal{F}, \mathbf{P})$ where $\Omega = [0,1]$, $\mathcal{F} = \mathcal{B}_{[0,1]}$ and $\mathbf{P}$ is the Lebesgue measure. Define the sequence of r.v.s $\{X_n, n \geq 1\}$ by

$$X_n = X_n(\omega) = \begin{cases} 0, & \text{if } 0 \leq \omega \leq 1 - n^{-1} \\ 1, & \text{if } 1 - n^{-1} < \omega \leq 1. \end{cases}$$

Obviously $X_n \xrightarrow{\text{a.s.}} 0$ as $n \to \infty$. However, for any $\varepsilon_n > 0$ with $\varepsilon_n \downarrow 0$ we have $\mathbf{P}[|X_n| > \varepsilon_n] = \mathbf{P}[X_n = 1] = n^{-1}$ for sufficiently large n. Thus

$$\sum_{n=1}^{\infty} \mathbf{P}[|X_n| > \varepsilon_n] = \infty.$$

Therefore condition (2) is sufficient but not necessary for the almost sure convergence of X_n to zero.

14.5. On the convergence of sequences of random variables in L^r-sense for different values of r

Suppose X and X_n, $n \geq 1$ are r.v.s in L^r for some fixed r, $r > 0$. Then $X, X_n \in \mathrm{L}^s$ for each s, $0 < s < r$. This follows from the well known Lyapunov inequality (see Feller 1971; Shiryaev 1995), $(\mathbf{E}[|X|^s])^{1/s} \leq (\mathbf{E}[|X|^r])^{1/r}$, $0 < s < r$, or from the elementary inequality $|x|^s \leq 1 + |x|^r$, $x \in \mathbb{R}^1$, $0 < s < r$ (used once before in Example 6.5). In other words

$$X_n \xrightarrow{\mathrm{L}^r} X \Rightarrow X_n \xrightarrow{\mathrm{L}^s} X \quad \text{for} \quad 0 < s < r.$$

Let us illustrate the fact that in general the converse is not true. Consider the sequence of r.v.s $\{X_n, n \geq 1\}$ where

$$\mathbf{P}[X_n = n] = n^{-(r+s)/2} = 1 - \mathbf{P}[X_n = 0], \quad n \geq 1,\ 0 < s < r.$$

Then we find

$$\mathbf{E}[X_n^s] = n^{(s-r)/2} \to 0 \quad \text{as} \quad n \to \infty$$

which implies that $X_n \xrightarrow{\mathrm{L}^r} 0$ as $n \to \infty$. However,

$$\mathbf{E}[X_n^r] = n^{(r-s)/2} \to \infty \quad \text{as} \quad n \to \infty$$

and therefore $X_n \xrightarrow{\mathrm{L}^s} 0 \not\Rightarrow X_n \xrightarrow{\mathrm{L}^r} 0$ for all $r > s$.

14.6. Sequences of random variables converging in probability but not in L^r-sense

(i) Let $\{X_n, n \geq 1\}$ be r.v.s such that

$$\mathbf{P}[X_n = \mathrm{e}^n] = \frac{1}{n}, \quad \mathbf{P}[X_n = 0] = 1 - \frac{1}{n}, \quad n \geq 1.$$

Then for any $\varepsilon > 0$ we have

$$\mathbf{P}[|X_n| < \varepsilon] = \mathbf{P}[X_n = 0] = 1 - \frac{1}{n} \to 1 \quad \text{as } n \to \infty$$

and hence $X_n \xrightarrow{\mathrm{P}} 0$ as $n \to \infty$. However, for each $r > 0$,

$$\mathbf{E}[X_n^r] = \mathrm{e}^{rn}\frac{1}{n} \to \infty \quad \text{as } n \to \infty$$

and therefore $X_n \overset{\mathrm{L}^r}{\not\to} 0$ as $n \to \infty$.

(ii) Consider the sequence $\{X_n, n \geq 1\}$ where X_n has a density

$$f_n(x) = \frac{n}{\pi(1 + n^2x^2)}, \quad x \in \mathbb{R}^1, \quad n \geq 1$$

(that is, X_n has a Cauchy distribution). Since for any $\varepsilon > 0$,

$$\mathbf{P}[|X_n| < \varepsilon] = \int_{-\varepsilon}^{\varepsilon} f_n(x)\,\mathrm{d}x = \frac{2}{\pi}\arctan(n\varepsilon) \to 1 \quad \text{as } n \to \infty$$

we conclude that $X_n \xrightarrow{\mathrm{P}} 0$ as $n \to \infty$. But for any $r \geq 1$, $\mathbf{E}[|X_n|^r] = \infty$ and thus the sequence $\{X_n\}$ cannot converge to zero as $n \to \infty$ in L^r-sense for $r \geq 1$. Moreover, since X_n, $n \geq 1$, do not belong to the space L^r it is not sensible to speak about L^r-convergence.

(iii) Define the sequences $\{Y_n, n \geq 2\}$ and $\{Z_n, n \geq 1]$ as follows:

$$\mathbf{P}[Y_n = 1] = 1/\log n = 1 - \mathbf{P}[Y_n = 0],$$

$$\mathbf{P}[Z_n = 0] = 1 - n^{-\alpha}, \quad \mathbf{P}[Z_n = \pm n] = 1/(2n^{\alpha}), \quad 0 < \alpha \leq 2.$$

Then, as $n \to \infty$, $Y_n \xrightarrow{\mathrm{P}} 0$ but $Y_n \overset{\mathrm{L}^r}{\not\to} 0$ for any $r > 0$; $Z_n \xrightarrow{\mathrm{P}} 0$ but $Z_n \overset{\mathrm{L}^2}{\not\to} 0$.

14.7. Convergence in L^r-sense does not imply almost sure convergence

(i) Consider again the sequence $\{X_n, n \geq 1\}$ defined in Example 14.3(i): namely, $X_n = X_n(\omega) = 1_{A_n}(\omega)$ for $n = 2^m + k$ and $A_n = [k2^{-m}, (k+1)2^{-m})$. Since $\omega \in \Omega = [0, 1]$ and $\mathbf{P}$ is the Lebesgue measure then $\mathbf{E}[|X_n|] = \mathbf{E}[X_n] = 2^{-m} \to 0$ as $n \to \infty$. Thus

$$X_n \xrightarrow{\mathrm{L}^1} 0 \quad \text{as } n \to \infty.$$

Nevertheless, as was shown in Example 14.3(i), $X_n \overset{\text{a.s.}}{\not\to} 0$ as $n \to \infty$.

(ii) Let $\{X_n, n \geq 1\}$ be a sequence of independent r.v.s defined by

$$\mathbf{P}[X_n = n^{1/(2r)}] = 1/n, \quad \mathbf{P}[X_n = 0] = 1 - 1/n, \quad n \geq 1$$

where $r > 0$ is arbitrary. It is easy to see that

$$\mathbf{E}[|X_n|^r] = \mathbf{E}[X_n^r] = (n^{1/(2r)})^r n^{-1} = n^{-1/2} \to 0 \quad \text{as } n \to \infty.$$

Therefore for any $r > 0$, $X_n \xrightarrow{L^r} 0$ as $n \to \infty$.

Let $M < N$ be positive integers. Since X_n are independent, we find

$$\mathbf{P}[\text{all } X_n = 0 \quad \text{for } M \le n \le N] = \prod_{n=M}^{N} (1 - 1/n).$$

The continuity of $\mathbf{P}$ ($B_N \downarrow B_0 \Rightarrow \mathbf{P}(B_N) \downarrow \mathbf{P}(B_0)$) implies that

$$\mathbf{P}[\cap_{n=M}^{\infty}(\omega : X_n(\omega) \le \varepsilon)] = \lim_{N\to\infty} \prod_{n=M}^{N} (1 - 1/n)$$

for arbitrary $\varepsilon > 0$ and integer M. Separately we can check that for arbitrary M, $\prod_{n=2}^{\infty}(1 - 1/n) = 0$ and $\prod_{n=M}^{\infty}(1 - 1/n) = 0$. Thus

$$\mathbf{P}[\cap_{n=M}^{\infty}(\omega : X_n(\omega) \le \varepsilon)] = 0.$$

Since the r.v.s X_n are non-negative this relation means that the sequence $\{X_n\}$ cannot converge almost surely.

(iii) Let $\{Y_n, n \ge 1]$ be a sequence of independent r.v.s given by

$$\mathbf{P}[Y_n = 0] = 1 - 1/n^{1/4}, \quad \mathbf{P}[Y_n = \pm 1] = 1/(2n^{1/4}), \quad n \ge 1.$$

Then it can be shown that $Y_n \xrightarrow{L^2} 0$ but $Y_n \overset{\text{a.s.}}{\not\to} 0$ as $n \to \infty$.

(iv) Let $\{S_n, n \ge 1\}$ be a symmetric Bernoulli walk, that is $S_n = \xi_1 + \ldots + \xi_n$ where ξ_j are i.i.d. r.v.s each taking the values $(+1)$ or (-1) with probability $\frac{1}{2}$. Define $X_n = X_n(\omega) = 1_{[S_n=0]}(\omega)$, $n \ge 1$. Then for every $r > 0$ we have

$$\lim_{n\to\infty} \mathbf{E}[X_n^r] = 0.$$

Thus $X_n \xrightarrow{L^r} 0$ as $n \to \infty$ for $r > 0$. However, the symmetric random walk $\{S_n\}$ is recurrent in the sense that S_n crosses the level zero for infinitely many values of n (for details see Feller 1968; Chung 1974; Shiryaev 1995). This means that $X_n = 1$ i.o. and therefore $X_n \overset{\text{a.s.}}{\not\to} 0$ as $n \to \infty$.

14.8. Almost sure convergence does not necessarily imply convergence in L^r-sense

(i) Define the sequence of r.v.s $\{X_n, n \ge 1\}$ as follows:

$$\mathbf{P}[X_n = 0] = 1 - 1/n^{\alpha}, \quad \mathbf{P}[X_n = n] = \mathbf{P}[X_n = -n] = 1/(2n^{\alpha}), \quad \alpha > 0.$$

Since $\mathbf{E}[|X_n|^{1/2}] = 1/n^{\alpha-1/2}$ we find that $\sum_{n=1}^{\infty} \mathbf{E}[|X_n|^{1/2}] < \infty$ for any $\alpha > \frac{3}{2}$. According to the Markov inequality we have $\mathbf{P}[|X_n| > \varepsilon] \le \varepsilon^{-1/2}\mathbf{E}[|X_n|^{1/2}]$ and hence $\sum_{n=1}^{\infty} \mathbf{P}[|X_n| > \varepsilon] < \infty$ for every $\varepsilon > 0$. Using the Borel–Cantelli lemma as in Example 14.4 we conclude that $X_n \overset{\text{a.s.}}{\not\to} 0$ as $n \to \infty$.

Further, $\mathbf{E}[|X_n|^2] = 1/n^{\alpha-2}$ and hence for any $\alpha \le 2$, $X_n \overset{\text{L}^2}{\not\to} 0$ as $n \to \infty$.

Therefore, if $\alpha \in [\frac{3}{2}, 2]$, then $X_n \xrightarrow{\text{a.s.}} 0$ but $X_n \overset{\text{L}^2}{\not\to} 0$ as $n \to \infty$.

(ii) Let $\{Y_n, n \ge 1\}$ be a sequence of r.v.s where Y_n takes the values e^n and 0, with probability n^{-2} and $1 - n^{-2}$ respectively. Since for any $\varepsilon > 0$, $\mathbf{P}[|X_n| > \varepsilon] = \mathbf{P}[X_n > \varepsilon] = \mathbf{P}[X_n = \mathrm{e}^n] = n^{-2}$ and

$$\sum_{n=1}^{\infty} \mathbf{P}[|X_n| > \varepsilon] = \sum_{n=1}^{\infty} \frac{1}{n^2} < \infty$$

we conclude as in case (i) above that $X_n \xrightarrow{\text{a.s.}} 0$ as $n \to \infty$. Obviously,

$$\mathbf{E}[|X_n|^r] = \mathbf{E}[X_n^r] = \mathrm{e}^{nr}/n^2 \to \infty \quad \text{as } n \to \infty$$

for any $r > 0$. Therefore, as $n \to \infty$, $X_n \xrightarrow{\text{a.s.}} 0$ but $X_n \overset{\text{L}^r}{\not\to} 0$ for all $r > 0$.

14.9. Weak convergence of the distribution functions does not imply convergence of the densities

Let F, F_n, $n \ge 1$ be d.f.s such that their densities f, f_n, $n \ge 1$ exist. According to the well known Scheffé theorem (see Billingsley 1968), if $f_n(x) \to f(x)$ as $n \to \infty$ for almost all $x \in \mathbb{R}^1$ then $F_n \xrightarrow{\text{d}} F$ as $n \to \infty$. It is natural to ask whether or not the converse is true. The example below shows that in general the answer is negative. Consider the function

$$F_n(x) = \begin{cases} 0, & \text{if } x \le 0 \\ x\left(1 - \dfrac{\sin 2n\pi x}{2n\pi x}\right), & \text{if } 0 < x \le 1 \\ 1, & \text{if } x \ge 1. \end{cases}$$

Then F_n is an absolutely continuous d.f. with density

$$f_n(x) = \begin{cases} 1 - \cos 2n\pi x, & \text{if } x \in [0, 1] \\ 0, & \text{otherwise.} \end{cases}$$

Also introduce the functions

$$F(x) = \begin{cases} 0, & \text{if } x \le 0 \\ x, & \text{if } 0 < x \le 1 \\ 1, & \text{if } x > 1, \end{cases} \qquad f(x) = \begin{cases} 1, & \text{if } x \in (0, 1] \\ 0, & \text{otherwise.} \end{cases}$$

Obviously F and f are the d.f. and the density corresponding to a uniform distribution on the interval (0, 1] respectively. It is easy to see that

$$F_n(x) \to F(x) \quad \text{as} \quad n \to \infty \text{ for all } x \in \mathbb{R}^1.$$

However,

$$f_n(x) \not\to f(x) \quad \text{as} \quad n \to \infty.$$

Therefore we have established that in general $F_n \xrightarrow{\text{d}} F \not\Rightarrow f_n \to f$.

14.10. The convergence $X_n \xrightarrow{\text{d}} X$ and $Y_n \xrightarrow{\text{d}} Y$ does not always imply that $X_n + Y_n \xrightarrow{\text{d}} X + Y$

Let X, X_n, $n \geq 1$ and Y, Y_n, $n \geq 1$ be r.v.s defined on the same probability space. Suppose $X_n \xrightarrow{\text{d}} X$ and $Y_n \xrightarrow{\text{d}} Y$ as $n \to \infty$. Does it follow from this that $X_n + Y_n \xrightarrow{\text{d}} X + Y$ as $n \to \infty$? There are cases when the answer to this question is positive, for example if X_n and Y_n, $n \geq 1$ are independent, or if the joint distribution of X_n, Y_n converges to that of X, Y (see Grimmett and Stirzaker 1982). The examples below aim to show that the answer is negative if we drop the independence condition.

(i) Let $\{X_n, n \geq 1\}$ be i.i.d. r.v.s such that $X_n = 1$ or 0 with probability $\frac{1}{2}$ each and put $Y_n = 1 - X_n$. Then $X_n \xrightarrow{\text{d}} X$ and $Y_n \xrightarrow{\text{d}} Y$ as $n \to \infty$ where each of X and Y takes the values 1 and 0 with equal probabilities. Further, since $X_n + Y_n = 1$, it is obvious that $X_n + Y_n$ does not tend in distribution to the sum $X + Y$ which is a r.v. with three possible values, 0, 1 and 2, with probabilities $\frac{1}{4}$, $\frac{1}{2}$ and $\frac{1}{4}$ respectively.

(ii) Suppose now the sequences of r.v.s $\{X_n, n \geq 1\}$ and $\{Y_n, n \geq 1\}$ are such that $X_n \xrightarrow{\text{d}} X$ and $Y_n \xrightarrow{\text{d}} Y$ where $X \sim \mathcal{N}(0,1)$, $Y \sim \mathcal{N}(0,1)$. If for each n, X_n and Y_n are independent, then $X_n + Y_n \xrightarrow{\text{d}} Z$ with $Z \sim \mathcal{N}(0,2)$. Moreover, in this case the distribution of (X_n, Y_n) converges to the standard two-dimensional normal distribution with zero mean and covariance matrix $\begin{pmatrix} 1 & 0 \\ 0 & 1 \end{pmatrix}$.

Let us now drop the condition that X_n and Y_n are independent. Again take $\{X_n, n \geq 1\}$ such that $X_n \xrightarrow{\text{d}} X$ with $X \sim \mathcal{N}(0,1)$ and let $Y_n = X_n$ for all $n \in \mathbb{N}$. Then $Y_n \xrightarrow{\text{d}} Y$ where $Y \sim \mathcal{N}(0,1)$. Obviously the sum $X_n + Y_n = 2X_n$ and it converges in distribution to a r.v. $\tilde{Z}$ where $\tilde{Z} \sim \mathcal{N}(0,4)$ but not to a r.v. distributed $\mathcal{N}(0,2)$ as expected.

14.11. The convergence in probability $X_n \xrightarrow{\text{P}} X$ does not always imply that $g(X_n) \xrightarrow{\text{P}} g(X)$ for any function g

The following result is well known and is used in many probabilistic problems (see Feller 1971; Billingsley 1995; Serfling 1980).

If X, X_n, $n \geq 1$ are r.v.s such that $X_n \xrightarrow{\mathrm{P}} X$ as $n \to 0$ and $g(x)$, $x \in \mathbb{R}^1$ is a continuous function, then $g(X_n) \xrightarrow{\mathrm{P}} g(X)$ as $n \to \infty$.

By a specific example we show that the continuity of g is an essential condition in the sense that it cannot be replaced by measurability only. To see this, consider the function

$$g(x) = \begin{cases} 0, & \text{if } x \leq 0 \\ 1, & \text{if } x > 0. \end{cases}$$

The sequence $\{X_n, n \geq 1\}$ can be taken arbitrarily but so as to satisfy the properties $X_n > 0$ for all $n \in \mathbb{N}$ and $X_n \xrightarrow{\mathrm{P}} 0$ as $n \to \infty$. For example, let X_n take the values 1 and n^{-1} with probabilities n^{-1} and $1 - n^{-1}$ respectively. Then obviously $X_n \xrightarrow{\mathrm{P}} X$ where $X = 0$ a.s. Moreover, for each n we have $g(X_n) = 1$. However, $g(X) = 0$ and hence $g(X_n)$ cannot converge in any reasonable sense to $g(X)$. In particular, $g(X_n) \stackrel{\mathrm{P}}{\not\to} g(X)$ as $n \to \infty$ despite the fact that $X_n \xrightarrow{\mathrm{P}} X$.

We come to the same conclusion by considering the function g defined above and the sequence of r.v.s $\{X_n, n \geq 1\}$ where $X_n \sim \mathcal{N}(0, \sigma^2/n)$, $\sigma^2 > 0$. Obviously $X_n \xrightarrow{\mathrm{P}} X$ as $n \to \infty$ with $X = 0$ a.s. Since X_n is symmetric, we have for each n,

$$g(X_n) = \begin{cases} 0, & \text{with probability } \frac{1}{2} \\ 1, & \text{with probability } \frac{1}{2}. \end{cases}$$

However, $g(X) = 0$ a.s. and hence $g(X_n) \stackrel{\mathrm{P}}{\not\to} g(X)$ as $n \to \infty$.

14.12. Convergence in variation implies convergence in distribution but the converse is not always true

Let X and Y be discrete r.v.s such that

$$\mathbf{P}[X = a_k] = p_k, \quad \mathbf{P}[Y = a_k] = q_k$$

where

$$a_k \in \mathbb{R}^1, \ p_k \geq 0, \ q_k \geq 0, \ k = 1, 2, \ldots, \quad \textstyle\sum_k p_k = 1, \quad \sum_k q_k = 1.$$

If F and G are the d.f.s of X and Y respectively, then the *distance in variation*, $v(F, G)$, is defined by

$$v(F, G) = \textstyle\sum_k |p_k - q_k|. \tag{1}$$

If X and Y are absolutely continuous r.v.s whose d.f.s and densities are F, G and f, g, then $v(F, G)$ is defined by

$$v(F, G) = \int_{-\infty}^{\infty} |f(x) - g(x)| \, \mathrm{d}x. \tag{2}$$

Suppose F, F_n, $n \geq 1$, are the d.f.s of the r.v.s X, X_n, $n \geq 1$, respectively. If $v(F_n, F) \to 0$ as $n \to \infty$ we write $F_n \xrightarrow{\text{v}} F$ and also $X_n \xrightarrow{\text{v}} X$ and say that the sequence $\{X_n\}$ *converges in variation* to X as $n \to \infty$. It is easy to see that convergence in variation implies convergence in distribution, that is, $F_n \xrightarrow{\text{v}} F \Rightarrow F_n \xrightarrow{\text{d}} F$. However, as we shall now see, the converse is not true.

(i) Let F_n be the d.f. of a r.v. X_n concentrated at the point $1/n$. Then $F_n \xrightarrow{\text{d}} F_0$ as $n \to \infty$ where F_0 is the d.f. of the r.v. $X_0 \equiv 0$, while the quantity $v(F_n, F_0)$ calculated by (1) does not tend to zero as $n \to \infty$.

(ii) Let $F(x)$, $x \in \mathbb{R}^1$ be a d.f. with density $f(x)$, $x \in \mathbb{R}^1$. Our goal is to construct a sequence of d.f.s $\{F_n, n \geq 1\}$ such that $F_n \xrightarrow{\text{w}} F$ but $F_n \not\xrightarrow{\text{v}} F$ as $n \to \infty$. Denote

$$I_n = \int_{-\infty}^{\infty} f(x) \cos^2 nx \,\mathrm{d}x, \quad J_n = \int_{-\infty}^{\infty} f(x) \sin^2 nx \,\mathrm{d}x, \quad n \geq 1.$$

The obvious identity $I_n + J_n = \int_{-\infty}^{\infty} f(x)\,\mathrm{d}x = 1$ implies that the numerical sequences $\{I_n, n \geq 1\}$ and $\{J_n, n \geq 1\}$ cannot simultaneously tend to zero as $n \to \infty$. Thus, we can assume that e.g. $I_n \not\to 0$ as $n \to \infty$. In such a case we introduce the function

$$f_n(x) = c_n f(x)(1 + \cos nx), \quad x \in \mathbb{R}^1$$

where $c_n^{-1} = \int_{-\infty}^{\infty} f(x)(1+\cos nx)\,\mathrm{d}x$. Then for each n the function f_n is a density and let F_n be the corresponding d.f. Let us try to find the limit of the sequence $\{F_n\}$ as $n \to \infty$. Since f is a density, then the well known Riemann–Lebesgue lemma (see e.g. Kolmogorov and Fomin 1970)

$$\int_B f(x) \cos nx \,\mathrm{d}x \to 0 \quad \text{as } n \to \infty$$

holds for any Borel set $B \in \mathcal{B}^1$. Hence

$$\int_B f_n(x)\,\mathrm{d}x \to \int_B f(x)\,\mathrm{d}x, \quad \text{as} \quad n \to \infty \quad \Rightarrow \quad F_n \xrightarrow{\text{w}} F \quad \text{as} \quad n \to \infty.$$

Let us now calculate the distance in variation $v(F_n, F)$. We have

$$\begin{aligned} v(F_n, F) &= \textstyle\int_{-\infty}^{\infty} |f_n(x) - f(x)|\,\mathrm{d}x \\ &= \textstyle\int_{-\infty}^{\infty} |c_n f(x) \cos nx - (1 - c_n) f(x)|\,\mathrm{d}x \\ &\geq |\textstyle\int_{-\infty}^{\infty} |c_n f(x) \cos nx|\,\mathrm{d}x - \int_{-\infty}^{\infty} |(1 - c_n) f(x)|\,\mathrm{d}x|. \end{aligned}$$

We find that $c_n \to 1$ as $n \to \infty$ and

$$\int_{-\infty}^{\infty} f(x) |\cos nx|\,\mathrm{d}x \geq \int_{-\infty}^{\infty} f(x) \cos^2 nx\,\mathrm{d}x = I_n \not\to 0.$$

Therefore $v(F_n, F) \not\to 0$ as $n \to \infty$, i.e. the sequence of d.f.s $\{F_n\}$ does not converge in variation to F despite the weak convergence established above.

14.13. There is no metric corresponding to almost sure convergence

It is well known that each of the following kinds of convergence: (i) in distribution; (ii) in probability; (iii) in L^r-sense, can be metrized (see Ash and Gardner 1975; Dudley 1976). It is therefore natural to ask whether almost sure convergence can be metrized. Let us show that the answer is negative.

Let $\mathcal{R}$ denote a set of r.v.s defined on the probability space $(\Omega, \mathcal{F}, \mathbf{P})$ and $d : \mathcal{R} \times \mathcal{R} \to \mathbb{R}^+$ a metric on $\mathcal{R}$, that is, d is non-negative, symmetric and satisfies the triangle inequality. Let us check the correctness of the following statement:

$$\text{For } X,\ X_1,\ X_2, \ldots \in \mathcal{R},\quad d(X_n, X) \to 0 \quad \text{iff} \quad X_n \xrightarrow{\text{a.s.}} X.$$

Suppose such a function d does exist. Let $\{X_n, n \geq 1\}$ be a sequence of r.v.s converging to some r.v. X in probability but not almost surely (see Example 14.3). Then for some $\delta > 0$ the inequality $d(X_n, X) \geq \delta$ will be satisfied for infinitely many n. Let Λ denote the set of these n. However, since $X_n \xrightarrow{\text{P}} X$ there exists a subsequence $\{X_{n_k}, n_k \in \Lambda\}$ of $\{X_n, n \in \Lambda\}$ converging to X almost surely. But this would mean that $d(X_{n_k}, X) \to 0$ as $n_k \to \infty$, which is impossible because $d(X_{n_k}, X) \geq \delta$ for each $n_k \in \Lambda$.

Thus the statement given above is incorrect and we conclude that a.s. convergence is not metrizable. Note, however, that this type of convergence can be metrized iff the probability space is atomic (see Thomasian 1957; Tomkins 1975a).

14.14. Complete convergence of sequences of random variables is stronger than almost sure convergence

The sequence of r.v.s $\{X_n, n \geq 1\}$ is called *completely convergent* to 0 if

$$\text{(1)} \qquad \lim_{n\to\infty} \sum_{m=n}^{\infty} \mathbf{P}[|X_m| > \varepsilon] = 0 \quad \text{for every} \quad \varepsilon > 0.$$

In this case the following notation is used: $X_n \xrightarrow{\text{c}} 0$.

In order to compare this mode of convergence with a.s. convergence, recall that

$$\text{(2)} \qquad X_n \xrightarrow{\text{a.s.}} 0 \iff \lim_{n\to\infty} \mathbf{P}[\cup_{m=n}^{\infty}\{|X_m| > \varepsilon\}] = 0.$$

Since the probability $\mathbf{P}$ is semi-additive, we obtain

$$\mathbf{P}[\cup_{m=n}^{\infty}\{|X_m| > \varepsilon\}] \leq \textstyle\sum_{m=n}^{\infty} \mathbf{P}[|X_m| > \varepsilon]$$

which immediately implies that $X_n \xrightarrow{\text{c}} 0 \Rightarrow X_n \xrightarrow{\text{a.s.}} 0$. However, the converse is not always true. To see this, consider the probability space $(\Omega, \mathcal{F}, \mathbf{P})$ where $\Omega = [0, 1]$, $\mathcal{F} = \mathcal{B}_{[0,1]}$ and $\mathbf{P}$ is the Lebesgue measure. Take the sequence $\{X_n, n \geq 1\}$ where

$$X_n = X_n(\omega) = \begin{cases} 1, & \text{if } 0 \leq \omega < \frac{1}{n} \\ 0, & \text{if } \frac{1}{n} \leq \omega \leq 1. \end{cases}$$

Then clearly this sequence converges to zero almost surely but not completely.

These two kinds of convergence are equivalent if the r.v.s X_n, $n \geq 1$, are independent (Hsu and Robbins 1947).

14.15. The almost sure uniform convergence of a random sequence implies its complete convergence, but the converse is not true

Recall that the sequence of r.v.s $\{X_n, n \geq 1\}$ is said to *converge almost surely uniformly* to a r.v. X if there exists a set $A \in \mathcal{F}$ with $\mathbf{P}(A) = 0$ such that $X_n = X_n(\omega)$ converge uniformly (in ω) to X on the complement A^c. Note that almost sure uniform convergence implies complete convergence discussed in Example 14.14 (see Lukacs 1975). Thus we come to the question of the validity of the converse statement. To find the answer we consider an example.

Let the probability space $(\Omega, \mathcal{F}, \mathbf{P})$ be given by $\Omega = [0, 1]$, $\mathcal{F} = \mathcal{B}_{[0,1]}$ and $\mathbf{P}$ is the Lebesgue measure. Consider the sequence $\{X_n, n \geq 1\}$ of r.v.s such that

$$X_n = X_n(\omega) = \begin{cases} 1 - 2n^2\omega, & \text{if } 0 \leq \omega \leq 1/(2n^2) \\ 0, & \text{if } 1/(2n^2) < \omega \leq 1 - 1/(2n^2) \\ 1 - 2n^2 + 2n^2\omega, & \text{if } 1 - 1/(2n^2) < \omega \leq 1. \end{cases}$$

For arbitrary $\varepsilon > 0$, $0 \leq X_n < \varepsilon$ iff $\omega \in ((1-\varepsilon)/(2n^2), 1 - (1-\varepsilon)/(2n^2))$. Hence

$$\mathbf{P}[|X_n| \geq \varepsilon] = \mathbf{P}[X_n \geq \varepsilon] = (1-\varepsilon)/n^2$$

so that

$$\sum_{n=1}^{\infty} \mathbf{P}[|X_n| \geq \varepsilon] = (1-\varepsilon) \sum_{n=1}^{\infty} (1/n^2) < \infty.$$

This means that the sequence $\{X_n\}$ converges completely to zero.

Now let us introduce the sets $B_n = [0, \frac{1}{4}n^2) \cup (1 - \frac{1}{4}n^2, 1]$, $n \geq 1$. Clearly $\mathbf{P}(B_n) = 1/(2n^2)$. Suppose for some set A with $\mathbf{P}(A) = 0$, X_n converges to zero almost surely uniformly on A^c. Then there exists a number $n_\varepsilon \in \mathbb{N}$ (independent of ω) such that $|X_n| \leq \varepsilon < \frac{1}{2}$ on A^c provided $n \geq n_\varepsilon$. However, we have $B_n \cap A^c = \emptyset$ and $B_{n_\varepsilon} \subset A$. Hence $\mathbf{P}(A) \geq \mathbf{P}(B_{n_\varepsilon}) = \frac{1}{2}n_\varepsilon^{-2}$. This contradiction shows that the sequence $\{X_n\}$ defined above does not converge almost surely uniformly to zero.

14.16. Converging sequences of random variables such that the sequences of the expectations do not converge

If the sequence of r.v.s $\{X_n\}$ converges in probability or a.s. to some r.v. X, then under additional assumptions we can show that the sequence of the expectations $\{\mathbf{E}X_n\}$ will tend to $\mathbf{E}X$. However, in general such a statement is not true without appropriate assumptions.

(i) Let $\{X_n, n \geq 1\}$ be r.v.s defined by

$$\mathbf{P}[X_n = -n - 4] = 1/(n+4), \; \mathbf{P}[X_n = -1] = 1 - 4/(n+4),$$
$$\mathbf{P}[X_n = n + 4] = 3/(n+4).$$

Obviously for any $\varepsilon > 0$ we have

$$\mathbf{P}[|X_n - (-1)| > \varepsilon] = 4/(n+4)$$

and hence $X_n \xrightarrow{\mathrm{P}} (-1)$ as $n \to \infty$. On the other hand,

$$\mathbf{E}X_n = 1 + 4/(n+1) \quad \text{and} \quad \lim_{n\to\infty} \mathbf{E}X_n = 1.$$

Therefore

$$\lim_{n\to\infty} \mathbf{E}X_n = 1 \neq -1 = \mathbf{E}\left[\mathrm{P} - \lim_{n\to\infty} X_n\right]$$

and the convergence in probability of X_n to X is not sufficient to ensure that $\mathbf{E}X_n \to \mathbf{E}X$. This can be explained by referring to the standard result (see Lukacs 1975; Chow and Teicher 1978): if $X_n, n \geq 1$ and X are L^r r.v.s and $X_n \xrightarrow{\mathrm{L}^r} X$ then $\mathbf{E}[|X_n|^k] \to \mathbf{E}[|X|^k]$ for each $0 < k \leq r$.

(ii) Consider the sequence $\{Y_n, n \geq 1\}$ of r.v.s where

$$Y_n(\omega) = \begin{cases} n^2, & \text{if } 0 \leq \omega \leq n^{-1} \\ 0, & \text{if } n^{-1} < \omega \leq 1 \end{cases}$$

and also the r.v. $Y(\omega) \equiv 0$, $\omega \in [0,1]$. Then for every $\omega \in [0,1]$ we have $Y_n(\omega) \to Y(\omega)$ as $n \to \infty$. However, $\mathbf{E}Y_n = n$ and $\mathbf{E}Y_n \not\to \mathbf{E}Y = 0$ as $n \to \infty$.

Let us note that in case (ii) $\mathbf{E}Y_n$ is unbounded, while in case (i) $\mathbf{E}X_n$ is bounded. According to Billingsley (1995), if $\{Z_n\}$ is uniformly bounded and $\lim_{n\to\infty} Z_n = Z$ on a set of probability 1, then $\mathbf{E}Z = \lim_{n\to\infty} \mathbf{E}Z_n$. Both cases (i) and (ii) show that uniform boundedness is essential.

14.17. Weak L^1-convergence of random variables is weaker than both weak convergence and convergence in L^1-sense

Recall that the sequence $\{X_n, n \geq 1\}$ of r.v.s in the space L^1 is said to *converge weakly in* L^1 to the r.v. X iff for any bounded r.v. Y we have

$$\lim_{n\to\infty} \mathbf{E}[X_n Y] = \mathbf{E}[XY]. \tag{1}$$

In this case the following notation is used: $X_n \xrightarrow{\mathrm{w,L^1}} X$ as $n \to \infty$.

Clearly the limit X belongs to L^1 and it is unique up to equivalence (see Neveu 1965; Chung 1974).

It is of general interest to clarify the connection between this mode of convergence and the others discussed in the previous examples. In particular, if $X_n \xrightarrow{\text{w,L}^1} X$, does it follow that $X_n \xrightarrow{\text{w}} X$ or that $X_n \xrightarrow{\text{L}^1} X$?

Remark. Here the notation $\xrightarrow{\text{w}}$ is used to denote the so-called weak convergence of X_n to X as $n \to \infty$ which in this case is equivalent to convergence in distribution. In a more general context weak convergence will be considered in Section 16.

To answer these questions consider the probability space $(\Omega, \mathcal{F}, \mathbf{P})$ where $\Omega = [0,1]$, $\mathcal{F} = \mathcal{B}_{[0,1]}$ and **P** is the Lebesgue measure, and take the following sequence of r.v.s $X_n(\omega) = \sin 2\pi n\omega$, $n \geq 1$. Note that $\{X_n\}$ is not convergent in either sense—weak or L^1-sense. Nevertheless we shall show that $X_n \xrightarrow{\text{w,L}^1} 0$ in the sense of definition (1).

Let Y be any bounded r.v., that is $Y = Y(\omega)$, $\omega \in [0,1]$ is an $\mathcal{F}$-measurable function. Then there is a sequence of stepwise functions $\{Y^{(m)}(\omega), m \geq 1\}$ such that $Y^{(m)} \xrightarrow{\text{a.s.}} Y$ as $m \to \infty$ (see Loève 1978). By the Egorov theorem (see Kolmogorov and Fomin 1970; Royden 1968) for any $\varepsilon > 0$ we can find an open set $\Lambda_\varepsilon \subset [0,1]$ such that the convergence $Y^{(m)} \to Y$ as $m \to \infty$ is uniform for $\omega \in \Lambda_\varepsilon^c = [0,1] \backslash \Lambda_\varepsilon$. Here we can also use the Lusin theorem (see Kolmogorov and Fomin 1970) on the existence of a continuous function Y^* coinciding with Y on the complement of a set of ε-measure. In both cases, for stepwise or continuous Y, we have

$$\mathbf{E}[X_n Y^*] = \int_0^1 Y^*(\omega)\ \sin 2\pi n\omega \, d\omega \to 0 \quad \text{as} \quad n \to \infty.$$

Since Y and Y^* are bounded and Y^* is close to Y, the difference $|\mathbf{E}[X_n Y^*] - \mathbf{E}[X_n Y]|$ can be made arbitrarily small. Hence $\mathbf{E}[X_n Y] \to 0$ as $n \to \infty$ for any bounded r.v. Y. Therefore $X_n \xrightarrow{\text{w,L}^1} 0$ as $n \to \infty$. However, as noted above, neither of the relations $X_n \xrightarrow{\text{w}} 0$ or $X_n \xrightarrow{\text{L}^1} 0$ is true.

14.18. A converging sequence of random variables whose Cesàro means do not converge

Let $\{X_n, n \geq 1\}$ be a sequence of r.v.s. Then the following implication holds:

$$(1) \qquad X_n \xrightarrow{\text{a.s.}} 0 \quad \text{as} \quad n \to \infty \Rightarrow \frac{1}{n}(X_1 + \ldots + X_n) \xrightarrow{\text{a.s.}} 0 \quad \text{as} \quad n \to \infty.$$

This follows from the standard theorem in analysis about the Cesàro means.

Our aim now is to show that almost sure convergence in (1) cannot be replaced by convergence in probability. Indeed, consider the sequence of independent r.v.s $\{\xi_n, n \geq 1\}$ where ξ_n has a d.f. F_n given by

$$F_n(x) = \begin{cases} 0, & \text{if } x \leq 0 \\ 1 - 1/(x+n), & \text{if } x > 0. \end{cases}$$

Then for every fixed $\varepsilon > 0$ we have

$$\mathbf{P}[|\xi_n| > \varepsilon] = 1 - F_n(\varepsilon) = 1/(\varepsilon + n)$$

which means that $\xi_n \xrightarrow{\mathrm{P}} 0$ as $n \to \infty$. Let us show that the Cesàro means

$$\eta_n := \frac{1}{n}(\xi_1 + \ldots + \xi_n) \overset{\mathrm{P}}{\not\to} 0 \quad \text{as} \quad n \to \infty.$$

Denoting $M_n = \max\{\xi_1, \ldots, \xi_n\}$ and taking into account the independence of the variables ξ_j we can easily show that for any $x > 0$

$$\mathbf{P}[M_n \le x] = \left(1 - \frac{1}{x+1}\right)\left(1 - \frac{1}{x+2}\right)\ldots\left(1 - \frac{1}{x+n}\right) < \left(1 - \frac{1}{x+n}\right)^n.$$

Therefore

$$\mathbf{P}[M_n/n \le \varepsilon] < \left(1 - \frac{1}{\varepsilon n + n}\right)^n. \tag{2}$$

Since $[M_n/n > \varepsilon] \subset [\eta_n > \varepsilon]$,

$$\mathbf{P}[M_n/n > \varepsilon] \le \mathbf{P}[\eta_n > \varepsilon] \quad \Rightarrow \quad \mathbf{P}[M_n/n \le \varepsilon] \ge \mathbf{P}[\eta_n \le \varepsilon].$$

Combining the last relation with (2) we see that

$$\mathbf{P}[\eta_n \le \varepsilon] < \left(1 - \frac{1}{(\varepsilon+1)n}\right)^n$$

and hence

$$\lim_{n\to\infty} \mathbf{P}[\eta_n > \varepsilon] \ge 1 - \lim_{n\to\infty} \mathbf{P}[\eta_n \le \varepsilon] \ge 1 - \exp(-(\varepsilon+1)^{-1}) > 0.$$

This means that η_n does not converge to zero in probability. Therefore in general $\xi_n \xrightarrow{\mathrm{P}} 0 \not\Rightarrow \frac{1}{n}(\xi_1 + \ldots + \xi_n) \xrightarrow{\mathrm{P}} 0$ as $n \to \infty$.

Finally, to indicate one additional case leading to the same result. Let $\{X_n, n \ge 1\}$ be independent r.v.s, where X_n takes the values 2^n and 0, with probabilities n^{-1} and $1 - n^{-1}$ respectively. Then $X_n \xrightarrow{\mathrm{P}} 0$ but $\frac{1}{n}(X_1 + \cdots + X_n) \overset{\mathrm{P}}{\not\to} 0$ as $n \to \infty$ (the details are left to the reader).

SECTION 15. LAWS OF LARGE NUMBERS

Let $\{X_n, n \ge 1\}$ be a sequence of r.v.s defined on the probability space $(\Omega, \mathcal{F}, \mathbf{P})$. Define $S_n = X_1 + \cdots + X_n$, $a_k = \mathbf{E}X_k$, $A_n = \mathbf{E}S_n = a_1 + \cdots + a_n$.

We say that the sequence $\{X_n\}$ satisfies the **weak law of large numbers** (WLLN) (or that $\{X_n\}$ obeys the WLLN) if $\frac{1}{n}S_n - \frac{1}{n}A_n \xrightarrow{\text{P}} 0$ as $n \to \infty$, that is if for any $\varepsilon > 0$ we have

$$\lim_{n\to\infty} \mathbf{P}\left[\left|\frac{1}{n}S_n - \frac{1}{n}A_n\right| \geq \varepsilon\right] = 0.$$

Further, if $\frac{1}{n}S_n - \frac{1}{n}A_n \xrightarrow{\text{a.s.}} 0$ as $n \to \infty$, that is if

$$\mathbf{P}\left[\omega : \lim_{n\to\infty}\left(\frac{1}{n}S_n(\omega) - \frac{1}{n}A_n\right) = 0\right] = 1$$

we say that the sequence $\{X_n\}$ satisfies the **strong law of large numbers** (SLLN) (or that $\{X_n\}$ obeys the SLLN).

Let us formulate some of the basic results concerning the WLLN and the SLLN. It is obvious that either $\{X_n\}$ is a sequence of identically distributed r.v.s or these variables are arbitrarily distributed.

Khintchine theorem. *Suppose that $\{X_n, n \geq 1\}$ is a sequence of i.i.d. r.v.s with $\mathbf{E}[|X_1|] < \infty$. Then this sequence satisfies the WLLN and $\frac{1}{n}S_n \xrightarrow{\text{P}} a$ as $n \to \infty$ where $a = \mathbf{E}X_1$.*

Kolmogorov theorem 1. *Let $\{X_n, n \geq 1\}$ be a sequence of i.i.d. r.v.s. The existence of $\mathbf{E}[|X_1|]$ is a necessary and sufficient condition for the sequence $\{X_n\}$ to satisfy the SLLN and $\frac{1}{n}S_n \xrightarrow{\text{a.s.}} a$ as $n \to \infty$ where $a = \mathbf{E}X_1$.*

Markov theorem. *Suppose $\{X_n, n \geq 1\}$ is an arbitrary sequence of r.v.s such that the following condition holds:*

$$(1) \qquad \frac{1}{n^2}\mathbf{V}[X_1 + \cdots + X_n] \to 0 \quad \text{as } n \to \infty \quad \textit{(Markov condition)}.$$

Then $\{X_n\}$ satisfies the WLLN.

Kolmogorov theorem 2. *Let $\{X_n, n \geq 1\}$ be a sequence of independent r.v.s with $\sigma_n^2 = \mathbf{V}X_n < \infty$, $n \geq 1$. Suppose the following condition is fulfilled:*

$$(2) \qquad \sum_{n=1}^{\infty} \frac{\sigma_n^2}{n^2} < \infty \quad \textit{(Kolmogorov condition)}.$$

Then the given sequence satisfies the SLLN.

In the examples below we refer to (1) and (2) as the Markov condition and the Kolmogorov condition respectively.

A detailed presentation of the laws of large numbers can be found in the books by Doob (1953), Gnedenko (1962), Fisz (1963), Révész (1967), Feller (1971), Chung (1974), Petrov (1975), Laha and Rohatgi (1979), Billingsley (1995) and Shiryaev (1995).

In this section we consider examples which illustrate the importance of the conditions ensuring the validity of the WLLN or of the SLLN as well as the relationship between these two laws and some other related topics.

15.1. The Markov condition is sufficient but not necessary for the weak law of large numbers

(i) Let $\{X_n, n \geq 1\}$ be a sequence of independent r.v.s such that X_n has a χ_n^2-*distribution with n degrees of freedom*: that is, X_n has a density

$$f_n(x) = \begin{cases} [2\Gamma(n/2)]^{-1}(x/2)^{(n-2)/2}\exp(-x/2), & \text{if } x > 0 \\ 0, & \text{otherwise.} \end{cases}$$

Then $\mathbf{E}X_n = n$, $\mathbf{V}X_n = 2n$ and clearly the Markov condition is not satisfied. Hence we cannot apply the Markov theorem to determine whether the sequence $\{X_n\}$ satisfies the WLLN.

We use the following result (see Feller (1971) or Shiryaev (1995)).

If $\{\xi_n, n \geq 1\}$ is a sequence of r.v.s and ξ_n has a ch.f. ϕ_n, then

$$\xi_n \xrightarrow{\text{P}} 0 \quad \text{as} \quad n \to \infty \iff \phi_n(t) \to 1 \quad \text{as} \quad n \to \infty \quad \text{for all} \quad t \in \mathbb{R}^1.$$

The ch.f. ψ_n of X_n is $\psi_n(t) = (1 - 2it)^{-n/2}$. Then calculating the ch.f. $\tilde{\psi}_n$ of $(S_n - \mathbf{E}S_n)/n$ where $S_n = X_1 + \cdots + X_n$ and $\mathbf{E}S_n = \frac{1}{2}n(n+1)$ we find that $\tilde{\psi}_n \to 1$ as $n \to \infty$ for all $t \in \mathbb{R}^1$ and in view of the above result we conclude that the sequence $\{X_n\}$ does satisfy the WLLN.

Note that analogous reasoning leads to the same conclusion if we consider the sequence of discrete independent r.v.s $\{Y_n, n \geq 1\}$ where $\mathbf{P}[Y_n = \pm 1] = \frac{1}{2}(1 - 2^{-n})$ and $\mathbf{P}[Y_n = \pm 2^n] = 2^{-(n+1)}$.

Therefore the Markov condition is not necessary for the WLLN.

(ii) Let $\{Y_n, n \geq 1\}$ be independent r.v.s where Y_n has a density

$$f_n(x) = (\sqrt{2}\sigma_n)^{-1}\exp(-\sqrt{2}|x|/\sigma_n), \quad x \in \mathbb{R}^1.$$

It is easy to show that $\mathbf{E}Y_n = 0$ and $\mathbf{V}Y_n = \sigma_n^2$. Let us choose σ_n^2 in the following special way: $\sigma_n^2 = n^{1+\delta}$ where $0 \leq \delta < 1$. Then the Markov condition is not fulfilled. Nevertheless, as will be shown, the sequence $\{Y_n\}$ satisfies the WLLN. However, to prove this statement we need the following result (see Feller 1971): let η_n be independent r.v.s and let

$$\text{(1)} \qquad \mathbf{P}\left[\frac{1}{n}|\eta_k| > \varepsilon\right] < \delta$$

for any positive ε, δ, $k = 1, 2, \ldots, n$ and all sufficiently large n. Denote by $\{\tilde{\eta}_n\}$ the truncated sequence with some constant $c > 0$, that is $\tilde{\eta}_k = \eta_k$ if $|\eta_k| \leq c$ and $\tilde{\eta}_k = c$ if $|\eta_k| > c$. Then $\{\eta_k\}$ obeys the WLLN iff the following two conditions hold for any $\varepsilon > 0$ and $c > 0$:

$$\text{(2)} \qquad \lim_{n\to\infty} \sum_{k=1}^{n} \mathbf{P}\left[\frac{1}{n}|\tilde{\eta}_k| > \varepsilon\right] = 0$$

and

$$\lim_{n\to\infty} \frac{1}{n^2} \sum_{k=1}^{n} \mathbf{V}\tilde{\eta}_k = 0. \tag{3}$$

Now for any fixed $\varepsilon > 0$ we can easily show that

$$\mathbf{P}\left[\frac{1}{n}|Y_k| > \varepsilon\right] = \exp[-\sqrt{2}\varepsilon n/k^{(1+\delta)/2}], \quad k = 1, 2, \dots, n$$

and for sufficiently large n the right-hand side can be made arbitrarily small. Thus condition (1) holds.

For given $\varepsilon > 0$ and constant $c > 0$ let N be an integer satisfying the relations: $\varepsilon N \le c$ and $\varepsilon(N+1) > c$. Choose $n > N$. Then

$$\lim_{n\to\infty} \sum_{k=1}^{n} \mathbf{P}[|\tilde{Y}_k| > \varepsilon n] = \lim_{n\to\infty} \sum_{k=1}^{N} \exp[-\sqrt{2}\varepsilon n/k^{(1+\delta)/2}]. \tag{4}$$

Since the sum on the right-hand side of (4) contains a finite number of terms and each term tends to zero as $n \to \infty$, (4) implies (2).

It then remains for us to check condition (3). A direct calculation shows that

$$\mathbf{V}\tilde{Y}_k = k^{1+\delta}[1 - \exp(-c\sqrt{2}/k^{(1+\delta)/2})] - \sqrt{2}k^{(1+\delta)/2}\exp(-c\sqrt{2}/k^{(1+\delta)/2}).$$

Using a Taylor expansion we find

$$\mathbf{V}\tilde{Y}_k = c^2 + (c^3/k^{(1+\delta)/2})t_k$$

where t_k includes higher-order terms (their exact expressions are not important). From this we can easily derive (3).

Therefore according to the Feller theorem cited above the sequence $\{Y_n\}$ satisfies the WLLN. Again, as in case (i), the Markov condition is not satisfied.

15.2. The Kolmogorov condition for arbitrary random variables is sufficient but not necessary for the strong law of large numbers

Consider the sequence $\{X_n, n \ge 1\}$ of independent r.v.s where

$$\mathbf{P}[X_n = \pm 1] = \tfrac{1}{2}(1 - 2^{-n}), \quad \mathbf{P}[X_n = 2^n] = \mathbf{P}[X_n = -2^n] = 2^{-(n+1)}.$$

Obviously $\mathbf{E}X_n = 0$, $\sigma_n^2 = \mathbf{V}X_n = 1 - 2^{-n} + 2^n$ so that $\sum_{n=1}^{\infty} \sigma_n^2/n^2$ diverges. Thus the Kolmogorov condition, $\sum_{n=1}^{\infty} \sigma_n^2/n^2 < \infty$ is not satisfied. Nevertheless we shall show that $\{X_n\}$ obeys the SLLN.

Recall that two sequences of r.v.s $\{\xi_n\}$ and $\{\eta_n\}$ are said to be *equivalent in the sense of Khintchine* if $\sum_{n=1}^{\infty} \mathbf{P}[\xi_n \ne \eta_n] < \infty$. According to Révész (1967) two such sequences simultaneously satisfy or do not satisfy the SLLN.

Introduce the sequence $\{Y_n, n \geq 1\}$ where

$$\mathbf{P}[Y_n = 1] = \mathbf{P}[Y_n = -1] = \tfrac{1}{2}(1 - 2^{-n}), \quad \mathbf{P}[Y_n = 0] = 2^{-n}.$$

Clearly $\mathbf{E}X_n = \mathbf{E}Y_n$ and $\mathbf{P}[X_n \neq Y_n] = 2^{-n}$ for $n \in \mathbb{N}$. Since the series $\sum_{n=1}^{\infty} \mathbf{P}[X_n \neq Y_n] = \sum_{n=1}^{\infty} 2^{-n}$ is convergent, the sequences $\{X_n\}$ and $\{Y_n\}$ are equivalent in the sense of Khintchine. Further, $\mathbf{V}Y_n = 1 - 2^{-n}$ so that $\sum_{n=1}^{\infty} \mathbf{V}Y_n/n^2 < \infty$. Thus the Kolmogorov condition is satisfied for the sequence $\{Y_n\}$ and therefore $\{Y_n\}$ obeys the SLLN. By the above result it follows that the sequence $\{X_n\}$ also obeys the SLLN.

Thus we have shown that the Kolmogorov condition for arbitrarily distributed r.v.s is not necessary for the validity of the SLLN.

15.3. A sequence of independent discrete random variables satisfying the weak but not the strong law of large numbers

Let $\{X_n, n \geq 2\}$ be independent r.v.s such that

$$\mathbf{P}[X_n = \pm n] = \frac{1}{2n \log n}, \quad \mathbf{P}[X_n = 0] = 1 - \frac{1}{n \log n}, \quad n = 2, 3, \ldots .$$

Consider the events $A_n = \{|X_n| \geq n\}$, $n \geq 2$. Then

$$\mathbf{P}(A_n) = \frac{1}{n \log n} \quad \Rightarrow \quad \sum_{n=2}^{\infty} \mathbf{P}(A_n) = \infty.$$

The divergence of the series $\sum_{n=2}^{\infty} \mathbf{P}(A_n)$, the mutual independence of the variables X_n and the Borel–Cantelli lemma allow us to conclude that the event $[A_n \text{ i.o.}]$ has probability 1. In other words,

$$\mathbf{P}[|X_n| \geq n \text{ i.o.}] = 1 \ \Rightarrow \ \mathbf{P}\left[\lim_{n\to\infty} \frac{S_n}{n} \neq 0\right] = 1.$$

Therefore the sequence $\{X_n, n \geq 2\}$ cannot satisfy the SLLN.

Now we shall show that $\{X_n\}$ obeys the WLLN. Obviously $\mathbf{V}X_k = k/\log k$. Since the function $x/\log x$ has a local minimum at $x = \mathrm{e}$ and $\sum_{k=3}^{n} k/\log k$ is a lower Riemann sum for the integral $\int_3^{n+1} (x/\log x)\,\mathrm{d}x$, we easily obtain that

$$\begin{aligned}
\frac{1}{n^2} \sum_{k=2}^{n} \mathbf{V}X_k &\leq \frac{1}{n^2}\left[\frac{2}{\log 2} + \int_3^{n+1} \frac{x}{\log x}\,\mathrm{d}x\right] \\
&\leq \frac{2}{n^2 \log n} + \frac{(n-2)(n+1)}{n^2 \log n} \to 0 \quad \text{as } n \to \infty.
\end{aligned}$$

Thus the Markov condition for $\{X_n\}$ is satisfied and therefore the sequence $\{X_n\}$ obeys the WLLN.

Finally, let us indicate another sequence whose properties are close to those of $\{X_n\}$. Let $\{Y_n, n \geq 2\}$ be a sequence of i.i.d. r.v.s such that

$$\mathbf{P}[Y_2 = \pm n] = \frac{C}{n^2 \log n}, \; n = 2, 3, \ldots, \; C = \frac{1}{2}\left(\sum_{n=2}^{\infty} \frac{1}{n^2 \log n}\right)^{-1}.$$

It can be shown that this sequence obeys the WLLN but does not satisfy the SLLN. The easiest way to do this is to use ch.f.s showing, for example, that $\psi_n(t) \to 1$ as $n \to \infty$ where ψ_n is the ch.f. of $\frac{1}{n}(Y_2 + \cdots + Y_{n+1})$.

15.4. A sequence of independent absolutely continuous random variables satisfying the weak but not the strong law of large numbers

Let $\{X_n, n \geq 1\}$ be independent r.v.s where the density of X_n is given by

$$f_n(x) = \frac{1}{\sqrt{2}\sigma_n} \exp(-\sqrt{2}|x|/\sigma_n), \quad x \in \mathbb{R}^1.$$

Then $\mathbf{V}X_n = \sigma_n^2$ and define σ_n^2 as follows: $\sigma_n^2 = 2n^2/(\log n)^2, n \geq 2$.

First we shall establish that $\{X_n\}$ does not obey the SLLN. Indeed, the probability of the event $A_n = \{|X_n| \geq n\}$ is

$$\mathbf{P}(A_n) = \frac{2}{\sqrt{2}\sigma_n} \int_n^{\infty} \exp(-\sqrt{2}x/\sigma_n)\, dx = \exp\left[-\frac{1}{2}\sqrt{2}(\log n)^2/n\right].$$

Since $(\log n)^2/n \to 0$ as $n \to \infty$, $\sum_{n=2}^{\infty} \mathbf{P}(A_n) = \infty$. Using similar reasoning to that in Example 15.3, we conclude that $\{X_n\}$ does not obey the SLLN.

Our purpose now is to show that $\{X_n\}$ satisfies the WLLN. However, one can check that the Markov condition for $\{X_n\}$ does not hold. Then the proof uses the Feller theorem cited in Example 15.1. First, we can see that

$$\text{(1)} \qquad \mathbf{P}\left[\frac{1}{n}|X_k| > \varepsilon\right] = \exp\left(-n\varepsilon\frac{\log k}{k}\right), \quad k = 2, 3, \ldots, n$$

and clearly this probability can be made arbitrarily small for large n.

For any truncation level $c > 0$ and $\varepsilon > 0$ we introduce the variables $\tilde{X}_k$ where $\tilde{X}_k = X_k$, if $|X_k| \leq c$ and $\tilde{X}_k = c$, if $|X_k| > c$. Using (1) we obtain

$$\sum_{k=2}^{n} \mathbf{P}\left[\frac{1}{n}|\tilde{X}_k| > \varepsilon\right] \to 0 \quad \text{as } n \to \infty.$$

Similarly to Example 15.1 we can verify that

$$\frac{1}{n^2}\sum_{k=2}^{n} \mathbf{V}\tilde{X}_k \to 0 \quad \text{as} \quad n \to \infty.$$

Thus, by the Feller theorem, the sequence $\{X_n\}$ satisfies the WLLN.

15.5. The Kolmogorov condition $\sum_{n=1}^{\infty}\sigma_n^2/n^2<\infty$ is the best possible condition for the strong law of large numbers

Let $\{X_n, n\geq 1\}$ be a sequence of independent r.v.s with finite variances σ_n^2 and $\{b_n, n\geq 1\}$ be a non-decreasing sequence of positive constants with $b_n\to\infty$. We say that the sequence $\{X_n\}$ obeys the SLLN with respect to $\{b_n\}$ if $b_n^{-1}S_n - b_n^{-1}\mathbf{E}S_n \xrightarrow{\text{a.s.}} 0$ as $n\to\infty$ where $S_n = X_1+\cdots+X_n$.

According to the Kolmogorov theorem the condition $\sum_{n=1}^{\infty}\sigma_n^2/b_n^2<\infty$ implies that $\{X_n\}$ satisfies the SLLN with respect to $\{b_n\}$. Note that in the classical Kolmogorov theorem $b_n=n, n\geq 1$.

It is of general interest to understand the importance of the condition $\sum_{n=1}^{\infty}\sigma_n^2/b_n^2<\infty$ in the SLLN. We shall now show that this condition is the best possible in the following sense. For simplicity we confine ourselves to the case $b_n=n, n\geq 1$. So, let $\{\sigma_n^2\}$ be a sequence of positive numbers with

$$\sum_{n=1}^{\infty}\sigma_n^2/n^2=\infty. \tag{1}$$

We aim to construct a sequence $\{Y_n, n\geq 1\}$ of independent r.v.s with $\mathbf{V}Y_n=\sigma_n^2$ such that $\{Y_n\}$ does not satisfy the SLLN. Let us describe the sequence $\{Y_n\}$. If $\sigma_n^2/n^2\leq 1$ then the r.v. Y_n takes the values $(-n)$, 0 and n with probabilities $\sigma_n^2/(2n^2)$, $1-\sigma_n^2/n^2$ and $\sigma_n^2/(2n^2)$ respectively. If $\sigma_n^2/n^2>1$ then $Y_n=\pm\sigma_n$ with probability $\frac{1}{2}$ each.

Clearly $\mathbf{E}Y_n=0$, $\mathbf{V}Y_n=\sigma_n^2$. For any $\varepsilon>0$ we have

$$\mathbf{P}[|Y_n|/n>\varepsilon]=\mathbf{P}[Y_n\neq 0]=\begin{cases}\sigma_n^2/n^2, & \text{if } \sigma_n^2/n^2\leq 1\\ 1, & \text{if } \sigma_n^2/n^2>1.\end{cases}$$

Suppose the sequence $\{Y_n\}$ does obey the SLLN. Then necessarily $Y_n/n\xrightarrow{\text{a.s.}}0$ as $n\to\infty$. From (1) it is easy to derive that $\sum_{n=1}^{\infty}\mathbf{P}[|Y_n|>\varepsilon n]=\infty$. By the Borel–Cantelli lemma the events $[|Y_n|>\varepsilon n]$ occur infinitely often, so the convergence $Y_n/n\xrightarrow{\text{a.s.}}0$ as $n\to\infty$ is not possible.

Therefore $\{Y_n\}$ does not obey the SLLN.

15.6. More on the strong law of large numbers without the Kolmogorov condition

Consider the sequence $\{X_n, n\geq 2\}$ of independent r.v.s where

$$\mathbf{P}[X_n=\pm(n/\log n)^{1/2}]=\tfrac{1}{2}. \tag{1}$$

It is easy to check that the Kolmogorov condition $\sum_{n=2}^{\infty}\sigma_n^2/n^2<\infty$ is not satisfied. However, Example 15.2 shows that the SLLN can also hold without this condition. In our specific case the most suitable result which can be applied is the following theorem (see Révész 1967): let $\{\xi_n\}$ be independent r.v.s with $\mathbf{E}\xi_n = 0$ and let for some $r \geq 1$

$$\mathbf{E}[|\xi_n|^{2r}] < \infty \quad \text{and} \quad \sum_{n=1}^{\infty} \mathbf{E}[|\xi_n|^{2r}]/n^{r+1} < \infty.$$

Then the sequence $\{\xi_n\}$ satisfies the SLLN.

Clearly for the sequence $\{X_n\}$ defined by (1) it is sufficient to take $r = 2$ and verify directly the conditions in the Révész theorem. Thus we arrive at the conclusion that $\{X_n\}$ obeys the SLLN.

15.7. Two 'near' sequences of random variables such that the strong law of large numbers holds for one of them and does not hold for the other

Consider two sequences of r.v.s, $\{X_n, n \geq 2\}$ and $\{Y_n, n \geq 2\}$ where

$$\mathbf{P}[X_n = n/\log n] = \mathbf{P}[X_n = -n/\log n] = \log n/(2n), \; \mathbf{P}[X_n = 0] = 1 - \log n/n,$$

$$\mathbf{P}[Y_n = \beta n] = \mathbf{P}[Y_n = -\beta n] = 1/(2\beta^2 n \log n), \; \mathbf{P}[Y_n = 0] = 1 - 1/(\beta^2 n \log n)$$

with $0 < \beta < 1$. Obviously X_n and Y_n are symmetric r.v.s with

$$\mathbf{E}X_n = \mathbf{E}Y_n = 0, \quad \mathbf{V}X_n = \mathbf{V}Y_n = n/\log n$$

and both satisfy the inequalities

$$|X_n| < n \quad \text{a.s.}, \quad |Y_n| < n \quad \text{a.s.}, \quad n = 3, 4, \ldots .$$

We are interested to know whether or not these sequences satisfy the SLLN. We shall show that $\{X_n\}$ obeys the SLLN while $\{Y_n\}$ does not. For this purpose we introduce H_r where

$$H_r = 2^{-2r} \sum_{n=2^r+1}^{2^{r+1}} \mathbf{V}X_n.$$

For any choice of $\varepsilon > 0$ we have $\exp(-\varepsilon/H_r) < \exp(-\varepsilon r \log 2/2)$ implying

$$\sum_{r=1}^{\infty} \exp(-\varepsilon/H_r) < \infty.$$

However, this condition is sufficient to conclude that the sequence $\{X_n\}$ obeys the SLLN (see Prohorov 1950).

Suppose now that $\{Y_n\}$ also satisfies the SLLN. Then necessarily

$$\mathbf{P}[Y_n/n \to 0] = 1.$$

It can easily be seen from the definition of $\{Y_n\}$ that

$$\sum_{n=2}^{\infty} \mathbf{P}[|Y_n|/n = \beta] = \infty.$$

Then by the Borel–Cantelli lemma, the events $[|Y_n| = n\beta]$ occur infinitely often. This, however, contradicts the above relation, namely that $Y_n/n \xrightarrow{\text{a.s.}} 0$ as $n \to \infty$, and therefore the sequence $\{Y_n\}$ does not obey the SLLN.

15.8. The law of large numbers does not hold if almost sure convergence is replaced by complete convergence

Let $\{X_n, n \geq 1\}$ be a sequence of i.i.d. r.v.s, $F(x)$, $x \in \mathbb{R}^1$ their common d.f. and $\mathbf{E}X_1 = \int_{-\infty}^{\infty} x \, \mathrm{d}F(x) = 0$. Suppose that $\{X_n\}$ satisfies the SLLN. Then

$$(1) \qquad Y_n := \frac{1}{n}(X_1 + \cdots + X_n) \xrightarrow{\text{a.s.}} 0 \quad \text{as} \quad n \to \infty.$$

It is natural to ask whether the conditions for the SLLN could guarantee that in (1) almost sure convergence can be replaced by a stronger kind of convergence, in particular by complete convergence (see Example 14.14).

Under the conditions

$$(2) \qquad \int_{-\infty}^{\infty} x \, \mathrm{d}F(x) = 0, \quad \sigma^2 = \int_{-\infty}^{\infty} x^2 \, \mathrm{d}F(x) < \infty$$

Hsu and Robbins (1947) have shown the convergence of the series $\sum_{n=1}^{\infty} \mathbf{P}[|Y_n| > \varepsilon]$ for any $\varepsilon > 0$. Therefore if condition (2) is satisfied, the sequence $\{Y_n\}$ converges completely. Thus instead of (1) we have $Y_n \xrightarrow{\text{c}} 0$ as $n \to \infty$.

Suppose now that condition (2) is relaxed a little as follows:

$$(3) \quad \int_{-\infty}^{\infty} x \, \mathrm{d}F(x) = 0, \quad \int_{-\infty}^{\infty} |x|^{\alpha} \, \mathrm{d}F(x) < \infty, \quad \int_{-\infty}^{\infty} x^2 \, \mathrm{d}F(x) = \infty$$

where $\alpha =$ constant and $\frac{1}{2}(1 + \sqrt{5}) \leq \alpha < 2$. Then the sequence $\{X_n\}$ satisfies the SLLN. However, the series $\sum_{n=1}^{\infty} \mathbf{P}[|Y_n| > \varepsilon]$ diverges for every $\varepsilon > 0$ and hence the relation $Y_n \xrightarrow{\text{c}} 0$ fails to hold. Therefore there are sequences of i.i.d. r.v.s such that the corresponding arithmetic means $\{Y_n\}$ converge a.s. but not completely.

Finally, it remains for us to indicate a particular case when conditions (3) are satisfied. For example, take X_1 to be absolutely continuous with density $f(x) = |x|^{-3}$ for $|x| \geq 1$ and $f(x) = 0$ otherwise.

15.9. The uniform boundedness of the first moments of a tight sequence of random variables is not sufficient for the strong law of large numbers

Recall that the sequence $\{X_n, n \geq 1\}$ of real-valued r.v.s is said to be *tight* if for each $\varepsilon > 0$ there exists a compact interval $K_\varepsilon \subset \mathbb{R}^1$ such that $\mathbf{P}[X_n \in K_\varepsilon] > 1 - \varepsilon$ for all n.

Let $\{X_n, n \geq 1\}$ be a sequence of independent r.v.s. According to a result derived by Taylor and Wei (1979), if $\{X_n\}$ is a tight sequence and the rth moments with $r > 1$ are uniformly bounded ($\mathbf{E}[|X_n|^r] < M = \text{constant} < \infty$) then $\{X_n\}$ satisfies the SLLN. Is it then possible to weaken the assumption for r, $r > 1$, replacing it by $r = 1$ in the above result?

By a specific example we show that the answer to this question is negative. Let $\{X_n, n \geq 1\}$ be a sequence of independent r.v.s such that

$$\mathbf{P}[X_n = \pm n] = \tfrac{1}{2}[n \log(n+2)]^{-1}, \quad \mathbf{P}[X_n = 0] = 1 - [n \log(n+2)]^{-1}.$$

Then $\mathbf{E}X_n = 0$, $\mathbf{E}[|X_n|] = 1/\log(n+2)$. So $\mathbf{E}[|X_n|]$ are uniformly bounded, and indeed, $\mathbf{E}[|X_n|] \to 0$. Taking into account the relation $\mathbf{P}[|X_n| \geq n] = 1/[n \log(n+2)]$, we conclude that the sequence $\{X_n\}$ is tight.

Further, $\sum_{n=1}^{\infty} \mathbf{P}[|X_n| \geq n] = \infty$ and the Borel–Cantelli lemma implies that the event $[|X_n| \geq n \text{ i.o.}]$ has probability 1. However, this means that the SLLN cannot be valid for the sequence $\{X_n\}$.

15.10. The arithmetic means of a random sequence can converge in probability even if the strong law of large numbers fails to hold

Let $\{X_n, n \geq 1\}$ be a sequence of i.i.d. r.v.s such that $\mathbf{E}[|X_1|] = \infty$. According to the Kolmogorov theorem this sequence does not satisfy the SLLN, i.e. $Y_n = S_n/n$, where $S_n = X_1 + \cdots + X_n$, is not a.s. convergent as $n \to \infty$. However we can still ask about the convergence of Y_n, the arithmetic means, in a weaker sense, e.g. in probability. This possibility is considered in the next example.

Consider the sequence $\{\xi_n, n \geq 1\}$ of i.i.d. r.v.s where

$$\mathbf{P}[\xi_1 = (-1)^{k-1}k] = \frac{6}{\pi^2 k^2}, \quad k = 1, 2, \ldots$$

The divergence of the harmonic series implies that $\mathbf{E}[|\xi_1|] = \infty$. Hence $\{\xi_n\}$ does not satisfy the SLLN.

Let us show now that the arithmetic means $(\xi_1 + \cdots + \xi_n)/n$ converge in probability to a fixed number as $n \to \infty$. Our reasoning is based on the following general and very useful result (see e.g. Feller 1971 or Shiryaev 1995): if the ch.f. $\psi(t) = \mathbf{E}[\mathrm{e}^{it\xi_1}]$, $t \in \mathbb{R}^1$ of ξ_1 is differentiable at $t = 0$ and $\psi'(0) = ic$, where $i = \sqrt{-1}$ and $c \in \mathbb{R}^1$, then $(\xi_1 + \cdots + \xi_n)/n \xrightarrow{\mathrm{P}} c$ as $n \to \infty$. Thus we first have to find the ch.f. ψ of the r.v. ξ_1 defined above. We have

$$\psi(t) = \frac{6}{\pi^2} \sum_{j=1}^{\infty} \frac{(\mathrm{e}^{it})^{2j-1}}{(2j-1)^2} + \frac{6}{\pi^2} \sum_{k=1}^{\infty} \frac{(\mathrm{e}^{it})^{2k}}{(2k)^2}.$$

If we introduce the functions

$$h_1(u) = \sum_{j=1}^{\infty} \frac{u^{2j-1}}{(2j-1)^2}, \ |u| \leq 1 \quad \text{and} \quad h_2(u) = \sum_{k=1}^{\infty} \frac{u^{2k}}{(2k)^2}, \ |u| \leq 1$$

we easily find that they both are differentiable and

$$h_1'(u) = \frac{1}{2u}\log\frac{1+u}{1-u}, \quad h_2'(u) = -\frac{1}{2u}\log(1-u^2).$$

Hence $h_1'(u) - h_2'(u) = (1/u)\ln(1+u)$ which implies that $\psi'(0)$ exists and

$$\psi'(0) = i\frac{6}{\pi^2}[h_1'(1) - h_2'(1)] = i\frac{6\log 2}{\pi^2}.$$

Thus we arrive at the final conclusion that

$$\frac{1}{n}(\xi_1 + \cdots + \xi_n) \xrightarrow{\mathrm{P}} \frac{6\log 2}{\pi^2} \quad \text{as} \quad n \to \infty.$$

Note that in this case the sequence $\{\xi_n\}$ satisfies the so-called generalized law of large numbers (see Example 15.12).

15.11. The weighted averages of a sequence of random variables can converge even if the law of large numbers does not hold

Let $\{X_k, k \geq 1\}$ be a sequence of non-degenerate i.i.d. r.v.s, $\{c_k, k \geq 1\}$ a sequence of positive numbers and let $S_n = \sum_{k=1}^n c_k X_k$ and $C_n = \sum_{k=1}^n c_k$, $n \geq 1$. The ratios S_n/C_n, $n \geq 1$ are called *weighted averages* generated by $\{X_k, c_k, k \geq 1\}$. We say that the weak (strong) law holds for the weighted averages of $\{X_k, c_k, k \geq 1\}$ iff S_n/C_n converges in probability (a.s.) to some constant as $n \to \infty$.

Without any loss of generality we can suppose that $\mathbf{E}X_k = 0$ for all $k \geq 1$. We now want to see whether S_n/C_n converges to 0 as $n \to \infty$.

Obviously if all $c_k \equiv 1$ then $S_n = X_1 + \cdots + X_n$, $C_n = n$ and we are in the framework of the classical laws of large numbers.

Our aim now is to show that there is a sequence of i.i.d. r.v.s $\{X_k\}$ and a sequence of weights $\{c_n\}$ such that

$$\frac{S_n}{C_n} \xrightarrow{\text{a.s.}} 0 \quad \text{while} \quad \frac{1}{n}(X_1 + \cdots + X_n) \overset{\text{a.s.}}{\not\to} 0 \quad \text{as} \quad n \to \infty.$$

In other words, the strong law holds for weighted averages but the classical SLLN is not valid. An analogous conclusion can be drawn about the weak law for weighted averages and the classical WLLN.

Firstly, consider the strong law. By assumption the variables X_n are identically distributed and in this case the SLLN does not hold iff $\mathbf{E}[|X_1|] = \infty$. Further, we need the following result (see Wright *et al*): let $g(x)$, $x \in \mathbb{R}^+$ be a non-negative measurable function with $g(x) \to \infty$ as $x \to \infty$. Then there exists a sequence $\{X_k, c_k, k \geq 1\}$ whose weighted averages $S_n/C_n, n \geq 1$, satisfy the strong law and $\mathbf{E}[g(X_1^+)] = \mathbf{E}[g(X_1^-)] = \infty$.

Actually this result contains all that we wanted, namely a sequence of i.i.d. r.v.s $\{X_k, k \geq 1\}$ with $\mathbf{E}|X_1| = \infty$ and a sequence of weights $\{c_k, k \geq 1\}$ such that

$S_n/C_n \xrightarrow{\text{a.s.}} 0$ as $n \to \infty$, although the sequence $\{X_k\}$ does not obey the classical SLLN.

A similar conclusion can be obtained by using a result of Chow and Teicher (1971) which states that there is a r.v. X with $\mathbf{E}|X| = \infty$ such that the sequence $\{X_k\}$ of independent copies of X together with a suitable sequence of weights $\{c_k\}$ generates weighted averages S_n/C_n which converge a.s. as $n \to \infty$. Obviously it is impossible in this case to take $c_k \equiv 1$ since the classical SLLN for $\{X_k\}$ is not satisfied. In this connection Chow and Teicher (1971) give two specific examples. The first one arises in the so-called St Petersburg game ($X_k = 2^k$ with probability 2^{-k}, and 0 otherwise), while in the second case X has a Cauchy distribution.

It is of general interest to compare some consequences of the results cited above. In particular, let us look at the value of $\mathbf{E}[|X|^r]$ for different r. Both examples considered by Chow and Teicher are such that

$$\lim_{x\to\infty} x^r \mathbf{P}[|X| \geq x] = 0 \quad \text{for all} \quad 0 < r < 1$$

which implies that $\mathbf{E}[|X|^r] < \infty$ for all $0 < r < 1$. In the result of Wright *et al* (1977) we can take the function $g(x) = (\log x)^+$ and choose a sequence $\{X_k\}$ of i.i.d. r.v.s such that $\mathbf{E}[|X|^r] = \infty$ for all $r > 0$ and find weights $\{c_k\}$ such that $S_n/C_n \xrightarrow{\text{a.s.}}$ constant. Clearly the SLLN fails to hold.

Consider now the weak law. It is easy to see that if the weak law holds for the weighted averages of $\{X_k, c_k\}$ then $\{c_k\}$ must satisfy the condition

$$(1) \qquad C_n \to \infty, \quad c_n/C_n \to 0 \quad \text{as} \quad n \to \infty.$$

According to Jamison *et al* (1965), the weak law holds for any sequence of weights $\{c_k\}$ satisfying (1) if $\int_{|x|\leq T} x \, \mathrm{d}F(x) \to a =$ constant as $T \to \infty$ and

$$(2) \qquad \lim_{T\to\infty} T\mathbf{P}[|X| \geq T] = 0$$

where F is the d.f of X. This result and a statement by Loève (1978) allow us to conclude that if X has a fixed distribution (we consider only the case of i.i.d.) then the weak law holds for $\{X_k, c_k\}$ for any $\{c_k\}$ iff $\{X_k\}$ satisfies the classical WLLN (when all $c_k \equiv 1$).

However, using the result of Wright *et al* with $g(x) = x^r$ and $0 < r < 1$, one can obtain a sequence $\{X_k, c_k\}$ for which the weak law holds but condition (2) is not satisfied. In such a case the weak law does not hold for the sequence $\{X_k, 1\}$. Obviously this means that the sequence $\{X_k\}$ does not obey the classical WLLN in spite of the fact that for some weights $\{c_k\}$, the weighted averages S_n/C_n converge in probability.

15.12. The law of large numbers with a special choice of norming constants

Let $\{X_n, n \geq 1\}$ be a sequence of independent r.v.s and $S_n = X_1 + \cdots + X_n$. If for some number sequences $\{a_n, n \geq 1\}$ and $\{b_n, n \geq 1\}$, with all $b_n > 0$, the following relation holds:

$$(1) \qquad (S_n - a_n)/b_n \to 0 \quad \text{as} \quad n \to \infty$$

and we say that $\{X_n\}$ satisfies a *generalized law of large numbers* (LLN). This law is weak or strong depending on the type of convergence in (1). If $a_n = \mathbf{E}S_n$ and $b_n = n$ we obtain the scheme of the classical LLN. There are sequences of r.v.s for which the classical LLN does not hold, but for some choice of $\{a_n\}$ and $\{b_n\}$ the generalized LLN holds. Let us consider an example.

In the well known St Petersburg game (also mentioned in Example 15.11), a player wins 2^k roubles if heads first appears at the kth toss of a symmetric coin, $k = 1, 2, \ldots$. Thus we get a sequence of independent r.v.s $\{X_k, k \geq 1\}$ where $\mathbf{P}[X_k = 2^k] = 2^{-k} = 1 - \mathbf{P}[X_k = 0]$. It is easy to check that $\{X_k\}$ does not obey the WLLN. However, we can hope that a relation like (1) will hold.

Using game terminology, suppose that a player pays variable entrance fees with a cumulative fee $b_n = n \log n$ for the first n games. Then the game becomes 'fair' in the sense that

$$(2) \qquad \lim_{n\to\infty} S_n/b_n = 1 \quad \text{in probability.}$$

It is natural to ask whether this game is 'fair' in a strong sense, that is, whether (2) is satisfied with probability 1. Actually we shall show that the St Petersburg game with $b_n = n \log n$ is 'fair' in a weak but not in a strong sense. In other words, it will be shown that $\{X_k\}$ obeys the weak but not the strong generalized LLN with $a_n = b_n = n \log n$, $n \geq 2$.

The result that $S_n/b_n \xrightarrow{\text{P}} 1$ as $n \to \infty$ is left to the reader as a useful exercise. Further, it is easy to see that $\mathbf{P}[X_n > c] \geq 1/c$ for any $c > 1$ and every $n \geq 2$. Hence for $c =$ constant > 1 and $n \geq 2$ we have

$$\mathbf{P}[X_n > cb_n] \geq 1/(cb_n) = 1/(cn \log n) \quad \text{and} \quad \sum_{n=2}^{\infty} \mathbf{P}[X_n > cb_n] = \infty.$$

This and the Borel–Cantelli lemma imply that $\mathbf{P}[X_n/b_n > c \text{ i.o.}] = 1$. Thus

$$\mathbf{P}[\overline{\lim} X_n/b_n = \infty] = 1 \quad \text{and} \quad \mathbf{P}[\overline{\lim} S_n/b_n = \infty] = 1.$$

Therefore

$$\mathbf{P}[\lim_{n\to\infty} S_n/b_n = 1] = 0$$

showing that (2) is satisfied for convergence in probability but not a.s.

SECTION 16. WEAK CONVERGENCE OF PROBABILITY MEASURES AND DISTRIBUTIONS

In Section 14 we introduced the notion of convergence in distribution and illustrated it by examples. In particular, we mentioned that this kind of convergence is close to so-called weak convergence. In this section we define weak convergence and clarify its relationship with other kinds of convergence.

Let $F_n, n \geq 1$, and F be d.f.s over the real line $\mathbb{R}^1$. Denote by $\mathbf{P}_n$ and $\mathbf{P}$ the probability measures over $(\mathbb{R}^1, \mathcal{B}^1)$ generated by F_n and F respectively. Recall that $\mathbf{P}_n$ and $\mathbf{P}$ are determined uniquely by the relations $\mathbf{P}_n(-\infty, x] = F_n(x)$ and $\mathbf{P}(-\infty, x] = F(x)$, $x \in \mathbb{R}^1$. Since F is continuous at the point x iff $\mathbf{P}(\{x\}) = 0$, then convergence in distribution $F_n \xrightarrow{\text{d}} F$ means that $\mathbf{P}_n(-\infty, x] \to \mathbf{P}(-\infty, x]$ for every x such that $\mathbf{P}(\{x\}) = 0$. Let us consider a more general situation.

For any Borel set A in $\mathbb{R}^1$ (that is $A \in \mathcal{B}^1$), ∂A will denote the boundary of A. Suppose $\mathbf{P}$ and $\mathbf{P}_n$, $n \geq 1$, are probability measures on $(\mathbb{R}^1, \mathcal{B}^1)$. We say that the sequence $\{\mathbf{P}_n\}$ **converges weakly** to $\mathbf{P}$ and write $\mathbf{P}_n \xrightarrow{\text{w}} \mathbf{P}$, if for any $A \in \mathcal{B}^1$ with $\mathbf{P}(\partial A) = 0$ we have

$$\mathbf{P}_n(A) \to \mathbf{P}(A) \quad \text{as} \quad n \to \infty.$$

Now we formulate the following fundamental result.

Theorem 1. *The following statements are equivalent:*

(a) $\mathbf{P}_n \xrightarrow{\text{w}} \mathbf{P}$;
(b) $\overline{\lim}_{n\to\infty} \mathbf{P}_n(A) \leq \mathbf{P}(A)$ *for any closed set* $A \in \mathcal{B}^1$;
(c) $\underline{\lim}_{n\to\infty} \mathbf{P}_n(A) \geq \mathbf{P}(A)$ *for any open set* $A \in \mathcal{B}^1$;
(d) *For every continuous and bounded function* g *on* $\mathbb{R}^1$ *we have*

$$\int_{\mathbb{R}^1} g(x)\mathbf{P}_n(\mathrm{d}x) \to \int_{\mathbb{R}^1} g(x)\mathbf{P}(\mathrm{d}x) \quad \textit{as} \quad n \to \infty.$$

Weak convergence can be studied in much more general situations not just for probability measures defined on the real line $\mathbb{R}^1$. However, convergence in distribution treated in Section 14 is equivalent to weak convergence discussed above. If we work with probability measures, the term weak convergence is preferable, while for d.f.s both terms, weak convergence and convergence in distribution, are used, as well as both notations, $F_n \xrightarrow{\text{w}} F$ and $F_n \xrightarrow{\text{d}} F$.

We now formulate another fundamental result connecting the weak convergence of d.f.s with the pointwise convergence of the corresponding ch.f.s.

Theorem 2. *(Continuity theorem.) Let* $\{F_n, n \geq 1\}$ *be a sequence of d.f.s on* $\mathbb{R}^1$ *and* $\{\phi_n, n \geq 1\}$ *be the corresponding sequence of the ch.f.s.*
(a) *If* $F_n \xrightarrow{\text{w}} F$ *for a d.f.* F*, then* $\phi_n(t) \to \phi(t)$, $t \in \mathbb{R}^1$ *where* ϕ *is the ch.f. of* F.
(b) *If* $\lim_{n\to\infty} \phi_n(t)$ *exists for each* $t \in \mathbb{R}^1$ *and* $\phi(t) := \lim_{n\to\infty} \phi_n(t)$ *is continuous at* $t = 0$*, then* ϕ *is the ch.f. of a d.f.* F *and* $F_n \xrightarrow{\text{w}} F$ *as* $n \to \infty$.

We refer the reader to the books by Billingsley (1968, 1995), Chung (1974) or Shiryaev (1995) for a detailed proof of Theorems 1 and 2 and of several others.

In this section we have included examples illustrating some aspects of the weak convergence of probability measures, distributions, and densities.

16.1. Defining classes and classes defining convergence

Let $(\Omega, \mathcal{F})$ be a measurable space and **P**, **Q** probabilities on this space. The class of events $\mathcal{A} \subset \mathcal{F}$ is said to be a *defining class*, if

$$\mathbf{P} = \mathbf{Q} \quad \text{on} \quad \mathcal{A} \Rightarrow \mathbf{P} = \mathbf{Q} \quad \text{on} \quad \mathcal{F}.$$

We say that $\mathcal{A} \subset \mathcal{F}$ is a *class defining convergence* if

$$\begin{aligned} &\mathbf{P}_n(A) \to \mathbf{P}(A) \quad \text{for all sets} \quad A \in \mathcal{A} \quad \text{with} \quad \mathbf{P}(\partial A) = 0 \\ \Rightarrow\ &\mathbf{P}_n(A) \to \mathbf{P}(A) \quad \text{for all sets} \quad A \in \mathcal{F} \quad \text{with} \quad \mathbf{P}(\partial A) = 0 \end{aligned}$$

that is, that $\mathbf{P}_n \xrightarrow{\text{w}} \mathbf{P}$ as $n \to \infty$.

Let us illustrate the relationship between these two notions.

(i) Obviously every class defining convergence is a defining class. However, the converse is not always true.

Let $\Omega = [0, 1)$, $\mathcal{F} = \mathcal{B}_{[0,1)}$ and $\mathcal{A} \subset \mathcal{F}$ be the field of all finite sums of disjoint subintervals of the type $[a, b)$ where $0 < a < b < 1$. Then $\mathcal{A}$ is a defining class but not a class defining convergence. To see this it is enough to consider the probabilities $\mathbf{P}_n$ and **P** concentrated at the points $1 - 1/n$ and 0 respectively.

(ii) Let $\{\mathbf{P}_n, n \geq 1\}$, **P** and **Q** be probabilities on $(\Omega, \mathcal{F})$ where $\Omega = \mathbb{R}^1$, $\mathcal{F} = \mathcal{B}^1$ and let $\mathcal{A} \subset \mathcal{F}$ be a defining class. Suppose two conditions are satisfied:

(1) $$\mathbf{P}_n(A) \to \mathbf{Q}(A) \quad \text{as} \quad n \to \infty \quad \text{for all} \quad A \in \mathcal{A}$$

and

(2) $$\mathbf{P}_n \xrightarrow{\text{w}} \mathbf{P} \quad \text{as} \quad n \to \infty.$$

Since $\mathcal{A}$ is a defining class, from (1) and (2) we could expect that $\mathbf{P} = \mathbf{Q}$. However, this is not the case. Define $\mathbf{P}_n$, **P** and **Q** as follows:

$$\begin{gathered} \mathbf{P}_n\left(\left\{\tfrac{1}{n}\right\}\right) = \mathbf{P}_n\left(\left\{1 + \tfrac{1}{n}\right\}\right) = \tfrac{1}{2}, \\ \mathbf{P}(\{0\}) = \mathbf{P}(\{1\}) = \tfrac{1}{2}, \quad \mathbf{Q}(\{0\}) = 1. \end{gathered}$$

It is easy to see that $\mathbf{P}_n \xrightarrow{\text{w}} \mathbf{P}$ as $n \to \infty$. Further, let B consist of the points 0, 1, $\frac{1}{n}$, $1 + \frac{1}{n}$ where $n \geq 1$. Denote by $\mathcal{A}$ the field containing all $A \in \mathcal{F}$ such that either AB is finite and $0 \notin A$, or $A^c B$ is finite and $0 \notin A^c$. Then $\mathcal{A}$ is a defining class and $\mathbf{P}_n(A) \to \mathbf{Q}(A)$ as $n \to \infty$ for every $A \in \mathcal{A}$. So (1) and (2) are satisfied, but $\mathbf{P} \neq \mathbf{Q}$.

(iii) Let $\mathbb{C}[0, 1]$ be the space of all continuous functions on $[0, 1]$ and $\mathcal{C}$ its Borel σ-field. For $k \in \mathbb{N}$ and $t_1, \ldots, t_k \in [0, 1]$ let

$$\pi_{t_1 \ldots t_k} : \mathbb{C}[0, 1] \longmapsto \mathbb{R}^k$$

map the point (function) $x \in \mathbb{C}[0,1]$ into the point $(x(t_1), \ldots, x(t_k)) \in \mathbb{R}^k$. The finite-dimensional sets (cylinders) in $\mathbb{C}[0,1]$ are defined as sets of the form $\pi_{t_1 \ldots t_k}^{-1} H$ where $H \in \mathcal{B}^k$. Denote by $\mathcal{A}$ the class of all such sets. Since the σ-field $\mathcal{C}$ is generated by $\mathcal{A}$, $\mathcal{A}$ is a defining class. This leads to the following question: does $\mathcal{A}$ form a class defining convergence? The answer is positive if we consider the space $(\mathbb{R}^\infty, \mathcal{B}^\infty)$ and the class $\mathcal{A}$ consisting of the finite-dimensional sets in $\mathbb{R}^\infty$.

However, as we shall show now, in the space $\mathbb{C}[0,1]$, $\mathcal{A}$ need not be a class defining convergence. To see this, consider the probability measures $\mathbf{P}$ and $\mathbf{P}_n$ where $\mathbf{P}$ is concentrated on the function $x \equiv 0$ (that is, $x(t) = 0$, $t \in [0,1]$) and $\mathbf{P}_n$ is concentrated on the function x_n defined by

$$x_n(t) = \begin{cases} nt, & \text{if } 0 \le t \le \frac{1}{n} \\ 2 - nt, & \text{if } \frac{1}{n} < t \le \frac{2}{n} \\ 0, & \text{if } \frac{2}{n} < t \le 1. \end{cases}$$

Since x_n does not converge to 0 uniformly in $\mathbb{C}[0,1]$, the measures $\mathbf{P}_n$ cannot converge weakly to $\mathbf{P}$ as $n \to \infty$. For example, if $A = S(0, \frac{1}{2})$ is the ball in $\mathbb{C}[0,1]$ with centre at 0 and radius $\frac{1}{2}$, then $\mathbf{P}(\partial A) = 0$ but $\mathbf{P}_n(A) = 0 \not\to \mathbf{P}(A) = 1$.

The relation $\mathbf{P}_n(A) \to \mathbf{P}(A)$ holds for any finite-dimensional set A in $\mathbb{C}[0,1]$ with $\mathbf{P}(\partial A) = 0$. This follows from the equality $\mathbf{P}_n(A) = \mathbf{P}(A)$ which is satisfied for any A of the form $\pi_{t_1 \ldots t_k}^{-1} H$, $H \in \mathcal{B}^k$ and $n \ge n_0$ where $n_0 = [2/t_{\min}] + 1$ with $t_{\min} = \min\{t_j, t_j \ne 0\}$.

This example shows that weak convergence in the space $\mathbb{C}[0,1]$ cannot be characterized by convergence for all finite-dimensional sets (as in $\mathbb{R}^\infty$).

16.2. In the case of convergence in distribution, do the corresponding probability measures converge for all Borel sets?

Let $F_0(x)$, $F_n(x)$, $n \ge 1$, be d.f.s and μ_0, μ_n, $n \ge 1$, their probability measures on $(\mathbb{R}^1, \mathcal{B}^1)$. Suppose $F_n \xrightarrow{\text{d}} F_0$ as $n \to 0$. It follows that

$$\mu_n((-\infty, x]) \to \mu_0((-\infty, x])$$

for every $x \in \mathbb{R}^1$ which is a continuity point of F_0. However this is a convergence of μ_n to μ_0 but for a special kind of sets, namely for infinite intervals which of course belong to $\mathcal{B}^1$. Thus we arrive at the following question.

Is it true that $F_n \xrightarrow{\text{d}} F_0$ imply $\mu_n(B) \to \mu_0(B)$ for all $B \in \mathcal{B}^1$?

In fact, the negative answer to this question is contained in the definition of convergence in distribution. Perhaps the easiest illustration is to take $F_n(x) = 1_{[1/n, \infty)}(x)$, $n \ge 1$, and $F_0(x) = 1_{[0,\infty)}(x)$, $x \in \mathbb{R}^1$. Then obviously $F_n(x) \to F_0(x)$ as $n \to \infty$ for all x except the only point $x = 0$ where F_0 has a jump (of size 1). Thus in this completely degenerate case we obviously have $F_n \xrightarrow{\text{d}} F_0$ as $n \to \infty$. Taking, for example, the Borel set $(-\infty, 0]$, we find

$$\mu_n((-\infty, 0]) = F_n(0) = 0 \not\to \mu_0((-\infty, 0]) = F_0(0) = 1 \text{ as } n \to \infty.$$

In the above case the limiting function F_0 is discontinuous. Let us assume now that F_0 is continuous everywhere on $\mathbb{R}^1$. Of course, if $F_n \xrightarrow{\text{d}} F_0$ and B is a Borel set with $\mu_0(\partial B) = 0$, then $\mu_n(B) \to \mu_0(B)$ as $n \to \infty$. Let us illustrate what we expect if $\mu_0(\partial B) \neq 0$.

Consider the r.v.s X_0 and X_n, $n \geq 1$, where X_0 is uniformly distributed on the interval (0,1) and X_n is defined by $\mathbf{P}[X_n = \frac{k}{n}] = \frac{1}{n}$ for $k = 0, 1, \ldots, n-1$, $n \geq 1$ (uniform discrete distribution). If F_0 and F_n are the d.f.s of X_0 and X_n respectively, we have (by $[a]$ standing for the integer part of a)

$$F_0(x) = \begin{cases} 0, & \text{if } x \leq 0 \\ x, & \text{if } 0 < x \leq 1 \\ 1, & \text{if } x > 1, \end{cases} \qquad F_n(x) = \begin{cases} 0, & \text{if } x \leq 0 \\ [nx]/n, & \text{if } 0 < x \leq 1 \\ 1, & \text{if } x > 1. \end{cases}$$

Since $|[nx]/n - x| \leq 1/n$ for any $x \in \mathbb{R}^1$ and any $n \geq 1$ we conclude that $X_n \xrightarrow{\text{d}} X_0$ as $n \to \infty$ (equivalently, that $F_n \xrightarrow{\text{d}} F_0$). Denote by P_0 and P_n the measures on $\mathbb{R}^1$ induced by F_0 and F_n and let Q be the set of all rational numbers in $\mathbb{R}^1$. Then $P_n(Q) = 1$ for each n, $P_0(Q) = 0$ and hence

$$\lim_{n\to\infty} P_n(Q) = 1 \neq 0 = P_0(Q).$$

In this example $P_0(\partial Q) = 1$, that is the crucial condition $P_0(\partial B) = 0$ is not satisfied for $B = Q$.

Note that the limiting function F_0 is not only continuous, it is absolutely continuous with a finite support (uniform distribution on (0,1)). A conclusion similar to the above concerning the eventual convergence of P_n to P_0 can also be derived for absolutely continuous F_0 having the whole real line $\mathbb{R}^1$ as its support. Consider a sequence of independent Bernoulli r.v.s $\xi_1, \xi_2, \ldots$: $\mathbf{P}[\xi_i = 1] = p$, $\mathbf{P}[\xi_i = 0] = q$, $q = 1 - p$, $0 < p < 1$. Denote by G_n the d.f. of the quantity $\tilde{S}_n = (S_n - np)/(npq)^{1/2}$ where $S_n = \xi_1 + \cdots + \xi_n$ and let θ be a r.v. distributed normally $\mathcal{N}(0, 1)$. Then $\tilde{S}_n \xrightarrow{\text{d}} \theta$, or equivalently, $G_n \xrightarrow{\text{d}} \Phi$ as $n \to \infty$ (Φ is the standard normal d.f.). If P_0 and P_n are the measures on $\mathbb{R}^1$ induced by Φ and G_n and the Borel set B is defined by $B = \bigcup_{k=0}^{n} \{(k - np)/(npq)^{1/2}\}$, then obviously

$$P_n(B) = \mathbf{P}[\tilde{S}_n \in b] = 1 \not\to P_0(B) = \mathbf{P}[\theta \in B] = 0 \text{ as } n \to \infty.$$

Once again this is due to the fact that the condition $P_0(\partial B) = 0$ is not satisfied.

16.3. Weak convergence of probability measures need not be uniform

Let $F_0(x)$, $F_n(x)$, $x \in \mathbb{R}^1$, $n \geq 1$ be d.f.s and μ_0, μ_n, $n \geq 1$, their corresponding probability measures on $(\mathbb{R}^1, \mathcal{B}^1)$. Let us suppose that

(1) $$\lim_{n\to\infty} \mu_n(B) = \mu_0(B) \quad \text{for all } B \in \mathcal{B}^1.$$

It is natural to ask if (1) holds uniformly in B. The example below shows that in general the answer is negative even for absolutely continuous d.f.s. Indeed, if F_0, F_n, $n \geq 1$, have densities f_0, f_n, $n \geq 1$, respectively, then (1) can be written in the form

$$\lim_{n\to\infty} \int_B f_n(x)\,\mathrm{d}x = \int_B f_0(x)\,\mathrm{d}x, \quad B \in \mathcal{B}^1. \tag{2}$$

Consider now the following functions:

$$f_n(x) = \begin{cases} 1+\sin(2\pi n x), & \text{if } x \in [0,1] \\ 0, & \text{if } x \notin [0,1], \end{cases} \qquad f_0(x) = \begin{cases} 1, & \text{if } x \in [0,1] \\ 0, & \text{if } x \notin [0,1]. \end{cases}$$

It is easy to see that f_0 and f_n for each $n \geq 1$ are density functions. Clearly f_0 is a uniform density on [0,1]. If F_0 and F_n, $n \geq 1$, are the d.f.s of f_0 and f_n, $n \geq 1$, then $F_n \xrightarrow{\mathrm{d}} F_0$ as $n \to \infty$. Moreover, applying the Riemann–Lebesgue theorem (see Rudin 1966; Royden 1968), we conclude that relation (2), and hence (1), is satisfied for this choice of f_0, f_n, $n \geq 1$, and for all $B \in \mathcal{B}^1$.

Consider now the sets $B_n = \{x \in [0,1] : f_n(x) \geq 1\}$, $n \geq 1$. Then

$$\int_{B_n} f_0(x)\,\mathrm{d}x = \frac{1}{2}, \quad \int_{B_n} f_n(x)\,\mathrm{d}x = \frac{1}{2} + \frac{1}{\pi}, \qquad n = 1, 2, \ldots .$$

Therefore in general the convergence in (1) and (2) can be non-uniform.

16.4. Two cases when the continuity theorem is not valid

Let F_0, F_n, $n \geq 1$, be d.f.s with ch.f.s ϕ_0, ϕ_n, $n \geq 1$, respectively. The continuity theorem states that as $n \to \infty$,

$$F_n \xrightarrow{\mathrm{d}} F_0 \iff \phi_n(t) \to \phi_0(t) \quad \text{where } \phi_0 \text{ is continuous at 0.}$$

Let us show that the continuity of ϕ_0 at 0 is essential.

(i) Consider the sequence of r.v.s $\{X_n, n \geq 1\}$ where $X_n \sim \mathcal{N}(0, n)$. Then the ch.f. ϕ_n of X_n is given by $\phi_n(t) = \exp(-\frac{1}{2}nt^2)$, $t \in \mathbb{R}^1$. Obviously we have $\phi_n(t) \to \tilde{\phi}(t)$ as $n \to \infty$ where

$$\tilde{\phi}(t) = \begin{cases} 0, & \text{if } t \neq 0 \\ 1, & \text{if } t = 0. \end{cases}$$

Thus the limiting function $\tilde{\phi}$ is discontinuous at 0 and hence the continuity theorem does not hold. On the other hand, we have

$$F_n(x) = \mathbf{P}[X_n \leq x] = \mathbf{P}[n^{-1/2}X_n \leq n^{-1/2}x] = \Phi(n^{-1/2}x) \to \tfrac{1}{2} \text{ as } n \to \infty.$$

Clearly $\lim_{n\to\infty} F_n(x) = \tilde{F}(x)$ exists for all $x \in \mathbb{R}^1$ but $\tilde{F}(x) \equiv \frac{1}{2}$ is not a d.f.

(ii) Consider the family of functions $\{F_n, n \geq 1\}$ where

$$F_n(x) = \begin{cases} 0, & \text{if } x < -n \\ (n+x)/(2n), & \text{if } -n \leq x \leq n \\ 1, & \text{if } x \geq n. \end{cases}$$

Then for each n, F_n is a d.f. and clearly for all $x \in \mathbb{R}^1$ we have $\lim_{n\to\infty} F_n(x) = \frac{1}{2}$. Thus the sequence $\{F_n\}$ is convergent but its limit, the constant $\frac{1}{2}$, is not a d.f. A simple explanation of this fact can be given if we consider the ch.f. ϕ_n of F_n. Since $\phi_n(t) = (\sin nt)/(nt)$ then

$$\lim_{n\to\infty} \phi_n(t) = \tilde{\phi}(t) = \begin{cases} 1, & \text{if } t = 0 \\ 0, & \text{if } t \neq 0. \end{cases}$$

Again, as in case (i), the limiting function $\tilde{\phi}$ is discontinuous at 0 and therefore the continuity theorem cannot be applied.

16.5. Weak convergence and Lévy metric

For given two d.f.s $F(x)$, $X \in \mathbb{R}^1$ and $G(x)$, $x \in \mathbb{R}^1$ the following quantity

$$L(F,G) = \inf\{\varepsilon > 0 : F(x-\varepsilon) - \varepsilon \leq G(x) \leq F(x+\varepsilon) + \varepsilon, x \in \mathbb{R}^1\}$$

is called a *Lévy metric* (distance) between F and G. Note that $L(\cdot,\cdot)$ is a metric in the space of all d.f.s and plays an essential role in probability theory; e.g. the following result is frequently used. Let F and F_n, $n \geq 1$ be d.f.s. Then, as $n \to \infty$

$$F_n \xrightarrow{\text{w}} F \iff L(F_n, F) \to 0.$$

Consider now the sequence $\{X_n, n \geq 1\}$ of independent r.v.s. Denote $S_n = X_1 + \cdots + X_n$, $s_n^2 = \mathbf{V}S_n$ and let F_n be the d.f. of S_n/s_n. Suppose the variables X_n are such that $F_n \xrightarrow{\text{w}} G$ as $n \to \infty$, where G is a d.f. (Actually, G belongs to the class of infinitely divisible distributions.) This is equivalent to saying that for any $\varepsilon > 0$ there is an index n_ε such that for all $n \geq n_\varepsilon$ we have $L(F_n, G) < \varepsilon$. Since the quantity $L(F_n, G)$ is 'small', we can suggest that another related quantity, $L(\tilde{F}_n, \tilde{G}_n)$, is also 'small'. Here $\tilde{F}_n$ is the d.f. of S_n (without normalization!) and $\tilde{G}_n(x) = G(xs_n)$. In several cases such a statement is true, but not always, as in the next example.

Let X_{nj}, $j = 1, \ldots, n$, $n \geq 1$ be independent r.v.s where

$$\mathbf{P}[X_{nj} = \pm 1] = \frac{1}{2}\left(1 - \frac{1}{n}\right), \quad \mathbf{P}[X_{nj} = \pm n\sqrt{5}] = \frac{1}{2n}, \quad j = 1, \ldots, n.$$

If $S_n = X_{n1} + \cdots + X_{nn}$, then $\mathbf{E}S_n = 0$ and $s_n^2 = \mathbf{V}S_n = 5n^2 + n - 1 \to \infty$ as $n \to \infty$. For the normalized variable $\eta_n = S_n/(n\sqrt{5})$ we have $\mathbf{E}\eta_n = 0$ and

$\mathbf{V}\eta_n = 1 + (n-1)/(5n^2)$ implying that $\mathbf{V}\eta_n \to 1$ as $n \to \infty$. Let us find the limit of the d.f. $F_n(x) = \mathbf{P}[\eta_n \le x]$ as $n \to \infty$. In this case the best way is to find the ch.f. $\psi_n(t) = \mathbf{E}[\mathrm{e}^{it\eta_n}]$. By using the structure of the variables X_{nj} and the properties of ch.f.s we find that

$$\lim_{n\to\infty} \psi_n(t) = \psi(t) = \exp(\cos t - 1), \quad t \in \mathbb{R}^1.$$

However $\psi(t) = \exp(\cos t - 1)$ is a ch.f. corresponding to a concrete r.v., say η_0 and $\eta_0 = \xi_1 - \xi_2$ with ξ_1 and ξ_2 independent r.v.s each having a Poisson distribution with parameter $\frac{1}{2}$. Hence by the continuity theorem, we have

$$F_n \xrightarrow{\mathrm{w}} G \text{ as } n \to \infty$$

with $G(x) = \mathbf{P}[\eta_0 \le x]$, $x \in \mathbb{R}^1$, or equivalently $\lim_{n\to\infty} L(F_n, G) = 0$.

Thus the quantity $L(F_n, G)$ is 'small' and we want to see if $L(\tilde{F}_n, \tilde{G}_n)$ is also 'small'. Recall that $\tilde{F}_n$ is the d.f. of S_n itself, while $\tilde{G}_n(x) = G(xs_n)$. Note first that $\tilde{F}_n$ and $\tilde{G}_n$ correspond to discrete r.v.s. Specifically, the values of S_n are in the set $\{\pm j, \pm k\sqrt{5} \pm l : j, k, l = 1, \ldots, n\}$ and the d.f. $\tilde{F}_n$ has jumps at all points of this set. Further, $\tilde{G}_n(x) = \mathbf{P}[\zeta_n \le x]$, where $\zeta_n = \eta_0 \cdot n\sqrt{5}$ and it is obvious that ζ_n takes its values in the set $\{0, \pm k \cdot n\sqrt{5} : k = 1, 2, \ldots\}$ at each point of which $\tilde{G}_n$ has a jump. In particular, $\mathbf{P}[\eta_0 = 0] > 0$ which implies that for an odd index n we can find a number $c > 0$ (expressed through $\mathbf{P}[\eta_0 = 0]$) such that $L(\tilde{F}_n, \tilde{G}_n) \ge c$. Hence we conclude that in this case the quantity $L(\tilde{F}_n, \tilde{G}_n)$ is not small.

16.6. A sequence of probability density functions can converge in the mean of order 1 without being converging everywhere

Let $f_0(x), f_1(x), f_2(x), \ldots$, $x \in \mathbb{R}^1$ be probability density functions. Here we consider two kinds of convergence of f_n to f_0: convergence almost everywhere and convergence in the mean of order 1 which are expressed respectively by

(1) $$\lim_{n\to\infty} f_n(x) = f_0(x) \text{ almost everywhere}$$

and

(2) $$\lim_{n\to\infty} \int_{-\infty}^{\infty} |f_n(x) - f_0(x)| \, \mathrm{d}x = 0.$$

Let us compare (1) and (2). According to a result by Robbins (1948), (1)$\Rightarrow$(2). However, the converse is not always true. Indeed, let

$$f_n(x) = \begin{cases} n/(n-1), & \text{if } (k-1)/n < x < k/n - 1/n^2, k = 1, 2, \ldots, n \\ 0, & \text{otherwise} \end{cases}$$

and let f_0 be the uniform density on the interval (0,1). It is easy to see that for every n, f_n is a density and if $B_n = \{x \in (0,1) : f_n(x) > 0\}$, then

$$|f_n(x) - f_0(x)| = \begin{cases} 1/(n-1), & \text{if } x \in B_n \\ 1, & \text{if } x \in B_n^c \cap (0,1). \end{cases}$$

Since the sets B_n and $B_n^c(0,1)$ have Lebesgue measures $(n-1)/n$ and $1/n$ respectively, we obtain the relation

$$\int_0^1 |f_n(x) - f_0(x)|\,dx = \frac{2}{n} \Rightarrow \lim_{n\to\infty} \int_0^1 |f_n(x) - f_0(x)|\,dx = 0$$

that is, f_n converges to f_0 in the mean of order 1. It now remains to show that

$$f_n(x) \not\to f_0(x) \equiv 1, \quad x \in (0,1).$$

For any fixed irrational number z there exist infinitely many rational numbers j/k such that $j/k - 1/k^2 < z < j/k$. This fact and the definition of f_n imply that $f_n(x) = 0$ for infinitely many n and for any fixed irrational $x \in (0,1)$. Furthermore, if x is a rational number in (0,1), then $x = j/k$ for some positive integers j and k with $j < k$, and moreover

$$f_n(x) = 0 \text{ for } n = lk,\ l = 1, 2, \ldots.$$

Thus for any $x \in (0,1)$ the densities $f_n(x)$ cannot converge to $f_0(x) \equiv 1$.

16.7. A version of the continuity theorem for distribution functions which does not hold for some densities

Let X_n be a r.v. with d.f. F_n, density f_n and ch.f. ϕ_n, $n \geq 1$. The continuity theorem provides necessary and sufficient conditions for the weak convergence of $\{F_n\}$ in terms of $\{\phi_n\}$. Now we want to find conditions which relate the ch.f.s $\{\phi_n\}$ and the densities $\{f_n\}$.

For some r.v. X_0 with d.f. F_0, density f_0 and ch.f. ϕ_0 we introduce the following three conditions:

(1) $$\lim_{n\to\infty} f_n(x) = f_0(x) \text{ for almost all } x \in \mathbb{R}^1,$$

(2) $$F_n \xrightarrow{w} F_0 \text{ as } n \to \infty,$$

(3) $$\lim_{n\to\infty} \phi_n(t) = \phi_0(t) \text{ for all } t \in \mathbb{R}^1 \text{ and } \phi_0 \text{ is continuous at } 0.$$

By the continuity theorem we have (2) $\Longleftrightarrow$ (3). According to the Scheffé theorem (see Example 14.9), (1) $\Rightarrow$ (2). Example 14.9 also shows that in general (2) $\not\Rightarrow$ (1).

Thus we conclude that $(1) \Rightarrow (3)$ and can expect that in general $(3) \not\Rightarrow (1)$. Let us illustrate by an example that indeed $(3) \not\Rightarrow (1)$.

Consider the standard normal density $\varphi(x) = (2\pi)^{-1/2}\exp(-\frac{1}{2}x^2)$ and its ch.f. $\phi_0(t) = \exp(-\frac{1}{2}t^2)$. Define the functions

$$(4) \qquad f_\lambda(x) = \varphi(x)(1 - \cos\lambda x)/(1 - \phi_0(\lambda)), \quad x \in \mathbb{R}^1,$$

$$(5) \qquad \psi_\lambda(t) = [2\phi_0(t) - \phi_0(t+\lambda) - \phi_0(t-\lambda)]/[2(1 - \phi_0(\lambda))], \quad t \in \mathbb{R}^1$$

where λ is any real number (e.g. take $\lambda = n$). It is not difficult to check that for each λ, $f_\lambda(x)$, $x \in \mathbb{R}^1$ is a probability density function, $\psi_\lambda(t)$, $t \in \mathbb{R}^1$ is a ch.f., and moreover, ψ_λ corresponds to f_λ. Further, we find

$$(6) \qquad \lim_{\lambda\to\infty} \psi_\lambda(t) = \phi_0(t) = \exp(-\tfrac{1}{2}t^2) \text{ for all } t \in \mathbb{R}^1$$

where the limiting function ϕ_0 is continuous at 0 and thus (3) is satisfied.

However,

$$(7) \qquad \lim_{\lambda\to\infty} f_\lambda(x) \neq \varphi(x) = (2\pi)^{-1/2}\exp(-\tfrac{1}{2}x^2)$$

and hence condition (1) does not hold.

Comparing (6) and (7) we see that in general the pointwise convergence of the ch.f.s ϕ_n given by (3) is not enough to ensure the convergence (1) of the densities f_n. At this point the following result may be useful (see Feller 1971).

Let ϕ_n and ϕ be absolutely integrable ch.f.s such that

$$(8) \qquad \lim_{n\to\infty} \int_{-\infty}^{\infty} |\phi_n(t) - \phi(t)|\,\mathrm{d}t = 0.$$

Then the corresponding d.f.s F_n and F have bounded continuous densities f_n and f respectively, and (8) implies that

$$(9) \qquad \lim_{n\to\infty} f_n(x) = f(x) \text{ uniformly in } x,\ x \in \mathbb{R}^1.$$

Obviously, in the above specific example, condition (9) is not satisfied (see (7)). It is easy to see that the pointwise convergence given by (6) does not imply the integral convergence (8).

16.8. Weak convergence of distribution functions does not imply convergence of the moments

Let F and F_n, $n \geq 1$ be d.f.s. Denote by m_k and $m_k^{(n)}$ their kth moments:

$$m_k = \int_{-\infty}^{\infty} x^k\,\mathrm{d}F(x), \quad m_k^{(n)} = \int_{-\infty}^{\infty} x^k\,\mathrm{d}F_n(x), \quad k = 1, 2, \ldots.$$

According to the Fréchet–Shohat theorem (see Feller 1971), if $m_k^{(n)} \to m_k$ as $n \to \infty$ for all k and the moment sequence $\{m_k\}$ determines F uniquely, then

$$F_n \xrightarrow{\text{w}} F \text{ as } n \to \infty. \tag{1}$$

(For such results also see the works of Kendall and Rao 1950, Lukacs 1970.)

Now let us answer the converse question: does the weak convergence (1) imply convergence of the moments $m_k^{(n)}$ to m_k? By two examples we show that (1) can hold even if $m_k^{(n)} \not\to m_k$ as $n \to \infty$ for any k.

(i) Consider the family of d.f.s $\{F_n, n \geq 1\}$ where

$$F_n(x) = \left(1 - \frac{1}{n}\right) \frac{1}{\sqrt{2\pi}} \int_{-\infty}^{x} \mathrm{e}^{-u^2/2}\, \mathrm{d}u + \frac{1}{2n}(1 + 1_{[n,\infty)}(x)), \;\; x \in \mathbb{R}^1.$$

It is easy to see that

$$\lim_{n\to\infty} F_n(x) = \Phi(x) \text{ for all } x \in \mathbb{R}^1$$

where Φ is the standard normal d.f., that is $F_n \xrightarrow{\text{w}} \Phi$ as $n \to \infty$.

However, the moments $m_k^{(n)}$ of any order k of F_n tend to infinity as $n \to \infty$ and hence $m_k^{(n)}$ cannot converge to the moments m_k of $\mathcal{N}(0,1)$. Recall that here $m_{2k-1} = 0$, $m_{2k} = (2k-1)!!$, $k = 1, 2, \ldots$.

(ii) Let F_n be the d.f. of a r.v. X_n distributed uniformly on the interval $[0, n]$ and F_0 be the d.f. of a degenerate r.v. X_0, for example, $X_0 = 0$. Define

$$G_n(x) = \frac{1}{n} F_n(x) + \left(1 - \frac{1}{n}\right) F_0(x), \; x \in \mathbb{R}^1, \; n \geq 1.$$

Then $\{G_n, n \geq 1\}$ is a sequence of d.f.s. The limit behaviour of $\{G_n\}$ can easily be investigated in terms of the corresponding ch.f.s $\{\psi_n\}$. Since

$$\psi_n(t) = \int_{-\infty}^{\infty} \mathrm{e}^{itx}\, \mathrm{d}G_n(x) = \frac{1}{n} \frac{\mathrm{e}^{itn} - 1}{itn} + \left(1 - \frac{1}{n}\right)$$

we find that $\lim_{n\to\infty} \psi_n(t) = 1$, $t \in \mathbb{R}^1$ which implies that

$$\lim_{n\to\infty} G_n(x) = F_0(x)$$

for all $x \in \mathbb{R}^1$ except $x = 0$ (the value of X_0; the only point of jump of F_0).

It remains for us to clarify whether the moments $m_k^{(n)}$ of G_n converge to the moments m_k of F_0. We have

$$m_k^{(n)} = \int_{-\infty}^{\infty} x^k\, \mathrm{d}G_n(x) = \frac{n^k}{k+1} \to \infty \text{ as } n \to \infty$$

for every k, $k = 1, 2, \ldots$, while the moments m_k of F_0 are all zero.

16.9. Weak convergence of a sequence of distributions does not always imply the convergence of the moment generating functions

Recall first a version of the continuity theorem. Suppose $\{F_n, n = 1, 2, \ldots\}$ are d.f.s and $\{M_n, n = 1, 2, \ldots\}$, the corresponding m.g.f.s $M_n(z)$ exist for all $|z| \leq r_0$ and all n. If F and M is another pair of a d.f. and m.g.f. such that $M_n(z) \to M(z)$ as $n \to \infty$ for all $|z| \leq r_1$ where $r_1 < r_0$, then $F_n \xrightarrow{\text{w}} F$.

Thus under general conditions the convergence of the m.g.f.s implies the weak convergence of the corresponding d.f.s and this motivates us to ask the inverse question: if $F_n \xrightarrow{\text{w}} F$, does it follow that $M_n \to M$ as $n \to \infty$?

Intuitively we may guess that the answer is 'no' rather than 'yes', simply because when talking about a m.g.f. we assume at least the existence of moments of any order. The latter is not necessary for the weak distribution.

A simple example shows that the answer to the above question is negative. Consider the d.f.s F and F_n, $n = 1, 2, \ldots$, defined by

$$F(x) = \begin{cases} 0, & \text{if } x < 0 \\ 1, & \text{if } x \geq 0, \end{cases} \qquad F_n(x) = \begin{cases} 0, & \text{if } x < -n \\ \frac{1}{2} + c_n \arctan(nx), & \text{if } -n \leq x < n \\ 1, & \text{if } x \geq n \end{cases}$$

where $c_n = 1/[2\arctan(n^2)]$.

It is easy to check that $F_n(x) \to F(x)$ as $n \to \infty$ at all points of continuity of F. Hence $F_n \xrightarrow{\text{w}} F$. Since F is a degenerate distribution concentrated at 0, then its m.g.f. $M_n(z) = 1$ for all z. Further, the m.g.f. $M_n(z)$ of F_n,

$$M_n(z) = \int_{-n}^{n} c_n e^{zx} \frac{n}{1 + n^2x^2}\, dx$$

exists for all z. It is almost obvious that $M_n(z) \to M(z)$ as $n \to \infty$ only for $z = 0$. If $z \neq 0$, $M_n(z) \not\to M(z)$ as $n \to \infty$ since $|M_n(z)| \to \infty$ as $n \to \infty$.

16.10. Weak convergence of a sequence of distribution functions does not always imply their convergence in the mean

Let $F_0, F_1, F_2, \ldots$ be d.f.s. Suppose for some $\beta > 0$ the following relation holds:

$$\text{(1)} \qquad \lim_{n\to\infty} \int_{-\infty}^{\infty} |F_n(x) - F_0(x)|^\beta\, dx = 0.$$

From here it is easy to derive that $F_n \xrightarrow{\text{w}} F$ as $n \to \infty$.

Now let us analyse (1) but in the opposite direction. Firstly, suppose that $F_n \xrightarrow{\text{w}} F_0$. The question is, under what additional assumptions we can obtain a relation like (1) with a suitable $\beta > 0$? One possible answer is contained in the following result (see Laube 1973). If $F_n \xrightarrow{\text{w}} F_0$ and for some $\gamma > 0$,

$$\text{(2)} \qquad \sup_{n\geq 1} \int_{-\infty}^{\infty} |x|^\gamma\, dF_n(x) < \infty$$

then F_n tends to F_0 in the mean of order $\beta > 1/\gamma$, that is (2) and the weak convergence of F_n to F_0 imply (1) with $\beta > 1/\gamma$.

Our aim now is to show that (1) need not be true if we take $\beta = 1/\gamma$. To see this, consider the following d.f.s:

$$F_0(x) = 1_{[0,\infty)}(x),\ x \in \mathbb{R}^1,$$

$$F_n(x) = \frac{1}{n}1_{[-n,0)}(x) + 1_{[0,\infty)}(x),\ x \in \mathbb{R}^1,\ n = 1, 2, \dots .$$

Then it can be easily seen that

$$\int_{-\infty}^{\infty} |x|\,\mathrm{d}F_n(x) = 1, \quad \lim_{n\to\infty} F_n(x) = 1_{[0,\infty)}(x) = F_0(x),\ \ x \in \mathbb{R}^1.$$

Obviously condition (2) is valid for $\gamma = 1$. However, relation (1) does not hold for $\beta = 1$, that is, for $\beta = 1/\gamma$, since

$$\int_{-\infty}^{\infty} |F_n(x) - F_0(x)|\,\mathrm{d}x = 1 \ \text{ for all } n.$$

Finally, note that relations like (1) can be used to obtain estimates for the global convergence behaviour in the central limit theorem (CLT) (see Laube 1973).

SECTION 17. CENTRAL LIMIT THEOREM

Let $\{X_n, n \geq 1\}$ be a sequence of independent r.v.s defined on the probability space $(\Omega, \mathcal{F}, \mathbf{P})$. As usual, denote

$$\begin{aligned} &S_n = X_1 + \cdots + X_n, \quad && a_k = \mathbf{E}X_k, \quad A_n = \mathbf{E}S_n = a_1 + \cdots + a_n, \\ &\sigma_k^2 = \mathbf{V}X_k, && s_n^2 = \mathbf{V}S_n = \sigma_1^2 + \cdots + \sigma_n^2. \end{aligned}$$

We say that the sequence $\{X_n\}$ satisfies the **central limit theorem** (CLT) (or, that $\{X_n\}$ obeys the CLT) if

$$\lim_{n\to\infty} \mathbf{P}[(S_n - A_n)/s_n \leq x] = \Phi(x) = \frac{1}{\sqrt{2\pi}} \int_{-\infty}^{x} \mathrm{e}^{-u^2/2}\,\mathrm{d}u \text{ for all } x \in \mathbb{R}^1.$$

Let F_k denote the d.f. of X_k. Clearly, we can suppose that $\mathbf{E}X_k = 0$ for all $k \geq 1$. Now introduce the following three conditions:

(L) $$\lim_{n\to\infty} \frac{1}{s_n^2} \sum_{k=1}^{n} \int_{|u|\geq\varepsilon s_n} u^2\,\mathrm{d}F_k(u) = 0 \text{ for each } \varepsilon > 0$$

(*Lindeberg condition*);

(F) $$\lim_{n\to\infty} \max_{1\leq k\leq n} \frac{\sigma_k^2}{s_n^2} = 0$$

(*Feller condition*);

(UAN) $$\lim_{n\to\infty} \max_{1\le k\le n} \mathbf{P}[|X_{kn} \ge \varepsilon] = 0 \ \text{ where } X_{kn} = X_k/s_n.$$

(*uniform asymptotic negligibility condition* (u.a.n. condition)).

Now we shall formulate in a compact form two fundamental results.

Lindeberg theorem.

$$\text{(L)} \Rightarrow \text{(CLT)}$$

Lindeberg–Feller theorem. If (F), then

$$\text{(L)} \iff \text{(CLT)}$$

or if (UAN), then

$$\text{(L)} \iff \text{(CLT)}.$$

The proof of these theorems and several other related topics can be found in many books. We refer the reader to the books by Gnedenko (1962), Fisz (1963), Breiman (1968), Billingsley (1968, 1995), Thomasian (1969), Rényi (1970), Feller (1971), Ash (1972), Chung (1974), Chow and Teicher (1978), Loève (1978), Laha and Rohatgi (1979) and Shiryaev (1995).

The examples below demonstrate the range of validity of the CLT and examine the importance of the conditions under which the CLT does hold. Some related questions are also considered.

17.1. Sequences of random variables which do not satisfy the central limit theorem

(i) Let $X_1, X_2, \ldots$ be independent r.v.s defined as follows: $\mathbf{P}[X_1 = \pm 1] = \frac{1}{2}$ and for $k \ge 2$ and some c, $0 < c < 1$,

$$\mathbf{P}[X_k = \pm 1] = \frac{1}{2}(1-c),\ \mathbf{P}[X_k = \pm k] = \frac{1}{2k^2}c,\ \mathbf{P}[X_k = 0] = \left(1 - \frac{1}{k^2}\right)c.$$

First let us check if the Lindeberg condition is satisfied. We have

$$\frac{1}{s_n^2}\sum_{k=1}^n \int_{|x|\ge\varepsilon s_n} x^2\,\mathrm{d}F_k(x) = \frac{1}{n}\sum_{k=1}^n \mathbf{E}[X_k^2 I(|X_k| \ge \varepsilon\sqrt{n})].$$

If n is large enough and such that $\varepsilon\sqrt{n} > 1$, $\varepsilon > 0$ is fixed, then we find

$$\frac{1}{n}\sum_{k=[\varepsilon\sqrt{n}]}^{n} k^2\mathbf{P}[|X_k| = k] \approx \frac{1}{n}(n - \varepsilon\sqrt{n})c \to c > 0.$$

Therefore the given sequence $\{X_k\}$ does not satisfy the Lindeberg condition. However, this does not mean that the CLT fails to hold for the sequence $\{X_k\}$ because the Lindeberg condition is only a sufficient condition. Actually the sequence $\{X_k\}$ does not obey the CLT. This follows from the fact that X_k/s_n satisfy the u.a.n. condition. Indeed,

$$\mathbf{P}[|X_k/s_n| \geq \varepsilon] = \mathbf{P}[|X_k| \geq \varepsilon\sqrt{n}] = \begin{cases} 0, & \text{if } k < \varepsilon\sqrt{n} \\ \frac{1}{k^2}c, & \text{if } k \geq \varepsilon\sqrt{n}. \end{cases}$$

Thus

$$\max_{1\leq k\leq n} \mathbf{P}[|X_k/s_n| \geq \varepsilon] \leq \frac{1}{\varepsilon^2 n}c \to 0 \text{ as } n \to \infty.$$

Now $S_n/s_n \xrightarrow{\mathrm{d}} \xi$ where $\xi \sim \mathcal{N}(0,1)$; this and the u.a.n. condition would imply the Lindeberg condition which, as we have seen above, is not satisfied.

Thus our final conclusion is that the Lindeberg condition is not satisfied and the CLT does not hold.

(ii) Let the r.v. Y take two values, 1 and -1, with probability $\frac{1}{2}$ each, and let $\{Y_k, k \geq 1\}$ be a sequence of independent copies of Y. Define a new sequence $\{X_k, k \geq 1\}$ where $X_k = \sqrt{15}Y_k/4^k$ and let $S_n = X_1 + \cdots + X_n$. Since $\mathbf{E}Y = 0$ and $\mathbf{V}X = 1$ we easily find that

$$\mathbf{E}S_n = 0 \text{ and } s_n^2 = \mathbf{V}S_n = 1 - \left(\tfrac{1}{16}\right)^n.$$

Thus $s_n^2 \approx 1$ for large n (this is why the factor $\sqrt{15}$ was involved).

On the other hand it is obvious that

$$\mathbf{P}[|S_n| \leq \tfrac{1}{2}] = 0 \text{ for every } n \geq 1.$$

Therefore the probabilities $\mathbf{P}[S_n \leq x]$ cannot converge to the standard normal d.f. $\Phi(x)$ for all x, so the sequence $\{X_k\}$ does not obey the CLT. Note that in this example X_1 'dominates' the other terms.

(iii) Suppose that for each n,

$$S_n = X_{n1} + X_{n2} + \cdots + X_{nn}$$

where $X_{n1}, \ldots, X_{nn}$ are independent r.v.s and each has a Poisson distribution with mean $1/(2n)$. We could expect that the distribution of the normalized quantity $(S_n - \mathbf{E}S_n)/\sqrt{\mathbf{V}S_n}$ will tend to the standard normal d.f. Φ. However, this is not the case, in spite of the fact that

$$\mathbf{P}[X_{nk} = 0] = \mathrm{e}^{-1/(2n)} \approx 1 \quad \text{for large } n$$

that is each X_{nk} is 'almost' zero. It is enough to note that for each n the sum S_n has a Poisson distribution with parameter $\frac{1}{2}$. In particular, $\mathbf{P}[S_n = 0] = \mathrm{e}^{-1/2}$ implying that the distribution of $(S_n - \mathbf{E}S_n)/\sqrt{\mathbf{V}S_n}$ cannot be close to Φ.

17.2. How is the central limit theorem connected with the Feller condition and the uniform negligibility condition?

Let $\{X_n, n \geq 1\}$ be a sequence of independent r.v.s such that $X_n \sim \mathcal{N}(0, \sigma_n^2)$ where $\sigma_1^2 = 1$ and $\sigma_k^2 = 2^{k-2}$ for $k \geq 2$. Then $S_n = X_1 + \cdots + X_n$ has variance $s_n^2 = 2^{n-1}$. Since $X_k/s_k \sim \mathcal{N}(0, \frac{1}{2})$ we find that

$$S_n/s_n \sim \mathcal{N}(0,1) \text{ for each } n$$

and therefore the CLT for $\{X_k\}$ is satisfied trivially. Further,

$$\lim_{n\to\infty} \max_{1\leq k\leq n} \frac{\sigma_k^2}{s_n^2} = \lim_{n\to\infty} \frac{2^{n-2}}{2^{n-1}} = \frac{1}{2} \neq 0$$

and moreover

$$\max_{1\leq k\leq n} \mathbf{P}[|X_k|/s_n \geq \varepsilon] \geq \mathbf{P}[|X_n|/s_n \geq \varepsilon] = 1 - \frac{1}{\sqrt{\pi}} \int_{-\varepsilon}^{\varepsilon} \mathrm{e}^{-u^2}\, du > 0.$$

Hence neither the Feller condition nor the u.a.n. condition holds. This implies that the Lindeberg condition also does not hold. However, despite these facts the sequence $\{X_n\}$ obeys the CLT.

17.3. Two 'equivalent' sequences of random variables such that one of them obeys the central limit theorem while the other does not

Consider again the sequence of independent r.v.s $\{X_n, n \geq 1\}$ from Example 17.1: namely, $\mathbf{P}[X_1 = \pm 1] = \frac{1}{2}$ and for $k \geq 2$ and $0 < c < 1$,

$$\mathbf{P}[X_k = \pm 1] = \frac{1}{2}(1-c),\ \mathbf{P}[X_k = \pm k] = \frac{1}{2k^2}c,\ \mathbf{P}[X_k = 0] = \left(1 - \frac{1}{k^2}\right)c.$$

Using truncation we define the sequence $\{\tilde{X}_{nk}, k = 1, \ldots, n, n \geq 1\}$ by

$$\tilde{X}_{nk} = \begin{cases} X_k, & \text{if } |X_k| \leq \sqrt{n} \\ 0, & \text{if } |X_k| > \sqrt{n}. \end{cases}$$

Denote $\tilde{S}_n = \tilde{X}_{n1} + \cdots + \tilde{X}_{nn}$, $\tilde{s}_n^2 = \mathbf{V}\tilde{S}_n$. Since $\mathbf{V}\tilde{X}_{nk} = 1$ if $k \leq \sqrt{n}$ and $\mathbf{V}\tilde{X}_{nk} = 1 - c$ if $k > \sqrt{n}$, we find that $\tilde{s}_n^2 = [\sqrt{n}] + (1-c)(n - [\sqrt{n}]) \approx n(1-c)$ and thus

$$\frac{1}{\tilde{s}_n^2} \sum_{k=1}^{n} \mathbf{E}[\tilde{X}_{nk}^2 I(|\tilde{X}_{nk}| \geq \varepsilon \tilde{s}_n)] \approx \frac{1}{n(1-c)}(\sqrt{n} - \varepsilon\sqrt{n}(1-c))c \to 0.$$

Therefore the Lindeberg condition holds and $\tilde{S}_n/\tilde{s}_n \xrightarrow{\mathrm{d}} \eta$ where η is a r.v. distributed $\mathcal{N}(0,1)$. So the sequence $\{\tilde{X}_{nk}\}$ obeys the CLT.

We shall show that the sequences $\{S_n\}$ and $\{\tilde{S}_n\}$ (not $\{X_n\}$ and $\{X_{nk}\}$) are 'equivalent' in the following sense:

(1) $$\mathbf{P}[S_n \neq \tilde{S}_n] \to 0 \quad \text{as} \quad n \to \infty.$$

Indeed,

$$\begin{aligned}\mathbf{P}[S_n \neq \tilde{S}_n] &\leq \sum_{k=1}^{n} \mathbf{P}[X_k \neq \tilde{X}_{nk}] \\ &\leq \sum_{k=1}^{n} \mathbf{P}[|X_k| > \sqrt{n}] \leq \sum_{k=[\sqrt{n}]}^{n} \mathbf{P}[|X_k| = k].\end{aligned}$$

Therefore

$$\mathbf{P}[|X_k| = k] = \frac{1}{k^2}\left(1 - \frac{1}{c}\right) \text{ and } \sum_{k=1}^{\infty} \frac{1}{k^2} < \infty \Rightarrow \mathbf{P}[S_n \neq \tilde{S}_n] \to 0 \text{ as } n \to \infty.$$

However (see Example 17.1) the sequence $\{X_n\}$ does not obey the CLT.

Thus we have constructed two sequences, $\{X_n\}$ and $\{\tilde{X}_{nk}\}$, which are equivalent in the sense of (1) and such that the CLT holds for $\{\tilde{X}_{nk}\}$ but does not hold for $\{X_n\}$. Note again that the Lindeberg condition is valid for $\{\tilde{X}_{nk}\}$ but not for $\{X_n\}$.

17.4. If the sequence of random variables $\{X_n\}$ satisfies the central limit theorem, what can we say about the variance of $S_n/\sqrt{\mathbf{V}S_n}$?

Consider two sequences, $\{X_k, k \geq 1\}$ and $\{Y_k, k \geq 1\}$, each consisting of independent r.v.s and such that

$$\mathbf{P}[X_k = \pm 1] = \tfrac{1}{2}(1 - k^{-2}), \quad \mathbf{P}[X_k = \pm k] = \tfrac{1}{2}k^{-2}, \mathbf{P}[Y_k = \pm 1] = \tfrac{1}{2}.$$

Denote

$$S_n = Y_1 + \cdots + Y_n, \quad \tilde{S}_n = X_1 + \cdots + X_n.$$

Obviously the sequence $\{Y_n\}$ obeys the CLT: that is, $S_n/\sqrt{n} \xrightarrow{\text{d}} \xi$ where $\xi \sim \mathcal{N}(0, 1)$. The truncation principle (see Gnedenko 1962; Feller 1971), when applied to the sequence $\{X_k\}$, shows that $\tilde{S}_n/\sqrt{n}$ has the same asymptotic behaviour as that of $S_n/\sqrt{n}$. Thus we conclude that $\tilde{S}_n/\sqrt{n} \xrightarrow{\text{d}} \eta$ as $n \to \infty$ where $\eta \sim \mathcal{N}(0, 1)$.

Then we can expect intuitively that

$$\mathbf{V}[S_n/\sqrt{n}] \to 1 \quad \text{and} \quad \mathbf{V}[\tilde{S}_n/\sqrt{n}] \to 1 \quad \text{as } n \to \infty.$$

For the sequence $\{Y_k\}$ we have $\mathbf{E}Y_k = 0$, $\mathbf{V}Y_k = 1$. Thus for each n,

$$1 = \mathbf{V}[S_n/\sqrt{n}] \to 1 \text{ as } n \to \infty.$$

On the other hand, for $\{X_k\}$ we find $\mathbf{E}X_k = 0$, $\mathbf{V}X_k = 2 - 1/k^2$ and

$$\mathbf{V}[\tilde{S}_n/\sqrt{n}] = \frac{1}{n}\sum_{k=1}^{n}\left(2 - \frac{1}{k^2}\right) = 2 - \frac{1}{n}\sum_{k=1}^{n}\frac{1}{k^2} \to 2 \text{ as } n \to \infty$$

(since $\sum_{k=1}^{\infty}(1/k^2) < \infty$), that is $\mathbf{V}[\tilde{S}_n/\sqrt{n}] \not\to 1$ as we assumed.

Therefore the CLT does not ensure in general the convergence of the moments of the normed sum $S_n/\sqrt{n}$ to the moments of the normal distribution $\mathcal{N}(0,1)$. For the convergence of the moments we need some additional integrability conditions. In particular,

$$\mathbf{E}[|S_n/\sqrt{n}|^{2+\delta}] < \infty,\ \delta > 0 \Rightarrow \mathbf{V}[S_n/\sqrt{n}] \to \mathbf{V}\xi.$$

17.5. Not every interval can be a domain of normal convergence

Suppose $\{X_n, n \geq 1\}$ is a sequence of i.i.d. r.v.s which satisfies the CLT. Denote by F_n the d.f. of $(S_n - \mathbf{E}S_n)/(\mathbf{V}S_n)^{1/2}$ where $S_n = X_1 + \cdots + X_n$. The uniform convergence $F_n(x) \to \Phi(x)$, $x \in \mathbb{R}^1$ implies

$$(1) \qquad \lim_{n\to\infty}\frac{1 - F_n(x)}{1 - \Phi(x)} = 1 \quad \text{uniformly in } x \text{ on any finite interval of } \mathbb{R}^1.$$

Note that (1) will hold uniformly on intervals of the type $[0, b_n]$ whose length b_n increases with n. In general, intervals for which (1) holds are called *domains of normal convergence*. Obviously such intervals exist, but we now show that not every interval can be a domain of normal convergence.

Consider $X_1, X_2, \ldots$ to be independent Bernoulli r.v.s with parameter p: that is, $\mathbf{P}[X_1 = 1] = p = 1 - \mathbf{P}[X_1 = 0]$. Obviously the sequence $\{X_n\}$ obeys the CLT. If $S_n = X_1 + \cdots + X_n$ then $\mathbf{E}S_n = np$, $s_n^2 = \mathbf{V}S_n = np(1-p)$ and

$$\begin{aligned} 1 - F_n(x) &= \mathbf{P}\left[(np(1-p))^{-1/2}\left(\sum_{k=1}^{n} X_k - np\right) > x\right] \\ &= \mathbf{P}\left[\sum_{k=1}^{n} X_k > x(np(1-p))^{1/2} + np\right]. \end{aligned}$$

Hence for an arbitrary $x > (n(1-p)/p)^{1/2}$ we obtain the equality

$$[1 - F_n(x)]/[1 - \Phi(x)] = 0$$

which clearly contradicts (1). Therefore (1) cannot hold for any interval of the type $[0, O(\sqrt{n})]$. In particular the interval $[0, c_p\sqrt{n}]$, where $c_p > ((1-p)/p)^{1/2}$ (p is fixed), cannot be a domain of normal convergence.

Finally note that intervals of the type $[0, o(\sqrt{n})]$ are domains of normal convergence. This follows from the well known Berry–Esseen estimates in the CLT (see Feller 1971; Chow and Teicher 1978; Shiryaev 1995).

17.6. The central limit theorem does not always hold for random sums of random variables

Let $\{X_n, n \geq 1\}$ be a sequence of r.v.s which satisfies the CLT. Take another sequence $\{\nu_n, n \geq 1\}$ of integer-valued r.v.s such that $\nu_n \xrightarrow{\text{a.s.}} \infty$ as $n \to \infty$ and define $T_n = S_{\nu_n} = X_1 + \cdots + X_{\nu_n}$ and $b_n^2 = \mathbf{V}T_n$. If

$$\lim_{n\to\infty} \mathbf{P}[(T_n - \mathbf{E}T_n)/b_n \leq x] = \Phi(x), \quad x \in \mathbb{R}^1$$

we say that the CLT holds for the *random sums* $\{S_{\nu_n}\}$ generated by $\{X_n\}$ and $\{\nu_n\}$.

In the next two examples we show that the CLT does not always hold for $\{S_{\nu_n}\}$. In both cases $\{X_n, n \geq 1\}$ is a sequence of i.i.d. r.v.s such that $\mathbf{P}[X_1 = \pm 1] = \frac{1}{2}$. Obviously if $\nu_n = n$ a.s. for each n, then $T_n = S_n = X_1 + \cdots + X_n$, $b^2 = n$ and $T_n/b_n \xrightarrow{\text{d}} \xi$ where $\xi \sim \mathcal{N}(0,1)$.

(i) Define the sequence $\{\nu_n, n \geq 0\}$ as follows:

$$\nu_0 = 0 \text{ and } \nu_n = \min\{k > \nu_{n-1} : S_k = (-1)^k\} \text{ for } n \geq 1.$$

Then $\nu_n \xrightarrow{\text{a.s.}} \infty$ as $n \to \infty$, $b_n^2 = \mathbf{V}T_n = n^2$ and clearly

$$\mathbf{P}[T_n/b_n = (-1)^n] = 1.$$

It follows that the distribution of T_n/b_n does not have a limit as $n \to \infty$ and hence the CLT cannot be valid for the random sums $\{S_{\nu_n}\}$.

(ii) Let $\{\nu_n, n \geq 1\}$ be independent r.v.s such that ν_n takes the values n and $2n$, with probabilities p and $q = 1 - p$ respectively. Suppose additionally that $\{\nu_n\}$ is independent of $\{X_n\}$. Then

$$b_n^2 = \mathbf{V}T_n = p\mathbf{E}[S_n^2] + q\mathbf{E}[S_{2n}^2] = (1+q)n.$$

It is easy to check that T_n/b_n does not converge in distribution to a r.v. $\xi \sim \mathcal{N}(0,1)$. More precisely, $\mathbf{P}[T_n/b_n \leq x]$ converges to the mixture of the distributions of two r.v.s, $\xi_1 \sim \mathcal{N}(0, (1+q)^{-2})$ and $\xi_2 \sim \mathcal{N}(0, 2(1+q)^{-2})$ with weights p and q respectively.

17.7. Sequences of random variables which satisfy the integral but not the local central limit theorem

Let $\{X_n, n \geq 1\}$ be a sequence of independent r.v.s. Denote by F_n and f_n respectively the d.f. and the density of $(S_n - \mathbf{E}S_n)/s_n$ where as usual $S_n = X_1 + \cdots + X_n$, $s_n^2 = \mathbf{V}S_n$.

Let us set down the following relations:

$$(1) \qquad \lim_{n\to\infty} F_n(x) = \Phi(x) = \frac{1}{\sqrt{2\pi}} \int_{-\infty}^{x} \mathrm{e}^{-u^2/2}\, \mathrm{d}u, \quad x \in \mathbb{R}^1;$$

(2) $$\lim_{n\to\infty} f_n(x) = \varphi(x) = (2\pi)^{-1/2}\mathrm{e}^{-x^2/2}, \quad x \in \mathbb{R}^1.$$

Recall that if (1) holds we say that the sequence $\{X_n\}$ obeys the *integral CLT*, while in case (2) we say that $\{X_n\}$ obeys the *local CLT* (for the densities). It is easy to see that (2)$\Rightarrow$(1). However, in general weak convergence does not imply convergence of the corresponding densities (see Example 14.9). Note that in (1) and (2) the limit distribution is $\mathcal{N}(0,1)$. Question: is the implication $(1) \Rightarrow (2)$ true? Two examples will be considered where $(1) \not\Rightarrow (2)$. In the first example the variables are identically distributed while in the second they have different distributions.

(i) Let X be a r.v. with density

(3) $$f(x) = \begin{cases} 0, & \text{if } |x| \ge \mathrm{e}^{-1} \\ 1/(2|x|\log^2|x|), & \text{if } |x| < \mathrm{e}^{-1} \end{cases} \text{ with } f(0) = c, 0 < c < \infty.$$

The density f is unbounded around $x = 0$, however, since X is a bounded r.v., the sequence $\{X_n, n \ge 1\}$ of independent copies of X satisfies the (integral) CLT. So the aim is to study the limit behaviour of the density f_n of $(X_1 + \cdots + X_n)/(\sigma\sqrt{n})$ where $\sigma^2 = \mathbf{V}X = \int_0^{\mathrm{e}^{-1}} (x/\log^2 x)\,\mathrm{d}x$, $0 < \sigma^2 < \infty$.

If g_2 is the density of the sum $X_1 + X_2$ then g_2 is expressed by the convolution

$$g_2(x) = \int_{-\mathrm{e}^{-1}}^{\mathrm{e}^{-1}} f(u)f(x-u)\,\mathrm{d}u.$$

Let us now try to find a lower bound for g_2. It is enough to consider x in a neighbourhood of 0; in particular we can assume that $|x| < \mathrm{e}^{-1}$, and, even more, that $0 < x < \mathrm{e}^{-1}$. Then $g_2(x) \ge \int_{-x}^{x} f(u)f(x-u)\,\mathrm{d}u$. Since $f(x-u)$ reaches its minimum in the domain $|u| \le x$ at $u = 0$, we have

$$g_2(x) \ge \frac{1}{2x\log^2 x}\int_{-x}^{x} \frac{1}{2|u|\log^2|u|}\,\mathrm{d}u = \frac{1}{2x|\log^3 x|}.$$

Analogously we establish that in a neighbourhood of 0 the density g_3 of the sum $X_1 + X_2 + X_3$ satisfies the inequality

$$g_3(x) > \frac{c_3}{x\log^4 x}, \quad c_3 = \text{constant} > 0.$$

In general, if g_n is the density of $X_1 + \ldots + X_n$ we find that around 0,

$$g_n(x) > \frac{c_n}{x|\log^{n+1} x|}, \quad c_n = \text{constant} > 0.$$

Thus for each n, $g_n(x)$ takes an infinite value at $x = 0$. Since f_n is obtained from g_n by suitable norming, then $f_n(x)$ cannot converge to $\varphi(x)$ as $n \to \infty$.

Therefore the sequence $\{X_n\}$ defined by the density (3) does not obey the local CLT although the integral CLT holds.

(ii) Let $\{X_n, n \geq 1\}$ be independent r.v.s where X_n has density

$$(4)\quad f_n(x) = \begin{cases} 2^n, & \text{if } -2^{-n-2} \leq x \leq 2^{-n-2} \text{ or } 1-2^{-n-3} < |x| < 1+2^{-n-3} \\ 0, & \text{otherwise.} \end{cases}$$

It is easy to see that $\mathbf{E}X_n = 0$, $\mathbf{V}X_n = \frac{1}{2} + 5/(3{\cdot}2^{2k+7})$. Then for an arbitrary $k \geq 1$, $\frac{1}{2} < \mathbf{V}X_k < 1$, the Lindeberg condition is satisfied and hence the sequence $\{X_n\}$ obeys the (integral) CLT.

Denote by $g_k(x)$, $x \in \mathbb{R}^1$ the density of the sum $S_k = X_1 + \cdots + X_k$. Then for $k = 2$, g_2 is the convolution of f_1 and f_2, that is

$$g_2(x) = (f_1 * f_2)(x) = \int_{-\infty}^{\infty} f_1(u) f_2(x-u)\, du.$$

Let us find the value of $g_2(x)$ at the point $x = \frac{1}{2}$. By (4) we have

$$\begin{aligned} f_1(u) &\neq 0, \quad \text{if} \quad -\tfrac{1}{8} \leq u \leq \tfrac{1}{8} \quad \text{or} \quad \tfrac{15}{16} < |u| < \tfrac{17}{16}, \\ f_2\left(\tfrac{1}{2} - u\right) &\neq 0, \quad \text{if} \quad \tfrac{7}{16} \leq u \leq \tfrac{9}{16} \quad \text{or} \quad \tfrac{15}{32} < |u| < \tfrac{17}{32}. \end{aligned}$$

Comparing the intervals where $f_1 \neq 0$ and $f_2 \neq 0$ we see that $g_2(\frac{1}{2}) = 0$. Analogously we find that

$$g_3\left(\tfrac{1}{2}\right) = (g_2 * f_3)(x)\big|_{x=\frac{1}{2}} = \int_{-\infty}^{\infty} g_2(u) f_3(x-u)\, du\Big|_{x=\frac{1}{2}} = 0$$

and, more generally, that $g_n(\frac{1}{2}) = 0$ for all $n \geq 2$. It is not difficult to see that $g_n(x) = 0$ for all x of the form $x = \frac{1}{2}(2m+1)$, $m = 0, \pm 1, \pm 2, \ldots$ and finally that $g_n(x) = 0$ for all $x = \frac{1}{2}(2m+1) + \delta$ where $m = 0, \pm 1, \pm 2, \ldots$ and $|\delta| < \frac{1}{4}$.

The sum $S_n = X_1 + \cdots + X_n$ has $\mathbf{E}S_n = 0$ and $\mathbf{V}S_n = s_n^2 = \frac{1}{2}n + \frac{5}{1152}(1-2^{-2n})$. Since the density g_n of S_n and the density p_n of S_n/s_n satisfy the relation $p_n(x) = s_n g_n(x s_n)$, we have to study the behaviour of the quantity $s_n g_n(x s_n)$ as $n \to \infty$. Again, take $x = \frac{1}{2}$. Then

$$s_n g_n\left(\tfrac{1}{2} s_n\right) = \left[\tfrac{1}{2}n + \tfrac{5}{1152}(1-2^{-2n})\right]^{1/2} \cdot g_n\left(\tfrac{1}{2}\left[\tfrac{1}{2}n + \tfrac{5}{1152}(1-2^{-2n})\right]^{1/2}\right).$$

If n is of the form $n = 2(2N+1)^2$, then the argument of g_n becomes

$$\tfrac{1}{2}(2N+1)[1 + \tfrac{5}{1152}(1 - 2^{-2.2(2N+1)^2})(2N+1)^{-2}].$$

For large N this expression takes the form $\frac{1}{2}(2N+1) + \delta$ with $|\delta| < \frac{1}{4}$. From the properties of g_n established above we conclude that

$$s_n g_n\left(\tfrac{1}{2} s_n\right) = 0 \ \text{ for sufficiently large } n.$$

This implies that

$$\lim_{n\to\infty} p_n\left(\tfrac{1}{2}\right) = 0.$$

However, $\varphi(\frac{1}{2}) \neq 0$ and thus relation (2) is not possible. Therefore the sequence $\{X_n\}$ defined by the densities (4) does not obey the local CLT.

General conditions ensuring convergence of both the d.f.s and the densities are described by Gnedenko and Kolmogorov (1954).

SECTION 18. DIVERSE LIMIT THEOREMS

In this section we have collected examples dealing with different kinds of limit behaviour of random sequences. The examples concern random series, conditional expectations, records and maxima of random sequences, versions of the law of the iterated logarithm and net convergence. The definitions of some of the notions are given in the examples themselves. For convenience we formulate one result here and give one definition.

Kolmogorov three-series theorem. *Let $\{X_n, n \geq 1\}$ be a sequence of independent r.v.s and $X_n^{(c)} = X_n I_{[|X_n| \leq c]}$ for some $c > 0$. A necessary condition for the convergence of $\sum_{n=1}^\infty X_n$ with probability 1 is that the series*

$$\sum_{n=1}^\infty \mathbf{E}[X_n^{(c)}], \quad \sum_{n=1}^\infty \mathbf{V}[X_n^{(c)}], \quad \sum_{n=1}^\infty \mathbf{P}[|X_n| \geq c]$$

converge for every $c > 0$. A sufficient condition is that these series are convergent for some $c > 0$.

The proof of this theorem and some useful corollaries can be found in the books by Breiman (1968), Chow and Teicher (1978), Shiryaev (1995).

Now let us define the so-called *net convergence* (see Neveu 1975). Let T be the set of all bounded stopping times with respect to the family $(\mathcal{F}_n, n \in \mathbb{N})$. Here $(\mathcal{F}_n)$ is a non-decreasing sequence of sub-σ-fields of $\mathcal{F}$ and a stopping time τ is a function with values in $[0, \infty]$ such that $[\tau = n] \in \mathcal{F}_n$ for each $n \in \mathbb{N}$. The family $(a_\tau, \tau \in \mathrm{T})$ of real numbers, called a *net*, is said to converge to the real number b provided for every $\varepsilon > 0$ there is $\tau_0 \in \mathrm{T}$ such that for all $\tau \in \mathrm{T}$ with $\tau \geq \tau_0$ we have $|a_\tau - b| < \varepsilon$.

Each of the examples given below contains appropriate references for further reading.

18.1. On the conditions in the Kolmogorov three-series theorem

(i) Let $\{X_n, n \geq 1\}$ be independent r.v.s with $\mathbf{E}X_n = 0$, $n \geq 1$. Then the condition $\sum_{n=1}^\infty \mathbf{V}X_n < \infty$ implies that $\sum_{n=1}^\infty X_n$ converges a.s. Note that this is one of the simplest versions of the Kolmogorov three-series theorem.

Let us show that the condition $\sum_{n=1}^{\infty} \mathbf{V}X_n < \infty$ is not necessary for the convergence of $\sum_{n=1}^{\infty} X_n$. Indeed, consider the sequence $\{X_n, n \geq 1\}$ of independent r.v.s where

$$\mathbf{P}[X_n = -n^4] = \mathbf{P}[X_n = n^4] = n^{-2}, \quad \mathbf{P}[X_n = 0] = 1 - 2n^{-2}.$$

Obviously $\sum_{n=1}^{\infty} \mathbf{V}X_n = \infty$ but nevertheless the series $\sum_{n=1}^{\infty} X_n$ is convergent a.s. according to the Borel–Cantelli lemma.

(ii) The Kolmogorov three-series theorem yields the following result (see Chow and Teicher 1978): if $\{X_n, n \geq 1\}$ are independent r.v.s with $\mathbf{E}X_n = 0$, $n \geq 1$ and

$$(1) \qquad \sum_{n=1}^{\infty} \mathbf{E}[X_n^2 I_{[|X_n| \leq 1]} + |X_n| I_{[|X_n|>1]}] < \infty$$

then the series $\sum_{n=1}^{\infty} x_n$ converges a.s.

Let us clarify the role of condition (1) in the convergence of $\sum_{n=1}^{\infty} X_n$. For this purpose consider the sequence $\{\xi_n, n \geq 1\}$ of i.i.d. r.v.s with $\mathbf{P}[\xi_1 = 1] = \mathbf{P}[\xi_1 = -1] = \frac{1}{2}$ and define $X_n = \xi_n/\sqrt{n}$, $n \geq 1$. It is easy to check that for any $r > 2$ the following condition holds:

$$(2) \qquad \sum_{n=1}^{\infty} \mathbf{E}[|X_n|^r] < \infty.$$

Condition (2) can be considered in some sense similar to (1). However the series $\sum_{n=1}^{\infty} X_n$ diverges a.s. This shows that the power 2 in the first term of the summands in (1) is essential.

Finally let us note that if condition (2) is satisfied for some $0 < r \leq 2$ then the series $\sum_{n=1}^{\infty} X_n$ does converge a.s. (see Loève 1978).

18.2. The independency condition is essential in the Kolmogorov three-series theorem

Let us start with a direct consequence of the Kolmogorov three-series theorem (sometimes called the 'two-series' theorem). If $X_n, n \geq 1$ are independent r.v.s and the series $\sum_{n=1}^{\infty} a_n$ and $\sum_{n=1}^{\infty} \sigma_n^2$, with $a_n = \mathbf{E}X_n$, $\sigma_n^2 = \mathbf{V}X_n$, are convergent, then the random series $\sum_{n=1}^{\infty} X_n$ is convergent with probability 1.

Our goal is to show that the independency property for X_n, $n \geq 1$, is essential for this and similar results to hold.

(i) Let ξ be a r.v. with $\mathbf{E}\xi = 0$ and $0 < \mathbf{V}\xi = b^2 < \infty$ (i.e. ξ is non-degenerate). Define $X_n = \xi/n, n \geq 1$. Then $a_n = \mathbf{E}X_n = 0$, $\quad \sigma_n^2 = \mathbf{V}X_n = b^2/n^2$ implying that

$$\sum_{n=1}^{\infty} a_n = 0 \text{ and } \sum_{n=1}^{\infty} \sigma_n^2 = b^2 \sum_{n=1}^{\infty} (1/n^2) < \infty.$$

Hence two of the conditions in the above result are satisfied and one condition, the independence of $X_n, n \geq 1$, is not. Nevertheless the question about the convergence of the random series $\sum_{n=1}^{\infty} X_n$ is reasonable. Since

$$\sum_{n=1}^{\infty} X_n(\omega) = \xi(\omega) \sum_{n=1}^{\infty} \frac{1}{n}$$

the series $\sum_{n=1}^{\infty} X_n(\omega)$ is convergent on the set $A = \{\omega : \xi(\omega) = 0\}$ and divergent on the set $A^{\rm c} = \{\omega : \xi(\omega) \neq 0\}$. If the non-degenerate r.v. ξ is such that $\mathbf{P}(A) = p$ where p is any number in [0,1), we get a random series $\sum_{n=1}^{\infty} X_n$ which is convergent with probability p (strictly less than 1) and divergent with probability $1 - p$ (strictly greater than 0).

(ii) In case (i) the dependence among the variables $X_n, n \geq 1$, is 'quite strong'—any two of them are functionally related. Let us see if the independence of $X_n, n \geq 1$, can be weakened and replaced e.g. by the exchangeability property. We use the following modification of the Kolmogorov three-series theorem. If X_n, $n \geq 1$, are i.i.d. r.v.s with $\mathbf{E}X_1 = 0$, $\mathbf{E}[X_1^2] < \infty$ and $c_n, n \geq 1$ are real numbers with $\sum_{n=1}^{\infty} c_n^2 < \infty$, then the random series $\sum_{n=1}^{\infty} c_n X_n$ is convergent with probability 1.

Let us now consider the sequence of i.i.d. r.v.s $\xi_n, n \geq 1$ with $\mathbf{E}\xi_1 = 0$ and $\mathbf{E}[\xi_1^2] < \infty$ and let η be another r.v. with $\mathbf{E}\eta = 0$, $0 < \mathbf{E}[\eta^2] < \infty$ and independent of $\{\xi_n, n \geq 1\}$. Define the sequence X_n, $n \geq 1$, by

$$X_n = \xi_n + \eta,\ n \geq 1.$$

Thus X_n, $n \geq 1$, is a sequence of identically distributed r.v.s with $\mathbf{E}X_1 = 0$ and $\mathbf{E}[X_1^2] < \infty$. Obviously the variables $X_n, n \geq 1$, are not independent. However X_n, $n \geq 1$, is an exchangeable sequence. (See also Example 13.8.) Our goal is to study the convergence of the series $\sum_{n=1}^{\infty} c_n X_n$ where c_n, $n \geq 1$, satisfy the condition $\sum_{n=1}^{\infty} c_n^2 < \infty$. Choose c_n, $n \geq 1$, such that $c_n > 0$ for any n and $\sum_{n=1}^{\infty} c_n = \infty$ (an easy case is $c_n = 1/n$). Since $c_n X_n = c_n \xi_n + c_n \eta$, we have

$$\sum_{n=1}^{\infty} c_n X_n = \sum_{n=1}^{\infty} c_n \xi_n + \eta \sum_{n=1}^{\infty} c_n.$$

The independence of ξ_n, $n \geq 1$, implies that the series $\sum_{n=1}^{\infty} c_n \xi_n$ is convergent a.s. Hence, in view of $\sum_{n=1}^{\infty} c_n = \infty$, the series $\sum_{n=1}^{\infty} c_n X_n$ is convergent on the set $A = \{\omega : \eta(\omega) = 0\}$ and divergent on $A^{\rm c} = \{\omega : \eta(\omega) \neq 0\}$. For preliminary given $p, p \in [0, 1)$, take the r.v. η such that $\mathbf{P}(A) = p$. Then the random series $\sum_{n=1}^{\infty} c_n X_n$ of exchangeable (but not independent) variables is convergent with probability $p < 1$ and divergent with probability $1 - p > 0$.

We have seen in both cases (i) and (ii) the role of the independence property for random series to converge with probability 1. The same examples lead to one additional conclusion. According to the Kolmogorov 0–1 law, if $X_n, n \geq 1$, are independent r.v.s, then the set $\{\omega : \sum_{n=1}^{\infty} X_n(\omega) \text{ converges}\}$ has probability 0 or 1. Hence, if $X_n, n \geq 1$, are not independent we can obtain $\mathbf{P}[\omega : \sum_{n=1}^{\infty} X_n(\omega) \text{ converges}] = p$ for arbitrarily given $p \in [0, 1)$.

18.3. The interchange of expectations and infinite summation is not always possible

Let us start with the formulation of a result showing that in some cases the operations of expectations and summation can be interchanged (see Chow and Teicher 1978). If $\{X_n, n \geq 1\}$ are non-negative r.v.s then

$$\mathbf{E}\left[\sum_{n=1}^{\infty} X_n\right] = \sum_{n=1}^{\infty} \mathbf{E}X_n. \tag{1}$$

Our aim now is to show that (1) is not true without the non-negativity of the variables X_n even if the series $\sum_{n=1}^{\infty} X_n$ is convergent.

Consider $\{\xi_n, n \geq 1\}$ to be i.i.d. r.v.s with $\mathbf{P}[\xi_1 = \pm 1] = \frac{1}{2}$ and define the stopping time $\tau = \inf\{n \geq 1 : \sum_{k=1}^{n} \xi_k = 1\}$ where $\inf\{\emptyset\} = \infty$. Then it is easy to check that $\mathbf{P}[\tau < \infty] = 1$. Setting $X_n = \xi_n I_{[\tau \geq n]}$, we get from the definition of τ that

$$\sum_{n=1}^{\infty} X_n = \sum_{n=1}^{\infty} \xi_n I_{[\tau \geq n]} = \sum_{n=1}^{\tau} \xi_n = 1 \Rightarrow \mathbf{E}\left[\sum_{n=1}^{\infty} X_n\right] = 1.$$

However, the event $[\tau \geq n] \in \sigma\{\xi_1, \dots, \xi_{n-1}\}$, the r.v.s ξ_n and $I_{[\tau \geq n]}$ are independent and from the properties of the expectation we obtain

$$\mathbf{E}X_n = \mathbf{E}\xi_n \mathbf{E}I_{[\tau \geq n]} = 0, \quad n \geq 1.$$

Thus $\sum_{n=1}^{\infty} \mathbf{E}X_n = 0$ and therefore (1) is not satisfied.

18.4. A relationship between a convergence of random sequences and convergence of conditional expectations

On the probability space $(\Omega, \mathcal{F}, \mathbf{P})$ we have given r.v.s X and $X_n, n \geq 1$, all in the space L^r (i.e. r-integrable) for some $r \geq 1$. Suppose $X_n \xrightarrow{\mathrm{L}^r} X$ as $n \to \infty$. Then for any sub-σ-field $\mathcal{A} \subset \mathcal{F}$ we have $\mathbf{E}[X_n|\mathcal{A}] \xrightarrow{\mathrm{L}^r} \mathbf{E}[X|\mathcal{A}]$ as $n \to \infty$ (e.g. see Neveu 1975 or Shiryaev 1995). This statement is a consequence of the Jensen inequality for conditional expectations. Obviously, we can ask the inverse question and the best way to answer it is to consider a specific example.

Let $X_n, n \geq 1$ be i.i.d. r.v.s with $\mathbf{P}[X = 2c] = \mathbf{P}[X = 0] = \frac{1}{2}$, $c \neq 0$ is a fixed real number. Take also (a trivial) r.v. $X = c : \mathbf{P}[X = c] = 1$. Then, if $\mathcal{A} = \sigma\{\emptyset, \Omega\}$, the trivial σ-field, we obviously get

$$\mathbf{E}[X_n|\mathcal{A}] = \mathbf{E}X_n = 2c \cdot \tfrac{1}{2} + 0 \cdot \tfrac{1}{2} = c = \mathbf{E}[X|\mathcal{A}] \text{ for any } n \geq 1.$$

Moreover because X, X_n are bounded, then for any $r \geq 1$, one has

$$\mathbf{E}[X_n|\mathcal{A}] \xrightarrow{L^r} \mathbf{E}[X|\mathcal{A}] \text{ as } n \to \infty.$$

However $\mathbf{E}[|X_n - X|^r] = \mathbf{E}[|X_n - c|^r] = \frac{1}{2}c^r + \frac{1}{2}c^r = c^r$ for all $n \geq 1$, $c \neq 0$ and hence $X_n \stackrel{L^r}{\not\to} X$ as $n \to \infty$.

18.5. The convergence of a sequence of random variables does not imply that the corresponding conditional medians converge

Let $(\Omega, \mathcal{F}, \mathbf{P})$ be a probability space, $\mathcal{F}_0 = \{\emptyset, \Omega\}$ the trivial σ-field and $\mathcal{D}$ a sub-σ-field of $\mathcal{F}$. If X is a r.v., then the *conditional median* of X with respect to $\mathcal{D}$ is defined as a $\mathcal{D}$-measurable r.v. M such that

$$\mathbf{P}[X \geq M|\mathcal{D}] \geq \tfrac{1}{2} \leq \mathbf{P}[X \leq M|\mathcal{D}] \text{ a.s.}$$

Usually the conditional median is denoted by $\mu(X|\mathcal{D})$ (see Example 6.10).

If $\{X_n, n \geq 1\}$ is a sequence of r.v.s which is convergent in a definite sense, then it is logical to expect that the corresponding sequence of the conditional medians also will be convergent. In this connection let us formulate the following result (see Tomkins 1975a). Let $\{X_n, n \geq 1\}$ and $\{M_n, n \geq 1\}$ be sequences of r.v.s such that for a given σ-field $\mathcal{D}$ we have $M_n = \mu(X_n|\mathcal{D})$ a.s. and there exist r.v.s X and M such that $X_n \xrightarrow{\text{a.s.}} X$ and $M_n \xrightarrow{\text{P}} M$ as $n \to \infty$. Then $M = \mu(X|\mathcal{D})$ a.s. We can now try to answer the question of whether the convergence of $\{X_n\}$ always implies convergence of the conditional medians $\{M_n\}$.

Let ξ be a r.v. distributed uniformly on the interval $(-\frac{1}{2}, \frac{1}{2})$, $\mathcal{D} = \mathcal{F}_0$ (the trivial σ-field) and define the sequence $\{X_n\}$ by

$$X_n = I_{[\xi \leq (-1)^n/n]},\ n \geq 1 \text{ and } X = I_{[\xi < 0]}.$$

It is easy to see that $X_n \xrightarrow{\text{a.s.}} X$ as $n \to \infty$. Moreover, X_n has a unique median M_n and $M_n = 0$ or 1 accordingly as n is odd or even. But clearly the sequence $\{M_n\}$ does not converge. (It would be useful for the reader to compare this example with the result cited above.)

18.6. A sequence of conditional expectations can converge only on a set of measure zero

If $(\Omega, \mathcal{F}, \mathbf{P})$ is a complete probability space, $(\mathcal{F}_n, n \in \mathbb{N})$ an increasing family of sub-σ-fields of $\mathcal{F}$, $\mathcal{F}_\infty = \lim_{n\to\infty} \mathcal{F}_n$ and X a positive r.v., the following result holds (see Neveu 1975):

(1) $\quad \mathbf{E}[X|\mathcal{F}_n] \xrightarrow{\text{a.s.}} \mathbf{E}[X|\mathcal{F}_\infty]$ outside the set $\{\omega : \mathbf{E}[X|\mathcal{F}_n] = \infty \text{ for all } n\}$.

We shall show that this result cannot be improved. More precisely, we give an example of an a.s. finite $\mathcal{F}_\infty$-measurable r.v. X such that $\mathbf{E}[X|\mathcal{F}_n] = \infty$ a.s. for all $n \in \mathbb{N}$. Clearly in such a case the convergence in (1) holds only on a set of measure zero.

Let $\Omega = [0,1], \mathcal{F} = \mathcal{B}_{[0,1]}$ and $\mathbf{P}$ be the Lebesgue measure. Consider the increasing sequence $(\mathcal{F}_n)$ of sub-σ-fields of $\mathcal{F}$ where $\mathcal{F}_n$ is generated by the dyadic partitions $\{[2^{-n}k, 2^{-n}(k+1)), 0 \le k < 2^n, n \in \mathbb{N}\}$. For each $n \in \mathbb{N}$ choose a positive measurable function $f_n : [0,1) \mapsto \mathbb{R}^+$ of period 2^{-n} with

$$\int_0^1 f_n(\omega)\,\mathrm{d}\omega = 1 \quad \text{and} \quad \int_0^1 1_{[f_n>0]}(\omega)\,\mathrm{d}\omega = 2^{-n}.$$

Since the sum $\sum_{n=1}^\infty 1_{[f_n>0]}$ is integrable, and hence a.s. finite, then $X = \sum_{n=1}^\infty f_n$ is a positive r.v. which is finite a.s. Thus the series $\sum_{n=1}^\infty f_n$ contains no more than a finite number of non-zero terms for almost all ω. On the other hand, for all $n \in \mathbb{N}$ and all k, $0 \le k < 2^n$, we have

$$\int_{2^{-n}k}^{2^{-n}(k+1)} X\,\mathrm{d}\omega \ge \sum_{m\ge n} \int_{2^{-n}k}^{2^{-n}(k+1)} f_m\,\mathrm{d}\omega = 2^{-n} \sum_{m\ge n} \int_0^1 f_m\,\mathrm{d}\omega = \infty$$

by the periodicity of f_m.

Therefore we have shown that $\mathbf{E}[X|\mathcal{F}_n] = \infty$ for all $n \in \mathbb{N}$ meaning that the a.s. convergence in (1) holds only on a set of measure zero.

18.7. When is a sequence of conditional expectations convergent almost surely?

Let $(\Omega, \mathcal{F}, \mathbf{P})$ be a probability space and $\{\mathcal{F}_n, n \ge 1\}$ an independent sequence of sub-σ-fields of $\mathcal{F}$, that is, for $k = 1, 2, \ldots$ and $A_j \in \mathcal{F}_j$, $1 \le j \le k$, we have

$$\mathbf{P}(A_1 A_2 \ldots A_k) = \mathbf{P}(A_1)\mathbf{P}(A_2)\ldots\mathbf{P}(A_k).$$

Let X be an integrable r.v. with $\mathbf{E}X = a$ and let $X_n = \mathbf{E}[X|\mathcal{F}_n]$. The following result is proved by Basterfield (1972):

$$\text{if } \mathbf{E}[|X|\log^+|X|] < \infty \text{ then } \mathbf{P}[X \to a \text{ as } n \to \infty] = 1$$

($\log x$ is defined for $x > 0$ and $\log^+ x = \log x$ if $x > 1$, and 0 if $0 < x \le 1$).

We aim to show that the assumption in this result cannot be weakened, e.g. it cannot be replaced by $\mathbf{E}[|X|] < \infty$. To see this, consider a sequence $\{A_n, n \ge 1\}$ of independent events with $\mathbf{P}(A_n) = 1/n$. Define the r.v. X by

$$X = \sum_{m=1}^\infty m!\xi_{m+2}$$

where ξ_m is the indicator function of the event $A_1 A_2 \dots A_m$. Since

$$\mathbf{E}\xi_m = \mathbf{P}(A_1 \dots A_m) = \mathbf{P}(A_1) \dots \mathbf{P}(A_m) = \frac{1}{m!}$$

we obtain

$$\mathbf{E}X = \sum_{m=1}^{\infty} \frac{m!}{(m+2)!} = \sum_{m=1}^{\infty} \left(\frac{1}{m+1} - \frac{1}{m+2} \right) = \frac{1}{2}.$$

Moreover, it is not difficult to verify that

$$\mathbf{E}[X \log^+ X] = \infty.$$

Consider now $\mathcal{F}_n = \{\emptyset, A_n, A_n^c, \Omega\}$ and $X_n = \mathbf{E}[X|\mathcal{F}_n]$. We need to check if $X_n \xrightarrow{\text{a.s.}} \frac{1}{2}$ as $n \to \infty$. Since $\mathbf{E}[X|A_n] = (1/\mathbf{P}(A_n)) \int_{A_n} X \, d\mathbf{P} = n \int_{A_n} X \, d\mathbf{P}$, replacing X by $\sum_{m=1}^{\infty} m! \xi_{m+2}$ we arrive at the equality

$$\mathbf{E}[X|A_n] = \tfrac{3}{2}.$$

However, $\sum_{n=1}^{\infty} \mathbf{P}(A_n) = \infty$ and, by the Borel–Cantelli lemma, almost all ω belong to infinitely many A_n. Therefore

$$\limsup_{n\to\infty} X_n = \tfrac{3}{2} \neq \tfrac{1}{2} = a = \mathbf{E}X.$$

Thus the condition $\mathbf{E}[|X| \log^+ |X|] < \infty$ cannot be replaced by $\mathbf{E}[|X|] < \infty$ so as to preserve the convergence $X_n = \mathbf{E}[X|\mathcal{F}_n] \xrightarrow{\text{a.s.}} a = \mathbf{E}X$ as $n \to \infty$.

18.8. The Weierstrass theorem for the unconditional convergence of a numerical series does not hold for a series of random variables

Let $\sum_{n=1}^{\infty} a_n$ be an infinite series of real numbers. This series is said to converge unconditionally if $\sum_{k=1}^{\infty} a_{n_k} < \infty$ for every rearrangement $\{n_1, n_2, \dots\}$ of $\{1, 2, \dots\}$. (By rearrangement we understand a one–one map of $\mathbb{N}$ onto $\mathbb{N}$.) We say that the series $\sum_{n=1}^{\infty} a_n$ converges absolutely if $\sum_{n=1}^{\infty} |a_n| < \infty$. According to the classical Weierstrass theorem these two concepts, unconditional convergence and absolute convergence, are equivalent.

Thus we arrive at the question: what happens when considering random series $\sum_{n=1}^{\infty} X_n(\omega)$, that is series of r.v.s?

Let $\{X_n, n \geq 1\}$ be a sequence of r.v.s defined on some probability space $(\Omega, \mathcal{F}, \mathbf{P})$. The series $\sum_{n=1}^{\infty} X_n$ is said to be *a.s. unconditionally convergent* if for every rearrangement $\{n_k\}$ of $\mathbb{N}$ we have $\sum_{k=1}^{\infty} X_{n_k} < \infty$ a.s. If $\sum_{n=1}^{\infty} |X_n| < \infty$ a.s., the given series is *a.s. absolutely convergent.*

So, bearing in mind the Weierstrass theorem, we could suppose that the concepts a.s. unconditional and a.s. absolute convergence are equivalent. However, as will be seen later, such a conjecture is not generally true.

Consider the sequence $\{r_n, n = 0, 1, 2, \ldots\}$ of the so-called Rademacher functions, that is $r_n(\omega) = \operatorname{sign}\sin(2^n \pi \omega)$, $0 \le \omega \le 1$, $n = 0, 1, \ldots$ (see Lukacs 1975). Actually r_n can also be written in the form

$$r_n(\omega) = \begin{cases} 1, & \text{if } 2k/2^n < \omega < (2k+1)/2^n \\ -1, & \text{if } (2k+1)/2^n < \omega < (2k+2)/2^n \\ 0, & \text{if } \omega = k/2^n,\ k = 0, 1, \ldots, 2^n. \end{cases}$$

Then $\{r_n\}$ is a sequence of independent r.v.s on the probability space $(\Omega, \mathcal{F}, \mathbf{P})$ with $\Omega = [0, 1]$, $\mathcal{F} = \mathcal{B}_{[0,1]}$ and $\mathbf{P}$ the Lebesgue measure. Moreover, r_n takes the values 1 and -1 with probability $\frac{1}{2}$ each, $\mathbf{E}r_n = 0$, $\mathbf{V}r_n = 1$.

Now take any numerical sequence $\{a_n\}$ such that

$$\sum_{n=1}^{\infty} a_n^2 < \infty \quad \text{but} \quad \sum_{n=1}^{\infty} |a_n| = \infty.$$

For example, $a_n = (-1)^n/(n+1)$. Using the sequence $\{r_n\}$ of the Rademacher functions and the numerical sequence $\{a_n\}$ we construct the series

$$(1) \qquad \sum_{n=1}^{\infty} a_n r_n(\omega).$$

Applying the Kolmogorov three-series theorem we easily conclude that this series is a.s. convergent. If $\{n_k\}$ is any rearrangement of $\mathbb{N}$ then the series $\sum_{k=1}^{\infty} a_{n_k} r_{n_k}(\omega)$ is also a.s. convergent. However, $\sum_{n=1}^{\infty} a_n r_n(\omega)$ is not absolutely convergent since $|r_n(\omega)| = 1$ and $\sum_{n=1}^{\infty} |a_n| = \infty$.

Therefore the series (1) is a.s. unconditionally convergent but not a.s. absolutely convergent, and so these two concepts of convergence of random series are not equivalent.

18.9. A condition which is sufficient but not necessary for the convergence of a random power series

Let $a_n = a_n(\omega)$, $n = 0, 1, 2, \ldots$, be a sequence of i.i.d. r.v.s. The *random power series*, that is a series of the type $\sum_{n=0}^{\infty} a_n(\omega) z^n$, is defined in the standard manner (see Lukacs 1975). As in the deterministic case (when a_n are numbers), one of the basic problems is to find the so-called *radius of convergence* $r = r(\omega)$. This r is a r.v. such that for all $|z| < r$ the series $\sum_{n=0}^{\infty} a_n(\omega) z^n$ is a.s. convergent. Moreover, $r(\omega) = (\limsup_{n\to\infty} \sqrt[n]{|a_n(\omega)|})^{-1}$.

Among the variety of results concerning random power series, we formulate here the following (see Lukacs 1975). If $\{a_n, n \ge 0\}$ are i.i.d. r.v.s and the d.f. $F(x)$, $x \in \mathbb{R}^+$ of $|a_1|$ satisfies the condition

$$(1) \qquad \int_1^{\infty} \log x \, \mathrm{d}F(x) < \infty$$

then the random power series $\sum_{n=0}^{\infty} a_n(\omega)z^n$ has a radius of convergence $r(\omega)$ such that $\mathbf{P}[r(\omega) \geq 1] = 1$.

Let us show by a concrete example that condition (1) is not necessary for the existence of r with $\mathbf{P}[r \geq 1] = 1$. Take ξ as a r.v. distributed uniformly on the interval [0,1]. Define a_n by $a_n(\omega) = \exp(1/\xi(\omega))$. Then the common d.f. of a_n is

$$F(x) = \begin{cases} 0, & \text{if } x < \mathrm{e} \\ 1 - (\log x)^{-1}, & \text{if } x \geq \mathrm{e}. \end{cases}$$

Clearly

$$\int_1^{\infty} \log x \, \mathrm{d}F(x) = \infty$$

and condition (1) is not satisfied. However, for any $\varepsilon > 0$ we have

$$\mathbf{P}[\limsup_{n\to\infty} [|a_n| \geq (1+\varepsilon)^n]] = \mathbf{P}[\limsup_{n\to\infty} [0, 1/(n\log(1+\varepsilon))]] = 0.$$

This relation, the definition of a radius of convergence and a result of Lukacs (1975) allow us to conclude that $\mathbf{P}[r(\omega) < x] = 0$ for all $x \in (0, 1]$. For $x = 1$ we get $\mathbf{P}[r(\omega) \geq 1] = 1$. Thus condition (1) is not necessary for the random power series $\sum_{n=0}^{\infty} a_n(\omega)z^n$ to have a radius of convergence $r \geq 1$.

18.10. A random power series without a radius of convergence in probability

As before consider a random power series and its partial sums

$$\sum_{n=0}^{\infty} a_n(\omega)z^n \quad \text{and} \quad U_N(z,\omega) = \sum_{n=0}^{N} a_n(\omega)z^n$$

where the coefficients $a_n(\omega)$, $n = 0, 1, \ldots$, are given r.v.s. If $U_n(z)$ are convergent in some definite sense as $N \to \infty$ then the random power series is said to converge in the same sense. There are several interesting results about the existence of the radii of convergence if we consider a.s. convergence, convergence in probability and L^r-convergence. Note that a.s. convergence was treated in Example 18.9.

Now we aim to show that no circle of convergence exists for convergence in probability of a random power series. For this let $\{a_n(\omega), n \geq 0\}$ be independent r.v.s with

$$\mathbf{P}[a_0 = 0] = 1, \quad \mathbf{P}[a_n = n^n] = 1/n, \quad \mathbf{P}[a_n = 0] = 1 - 1/n, \quad n \geq 1.$$

It is easy to check that the power series $\sum_{n=0}^{\infty} a_n(\omega)z^n$ is a.s. divergent, that is its radius of convergence is $r_0 = 0$. Clearly, this series cannot converge in probability or in L^r-sense.

From the definition of a_n we find that

$$\mathbf{P}[|a_n(\omega)z^n| > \varepsilon] = \mathbf{P}[a_n(\omega) > \varepsilon|z|^{-n}] = 1/n \Rightarrow a_n(\omega)z^n \xrightarrow{\mathrm{P}} 0 \text{ as } n \to \infty.$$

Define another power series, say $\sum_{n=0}^{\infty} b_n(\omega)z^n$, whose coefficients b_n are given by $b_0 = a_0$ and $b_n = a_n - a_{n-1}$ for $n \geq 1$. Obviously,

$$\lim_{N\to\infty} \sum_{n=0}^{N} b_n 1^n = \lim_{N\to\infty} a_N \stackrel{\mathrm{P}}{=} 0.$$

Furthermore, we have

$$\sum_{n=0}^{N} b_n(\omega)z^n = a_N(\omega)z^N + (1-z)\sum_{n=0}^{N-1} a_n(\omega)z^n. \tag{1}$$

It is clear that the series $\sum_{n=0}^{\infty} b_n(\omega)z^n$ converges in probability at least at two points, namely $z = 0$ and $z = 1$. If we suppose that it is convergent for a point z such that $z \neq 0$ and $z \neq 1$, we derive from (1) that

$$\sum_{n=0}^{N-1} a_n(\omega)z^n = \frac{1}{1-z}\left[\sum_{n=0}^{N} b_n(\omega)z^n - a_N(\omega)z^N\right]$$

which must also converge in probability as $N \to \infty$. However, this contradicts the fact that $r_0 = 0$.

Therefore in general the random power series has no circle of convergence in probability.

Finally, note that Arnold (1967) characterized probability spaces in which every random power series has a circle of convergence in probability. Such a property holds iff the probability space is atomic.

18.11. Two sequences of random variables can obey the same strong law of large numbers but one of them may not be in the domain of attraction of the other

Let $\{X_k, k \geq 1\}$ and $\{Y_k, k \geq 1\}$ each be a sequence of i.i.d. r.v.s. Omitting subscripts, we say that X and Y obey the same SLLN if for each number sequence $\{a_n, n \geq 1\}$ with $0 < a_n \uparrow \infty$ either

$$\sum_{k=1}^{n} X_k = o(a_n) \text{ and } \sum_{k=1}^{n} Y_k = o(a_n) \quad \text{a.s.} \tag{1}$$

or

$$\limsup_{n\to\infty} \frac{1}{a_n}\sum_{k=1}^{n} X_k = \infty \text{ and } \limsup_{n\to\infty} \frac{1}{a_n}\sum_{k=1}^{n} Y_k = \infty \text{ a.s.} \tag{2}$$

We also need the following result of Stout (1979): X and Y obey the same SLLN iff

$$\mathbf{P}[|Y| > x]/\mathbf{P}[|X| > x] = O(x) \text{ as } x \to \infty. \tag{3}$$

Note that the statement that two r.v.s X and Y, or more exactly two sequences $\{X_k, k \geq 1\}$ and $\{Y_k, k \geq 1\}$, obey the same SLLN is closely related to another statement involving the so-called *domain of attraction*. Let U and V be r.v.s with d.f.s G and H and ch.f.s ϕ and ψ respectively. Suppose H is a stable distribution with index γ (see Feller 1971; Zolotarev 1986). We say that U belongs to the domain of normal attraction of V if for suitable constants b_n and $c_n{:=}cn^{1/\gamma}$ the distribution of $(1/c_n)\sum_{k=1}^{n} U_k - b_n$ tends to H as $n \to \infty$, or in terms of the corresponding ch.f.s:

$$\lim_{n\to\infty} \exp(itb_n)\phi^n(t/c_n) = \psi(t), \quad t \in \mathbb{R}^1.$$

We write $U \in N(\gamma)$ to denote that U is in the domain of normal attraction of a stable law with index γ. Now let $X \in N(\gamma_x)$ and $Y \in N(\gamma_y)$ where $\gamma_x < 2, \gamma_y < 2$. Then, as a consequence of a result by Gnedenko and Kolmogorov (1954), we obtain that X and Y obey the same SLLN iff $\gamma_x = \gamma_y$.

Thus we come to the question: Can a r.v. Y fail to be in the domain of normal attraction of a stable law X and yet obey the same SLLN as X? By an example we show that the answer is positive. Consider a r.v. X with a Cauchy distribution and a r.v. Y whose d.f. F is given by

$$F(x) = \begin{cases} 1 - (x+3)^{-1}[2 + \sin(\log x)], & \text{if } x > 0 \\ 0, & \text{if } x \leq 0. \end{cases} \tag{4}$$

It is easy to check that X and Y satisfy condition (3) and hence X and Y obey the same SLLN in the sense of (1) and (2). According to a result by Gnedenko and Kolmogorov (1954) the r.v. Y is in the domain of attraction of a Cauchy distribution only if

$$\mathbf{P}[|Y| > x] = (\alpha + \beta(x))/x \tag{5}$$

for some $\alpha = \text{constant} > 0$ and $\beta(x) \to 0$ as $x \to \infty$. However, the d.f. F given by (4) does not satisfy (5).

Therefore Y is not in the domain of normal attraction of the Cauchy-distributed r.v. X despite the fact that X and Y obey the same SLLN.

18.12. Does a sequence of random variables always imitate normal behaviour?

Let $F(x)$, $x \in \mathbb{R}^1$ be a d.f. with zero mean and variance 1. Consider the sequence $\{X_n, n \geq 1\}$ of i.i.d. r.v.s whose d.f. is F, and another sequence of independent r.v.s $\{Y_n, n \geq 1\}$ each distributed $\mathcal{N}(0,1)$. We also have a non-decreasing sequence $\{a_n, n \geq 1\}$ of real numbers. As usual, let $S_n = X_1 + \cdots + X_n$. We say that the sequence $\{S_n, n \geq 1\}$ (generated by $\{X_n\}$) *imitates normal behaviour to within* $\{a_n\}$ if there is a probability space with r.v.s $\{X_n\}$ and $\{Y_n\}$ defined on it such that $\{X_n\}$ are i.i.d. with a common d.f. F, $\{Y_n\}$ are independent $\mathcal{N}(0,1)$ and

$$\frac{1}{a_n}[(X_1 + \cdots + X_n) - (Y_1 + \cdots + Y_n)] \xrightarrow{\text{a.s}} 0 \text{ as } n \to \infty. \tag{1}$$

Note that the first result of this type was obtained by Strassen (1964), who showed that every sequence $\{S_n\}$ with $\mathbf{E}X_1 = 0$ and $\mathbf{E}[X_1^2] = 1$ imitates normal behaviour to within $\{a_n\}$ with $a_n = (n \log\log n)^{1/2}$. He used this result to prove the *law of iterated logarithm* (LIL) for all such sequences. The question now is whether it is possible to choose a sequence $\{a_n\}$ 'smaller' than $\{(n \log\log n)^{1/2}\}$ and preserve the property described by (1). Some results in this direction can be found in Breiman (1967). Our aim now is to show that the condition on $\{a_n\}$, that is $a_n = (n \log\log n)^{1/2}$, cannot be weakened too much. More precisely, let us show that the sequence $\{S_n\}$ defined above does not imitate normal behaviour to within $\{b_n\}$ where $b_n = n^{1/2}$.

Firstly, define the sequence $\{n_k, k \geq 2\}$ by

$$n_{k+1} - n_k = n_{k+1}/g(n_{k+1}) \quad \text{where} \quad g(n_k) = \beta \log k,\ \beta > 2.$$

Thus the differences $n_{k+1} - n_k$, $k \geq 2$, are increasing.

Suppose now that $\{S_n\}$ imitates normal behaviour to within $\{b_n\}$. Then for $Z_n = \xi_1 + \cdots + \xi_n$ sums of independent $\mathcal{N}(0,1)$ r.v.s, the series

$$\sum_{k=2}^{\infty} \mathbf{P}[S_{n_{k+1}-n_k} > b_{n_{k+1}}] \quad \text{and} \quad \sum_{k=2}^{\infty} \mathbf{P}[Z_{n_{k+1}-n_k} > b_{n_{k+1}}]$$

must converge or diverge simultaneously as a consequence of (1). Take $X_k = \eta_k \theta_k$ where $\eta_1, \theta_1, \eta_2, \theta_2, \ldots$ are all mutually independent, $\theta_k \sim \mathcal{N}(0,1)$, $\mathbf{E}\eta_1 = 0$, $\mathbf{E}[\eta_1^2] = 1$ and the distribution of η_1 will be specified later. We have

$$\text{(2)} \qquad \mathbf{P}[Z_{n_{k+1}-n_k} > \sqrt{n_{k+1}}] \approx (g(n_{k+1}))^{-1/2} \exp[-\tfrac{1}{2} g(n_{k+1})].$$

For the sequence $\{S_n\}$ we find

$$\text{(3)} \qquad \mathbf{P}[S_{n_{k+1}-n_k} > \sqrt{n_{k+1}} \mid \eta_1, \eta_2, \ldots] \approx (u_k/n_{k+1})^{1/2} \exp[-n_{k+1}/(2u_k)]$$

where $u_k = \eta_1^2 + \cdots + \eta_{n_{k+1}-n_k}^2$. We can take η_1 distributed such that

$$\text{(4)} \qquad \mathbf{P}\left[\sum_{k=1}^{n} \eta_k^2 > n g(n)\right] \geq \mathbf{P}\left[\bigcup_{k=1}^{n} (\eta_k^2 > n g(n))\right] \geq \frac{c}{g(n)h(n)}$$

where $h(n) = (\log n)(\log\log n)^{1+\delta}$ and $\delta > 0$. From (3) and (4) it follows that

$$\text{(5)} \qquad \mathbf{P}[S_{n_{k+1}-n_k} > \sqrt{n_{k+1}}] \geq \tilde{c}/[\sqrt{g(n_{k+1})}\, h(n_{k+1})].$$

Taking (2) into account we find that

$$\sum_{k=2}^{\infty} \mathbf{P}[Z_{n_{k+1}-n_k} > \sqrt{n_{k+1}}] < \infty.$$

On the other hand, since $\log n_{k+1} \sim k/(\beta \log k)$, for any fixed δ, $0 < \delta < \frac{1}{2}$, we obtain from (5) that

$$\sum_{k=2}^{\infty} \mathbf{P}[S_{n_{k+1}-n_k} > \sqrt{n_{k+1}}] = \infty.$$

However, these two series must converge or diverge together. This contradiction shows that the sequence $\{S_n\}$ does not imitate normal behaviour to within $\{b_n\}$ where $b_n = \sqrt{n}$. Therefore in the result of Strassen (1964) the sequence $a_n = (n \log\log n)^{1/2}$ cannot be replaced by $b_n = n^{1/2}$.

18.13. On the Chover law of iterated logarithm

Let $\{X_n, n \geq 1\}$ be a sequence of i.i.d. r.v.s with a d.f. F. Denote

$$S_n = X_1 + \cdots + X_n, \quad \xi_n = S_n/b_n, \quad \eta_n = |\xi_n|^{1/\log\log n}, \quad n \geq 3$$

where $\{b_n, n \geq 1\}$ are norming constants, $b_n > 0, n \geq 1$. It is interesting to study the asymptotic behaviour of η_n as $n \to \infty$. For example, Vasudeva (1984) has proved the following result. Suppose there exists a sequence $\{b_n\}$ such that $\xi_n \xrightarrow{\text{d}} \xi$ as $n \to \infty$ where ξ is a stable r.v. with index γ, $0 < \gamma \leq 2$. Then $\eta_n \xrightarrow{\text{a.s.}} p$ where p is a definite number in the interval $[0, \infty)$.

Let us note that the a.s. convergence of $\{\eta_n\}$ is known as the *Chover law of iterated logarithm* (for references and details see Vasudeva 1984).

One can ask whether it is necessary to assume the weak convergence of $\{\xi_n\}$ in order to get the a.s. convergence of $\{\eta_n\}$. Our aim now is to describe a sequence of r.v.s $\{X_n, n \geq 1\}$ such that $\xi_n = (X_1 + \cdots + X_n)/b_n$, for given $\{b_n\}$, fails to converge weakly, but nevertheless

$$\eta_n = |\xi_n|^{1/\log\log n} \xrightarrow{\text{a.s.}} \text{constant} = \mathrm{e}^{1/\sqrt{2}} \quad \text{as } n \to \infty.$$

For this take the function $F(x)$, $x \in \mathbb{R}^1$ where

$$F(-x) = 1 - F(x) = \begin{cases} \frac{1}{2}, & \text{if } 0 < x \leq 1 \\ \frac{1}{2} x^{-\sqrt{2}}(1 + \frac{1}{12}\sin(\log x)), & \text{if } x \geq 1. \end{cases}$$

It is easy to see that F is a d.f. Let $\{X_n, n \geq 1\}$ be i.i.d. r.v.s whose common d.f. is F. Choose $b_n = n^{1/\sqrt{2}}$, $n \geq 1$, as a sequence of norming constants. Note that for all $x > 0$ we have

$$\frac{11}{12} x^{-\sqrt{2}} \leq 1 - F(x) + F(-x) \leq \frac{13}{12} x^{-\sqrt{2}}.$$

Then for $\eta_n = |S_n/b_n|^{1/\log\log n}$, $n \geq 3$, (with $b_n = n^{1/\sqrt{2}}$) we find the following two relations:

$$\mathbf{P}[|S_n| \geq b_n (\log n)^{(1-\varepsilon)/\sqrt{2}}] \geq c_1/(n(\log n)^{1-\varepsilon}),$$
$$\mathbf{P}[|S_n| \geq b_n (\log n)^{(1+\varepsilon)/\sqrt{2}}] \leq c_2/(n(\log n)^{1+\varepsilon})$$

valid for any $\varepsilon \in (0,1)$ and all $n \geq n_0$, where n_0 is a fixed natural number. By the Borel–Cantelli lemma and a result by Feller (1946) we find that

$$\mathbf{P}[|S_n| \geq b_n(\log n)^{(1-\varepsilon)/\sqrt{2}} \text{ i.o.}] = 1, \mathbf{P}[|S_n| \geq b_n(\log n)^{(1+\varepsilon)/\sqrt{2}} \text{ i.o.}] = 0.$$

Therefore

$$\mathbf{P}[\lim_{n\to\infty} \eta_n = \mathrm{e}^{1/\sqrt{2}}] = 1.$$

Applying a result of Zolotarev and Korolyuk (1961) we see that the sequence $\{\xi_n\}$ cannot converge weakly (to a non-degenerate r.v.) for any choice of the norming constants $\{b_n\}$.

18.14. On record values and maxima of a sequence of random variables

Let $\{X_n, n \geq 1\}$ be a sequence of i.i.d. r.v.s with a common d.f. F. Recall that X_k is said to be a *record value* of $\{X_n\}$ iff $X_k > \max\{X_1, \ldots, X_{k-1}\}$. By convention X_1 is a record. Define the r.v.s $\{\tau_n, n \geq 0\}$ by

$$\tau_0 = 0, \quad \tau_n = \min\{k : k > \tau_{n-1}, X_k > X_{\tau_{n-1}}\}.$$

Obviously the variables τ_n are the indices at which record values occur. Further, we shall analyse some properties of two sequences, $\{X_{\tau_n}, n \geq 1\}$ and the sequence of maxima $\{M_n, n \geq 1\}$ where $M_n = \max\{X_1, \ldots, X_n\}$.

The sequence of r.v.s $\{\xi_n, n \geq 1\}$ is called *stable* if there exist norming constants $\{b_n, n \geq 1\}$ such that $\xi_n/b_n \to 1$ as $n \to \infty$. If the convergence is with probability 1, then $\{\xi_n\}$ is a.s. stable, while if the convergence is in probability, we say that $\{\xi_n\}$ is stable in probability.

Let us formulate a result connecting the sequences $\{X_{\tau_n}\}$ and $\{M_n\}$ (see Resnik 1973): if $\{X_{\tau_n}\}$ is stable in probability, then the same holds for $\{M_n\}$.

Note firstly that the function $h(x) = -\log(1 - F(x))$ and its inverse function $h^{-1}(x) = \inf\{y : h(y) > x\}$ play an important role in studying records and maxima of random sequences. In particular the above result of Resnik has the following precise formulation: as $n \to \infty$

$$X_{\tau_n}/h^{-1}(n) \xrightarrow{\text{P}} 1 \Rightarrow M_n/h^{-1}(\log n) \xrightarrow{\text{P}} 1.$$

This naturally raises the question of whether the converse is true. By an example we show that the answer is negative. Take the function $h(x) = (\log x)^2$, $x \geq 1$, and let $\{X_n\}$ be i.i.d. r.v.s with d.f. F corresponding to this h. As above, $\{M_n\}$ and $\{X_{\tau_n}\}$ are the sequences of the maxima and the records respectively. Since the function $h^{-1}(\log y) = \exp[(\log y)^{1/2}]$ is slowly varying, according to Gnedenko (1943), $\{M_n\}$ is stable in probability. Moreover, from a result by Resnik (1973), $\{M_n\}$ is a.s. stable. Nevertheless, the sequence of records $\{X_{\tau_n}\}$ is not stable in probability since the function $h^{-1}((\log y)^2) = y$ (compare with the result cited above) is not slowly varying and this condition (see again Resnik 1973) is necessary for $\{X_{\tau_n}\}$ to be stable.

Chapter 4

Stochastic Processes

Courtesy of Professor A. T. Fomenko of Moscow University.

SECTION 19. BASIC NOTIONS ON STOCHASTIC PROCESSES

Let $(\Omega, \mathcal{F}, \mathbf{P})$ be a probability space, T a subset of $\mathbb{R}^+$ and $(E, \mathcal{E})$ a measurable space. The family $X = (X_t, t \in T)$ of random variables on $(\Omega, \mathcal{F}, \mathbf{P})$ with values in $(E, \mathcal{E})$ is called a **stochastic process** (or a random process, or simply a process). We call T the *parameter set* (or time set, or index set) and $(E, \mathcal{E})$ the *state space* of the process X. For every $\omega \in \Omega$, the mapping from T into E defined by $t \mapsto X_t(\omega)$ is called a *trajectory* (path, realization) of the process X. We shall restrict ourselves to the case of real-valued processes, that is processes whose state space is $E = \mathbb{R}^1$ and $\mathcal{E} = \mathcal{B}^1$. The index set T will be either discrete (a subset of $\mathbb{N}$) or some interval in $\mathbb{R}^+$.

Some classes of stochastic processes can be characterized by the family $\mathcal{P}$ of the *finite-dimensional distributions*. Recall that

$$\mathcal{P} = \{P_{t_1,\dots,t_n}(B_1, \dots, B_n),\ n \geq 1,\ t_1, \dots, t_n \in T,\ B_1, \dots, B_n \in \mathcal{B}^1\}$$

where

$$P_{t_1,\dots,t_n}(B_1, \dots, B_n) = \mathbf{P}[X_{t_1} \in B_1, \dots, X_{t_n} \in B_n].$$

The following fundamental result (**Kolmogorov theorem**) is important in this analysis. Let $\mathcal{P}$ be a family of finite-dimensional distributions,

$$\mathcal{P} = \{P_{t_1,\dots,t_n}(x_1, \dots, x_n),\ n \geq 1,\ t_1, \dots, t_n \in T,\ x_1, \dots, x_n \in \mathbb{R}^1\}$$

and $\mathcal{P}$ satisfies the compatibility (consistency) condition (see Example 2.3). Then there exists a probability space $(\Omega, \mathcal{F}, \mathbf{P})$ with a stochastic process $X = (X_t, t \in T)$ defined on it such that X has $\mathcal{P}$ as its family of the finite-dimensional distributions.

Let T be a finite or infinite interval of $\mathbb{R}^+$. We say that the process $X = (X_t, t \in T)$ is *continuous* (more exactly, almost surely continuous) if almost all of its trajectories $X_t(\omega), t \in T$ are continuous functions. In this case we say that X is a process in the space $\mathbb{C}(T)$ (of the continuous functions from T to $\mathbb{R}^1$). Further, X is said to be *right-continuous* if almost all of its trajectories are right-continuous functions: that is, for each t, $X_t = X_{t+}$ where $X_{t+} := \lim_{s \downarrow t} X_s$. In this case, if the left-hand limits exist for each time t, the process X is without discontinuity of second kind and we say that X is a process in the space $\mathbb{D}(T)$ (the space of all functions from T to $\mathbb{R}^1$ which are right-continuous and have left-hand limits). We can define the left-continuity of X analogously.

Let $(\mathcal{F}_t, t \in \mathbb{R}^+)$ be an increasing family of sub-σ-fields of the basic σ-field $\mathcal{F}$, that is $\mathcal{F}_t \subset \mathcal{F}$ for each $t \in \mathbb{R}^+$ and $\mathcal{F}_s \subset \mathcal{F}_t$ if $s < t$. The family $(\mathcal{F}_t, t \in \mathbb{R}^+)$ is called a *filtration* of Ω. For each $t \in \mathbb{R}^+$ define

$$\mathcal{F}_{t+} := \bigcap_{s>t} \mathcal{F}_s, \quad \mathcal{F}_{t-} := \bigvee_{s<t} \mathcal{F}_s.$$

(Recall that $\bigvee_{s<t} \mathcal{F}_s$ denotes the minimal σ-field including all $\mathcal{F}_s, s < t$.) For $t = 0$ we set $\mathcal{F}_{0-} = \mathcal{F}_0$ and $\mathcal{F}_\infty = \bigvee_{t \in \mathbb{R}^+} \mathcal{F}_t$. The filtration $(\mathcal{F}_t, t \in \mathbb{R}^+)$ is *continuous* if it is both left-continuous and right-continuous. We say that the process $X = (X_t, t \in \mathbb{R}^+)$ is *adapted* with the filtration $(\mathcal{F}_t, t \in \mathbb{R}^+)$ if for each $t \in \mathbb{R}^+$ the r.v. X_t is $\mathcal{F}_t$-measurable. In this case we write simply that X is $(\mathcal{F}_t)$ adapted.

The quadruple $(\Omega, \mathcal{F}, (\mathcal{F}_t, t \in \mathbb{R}^+), \mathbf{P})$ is called a *probability basis*. The phrase '$X = (X_t, t \in \mathbb{R}^+)$ is a process on $(\Omega, \mathcal{F}, (\mathcal{F}_t, t \in \mathbb{R}^+), \mathbf{P})$' means that $(\Omega, \mathcal{F}, \mathbf{P})$ is a probability space, $(\mathcal{F}_t, t \in \mathbb{R}^+)$ is a filtration, X is a process on $(\Omega, \mathcal{F}, \mathbf{P})$ and X is $(\mathcal{F}_t)$-adapted. If the filtration $(\mathcal{F}_t, t \in \mathbb{R}^+)$ is right-continuous and is completed by all subsets in Ω of $\mathbf{P}$-measure zero, we say that this filtration satisfies the *usual conditions*.

Other important notions and properties concerning the stochastic processes will be introduced in the examples themselves.

The reader will find systematic presentations of the basic theory of stochastic processes in many books (see Doob 1953, 1984; Blumenthal and Getoor 1968; Gihman and Skorohod 1974/1979; Ash and Gardner 1975; Prohorov and Rozanov 1969; Dellacherie 1972; Dellacherie and Meyer 1978, 1982; Loève 1978; Wentzell 1981; Métivier 1982; Jacod and Shiryaev 1987; Revuz and Yor 1991; Rao 1995).

19.1. Is it possible to find a probability space on which any stochastic process can be defined?

Let $(\Omega, \mathcal{F}, \mathbf{P})$ be a fixed probability space and $X = (X_t, t \in T \subset \mathbb{R}^+)$ be a stochastic process. It is quite natural to ask whether X can be defined on this space. Our motive for asking such a question will be clear if we recall the following result (see Ash 1972): there exists a universal probability space on which all possible random variables can be defined.

By the two examples considered below we show that some difficulties can arise when trying to extend this result to stochastic processes.

(i) Suppose the answer to the above question is positive and $(\Omega, \mathcal{F}, \mathbf{P})$ is such a space. Note that Ω is fixed, $\mathcal{F}$ is fixed and clearly the cardinality of $\mathcal{F}$ is less than or equal to 2^Ω. Choose the index set T such that its cardinality is greater than that of $\mathcal{F}$ and consider the process $X = (X_t, t \in T)$ where X_t are independent r.v.s with $\mathbf{P}[X_t = 0] = \mathbf{P}[X_t = 1] = \frac{1}{2}$ for each $t \in T$. However, if $t_1 \neq t_2$, $t_1, t_2 \in T$, the events $[X_{t_1} = 1]$ and $[X_{t_2} = 1]$ cannot be equivalent, since this would give

$$\tfrac{1}{2} = \mathbf{P}[X_{t_1} = 1] = \mathbf{P}[X_{t_1} = 1, X_{t_2} = 1] \neq \mathbf{P}[X_{t_1} = 1]\mathbf{P}[X_{t_2} = 1] = \tfrac{1}{4}.$$

This contradiction shows that $\mathcal{F}$ must contain events whose number is greater than or at least equal to the cardinality of T. But this contradicts the choice of T.

(ii) Let $\Omega = [0, 1]$, $\mathcal{F} = \mathcal{B}_{[0,1]}$ and $\mathbf{P}$ be the Lebesgue measure. We shall show that on this space $(\Omega, \mathcal{F}, \mathbf{P})$ there does not exist a process $X = (X_t, t \in [0, 1])$ such that the variables X_t are independent and X_t takes the values 0 and 1 with probability $\frac{1}{2}$ each. (Compare this with case (i) above.)

Suppose X does exist on $(\Omega, \mathcal{F}, \mathbf{P})$. Then $\mathbf{E}[X_t] < \infty$ for every $t \in [0,1]$. Let $\mathcal{I}$ be the countable set of simple r.v.s of the type $\sum_k c_k I_{A_k}$ where c_k are rational numbers and $\{A_k\}$ are finite partitions of the interval [0,1] into subintervals with rational endpoints. Since $\mathbf{E}[X_t] < \infty$ then according to Billingsley (1995) there is a r.v. $Y_t \in \mathcal{I}$ with $\mathbf{E}[|X_t - Y_t|] < \frac{1}{4}$. (Instead of $\frac{1}{4}$ we could take any $\varepsilon > 0$.) However, for arbitrary s, t we have $\mathbf{E}[|X_s - X_t|] = \frac{1}{2}$ implying that $\mathbf{E}[|Y_s - Y_t|] > 0$ for all $s \neq t$. But there are only countably many variables Y_t.

19.2. What is the role of the family of finite-dimensional distributions in constructing a stochastic process with specific properties?

Let $\mathcal{P} = \{P_n := P_{t_1,\dots,t_n}, n \in \mathbb{N}, t_1, \dots, t_n \in T, T \subset \mathbb{R}^+\}$ be a compatible family of finite-dimensional distributions. Then we can always find a probability space $(\Omega, \mathcal{F}, \mathbf{P})$ and a process $X = (X_t, t \in T)$ defined on it with just $\mathcal{P}$ as a family of its finite-dimensional distributions. Note, however, that this result says nothing about any other properties of X. Now we aim to clarify whether the compatibility conditions are sufficient to define a process satisfying some preliminary prescribed properties.

We are given the compatible family $\mathcal{P}$ and the following family of functions $\{X_t(\omega), \omega \in \Omega, t \in T\}$. Denote by $\mathcal{A}$ and $\tilde{\mathcal{A}}$ respectively the smallest Borel field and the smallest σ-field with respect to which every X_t is measurable.

Since every set in $\mathcal{A}$ has the form $A = \{\omega\colon (X_{t_1}(\omega), \dots, X_{t_n}(\omega)) \in B\}$ where $B \in \mathcal{B}^n$, by the relation

$$(1) \qquad \mathbf{P}(A) = \mathbf{P}\{\omega\colon (X_{t_1}(\omega), \dots, X_{t_n}(\omega)) \in B\} = \int_B \mathrm{d}P_{t_1,\dots,t_n}(x_1, \dots, x_n)$$

we define a probability measure $\mathbf{P}$ on $(\Omega, \mathcal{A})$ and this measure is additive. However, it can happen that $\mathbf{P}$ is not σ-additive and in this case we cannot extend $\mathbf{P}$ from $(\Omega, \mathcal{A})$ to $(\Omega, \tilde{\mathcal{A}})$ (recall that $\tilde{\mathcal{A}}$ is the σ-field generated by $\mathcal{A}$) to get a process $(X_t, t \in T)$ with the prescribed finite-dimensional distributions.

Let us illustrate this by an example. Take $T = [0,1]$ and $\Omega = \mathbb{C}[0,1]$ the space of all continuous functions on [0,1]. Let $X_t(\omega)$ be the coordinate function: $X_t(\omega) = \omega(t)$ where $\omega = \{\omega(t), t \in [0,1]\} \in \mathbb{C}[0,1]$. Suppose the family $\mathcal{P} = \{P_n\}$ is defined as follows:

$$(2) \qquad P_{t_1,\dots,t_n}(x_1, \dots, x_n) = \prod_{k=1}^{n} \int_{-\infty}^{x_k} g(u)\,\mathrm{d}u$$

where $g(u), u \in \mathbb{R}^1$ is any probability density function which is symmetric with respect to zero. It is easy to see that this family $\mathcal{P}$ is compatible. Further, the measure $\mathbf{P}$ which we want to find must satisfy the relation

$$\mathbf{P}[X_t > \varepsilon, X_s < -\varepsilon] = \left(\int_\varepsilon^\infty g(u)\,\mathrm{d}u \right)^2 \quad \text{for any } \varepsilon > 0.$$

Since $\Omega = \mathbb{C}[0,1]$, the sets $A_n = \{\omega\colon X_t(\omega) > \varepsilon, X_{t+1/n}(\omega) < -\varepsilon\}$ must tend to the empty set $\emptyset$ as $n \to \infty$ for every $\varepsilon > 0$. However,

$$\lim_{n\to\infty} \mathbf{P}(A_n) = \left(\int_\varepsilon^\infty g(u)\,\mathrm{d}u\right)^2 \neq 0.$$

Hence the measure $\mathbf{P}$ defined by (1) and (2) is not continuous at $\emptyset$ (or, equivalently, $\mathbf{P}$ is not σ-additive) and thus $\mathbf{P}$ cannot be extended to the σ-field $\tilde{\mathcal{A}}$. Moreover, for any probability measure $\mathbf{P}$ on $(\Omega, \mathcal{A})$ we have $\mathbf{P}(\Omega) = 1$ which means that with probability 1 every trajectory would be continuous. However, as we have shown, the family $\mathcal{P}$ is not consistent with this fact despite its compatibility.

In particular, note that (2) implies that $(X_t, t \in [0,1])$ is a set of r.v.s which are independent and each is distributed symmetrically with density g. The independence between X_t and X_s, even for very close s and t, is inconsistent with the desired continuity of the process.

This example and others show that the family, even though compatible, must satisfy additional conditions in order to obtain a stochastic process whose trajectories possess specific properties (see Prohorov 1956; Billingsley 1968).

19.3. Stochastic processes whose modifications possess quite different properties

Let $X = (X_t, t \in T)$ and $Y = (Y_t, t \in T)$ be two stochastic processes defined on the same probability space $(\Omega, \mathcal{F}, \mathbf{P})$ and taking values in the same state space $(E, \mathcal{E})$. We say that Y is a *modification* of X (and conversely) if for each $t \in T$, $\mathbf{P}[\omega\colon X_t(\omega) \neq Y_t(\omega)] = 0$. If we have

$$\mathbf{P}[\cup_{t\in T} : X_t(\omega) \neq Y_t(\omega)\}] = 0$$

then the processes X and Y are called *indistinguishable*.

The following examples illustrate the relationship between these two notions and show that two processes can have very different properties even if one of the processes is a modification of the other.

(i) Note firstly that if the parameter set T is countable, then X and Y are indistinguishable iff Y is a modification of X. Thus some differences can arise only if T is not countable.

So, define the probability space $(\Omega, \mathcal{F}, \mathbf{P})$ as follows: $\Omega = \mathbb{R}^+$, $\mathcal{F} = \mathcal{B}^+$ and $\mathbf{P}$ is any absolutely continuous probability distribution. Take $T = \mathbb{R}^+$ and consider the processes $X = (X_t, t \in \mathbb{R}^+)$ and $Y = (Y_t, t \in \mathbb{R}^+)$ where $X_t(\omega) \equiv 0$ and $Y_t(\omega) = 1_{\{t\}}(\omega)$. Obviously, Y is a modification of X and this fact is a consequence of the absolute continuity of $\mathbf{P}$. Nevertheless the processes X and Y are not indistinguishable, as is easily seen.

(ii) Let $\Omega = [0,1]$, $\mathcal{F} = \mathcal{B}_{[0,1]}$, $\mathbf{P}$ be the Lebesgue measure and $T = \mathbb{R}^+$. As usual, denote by $[t]$ the integer part of t. Consider two processes $X = (X_t, t \in \mathbb{R}^+)$ and

$Y = (Y_t, t \in \mathbb{R}^+)$ where

$$X_t(\omega) = 0 \text{ for all } \omega \text{ and all } t, \quad Y_t(\omega) = \begin{cases} 0, & \text{if } t - [t] \neq \omega \\ 1, & \text{if } t - [t] = \omega. \end{cases}$$

It is obvious that Y is a modification of X. Moreover, all trajectories of X are continuous while all trajectories of Y are discontinuous. (A similar fact holds for the processes X and Y in case (i).)

(iii) Let τ be a non-negative r.v. with an absolutely continuous distribution. Define the processes $X = (X_t, t \in \mathbb{R}^+)$ and $Y = (Y_t, t \in \mathbb{R}^+)$ where

$$X_t = X_t(\omega) = 1_{[\tau(\omega) \leq t]}(\omega), \quad Y_t = Y_t(\omega) = 1_{[\tau(\omega) < t]}(\omega), \quad X_0 = 0, \; Y_0 = 0.$$

It is easy to see that for each $t \in \mathbb{R}^+$ we have

$$\mathbf{P}[\omega : X_t(\omega) \neq Y_t(\omega)] = \mathbf{P}[\omega : \tau(\omega) = t] = 0.$$

Hence each of the processes X and Y is a modification of the other. But let us look at their trajectories. Clearly X is right-continuous with left-hand limits, while Y is left-continuous with right-hand limits.

19.4. On the separability property of stochastic processes

Let $X = (X_t, t \in T \subset \mathbb{R}^1)$ be a stochastic process defined on the probability space $(\Omega, \mathcal{F}, \mathbf{P})$ and taking values in the measurable space $(\mathbb{R}^1, \mathcal{B}^1)$. The process X is said to be *separable* if there exists a countable dense subset $S_0 \subset T$ such that for every closed set $B \in \mathcal{B}^1$ and every open set $I \in \mathbb{R}^1$,

$$\{\omega : X_t(\omega) \in B \text{ for all } t \in TI\} = \{\omega : X_s(\omega) \in B \text{ for all } s \in S_0 I\}.$$

Clearly, if the process X is separable, then any event associated with X can be represented by countably many operations like union and intersection. The last situation, as we know, is typical in probability theory. However, not every stochastic process is separable.

(i) Let τ be a r.v. distributed uniformly on the interval [0,1]. Consider the process $X = (X_t, t \in [0, 1])$ where

$$X_t = X_t(\omega) = \begin{cases} 1, & \text{if } \tau(\omega) = t \\ 0, & \text{if } \tau(\omega) \neq t. \end{cases}$$

If S is any countable subset of [0,1] we have

$$\mathbf{P}[X_t = 0 \text{ for } t \in S] = 1, \quad \mathbf{P}[X_t = 0 \text{ for } t \in [0, 1]] = 0.$$

Therefore the process X is not separable.

(ii) Consider the probability space $(\Omega, \mathcal{F}, \mathbf{P})$ where $\Omega = [0,1]$, $\mathcal{F}$ is the σ-field of Lebesgue-measurable sets of [0,1] and $\mathbf{P}$ is the Lebesgue measure. Let $T = [0,1]$ and Λ be a non-Lebesgue-measurable set contained in [0,1] (the construction of such sets is described by Halmos 1974). Define the function $X = (X_t, t \in T)$ by

$$X_t = X_t(\omega) = \begin{cases} 1, & \text{if } t \in \Lambda \text{ and } \omega = t \\ 0, & \text{otherwise.} \end{cases}$$

Then for each $t \in T$, $t \notin \Lambda$, we have $X_t(\omega) = 0$ for all $\omega \in \Omega$. Further, for each $t \in T, t \in \Lambda$, we have $X_t(\omega) = 0$ for all $\omega \in \Omega$ except for $\omega = t$ when $X_t(\omega) = 1$. Thus for every $t \in T$, $X_t(\omega)$ is $\mathcal{F}$-measurable and hence X is a stochastic process.

Let us note that for each $\omega \in \Omega$, $\omega \notin \Lambda$, we have $X_t(\omega) = 0$ for all $t \in T$, and for each $\omega \in \Omega$, $\omega \in \Lambda$, we have $X_t(\omega) = 0$ except for $t = \omega$ when $X_t(\omega) = 1$. Therefore every sample function of X is a Lebesgue-measurable function. Suppose now that the process X is separable. Then there would be a countable dense subset $S_0 \subset T$ such that for every closed B and open I,

$$\{\omega : X_t(\omega) \in B \text{ for all } t \in TI\} = \{\omega : X_s(\omega) \in B \text{ for all } s \in S_0 I\}$$

and both events belong to the σ-field $\mathcal{F}$. Take in particular $B = [0, \frac{1}{2}]$ and $I = \mathbb{R}^1$. Then the event

$$\{\omega : X_t(\omega) \in [0, \tfrac{1}{2}] \text{ for all } t \in T\} = [0,1] \setminus \Lambda$$

does not belong to $\mathcal{F}$. Hence the process X is not separable.

The processes considered in cases (i) and (ii) have modifications which are separable. For very general results concerning the existence of separable modifications of stochastic processes we refer the reader to the books by Doob (1953, 1984), Gihman and Skorohod (1974/1979), Yeh (1973), Ash and Gardner (1975) and Rao (1979, 1995).

19.5. Measurable and progressively measurable stochastic processes

Consider the process $X = (X_t, t \geq 0)$ defined on the probability basis $(\Omega, \mathcal{F}, (\mathcal{F}_t), \mathbf{P})$ and taking values in some measurable space $(E, \mathcal{E})$. Here $(\mathcal{F}_t)$ is a filtration satisfying the usual conditions (see the introductory notes). Recall that if for each t, X_t is $\mathcal{F}_t$-measurable, we say that the process X is $(\mathcal{F}_t)$-*adapted*. The process X is said to be *measurable* if the mapping $(t, \omega) \mapsto X_t(\omega)$ of $\mathbb{R}^+ \times \Omega$ to E is measurable with respect to the product σ-field $\mathcal{B}^+ \times \mathcal{F}$. Finally, the process X is called *progressively measurable* (or simply, a *progressive process*) if for each t, the map $(s, \omega) \mapsto X_s(\omega)$ of $[0, t] \times \Omega$ to E is $\mathcal{B}_{[0,t]} \times \mathcal{F}_t$-measurable.

Now let us consider examples to answer the following questions. (a) Does every process have a measurable modification? (b) What is the relationship between measurability and progressive measurability?

(i) Let $X = (X_t, t \in [0,1])$ be a stochastic process consisting of mutually independent r.v.s such that $\mathbf{E}X_t = 0$ and $\mathbf{E}[X_t^2] = 1$, $t \in [0,1]$. We want to know if this process is measurable. Suppose the answer is positive: that is there exists a (t,ω)-measurable family $(X_t(\omega))$ with these properties: $\mathbf{E}X_t = 0$ for $t \in [0,1]$, $\mathbf{E}[X_s X_t] = 0$ if $s \neq t$ and $\mathbf{E}[X_s X_t] = 1$ if $s = t$. It follows that for every subinterval I of $[0,1]$ we should have

$$\int_\Omega \int_I \int_I |X_s(\omega) X_t(\omega)| \mathbf{P}(\mathrm{d}\omega)\,\mathrm{d}s\,\mathrm{d}t < \infty.$$

Hence using the Fubini theorem we obtain

$$\mathbf{E}\left\{\left(\int_I X_t(\omega)\,\mathrm{d}t\right)^2\right\} = \mathbf{E}\left\{\int_I \int_I X_s(\omega) X_t(\omega)\,\mathrm{d}s\,\mathrm{d}t\right\} = \int_I \int_I \mathbf{E}[X_s X_t]\,\mathrm{d}s\,\mathrm{d}t = 0.$$

Thus for a set N_I with $\mathbf{P}(N_I) = 0$ we have $\int_I X_t(\omega)\,\mathrm{d}t = 0$ if $\omega \notin N_I$. Consider now all subintervals $I = [r', r'']$ with rational endpoints r', r'' and let $N = \cup_I N_I$. Then $\mathbf{P}(N) = 0$ and for all ω in the complement N^{c} of N we have $\int_a^b X_t(\omega)\,\mathrm{d}t = 0$ for any subinterval $[a,b]$ of $[0,1]$. This means that for $\omega \in N^{\mathrm{c}}$, $X_t(\omega) = 0$ for all t except possibly for a set of Lebesgue measure zero. Applying the Fubini theorem again we find that

$$\int_\Omega \int_0^1 X_t^2(\omega) \mathbf{P}(\mathrm{d}\omega)\,\mathrm{d}t = 0.$$

However, this is not possible, since

$$\int_\Omega \int_0^1 X_t^2(\omega) \mathbf{P}(\mathrm{d}\omega)\,\mathrm{d}t = \int_0^1 \mathbf{E}[X_t^2]\,\mathrm{d}t = 1.$$

This contradiction shows that the process X is not measurable. Moreover, the same reasoning shows that X does not have a measurable modification.

(ii) Consider now a situation which could be compared with case (i). Let $X = (X_t, t \in T \subset \mathbb{R}^1)$ be a second-order stochastic process ($\mathbf{E}[X_t^2] < \infty$ for all $t \in T$) and let $C(s,t) = \mathbf{E}[X_s X_t]$ be its covariance function. If X is a measurable process, then it follows from the Fubini theorem that $C(s,t)$ is a $\mathcal{B}_T \times \mathcal{B}_T$-measurable function. This fact leads naturally to the question: does the measurability of the covariance function $C(s,t)$ imply that the process X is measurable? The example below shows that the answer is negative.

Let $T = [0,1]$ and $X = (X_t, t \in [0,1])$ be a family of zero-mean r.v.s of unit variance and such that X_s and X_t for $s \neq t$ are uncorrelated: that is, $\mathbf{E}X_s = 0$ for $t \in [0,1]$, and $C(s,t) = \mathbf{E}[X_s X_t] = 0$ if $s \neq t$, $C(s,t) = 1$ if $s = t$, $s,t \in [0,1]$.

Since C is symmetric and non-negative definite, there exists a probability space $(\Omega, \mathcal{F}, \mathbf{P})$ and on it a real-valued process $X = (X_t, t \in [0,1])$ with C as its covariance function. Obviously, the given function C is $\mathcal{B}_{[0,1]} \times \mathcal{B}_{[0,1]}$-measurable.

Denote by $H(X)$ the closure in $L^2 = L^2(\Omega, \mathcal{F}, \mathbf{P})$ of the linear space generated by the r.v.s $\{X_t, t \in [0,1]\}$; $H(X)$ is called a linear space of the process X. According to Cambanis (1975) the following two statements are equivalent: (a) the process X has a measurable modification; (b) the covariance function C is $\mathcal{B}_I \times \mathcal{B}_I$-measurable and $H(X)$ is a separable space.

Now, since the values of X are orthogonal in L^2, that is $\mathbf{E}[X_s X_t] = 0$ for $s \neq t$, $s, t \in [0,1]$, the space $H(X)$ is not separable and therefore the process X does not have a measurable modification.

The same conclusion can be derived in the following way. Suppose X has a measurable modification, say $Y = (Y_t, t \in [0,1])$. Then

$$\mathbf{E}\left\{\int_0^1 Y_t^2 \,\mathrm{d}t\right\} = \int_0^1 C(t,t)\,\mathrm{d}t = 1 \tag{1}$$

and this relation implies that $\int_0^1 Y_t^2 \,\mathrm{d}t < \infty$ a.s. Let $\{\varphi_n, n \geq 1\}$ be a complete orthogonal system in the space $L^2[0,1] = L^2([0,1], \mathcal{B}_{[0,1]}, Leb)$ of all functions $f(t), t \in [0,1]$ which are $\mathcal{B}_{[0,1]}$-measurable and square-integrable: $\int_0^1 f^2(t)\,\mathrm{d}t < \infty$. Then (see Loève 1978)

$$Y_t = \sum_{n=1}^{\infty} \xi_n \varphi_n(t)$$

in $L^2[0,1]$ where $\xi_n = \int_0^1 Y_t \varphi_n(t)\,\mathrm{d}t$ a.s. Further, we have

$$\mathbf{E}[\xi_n^2] = \int_0^1 \int_0^1 C(s,t)\varphi_n(s)\varphi_n(t)\,\mathrm{d}s\,\mathrm{d}t = 0$$

that is $\mathbf{P}[\xi_n = 0] = 1$, and hence

$$\int_0^1 Y_t^2 \,\mathrm{d}t = \sum_{n=1}^{\infty} \xi_n^2 = 0 \quad \text{a.s.}$$

which contradicts equality (1). Therefore the process X with the covariance function C does not have a measurable modification.

(iii) Here we suggest a brief analysis of the usual and the progressive measurability of stochastic processes.

Let $X = (X_t, t \geq 0)$ be an $(\mathcal{F}_t)$-progressive process. Obviously X is $(\mathcal{F}_t)$-adapted and measurable. Is it then true that every $(\mathcal{F}_t)$-adapted and measurable process is progressive? The following example shows that this is not the case.

Let $\Omega = \mathbb{R}^+$, $\mathcal{F} = \mathcal{B}^+$ and $\mathbf{P}(\mathrm{d}x) = \mathrm{e}^{-x}\,\mathrm{d}x$ where $\mathrm{d}x$ corresponds to the Lebesgue measure. Define $\Delta = \{(x,x), x \in \mathbb{R}^+\}$ and let $\mathcal{F}_t$ for each $t \in \mathbb{R}^+$ be the σ-field generated by the points of $\mathbb{R}^+$ (this means that $A \in \mathcal{F}_t$ iff A or A^c is countable). Consider the process $X = (X_t, t \in \mathbb{R}^+)$ where

$$X_t(\omega) = 1_\Delta(t, \omega).$$

Then the process X is $(\mathcal{F}_t)$-adapted and $\mathcal{B}^+ \times \mathcal{F}$-measurable but is not progressively measurable.

It is useful to cite the following result: if X is an adapted and right-continuous stochastic process on the probability basis $(\Omega, \mathcal{F}, (\mathcal{F}_t), t \geq 0, \mathbf{P})$ and takes values in the metric space $(E, \mathcal{E})$, then X is progressively measurable. The proof of this result as well as a detailed presentation of many other results concerning measurability properties of stochastic processes can be found in the books by Doob (1953, 1984), Dellacherie and Meyer (1978, 1982), Dudley (1972), Elliott (1982), Rogers and Williams (1994) and Rao (1995).

19.6. On the stochastic continuity and the weak L^1-continuity of stochastic processes

Let $X = (X(t), t \in T)$ be a stochastic process where T is an interval in $\mathbb{R}^1$. We say that X is stochastically continuous (P-continuous) at a fixed point $t_0 \in T$ if for $t \in T$, $X(t) \xrightarrow{\mathrm{P}} X(t_0)$ as $t \to t_0$. The process X is said to be *stochastically continuous* if it is P-continuous at all points of T.

A second-order process $X = (X(t), t \in T \subset \mathbb{R}^1)$ is called *weakly* L^1*-continuous* if for every $t \in T$ and every r.v. ξ with $\mathbf{E}[\xi^2] < \infty$ we have

$$\lim_{s \to t} \mathbf{E}[X(s)\xi] = \mathbf{E}[X(t)\xi].$$

We now consider two specific examples. The first one shows that not every process is stochastically continuous, while the second examines the relationship between the two notions discussed above.

(i) Let the process $X = (X(t), t \in [0,1])$ consist of i.i.d. r.v.s with a common density $g(x)$, $x \in \mathbb{R}^1$. Let $t_0, t \in [0,1]$, $t \neq t_0$ and $\varepsilon > 0$. Then

$$p_\varepsilon := \mathbf{P}[|X(t) - X(t_0)| \geq \varepsilon] = \iint\limits_{|x-y| \geq \varepsilon} g(x)g(y)\,\mathrm{d}x\,\mathrm{d}y.$$

Obviously, if $\varepsilon \to 0$ then

$$p_\varepsilon \to \iint\limits_{x \neq y} g(x)g(y)\,\mathrm{d}x\,\mathrm{d}y = \int_{-\infty}^{\infty}\int_{-\infty}^{\infty} g(x)g(y)\,\mathrm{d}x\,\mathrm{d}y = 1.$$

This means that for some $\varepsilon_0 > 0$, $p_{\varepsilon_0} > \frac{1}{2}$, and hence

$$\mathbf{P}[|X(t) - X(t_0)| \geq \varepsilon_0] \not\to 0 \quad \text{as} \quad t \to t_0.$$

Therefore the process X is not stochastically continuous at each $t \in [0,1]$.

(ii) Let the probability space $(\Omega, \mathcal{F}, \mathbf{P})$ be defined by $\Omega = [0,1]$, $\mathcal{F} = \mathcal{B}_{[0,1]}$ with $\mathbf{P}$ the Lebesgue measure. Take the sequence of r.v.s $\{\eta_n, n \geq 1\}$ where

$\eta_n(\omega) = n^{3/4}1_{[0,1/n]}(\omega)$. Then for sufficiently small $\varepsilon > 0$ we have

$$\mathbf{P}[\omega: |\eta_n(\omega)| \geq \varepsilon] = \frac{1}{n} \to 0$$

and hence $\eta_n \xrightarrow{\mathrm{P}} 0$ as $n \to \infty$. However, $\mathbf{E}[\eta_n^2] = n^{1/2}$, the sequence $\{\eta_n\}$ is not bounded and consequently $\{\eta_n\}$ is not weakly L^1-convergent (see Masry and Cambanis 1973).

Our plan now is to use the sequence $\{\eta_n\}$ to construct a stochastic process which is stochastically but not weakly L^1-continuous.

Define the process $X = (X(t), t \in [0,1])$ by $X(t) = 0$ for $t = 0, \omega \in \Omega$ and

$$X(t) = (n+1)(1-nt)\eta_{n+1}(\omega) + n((n+1)t - 1)\eta_n(\omega)$$

for $t \in \left[\frac{1}{n+1}, \frac{1}{n}\right]$ and $\omega \in \Omega$, $n \geq 1$. Thus $X(0) = 0$, $X(1/n) = \eta_n$, $n \geq 1$, and for all $\omega \in \Omega$, $X(\cdot, \omega)$ is a linear function on every interval $[\frac{1}{n+1}, \frac{1}{n}]$, $n \geq 1$. Since $X(\frac{1}{n}) = \eta_n$ and the sequence $\{\eta_n\}$ does not converge weakly in L^1-sense, then the process X is not weakly L^1-continuous at the point $t = 0$.

It is easy to see that for all $\omega \in \Omega$ the process X is continuous on $(0,1]$ and hence X is stochastically continuous on the same interval $(0,1]$. Clearly, it remains for us to show that X is P-continuous at $t = 0$. Fix $\varepsilon > 0$ and $\delta > 0$. Since $\eta_n \xrightarrow{\mathrm{P}} 0$, there exists $N = N(\varepsilon, \delta)$ such that for all $n \geq N$, $\mathbf{P}[|\eta_n| \geq \varepsilon] < \frac{1}{2}\delta$. Now for all $t \in (0, N^{-1}]$ we have $t \in \left[\frac{1}{n+1}, \frac{1}{n}\right]$ for some concrete $n \geq N$ and it follows from the definition of X that

$$0 \leq |X(t)| \leq \max\{|\eta_n|, |\eta_{n+1}|\} \text{ for all } \omega \in \Omega.$$

Thus

$$\begin{aligned}\mathbf{P}[\omega: |X(t)| \geq \varepsilon] &\leq \mathbf{P}[\omega: \max\{|\eta_n|, |\eta_{n+1}|\} \geq \varepsilon] \\ &\leq \mathbf{P}[\omega: |\eta_n| \geq \varepsilon] + \mathbf{P}[\omega: |\eta_{n+1}| \geq \varepsilon] \\ &< \frac{1}{2}\delta + \frac{1}{2}\delta = \delta\end{aligned}$$

implying that

$$X(t) \xrightarrow{\mathrm{P}} 0 = X(0) \quad \text{as} \quad t \to 0$$

and hence the process X is stochastically continuous on the interval $[0,1]$.

Note that in this example the weak L^1-continuity of X is violated at the point $t = 0$ only. Using arguments from the paper by Masry and Cambanis (1973) we can construct a process which is stochastically continuous and weakly L^1-discontinuous at a finite or even at a countable number of points in $[0,1]$.

19.7. Processes which are stochastically continuous but not continuous almost surely

We know that convergence in probability is weaker than a.s. convergence (see Section 14). So it is not surprising that there are processes which are stochastically but not a.s. continuous. Consider the following two examples.

(i) Let $X = (X_t, t \in [0,1])$ be a stochastic process defined on the probability space $(\Omega, \mathcal{F}, \mathbf{P})$ where $\Omega = [0,1]$, $\mathcal{F} = \mathcal{B}_{[0,1]}$, $\mathbf{P}$ is the Lebesgue measure and

$$X_t = X_t(\omega) = \begin{cases} 1, & \text{if } t > \omega \\ 0, & \text{if } t \le \omega. \end{cases}$$

The state space of X consists of the values 1 and 0, and the finite-dimensional distributions are expressed as follows:

$$\mathbf{P}[X_{t_1} = 0, \dots, X_{t_{i-1}} = 0, X_{t_i} = 1, \dots, X_{t_n} = 1] = t_i - t_{i-1}$$

if $t_1 < t_2 < \dots < t_n$ and $1 < i \le n$;

$$\mathbf{P}[X_{t_1} = 0, \dots, X_{t_n} = 0] = 1 - t_n, \ \mathbf{P}[X_{t_1} = 1, \dots, X_{t_n} = 1] = t_1.$$

In all other cases $\mathbf{P}[X_{t_1} = k_1, \dots, X_{t_n} = k_n] = 0$ where $k_i = 0$ or 1. Clearly, the process X is stochastically continuous since for any $\varepsilon \in (0,1)$ we have

$$\mathbf{P}[|X_{t_2} - X_{t_1}| \ge \varepsilon] = \mathbf{P}[X_{t_1} = 0, X_{t_2} = 1] = t_2 - t_1.$$

However, almost all trajectories of X are discontinuous functions.

(ii) Consider the Poisson process $X = (X_t, t \ge 0)$ with a given parameter λ. That is, $X_0 = 0$ a.s., the increments of X are independent, and $X_t - X_s$ for $t > s$ has a Poisson distribution: $\mathbf{P}[X_t - X_s = k] = \mathrm{e}^{-\lambda(t-s)}[\lambda(t-s)]^k/k!$, $k = 0, 1, 2, \dots$. The definition immediately implies that for each fixed $t_0 \ge 0$

$$X_t \xrightarrow{\mathrm{P}} X_{t_0} \quad \text{as} \quad t \to t_0.$$

Hence X is stochastically continuous. However, it can he shown that every trajectory of X is a non-decreasing stepwise function with jumps of size 1 only. This and other results can be found e.g. in the book by Wentzell (1981).

Therefore the Poisson process is stochastically continuous but not a.s. continuous.

19.8. Almost sure continuity of stochastic processes and the Kolmogorov condition

Let $X = (X_t, t \in T \subset \mathbb{R}^+)$ be a real-valued stochastic process defined on some probability space $(\Omega, \mathcal{F}, \mathbf{P})$. Suppose X satisfies the following classical *Kolmogorov condition*:

(1) $\quad \mathbf{E}[|X_t - X_s|^p] \le K|t-s|^{1+q}, \ K = \text{constant}, \ p > 0, q > 0, \ t, s \in T.$

Then X is a.s. continuous. In other words, almost all of the trajectories of X are continuous functions. The same result can be expressed in another form: if condition (1) is valid a process $\tilde{X} = (\tilde{X}_t, t \in T)$ exists which is a.s. continuous and is a modification of the process X.

Since (1) is a sufficient condition for the continuity of a stochastic process, it is natural to ask whether this condition is necessary.

Firstly, if we consider the Poisson process (see Example 19.7) again we can easily see that condition (1) is not satisfied. Of course, we cannot make further conclusions from this fact. However, we know by other arguments that the Poisson process is not continuous.

Consider now the standard Wiener process $w = (w_t, t \geq 0)$. Recall that $w_0 = 0$ a.s., the increments of w are independent, and $w_t - w_s \sim \mathcal{N}(0, |t-s|)$. It is easy to check that w satisfies (1) with $p = 4$ and $q = 1$. Hence the Wiener process w is a.s. continuous.

Now based on the Wiener process we can construct an example of a process which is continuous but does not satisfy condition (1). Let $Y = (Y_t, t \geq 0)$ where $Y_t = \exp(w_t^3)$. This process is a.s. continuous. However, for any $p > 0$ the expectation $\mathbf{E}[|Y_t - Y_s|^p]$ does not exist and thus condition (1) cannot be satisfied. This example shows that the Kolmogorov condition (1) is not generally necessary for the continuity of a stochastic processes.

Important general results concerning continuity properties of stochastic processes, including some useful counterexamples, are given by Ibragimov (1983) and Balasanov and Zhurbenko (1985).

19.9. Does the Riemann or Lebesgue integrability of the covariance function ensure the existence of the integral of a stochastic process?

Suppose $X = (X_t, t \in [a,b] \subset \mathbb{R}^1)$ is a second-order real-valued stochastic process with zero mean and covariance function $\Gamma(s,t) = \mathbf{E}[X_s X_t]$, $s, t \in [a,b]$. We should like to analyse the conditions under which the integral $J = \int_a^b X_t \, \mathrm{d}t$ can be constructed. As usual, we consider integral sums of the type $J_N = \sum_{k=1}^N X_{s_k}(t_k - t_{k-1})$, $s_k \in (t_{k-1}, t_k)$ and define J as the limit of $\{J_N\}$ in a definite sense. One reasonable approach is to consider the convergence of the sequence $\{J_N\}$ in L^2-sense. In this case, if the limit exists, it is called an L^2-integral and is denoted as $(\mathrm{L}^2)\int_a^b X_t \, \mathrm{d}t$.

According to results which are generally accepted as classical (see Lévy 1965; Loève 1978), the integral $(\mathrm{L}^2)\int_a^b X_t \, \mathrm{d}t$ exists iff the Riemann integral $(\mathrm{R})\int_a^b \int_a^b \Gamma(s,t) \, \mathrm{d}s \, \mathrm{d}t$ exists. Note however that a paper by Wang (1982) provides some explanations of certain differences in the interpretation of the double Riemann integral. As an important consequence Wang has shown that the existence of $(\mathrm{R})\int_a^b \int_a^b \Gamma(s,t) \, \mathrm{d}s \, \mathrm{d}t$ is a sufficient but not necessary condition for the existence of $(\mathrm{L}^2)\int_a^b X_t \, \mathrm{d}t$. Let us consider this situation in more detail.

Starting with the points $a = x_0 < x_1 < \ldots < x_m = b$ on the axis Ox and $a = y_0 < y_1 < \ldots < y_n = b$ on the axis Oy we divide the square $[a,b] \times [a,b]$ into rectangles in the standard way. Define

$$\Delta_1 = \max_{1 \le i \le m} \Delta x_i,\ \Delta x_i = x_i - x_{i-1}, \quad \Delta_2 = \max_{1 \le j \le n} \Delta y_j,\ \Delta y_j = y_j - y_{j-1}.$$

Introduce the following two conditions:

(1) $$\lim_{\Delta_1 \to 0, \Delta_2 \to 0} \sum_{i=1}^{m} \sum_{j=1}^{n} \Gamma(u_i, v_j) \Delta x_i \Delta y_j \text{ exists;}$$

(2) $$\lim_{\Delta_1 \to 0, \Delta_2 \to 0} \sum_{i=1}^{m} \sum_{j=1}^{n} \Gamma(u_{ij}, v_{ij}) \Delta x_i \Delta y_j \text{ exists.}$$

(Here $u_i \in (x_{i-1}, x_i)$, $v_j \in (y_{j-1}, y_j)$ and $(u_{ij}, v_{ij} \in (x_{i-1}, x_i) \times (y_{j-1}, y_j)$.)

It is important to note that the Riemann integral (R)$\int_a^b \int_a^b \Gamma(x,y)\,\mathrm{d}x\,\mathrm{d}y$ exists iff condition (2) is fulfilled (not condition (1)). On the other hand, the integral (L^2)$\int_a^b X_t\,\mathrm{d}t$ exists iff condition (1) is fulfilled. Since (2) $\Rightarrow$ (1), then obviously the existence of (R)$\int_a^b \int_a^b \Gamma(x,y)\,\mathrm{d}x\,\mathrm{d}y$ is a sufficient condition for the existence of (L^2)$\int_a^b X_t\,\mathrm{d}t$.

Following Wang (1982) we describe a stochastic process $X = (X_t, t \in [0,1])$ such that its covariance function $\Gamma(s,t)$ is not Riemann-integrable but the integral (L^2)$\int_a^b X_t\,\mathrm{d}t$ does exist. In particular, this example will show that in general (1) $\not\Rightarrow$ (2), that is conditions (1) and (2) are not equivalent.

For the construction of the process X with the desired properties we need some notation and statements. Suppose that $[a,b] = [0,1]$. Let

$$A = \{(x,y) : \tfrac{1}{2} < x < y < 1\},$$
$$B = \{x : x \in (\tfrac{1}{2}, 1) \text{ where } x = (2^{2^k} + j)/2^{2^k+1}, 0 \le j \le 2^{2^k}, j, k \in \mathbb{N}\}$$

and $j(2^k) \equiv 2^k \pmod{j}$, $1 \le j(2^k) < 2^k$. Clearly, if j is odd, then $j(2^k)$ is odd.

For $x \in B$, $x = (2^{2^k} + j)/2^{2^k+1}$ we define the function

$$g(x) = j(2^k)/2^k + j/2^{2^{k+1}+k} = [2^{2^{k+1}} j(2^k) + j]/2^{2^{k+1}+k}.$$

Let us now formulate four statements numbered (I), (II), (III) and (IV). (For detailed proofs see Wang (1982).)

(I) If $x_1, x_2 \in B$ and $x_1 \ne x_2$, then $g(x_1) \ne g(x_2)$.

Now let

$$B' = \{x : x \in B, g(x) \notin B, x < g(x)\},\ D = \{(x,y) : x \in B', y = g(x), (x,y) \in A\}.$$

(II) We have $D \subset A$ and for arbitrary $\delta > 0$ and $(x_0, y_0) \in A$ there exists $(x, y) \in D$ such that $d[(x_0, y_0), (x, y)] < \delta$. (Here $d[\cdot, \cdot]$ is the usual Euclidean distance in the plane.)

Introduce the set

$$B'' = \{y : y = g(x), x \in B'\} \cap (\tfrac{1}{2}, 1).$$

Then $B''B' = \emptyset$. In the square $[\frac{1}{2}, 1] \times [\frac{1}{2}, 1]$ we define the function $\gamma(x, y)$ as follows.

1) If $(x, y) \in A = \{(x, y) : \frac{1}{2} < x < y < 1\}$ we put $\gamma(x, y) = 1$, if $(x, y) \in D$ and $\gamma(x, y) = 0$, otherwise.

2) If $\frac{1}{2} < x = y < 1$, let $\gamma(x, y) = 1$, if $x = y \in B' \cup B''$ and $\gamma(x, y) = 0$, otherwise.

3) If $\frac{1}{2} < y < x < 1$ we take $\gamma(x, y) = \gamma(y, x)$. For the boundary points of A, $x = \frac{1}{2}$, $x = 1$, $y = \frac{1}{2}$, $y = 1$, let $\gamma(x, y) = 0$.

(III) The Riemann integral $(\mathrm{R})\int_{\frac{1}{2}}^{1} \int_{\frac{1}{2}}^{1} \gamma(x, y)\, dx\, dy$ does not exist.

(IV) $\lim_{\Delta_1 \to 0, \Delta_2 \to 0} \sum_{i=1}^{m} \sum_{j=1}^{n} \gamma(u_i, v_j) \Delta x_i \Delta y_j$ exists and is zero.

So, having statements (I) and (II) we can now define a stochastic process whose covariance function equals $\gamma(s, t)$. For t in the interval $[\frac{1}{2}, 1]$ and $t \in B'$, let ξ_t be a r.v. distributed $\mathcal{N}(0, 1)$. If $t \in B''$ then there exists a unique $s \in B'$ such that $t = g(s)$; let $\xi_t = \xi_s$. If $t \notin B' \cup B''$, let $\xi_t \equiv 0$. Then it is not difficult to find that

$$\Gamma(s, t) := \mathbf{E}[\xi_s \xi_t] = \gamma(s, t)$$

where $\gamma(s, t)$ is exactly the function introduced above.

It remains for us to apply statements (III) and (IV). Obviously, (IV) implies the existence of $(\mathrm{L}^2)\int_{\frac{1}{2}}^{1} \xi_t\, dt$. However, (III) shows that the integral $(\mathrm{R})\int_{\frac{1}{2}}^{1} \int_{\frac{1}{2}}^{1} \gamma(s, t)\, ds\, dt$ does not exist.

Therefore the Riemann integrability of the covariance function is not necessary for the existence of the integral of the stochastic process.

As we have seen, the existence of the integral $(\mathrm{L}^2)\int_a^b X_t\, dt$ is related to the Riemann integrability of the covariance function of X. Thus we arrive at the question: is it possible to weaken this condition replacing it by the Lebesgue integrability? The next example gives the answer.

Define the process $X = (X_t, t \in [0, 1])$ as follows:

$$X_t = \begin{cases} 0, & \text{if } t \text{ is irrational} \\ \eta, & \text{if } t \text{ is rational} \end{cases}$$

where η is a r.v. distributed $\mathcal{N}(0, 1)$.

It is easy to see that $\Gamma(s, t) = \mathbf{E}[X_s X_t] = 1$ if both s and t are rational, and $\Gamma(s, t) = 0$ otherwise. Since $\Gamma(s, t) \neq 0$ over a set of plane Lebesgue measure zero, then $\Gamma(s, t)$ is Lebesgue-integrable on the square $[0, 1] \times [0, 1]$ and

$\int_0^1 \int_0^1 \Gamma(s,t)\,\mathrm{d}s\,\mathrm{d}t = 0$. However, the function $\Gamma(s,t)$ does not satisfy condition (1) which is necessary and sufficient for the existence of $(\mathrm{L}^2)\int_0^1 X_t\,\mathrm{d}t$. Hence the integral $(\mathrm{L}^2)\int_0^1 X_t\,\mathrm{d}t$ does not exist.

Therefore the Lebesgue integrability of the covariance function Γ of the process X is not sufficient to ensure the existence of the integral $(\mathrm{L}^2)\int_0^1 X_t\,\mathrm{d}t$.

19.10. The continuity of a stochastic process does not imply the continuity of its own generated filtration, and vice versa

Let $X = (X_t, t \geq 0)$ be a stochastic process defined on the probability space $(\Omega, \mathcal{F}, \mathbf{P})$. Denote by $\mathcal{F}_t^X = \sigma\{X_s, s \leq t\}$ the smallest σ-field generated by the process X up to time t. Clearly, $\mathcal{F}_t^X \subset \mathcal{F}_{t_1}^X$ if $t < t_1$. The family $(\mathcal{F}_t^X, t \geq 0)$ is called the own generated filtration of the process.

It is of general interest to clarify the relationship between the continuity of the process X and the continuity of the filtration $(\mathcal{F}_t^X)$. Recall the following well known result (see Liptser and Shiryaev 1977/78): if $X = w$ is the standard Wiener process, then the filtration $(\mathcal{F}_t^w, t \geq 0)$ is continuous.

Let us answer two questions. (a) Does the continuity of the process X imply that the filtration $(\mathcal{F}_t^X)$ is continuous? (b) Is it possible to have a continuous filtration $(\mathcal{F}_t^X)$ which is generated by a discontinuous process X?

(i) Let $\Omega = \mathbb{R}^1$, $\mathcal{F} = \mathcal{B}^1$ and $\mathbf{P}$ be an arbitrary probability measure on $\mathcal{B}^1$. Consider the process $X = (X_t, t \geq 0)$ where $X_t = X_t(\omega) = t\xi(\omega)$ and ξ is a r.v. distributed $\mathcal{N}(0,1)$. Obviously, the process X is continuous. Further, it is easy to see that for $t = 0$, $\mathcal{F}_0^X$ is the trivial σ-field $\{\emptyset, \Omega\}$. If $t > 0$, we have $\mathcal{F}_t^X = \mathcal{B}^1$. Thus $\mathcal{F}_0^X \neq \mathcal{F}_{0+}^X$. Therefore the filtration $(\mathcal{F}_t^X)$ is not right-continuous and hence not continuous despite the continuity of X.

(ii) Let $\Omega = [0,1]$, $\mathcal{F} = \mathcal{B}_{[0,1]}$ and $\mathbf{P}$ be the Lebesgue measure. Choose the function $h \in \mathbb{C}^\infty(\mathbb{R}^+)$ so that $h(x) = 0$ for $x \leq \frac{1}{2}$ and for $x > \frac{1}{2}$, $h(x) > 0$, and h is strictly increasing on the interval $[\frac{1}{2}, \infty)$. (It is easy to find examples of such functions.) Consider the process $X = (X_t, t \geq 0)$ where $X_t = X_t(\omega) = \omega h(t)$, $\omega \in \Omega$, $t \geq 0$ and let $(\mathcal{F}_t^X, t \geq 0)$ be its own generated filtration: $\mathcal{F}_t^X = \sigma\{X_s, s \leq t\}$. Then it is easy to check that

$$\mathcal{F}_t^X = \begin{cases} \{\emptyset, \Omega\}, & \text{if } 0 \leq t \leq \frac{1}{2} \\ \mathcal{B}^1, & \text{if } t > \frac{1}{2}. \end{cases}$$

Hence the filtration $(\mathcal{F}_t^X)$ is discontinuous even though the trajectories of X are in the space $\mathbb{C}^\infty$.

(iii) Now we aim to show that the filtration $(\mathcal{F}_t^X)$ of the process X can be continuous even if X has discontinuous trajectories.

Firstly, let $h : \mathbb{R}^+ \mapsto \mathbb{R}^1$ be any function. Then a countable dense set $D \subset \mathbb{R}^+$ exists such that for all $t \geq 0$ there is a sequence $\{t_n, n \geq 1\}$ in D with $t_n \to t$ and

$h(t_n) \to h(t)$ as $n \to \infty$. The reasoning is as follows. Let $(\Omega, \mathcal{F}, \mathbf{P})$ be a one-point probability space. Define $X_t(\omega) = h(t)$ for $\omega \in \Omega$ and all $t \geq 0$. Since the extended real line $\overline{\mathbb{R}^1}$ is a compact, the separability theorem (see Doob 1953, 1984; Gihman and Skorohod (1974/1979); Ash and Gardner 1975) implies that $(X_t, t \geq 0)$ has a separable version $(Y_t, t \geq 0)$ with $Y_t : \Omega \to \overline{\mathbb{R}^1}$, $t \geq 0$. But $Y_t = X_t$ a.s. and so $Y_t(\omega) = X_t(\omega) = h(t), t \geq 0$.

Thus we can construct a class of stochastic processes which are separable but whose trajectories need not possess any useful properties (for example, X can be discontinuous, and even non-measurable; of course, everything depends on the properties of the function h which, let us repeat, can be chosen arbitrarily).

Now take again the above one-point probability space $(\Omega, \mathcal{F}, \mathbf{P})$ and choose any function $h : \mathbb{R}^+ \mapsto \mathbb{R}^1$. Define the process $X = (X_t, t \geq 0)$ by $X_t = X_t(\omega) = h(t)$, $\omega \in \Omega$, $t \geq 0$. Then for all $t \geq 0$, $\mathcal{F}_t^X = \sigma\{X_s, s \leq t\}$ is a $\mathbf{P}$-trivial σ-field in the sense that each event $A \in \mathcal{F}_t^X$ has either $\mathbf{P}(A) = 1$ or $\mathbf{P}(A) = 0$. Therefore the filtration $(\mathcal{F}_t^X, t \geq 0)$ is continuous. By the above result the process X is separable but its trajectories are equal to h, and h is chosen arbitrarily. It is enough to take h as discontinuous.

Finally we can conclude that in general the continuity (and even the infinite smoothness) of a stochastic process does not imply the continuity of its own generated filtration (see case (i) and case (ii)). On the other hand, a discontinuous process can generate a continuous filtration (case (iii)).

The interested reader can find several useful results concerning fine properties of stochastic processes and their filtrations in the books by Dellacherie and Meyer (1978, 1982), Métivier (1982), Jacod and Shiryaev (1987), Revuz and Yor (1991) and Rao (1995).

SECTION 20. MARKOV PROCESSES

We recall briefly only a few basic notions concerning Markov processes. Some definitions will be given in the examples considered below. In a few cases we refer the reader to the existing literature.

Firstly, let $X = (X_t, t \in T \subset \mathbb{R}^+)$ be a family of r.v.s on the probability space $(\Omega, \mathcal{F}, \mathbf{P})$ such that for each t, X_t takes values in some countable set E. We say that X is a **Markov chain** if it satisfies the *Markov property*: for arbitrary $n \geq 1$, $t_1 < t_2 < \ldots < t_n < t, t_j, t \in T, i_1, \ldots, i_n, j \in E$,

(1) $$\mathbf{P}[X_t = j | X_{t_1} = i_1, \ldots, X_{t_{n-1}} = i_{n-1}, X_{t_n} = i_n] = \mathbf{P}[X_t = j | X_{t_n} = i_n].$$

The chain X is finite or infinite accordingly as the state space E is finite or infinite. If $T = \{0, 1, 2, \ldots\}$ we write $X = (X_n, n \geq 0)$ or $X = (X_n, n = 0, 1, \ldots)$ and say that X is a *discrete-time* Markov chain. If $T = \mathbb{R}^+$ or $T = [a, b] \subset \mathbb{R}^+$ we say that $X = (X_t, t \geq 0)$ or $X = (X_t, t \in [a, b])$ is a *continuous-time* Markov chain.

The probabilistic characteristics of any Markov chain can be found if we know the *initial distribution* $(r_j, j \in E)$ where $r_j = \mathbf{P}[X_0 = j]$, $r_j \geq 0$, $\sum_{j \in E} r_j = 1$ and the *transition probabilities* $p_{ij}(s,t) = \mathbf{P}[X_t = j | X_s = i]$, $t \geq s$, $i, j \in E$. The chain X is called *homogeneous* if $p_{ij}(s,t)$ depends on s and t only through $t - s$. In this case, if X is a discrete-time Markov chain, it is enough to know $(r_j, j \in E)$ and the 1-step *transition matrix* $P = (p_{ij})$ where $p_{ij} = \mathbf{P}[X_{n+1} = j | X_n = i]$, $n \geq 0$. The n-step transition probabilities form the matrix $P^{(n)} = (p_{ij}^{(n)})$ and satisfy the relation

$$(2) \qquad p_{ij}^{(m+n)} = \sum_{k \in E} p_{ik}^{(m)} p_{kj}^{(n)}$$

which is called the *Chapman–Kolmogorov equation.*

Note that the transition probabilities $p_{ij}(t)$ or $p_{ij}(s,t)$ of any continuous-time Markov chain satisfy the so-called forward and backward Kolmogorov equations.

In some of the examples below we assume that the reader is familiar with basic notions and results in the theory of Markov chains such as classification of the states, recurrence and transience properties, irreducibility, aperiodicity, infinitesimal matrix and Kolmogorov equations.

Now let us recall some more general notions. Let $X = (X_t, t \geq 0)$ be a real-valued process on the probability space $(\Omega, \mathcal{F}, \mathbf{P})$ and $(\mathcal{F}_t, t \geq 0)$ be its own generated filtration. We say that X is a **Markov process** with state space $(\mathbb{R}^1, \mathcal{B}^1)$ if it satisfies the *Markov property*: for arbitrary $\Gamma \in \mathcal{B}^1$ and $t \geq s$,

$$(3) \qquad \mathbf{P}[X_t \in \Gamma | \mathcal{F}_s] = \mathbf{P}[X_t \in \Gamma | X_s] \quad \text{a.s.}$$

This property can also be written in other equivalent forms.

The function $P(s, x; t, \Gamma)$ defined for $s, t \in \mathbb{R}^+$, $s \leq t$, $x \in \mathbb{R}^1$, $\Gamma \in \mathcal{B}^1$ is said to be a *transition function* if: (a) for fixed s, t and x, $P(s, x; t, \cdot)$ is a probability measure on $\mathcal{B}^1$; (b) $P(s, x; t, \Gamma)$ is $\mathcal{B}^1$-measurable in x for fixed s, t, Γ; (c) $P(s, x; s, \Gamma) = \delta_x(\Gamma)$ where $\delta_x(\Gamma)$ is the unit measure concentrated at x; (d) the following relation holds:

$$P(s, x; t, \Gamma) = \int_{\mathbb{R}^1} P(s, x; u, \mathrm{d}y) P(u, y; t, \Gamma), \quad s \leq u \leq t.$$

This relation, called the *Chapman–Kolmogorov equation*, is the continuous analogue of (2).

We say that $X = (X_t, t \geq 0)$ is a Markov process with transition function $P(s, x; t, \Gamma)$ if X satisfies (3) and

$$\mathbf{P}[X_t \in \Gamma | X_s] = P(s, X_s; t, \Gamma) \quad \text{a.s.}$$

The Markov process X is called *homogeneous* if its transition function $P(s, x; t, \Gamma)$ depends on s and t only through $t - s$. In this case we can introduce the function $P(t, x, \Gamma) = P(0, x; t, \Gamma)$ of three arguments, $t \geq 0$, $x \in \mathbb{R}^1$, $\Gamma \in \mathcal{B}^1$ and to express conditions (a)–(d) in a simpler form.

Note that the strong Markov property will be introduced and compared with the usual Markov property (3) in one of the examples.

Complete presentations of the theory of Markov chains in discrete and continuous time can be found in the books by Doob (1953), Chung (1960), Gihman and Skorohod (1974/1979), Isaacson and Madsen (1976) and Iosifescu (1980). Some important books are devoted to the general theory of Markov processes: see Dynkin (1961, 1965), Blumenthal and Getoor (1968), Rosenblatt (1971, 1974), Wentzell (1981), Chung (1982), Ethier and Kurtz (1986), Bhattacharya and Waymire (1990) and Rogers and Williams (1994).

In this section we have included examples which examine the relationships between some similar notions or illustrate some of the basic properties of the Markov chains and processes. Note especially that many other useful results and counterexamples can be found in the recent publications indicated in the Supplementary Remarks.

20.1. Non-Markov random sequences whose transition functions satisfy the Chapman–Kolmogorov equation

Here we consider a few examples to illustrate the difference between the Markov property, which defines a Markov process, and the Chapman–Kolmogorov equation, which is a consequence of the Markov property.

(i) Suppose an urn contains four balls numbered 1, 2, 3, 4. Randomly we choose one ball, note its number and return it to the urn. This procedure is repeated many times. Denote by ξ_n the number on the nth chosen ball. For $j = 1, 2, 3$ introduce the events $A_j^{(n)} = \{\text{either } \xi_n = j \text{ or } \xi_n = 4\}$ and let $X_{3(m-1)+j} = 1$ if $A_j^{(m)}$ occurs, and 0 otherwise, $m = 1, 2, \ldots$. Thus we have defined the random sequence $(X_n, n \geq 1)$ and want to establish whether it satisfies the Markov property and the Chapman–Kolmogorov equation.

If each of k_1, k_2, k_3 is 1 or 0, then

$$\mathbf{P}[X_n = k_1] = \mathbf{P}[X_n = k_2 | X_m = k_3] = \tfrac{1}{2}, \quad n > m.$$

Therefore for $l < m < n$ we have

$$\begin{aligned}\tfrac{1}{2} &= \mathbf{P}[X_n = k_2 | X_l = k_1] \\ &= \mathbf{P}[X_n = k_2 | X_m = 0]\mathbf{P}[X_m = 0 | X_l = k_1] \\ &+ \mathbf{P}[X_n = k_2 | X_m = 1]\mathbf{P}[X_m = 1 | X_l = k_1] = \tfrac{1}{2} \cdot \tfrac{1}{2} + \tfrac{1}{2} \cdot \tfrac{1}{2} = \tfrac{1}{2}.\end{aligned}$$

This means that the transition probabilities of the sequence $\{X_n, n \geq 1\}$ satisfy the Chapman–Kolmogorov equation. Further, the event $[X_{3m} = 1, X_{3m-1} = 1]$ means that $\xi_m = 4$ which implies that $X_{3m} = 1$. Thus

$$\mathbf{P}[X_{3m} = 1 | X_{3m-2} = 1, X_{3m-1} = 1] = 1, \quad m = 1, 2, \ldots.$$

This relation shows that the Markov property does not hold for the sequence $\{X_n, n \geq 1\}$. Therefore $\{X_n, n \geq 1\}$ is not a Markov chain despite the fact that its transition probabilities satisfy the Chapman–Kolmogorov equation.

(ii) In Example 7.1(iii) we constructed an infinite sequence of pairwise i.i.d r.v.s $\{X_n, n \geq 1\}$ where X_n takes the values 1, 2, 3 with probability $\frac{1}{3}$ each. Thus we have a random sequence such that $p_{ij} = \mathbf{P}[X_{n+1} = j | X_n = i] = \frac{1}{3}$ for all possible i, j. The Chapman–Kolmogorov equation is trivially satisfied with $p_{ij}^{(n)} = \frac{1}{3}, n \geq 1$. However, the sequence $\{X_n, n \geq 1\}$ is not Markovian. To see this, suppose that at time $n = 1$ we have $X_1 = 2$. Then a transition to state 3 at the next step is possible iff the initial state was 1. Hence the transitions following the first step depend not only on the present state but also on the initial state. This means that the Markov property is violated although the Chapman–Kolmogorov equation is satisfied.

(iii) Every $N \times N$ stochastic matrix P defines the transition probabilities of a Markov process with discrete time. Its n-step transition probabilities satisfy the Chapman–Kolmogorov equation which can be written as the semigroup relation $P^{m+n} = P^m P^n$. Now we are going to show that for $N \geq 3$ there is a non-Markov process with N states whose transition probabilities satisfy the same equation.

Let Ω_1 be the sample space whose points $(x^{(1)}, \ldots, x^{(N)})$ are the random permutations of $(1, \ldots, N)$ each with probability $1/N!$. Let i and ν be fixed numbers of the set $\{1, \ldots, N\}$ and Ω_2 be the set of the N points $(x^{(1)}, \ldots, x^{(N)})$ such that $x^{(i)} = \nu$. Each point in Ω_2 has probability $1/N$. Let Ω be the mixture of Ω_1 and Ω_2 with Ω_1 carrying weight $1 - 1/N$ and Ω_2 weight $1/N$. More formally, Ω contains $N!+N$ arrangements $(x^{(1)}, \ldots, x^{(N)})$ which represent either a permutation of $(1, \ldots, N)$ or the N-fold repetition of the integer ν, $\nu = 1, \ldots, N$. To each point of the first class we attribute probability $(1 - N^{-1})/N!$; to each point of the second class, probability N^{-2}. Then clearly

$$\mathbf{P}[x^{(i)} = \nu] = N^{-1}, \quad \mathbf{P}[x^{(i)} = \nu, x^{(j)} = \mu] = N^{-2}, \quad i \neq j.$$

Thus all transition probabilities of the sequence constructed above are the same, namely

$$\mathbf{P}[x^{(i)} = \nu | x^{(j)} = \mu] = N^{-1}.$$

If $x^{(1)} = 1, x^{(2)} = 1$, then $\mathbf{P}[x^{(3)} \neq 1] = 0$ which means that the Markov property is not satisfied. Nevertheless the Chapman–Kolmogorov equation is satisfied.

20.2. Non-Markov processes which are functions of Markov processes

If $X = (X_t, t \geq 0)$ is a Markov process with state space $(E, \mathcal{E})$ and g is a one–one mapping of E into E, then $Y = (g(X_t), t \geq 0)$ is again a Markov process. However, if g is not a one–one function, the Markov property may not hold. Let us illustrate this possibility by a few examples.

(i) Let $\{X_n, n = 0, 1, 2, \ldots\}$ be a Markov chain with state space $E = \{1, 2, 3\}$, transition matrix

$$P = \begin{pmatrix} 0 & \frac{1}{2} & \frac{1}{2} \\ \frac{1}{3} & \frac{1}{4} & \frac{5}{12} \\ \frac{2}{3} & \frac{1}{4} & \frac{1}{12} \end{pmatrix}$$

and initial distribution $r = (\frac{1}{3}, \frac{1}{3}, \frac{1}{3})$. It is easy to see that the chain $\{X_n\}$ is stationary.

Consider now the new process $\{Y_n, n = 0, 1, 2, \ldots\}$ where $Y_n = g(X_n)$ and g is a given function on E. Suppose the states i of X on which g equals some fixed constant are collapsed into a single state of the new process Y called an *aggregated process*. The collection of states on which g takes the value x will be called the set of states S_x. It is obvious that only non-empty sets of states are of interest.

For the Markov chain given above let us collapse the set of states S consisting of 1, 2 into one state. Then it is not difficult to find that

$$\mathbf{P}[X_{m+2} \in S, X_{m+1} \in S | X_m \in S] = \tfrac{29}{96} \neq (\mathbf{P}[X_{m+1} \in S | X_m \in S])^2 = \left(\tfrac{13}{24}\right)^2.$$

This relation implies that the new process Y is not Markov.

(ii) Let $\{X_n, n = 0, 1, \ldots\}$ be a stationary Markov chain with the following state space $E = \{1, 2, 3, 4\}$ and n-step transition matrices

$$P^{(n)} = \frac{1}{4}\begin{pmatrix} 1 & 1 & 1 & 1 \\ 1 & 1 & 1 & 1 \\ 1 & 1 & 1 & 1 \\ 1 & 1 & 1 & 1 \end{pmatrix} + \lambda^n \begin{pmatrix} 0 & 1 & -1 & 0 \\ 0 & 0 & 0 & 0 \\ 0 & -1 & 1 & 0 \\ 0 & 0 & 0 & 0 \end{pmatrix}$$

$$+ \left(\frac{\lambda'}{\sqrt{2}}\right)^n \frac{1}{\sqrt{2}} \begin{pmatrix} 1 & -1 & 1 & -1 \\ 0 & 0 & 0 & 0 \\ 0 & 0 & 0 & 0 \\ -1 & 1 & -1 & 1 \end{pmatrix}$$

where $n = 1, 2, \ldots$ and λ, λ' are real numbers sufficiently small in absolute value. Take the function g: $E \mapsto \{1, 2\}$ such that $g(1) = g(2) = 1$, $g(3) = g(4) = 2$, and consider the aggregated process $\{Y_n, n = 0, 1, \ldots\}$ where $Y_n = g(X_n)$. If $Q^{(n)}$ denotes the n-step transition matrix of Y, we find

$$Q^{(n)} = \begin{pmatrix} \frac{1}{2} & \frac{1}{2} \\ \frac{1}{2} & \frac{1}{2} \end{pmatrix} + \lambda^n \begin{pmatrix} \frac{1}{2} & -\frac{1}{2} \\ -\frac{1}{2} & \frac{1}{2} \end{pmatrix}.$$

It turns out that $Q^{(n)}$ does not depend on λ' and it is easy to check that $Q^{(n)}$, $n \geq 1$, satisfy the Chapman–Kolmogorov equation. However, the relation

$$\mathbf{P}[Y_0 = 1, Y_1 = 1, Y_2 = 1] = \tfrac{1}{8}(1 + 2\lambda + \lambda\lambda')$$

implies that the sequence $\{Y_n, n \geq 0\}$ is not Markov when $\lambda \neq \lambda'$.

(iii) Consider two Markov chains, X_1 and X_2, with the same state space E and initial distribution r, and with transition matrices P_1 and P_2 respectively. Define a new process, say X, with the same state space E, initial distribution r and n-step transition matrix $P^{(n)} = \frac{1}{2}P_1^{(n)} + \frac{1}{2}P_2^{(n)}$. Then it can be shown that the process X, which is called a mixture of X_1 and X_2, is not Markov.

(iv) Let $w = (w_t, t \geq 0)$ be a standard Wiener process. Consider the processes

$$|w| = (|w_t|, t \geq 0), \quad M = (M_t := \max_{0 \leq s \leq t} w_s, t \geq 0), \quad Y = M - w.$$

Then obviously the process M is not Markov. According to a result by Freedman (1971), see also Revuz and Yor (1991), Y is a Markov process distributed as $|w|$ where $|w|$ is called a Wiener process with a reflecting barrier at 0. Since the Wiener process w itself is a Markov process, we have the relation

$$M = Y + w.$$

Here the right-hand side is a sum of two Markov processes but the left-hand side is a process which is not Markov. In other words, the sum of two Markov processes need not be a Markov process. Note, however, that the sum of two independent Markov processes preserves this property.

20.3. Comparison of three kinds of ergodicity of Markov chains

Let $X = \{X_n, n = 0, 1, \ldots\}$ be a non-stationary Markov chain with state space E (E is a countable set, finite or infinite). The chain X is described completely by the initial distribution $f^{(0)} = (f_j^{(0)}, j \in E)$ and the sequence $\{P_n, n \geq 1\}$ of the transition matrices.

If $\lim_{n\to\infty} p_{ij}^{(k,k+n)} = \pi_j$ exists for all $j \in E$ independently of i, $\pi_i > 0$ and $\sum_{j\in E} \pi_j = 1$, we say that the chain X is *ergodic* and $(\pi_j, j \in E)$ is its *ergodic distribution.*

Introduce the following notation:

$$f^{(m)} = f^{(0)}P_1P_2\ldots P_m, \quad f^{(k,m)} = f^{(0)}P_{k+1}P_{k+2}\ldots P_m,$$

$$P^{(k,m)} = P_{k+1}P_{k+2}\ldots P_m.$$

The Markov chain X is called *weakly ergodic* if for all $k \in \mathbb{N}$,

$$\lim_{m\to\infty} \sup_{f^{(0)}, g^{(0)}} ||f^{(k,m)} - g^{(k,m)}|| = 0 \tag{1}$$

where $f^{(0)} = (f_j^{(0)}, j \in E)$ and $g^{(0)} = (g_j^{(0)}, j \in E)$ are arbitrary initial distributions of X.

The chain X is called *strongly ergodic* if there is a probability distribution $q = (q_j, j \in E)$ such that for all $k \in \mathbb{N}$,

$$\lim_{m\to\infty} \sup_{f^{(0)}} ||f^{(k,m)} - q|| = 0. \tag{2}$$

(In (1) and (2) the norm of the vector $x = (x_j, j \in E)$ is defined by $||x|| = \sum_{j\in E} |x_j|$.)

Now we can easily make a distinction between the ergodicity, the weak ergodicity and the strong ergodicity in the case of stationary Markov chains.

For every Markov chain we can introduce the so-called *δ-coefficient*. If $P = (p_{ij})$ is the transition matrix of the chain we put

$$\delta(P) = 1 - \inf_{i,k} \sum_{j\in E} \min\{p_{ij}, p_{kj}\}.$$

This coefficient is effectively used for studying Markov chains and will be used in the examples below.

Our aim is to compare the three notions of ergodicity introduced above. Obviously, strong ergodicity implies weak ergodicity. Thus the first question is whether the converse is true. According to a result by Isaacson and Madsen (1976), if the state space E is finite, the ergodicity and the weak ergodicity are equivalent notions. The second question is: what happens if E is infinite? The examples below will answer these and other related questions.

(i) Let $\{X_n\}$ be a non-stationary Markov chain with

$$P_{2n-1} = \begin{pmatrix} \frac{1}{2} & \frac{1}{2} \\ 1 & 0 \end{pmatrix}, \quad P_{2n} = \begin{pmatrix} 0 & 1 \\ 1 & 0 \end{pmatrix}, \qquad n = 1, 2, \ldots.$$

We can easily see that $\delta(P_{2n}) = 1$ and $\delta(P_{2n-1}) = \frac{1}{2}$. Hence for all k,

$$\delta(P^{(k,m)}) \leq \prod_{j=k+1}^{m} \delta(P_j) \leq (1/2)^{[(m-k)/2]} \to 0 \quad \text{as} \quad m \to 0.$$

However, the condition $\delta(P^{(k,m)}) \to 0$ for all k as $m \to \infty$ is necessary and sufficient for any Markov chain X to be weakly ergodic (see Isaacson and Madsen 1976). Therefore the Markov chain considered here is weakly ergodic. Let us determine whether X is strongly ergodic. Take $f^{(0)} = (0, 1)$ as an initial distribution. Then

$$\begin{aligned} f^{(2k)} &= f^{(0)}(P_1P_2)(P_3P_4)\ldots(P_{2k-1}P_{2k}) \\ &= (0, 1)\begin{pmatrix} \frac{1}{2} & \frac{1}{2} \\ 0 & 1 \end{pmatrix}^k = (0, 1) \end{aligned}$$

and

$$f^{(2k+1)} = f^{(0)}(P_1P_2)(P_3P_4)\dots(P_{2k-1}P_{2k})P_{2k+1}$$
$$= (0,1)\begin{pmatrix} \frac{1}{2} & \frac{1}{2} \\ 0 & 1 \end{pmatrix}^k = (0,1).$$

Hence $||f^{(2j)} - f^{(2j+1)}|| = 2$ for $j = k, k+1, \dots$ and the sequence $\{f^{(k)}\}$ does not converge. Therefore the chain X is not strongly ergodic.

(ii) Again, let $\{X_n\}$ be a non-stationary Markov chain with

$$P_{2n-1} = \begin{pmatrix} 1 - \frac{1}{2n-1} & \frac{1}{2n-1} \\ 1 - \frac{1}{2n-1} & \frac{1}{2n-1} \end{pmatrix}, \quad P_{2n} = \begin{pmatrix} \frac{1}{2n} & 1 - \frac{1}{2n} \\ \frac{1}{2n} & 1 - \frac{1}{2n} \end{pmatrix}, \quad n = 1, 2, \dots .$$

Then for any initial distribution $f^{(0)}$ we have

$$f^{(k,m)} = \begin{cases} \left(1 - \frac{1}{m}, \frac{1}{m}\right), & \text{if } m \text{ is odd} \\ \left(\frac{1}{m}, 1 - \frac{1}{m}\right), & \text{if } m \text{ is even.} \end{cases}$$

It is not difficult to check that condition (1) is satisfied while condition (2) is violated. Therefore this Markov chain is weakly, but not strongly, ergodic.

(iii) Let $\{X_n\}$ be a stationary Markov chain with infinite state space E and transition matrix

$$P = \begin{pmatrix} \frac{1}{2} & \frac{1}{2} & 0 & 0 & 0 & \dots \\ \frac{1}{2} & 0 & \frac{1}{2} & 0 & 0 & \dots \\ 0 & \frac{3}{4} & 0 & \frac{1}{4} & 0 & \dots \\ 0 & 0 & \frac{7}{8} & 0 & \frac{1}{8} & \dots \\ \dots & \dots & \dots & \dots & \dots & \dots \end{pmatrix}.$$

It can be shown that this chain is irreducible, positive recurrent and aperiodic. Hence it is ergodic (see Doob 1953; Chung 1960), that is independently of the initial distribution $f^{(0)}$, $\lim_{n\to\infty} f^{(0)}P^{(n)} = \pi$ exists and $\pi = (\pi_j, j \in E)$ is a probability distribution. However, $\delta(P^{(m)}) = 1$ for all m which implies that the chain is not weakly ergodic.

(iv) Since the condition $\delta(P^{(k,m)}) \to 0$ for all k as $m \to \infty$ is necessary and sufficient for the weak ergodicity of non-stationary Markov chains, it is natural to ask how this condition is expressed in the stationary case.

Let $\{X_n\}$ be a stationary Markov chain with transition matrix P. Then we have $P^{(k,m)} = P^{(m-k)}$ and for the δ-coefficient we find

$$\delta(P^{(k,m)}) = \delta(P^{(m-k)}) \le [\delta(P)]^{m-k}.$$

This means that the condition $\delta(P) < 1$ is sufficient for the chain to be weakly ergodic. However, this condition is not necessary. Indeed, let

$$P = \begin{pmatrix} 0 & 1 & 0 \\ \frac{1}{2} & 0 & \frac{1}{2} \\ \frac{1}{3} & \frac{1}{3} & \frac{1}{3} \end{pmatrix}.$$

The Markov chain with this P is irreducible, aperiodic and positive recurrent. Therefore (see Isaacson and Madsen 1976) the chain is weakly ergodic. At the same time $\delta(P) = 1$.

20.4. Convergence of functions of an ergodic Markov chain

Let $X = (X_n, n \geq 0)$ be a Markov chain with countable state space E and n-step transition matrix $(p_{ij}^{(n)})$. Let $\pi_j := \lim_{n\to\infty} p_{ij}^{(n)}$ be the ergodic distribution of X and $g : E \mapsto \mathbb{R}^1$ be a bounded and measurable function. We are interested in the conditions under which the following relation holds:

$$(1) \qquad \lim_{n\to\infty} \mathbf{E}[g(X_n)] = \sum_{j\in E} \pi_j g(j).$$

One of the possible answers, given by Holewijn and Hordijk (1975), can be formulated as follows. Let X be an irreducible, positive recurrent and aperiodic Markov chain with values in the space E. Denote by $(\pi_j, j \in E)$ its ergodic distribution. Suppose that the function g is non-negative and is such that $\sum_{j\in E} \pi_j g(j) < \infty$. Additionally, suppose that for some $i_0 \in E$, $\mathbf{P}[X_0 = i_0] = 1$. Then relation (1) does hold.

Our aim now is to show that the condition $X_0 = i_0$ is essential. In particular, this condition cannot be replaced by the assumption that X has some proper distribution over the whole space E at the time $n = 0$.

Consider the Markov chain $X = (X_n, n \geq 0)$ which takes values in the set $E = \{0, 1, 2, \ldots\}$ and has the following transition probabilities:

$$p_{0j} = q^j p, \text{ if } j = 0, 1, 2, \ldots, \quad p_{i,i-1} = 1, \text{ if } i = 1, 2, \ldots, \quad p_{ij} = 0, \text{ otherwise}$$

where $0 < p < 1$, $q = 1 - p$. A direct calculation shows that

$$p_{ij}^{(n)} = q^j p, \text{ if } i = 0, 1, 2, \ldots, \ j = 0, 1, \ldots, n-1,$$

$$p_{n-1+k,k-1}^{(n)} = 1, \quad \text{if } k = 1, 2, \ldots \quad \text{and} \quad p_{ij}^{(n)} = 0, \quad \text{otherwise.}$$

The chain X is irreducible, aperiodic and positive recurrent and its ergodic distribution $(\pi_j, j \in E)$ is given by

$$\pi_j := \lim_{n\to\infty} p_{ij}^{(n)} = q^j p.$$

Suppose now g is a function on E satisfying the following condition: $\sum_{j=0}^{\infty} q^j |g(j)| < \infty$. Suppose also that $(r_j, j \in E)$ is the initial distribution of the chain X. Then

$$\begin{aligned}\mathbf{E}[g(X_n)] &= \sum_{j=0}^{\infty} \left(\sum_{i=0}^{\infty} r_i p_{ij}^{(n)}\right) g(j)\\ &= (r_0 + \ldots + r_{n-1}) \sum_{j=0}^{\infty} pq^j g(j) + \sum_{j=0}^{\infty} r_{n+j} g(j).\end{aligned}$$

Clearly, $\mathbf{E}[g(X_n)]$ will converge to $\sum_{j=0}^{\infty} pq^j g(j)$ as $n \to \infty$ iff

$$\lim_{n\to\infty} \sum_{j=0}^{\infty} r_{n+j} g(j) = 0.$$

Now we can make our choice of g and $(r_j, j \in E)$. Let

$$g(j) = j^2, \quad \text{if } j = 0, 1, 2, \ldots \quad \text{and} \quad r_j = \begin{cases} 0, & \text{if } j = 0 \\ \dfrac{6}{\pi^2 j^2}, & \text{if } j = 1, 2, \ldots. \end{cases}$$

Then $\sum_{j=0}^{\infty} q^j |g(j)| < \infty$ and obviously for all $n \geq 0$ we get

$$\sum_{j=0}^{\infty} r_{n+j} g(j) = \infty.$$

Hence a relation like (1) is not possible.

Therefore in general we cannot replace the condition $\mathbf{P}[X_0 = i_0] = 1$ by another one assuming that X_0 is a non-degenerate r.v. with an arbitrary distribution over the whole state space E.

20.5. A useful property of independent random variables which cannot be extended to stationary Markov chains

It is well known that sequences of independent r.v.s obey several interesting properties (see Chung 1974; Stout 1974a; Petrov 1975). It turns out that the independence condition is essential for the validity of many of these properties.

Let us formulate the following result: if $\{X_n, n \geq 1\}$ is a sequence of i.i.d. r.v.s. and $\mathbf{E}X_1 = \infty$ then $\limsup_{n\to\infty}(X_n/n) = \infty$ a.s.

This result is a consequence of the Borel–Cantelli lemma (see Tanny 1974; O'Brien 1982). Our aim now is to clarify whether a similar result holds for a 'weakly' dependent random sequence. We shall treat the case of $\{X_n, n \geq 0\}$ forming a stationary Markov chain.

Let $X = (X_n, n \geq 0)$ be a stationary Markov chain with state space $E = \{1, 2, \ldots\}$ and

$$\begin{aligned}&\mathbf{P}[X_n = k] = 1/(k(k+1)), \quad k = 1, 2, \ldots, \ n = 0, 1, 2, \ldots\\ &\mathbf{P}[X_n = k+1 | X_{n-1} = k] = k/(k+2),\\ &\mathbf{P}[X_n = 1 | X_{n-1} = k] = 2/(k+2), \quad k = 1, 2, \ldots, \ n = 1, 2, \ldots.\end{aligned}$$

It is easy to see that $\mathbf{E}X_n = \infty$ for each n. However, we have $X_n \le X_0 + n$ a.s. for all n which implies that $\mathbf{P}[\limsup_{n\to\infty}(X_n/n) \le 1] = 1$. Hence $\limsup_{n\to\infty}(X_n/n)$ is not infinity as in the case of independent r.v.s. Using a result by O'Brien (1982) we conclude that $\limsup_{n\to\infty}(X_n/n) = 0$ a.s.

Therefore we have constructed a stationary Markov chain such that for each n, $\mathbf{E}X_n = \infty$ but with $\limsup_{n\to\infty}(X_n/n) = 0$ a.s.

20.6. The partial coincidence of two continuous-time Markov chains does not imply that the chains are equivalent

Let $X = (X_t, t \ge 0)$ and $\tilde{X} = (\tilde{X}_t, t \ge 0)$ be homogeneous continuous-time Markov chains with the same state space E, the same initial distribution and transition probabilities $p_{ij}(t)$ and $\tilde{p}_{ij}(t)$ respectively. If $p_{ij}(t) = \tilde{p}_{ij}(t)$, $i, j \in E$ for infinitely many t, but not for all $t \ge 0$, we say that X and $\tilde{X}$ coincide partially. If moreover we have $p_{ij}(t) = \tilde{p}_{ij}(t)$, $i, j \in E$ for all $t \ge 0$, then the processes X and $\tilde{X}$ are equivalent (stochastically equivalent) in the sense that each one is a modification of the other (see Example 19.3).

Firstly, let us note that the transition probabilities of any continuous-time Markov chain satisfy two systems of differential equations which are called Kolmogorov equations (see Chung 1960; Gihman and Skorohod 1974/1979). These equations are written in terms of the corresponding infinitesimal matrix $Q = (q_{ij})$ and under some natural conditions they uniquely define the transition probabilities $p_{ij}(t)$, $t \ge 0$.

Let X and $\tilde{X}$ be Markov chains each taking values in the set $\{1, 2, 3\}$. Suppose X and $\tilde{X}$ are defined by their infinitesimal matrices $Q = (q_{ij})$ and $\tilde{Q} = (\tilde{q}_{ij})$ respectively, where

$$Q = \begin{pmatrix} -1 & 1 & 0 \\ 0 & -1 & 1 \\ 1 & 0 & -1 \end{pmatrix}, \quad \tilde{Q} = \begin{pmatrix} -1 & \frac{1}{2} & \frac{1}{2} \\ \frac{1}{2} & -1 & \frac{1}{2} \\ \frac{1}{2} & \frac{1}{2} & -1 \end{pmatrix}.$$

Thus, knowing Q and $\tilde{Q}$ and using the Kolmogorov equations, we can show that the transition probabilities $p_{ij}(t)$ and $\tilde{p}_{ij}(t)$ have the following explicit form:

$$p_{11}(t) = p_{22}(t) = p_{33}(t) = \tfrac{1}{3} + \tfrac{2}{3}\mathrm{e}^{-3t/2}\cos(\sqrt{3}t/2),$$

$$p_{12}(t) = p_{23}(t) = p_{31}(t) = \tfrac{1}{3} + \tfrac{2}{3}\mathrm{e}^{-3t/2}\cos(\sqrt{3}t/2 - 2\pi/3),$$

$$p_{13}(t) = p_{21}(t) = p_{32}(t) = \tfrac{1}{3} + \tfrac{2}{3}\mathrm{e}^{-3t/2}\cos(\sqrt{3}t/2 + 2\pi/3),$$

$$\tilde{p}_{11}(t) = \tilde{p}_{22}(t) = \tilde{p}_{33}(t) = \tfrac{1}{3} + \tfrac{2}{3}\mathrm{e}^{-3t/2},$$

$$\tilde{p}_{ij}(t) = \tfrac{1}{3} - \tfrac{1}{3}\mathrm{e}^{-3t/2}, \quad \text{if } i \ne j,\ i, j = 1, 2, 3.$$

(The details are left to the reader.)

Obviously, $p_{ij}(t) = \tilde{p}_{ij}(t)$ for every $t = 4k\pi/\sqrt{3}$, $k \in \mathbb{N}$, but for all other t we have $p_{ij}(t) \neq \tilde{p}_{ij}(t)$. Therefore the processes X and $\tilde{X}$ are not equivalent, though they partially coincide.

20.7. Markov processes, Feller processes, strong Feller processes and relationships between them

Let $X = (X_t, t \geq s, \mathbf{P}_{s,x})$ be a Markov family: that is $(X_t, t \geq s)$ is a Markov process with respect to the probability measure $\mathbf{P}_{s,x}$, $\mathbf{P}_{s,x}[X_s = x] = 1$ and $P(s,x;t,\Gamma)$, $t \geq s$, $x \in \mathbb{R}^1$, $\Gamma \in \mathcal{B}^1$ is its transition function. As usual, $\mathbb{B} = \mathbb{B}(\mathbb{R}^1)$ and $\mathbb{C} = \mathbb{C}(\mathbb{R}^1)$ will denote the set of all bounded and measurable functions on $\mathbb{R}^1$, and the set of all bounded and continuous functions on $\mathbb{R}^1$ respectively. By the equality

$$P^{st}g(x) = \mathbf{E}_{s,x}[g(X_t)] = \int_{\mathbb{R}^1} g(y)P(s,x;t,\mathrm{d}y)$$

we define on $\mathbb{B}$ a semigroup of operators $\{P^{st}\}$. Obviously we have the inclusion $P^{st}\mathbb{B} \subset \mathbb{B}$ and, moreover, $P^{st}\mathbb{C} \subset \mathbb{B}$. A Markov process for which $P^{st}\mathbb{C} \subset \mathbb{C}$ is called a *Feller process*. This means that for each continuous and bounded function g, the function $P^{st}g(x)$ is continuous in x. In other words,

$$\int_{\mathbb{R}^1} g(y)P(s,x;t,\mathrm{d}y) \to \int_{\mathbb{R}^1} g(y)P(s,x_0;t,\mathrm{d}y) \quad \text{as} \quad x \to x_0,\ x_0 \in \mathbb{R}^1$$

which is equivalent to the weak continuity of $P(\cdot)$ with respect to the second argument (the starting point of the process).

Let us now introduce another notion. If for each $g \in \mathbb{B}$ the function $P^{st}g(x)$ is continuous in x, the Markov process is called a *strong Feller process*. Clearly, the assumption for a process to be strong Feller is more restrictive than that for a process to be Feller. Thus, every strong Feller process is also a Feller process. However the converse is not always true. There are Markov processes which are not Feller processes, although the condition for a process to be Feller seems very natural and not too strong.

(i) Let the family $X = (X_t, t \geq s, \mathbf{P}_{s,x})$ describe the motion for $t \geq s$ of a particle starting at time s from the position $X_s = x$: if $X_s < 0$, the particle moves to the left with unit rate; if $X_s > 0$ it moves to the right with unit rate; if $X_s = 0$, the particle moves to the left or to the right with probability $\frac{1}{2}$ for each of these two directions. Formally this can be expressed by:

$$\mathbf{P}_{s,x}[X_t = x + (t-s), t \geq s] = 1, \quad \text{if} \quad x > 0$$
$$\mathbf{P}_{s,x}[X_t = x - (t-s), t \geq s] = 1, \quad \text{if} \quad x < 0$$
$$\mathbf{P}_{s,x}[X_t = t-s, t \geq s] = \mathbf{P}_{s,x}[X_t = -(t-s), t \geq s] = \tfrac{1}{2}.$$

It is easy to see that $X = (X_t, t \geq s, \mathbf{P}_{s,x})$ is a Markov family. Further, if g is a continuous and bounded function, we find explicitly that

$$P^{st}g(x) = \begin{cases} g(x + (t-s)), & \text{if } x > 0 \\ g(x - (t-s)), & \text{if } x < 0 \\ \frac{1}{2}g(t-s) + \frac{1}{2}g(-(t-s)), & \text{if } x = 0. \end{cases}$$

Since $P^{st}g(x)$ has a discontinuity at $x = 0$, it follows from this that X is not a Feller process even though it is a Markov process.

(ii) It is easy to give an example of a process which is Feller but not strong Feller. Indeed, by the formula

$$P(t, x, \Gamma) = I_\Gamma(x + vt), \quad t \geq 0,\ x \in \mathbb{R}^1,\ \Gamma \in \mathcal{B}^1,\ v = \text{constant} > 0$$

we define a transition function which corresponds to a homogeneous Markov process. Actually, this P describes a motion with constant velocity v. All that remains is to check that the process is Feller but is not strong Feller.

20.8. Markov but not strong Markov processes

In the introductory notes to this section we defined the Markov property of a stochastic process. For the examples below we need another property called the strong Markov property. For simplicity we consider the homogeneous case.

Let $X = (X_t, t \geq 0)$ be a real-valued Markov process defined on the probability space $(\Omega, \mathcal{F}, \mathbf{P})$ and $(\mathcal{F}_t, t \geq 0)$ be its own generated filtration which is assumed to satisfy the usual conditions. Let τ be an $(\mathcal{F}_t)$-stopping time and $\mathcal{F}_\tau$ be the σ-field of all events $A \in \mathcal{F}$ such that $A \cap [\tau \leq t] \in \mathcal{F}_t$ for all $t \geq 0$.

Suppose the Markov process X is $(\mathcal{F}_t)$-progressive, and let η be an $\mathcal{F}_\tau$-measurable non-negative r.v. defined on the set $[\omega : \tau(\omega) < \infty]$. Then X is said to be a *strong Markov process* if, for any $\Gamma \in \mathcal{B}^1$,

$$\mathbf{P}[X_{\tau+\eta} \in \Gamma | \mathcal{F}_\tau] = \mathbf{P}[X_{\tau+\eta} \in \Gamma | X_\tau] \quad \text{a.s.}$$

This relation defines the *strong Markov property*. In terms of the transition function $P(t, x, \Gamma)$ it can be written in the form

$$\mathbf{P}[X_{\tau+\eta} \in \Gamma | \mathcal{F}_\tau] = P(\eta, X_\tau, \Gamma) \quad \text{a.s.} \tag{1}$$

If $(X_t, t \geq 0, \mathbf{P}_x)$ is a homogeneous Markov family (also see Example 20.5), the strong Markov property can be expressed by

$$\mathbf{P}_x\{A \cap [X_{\tau+\eta} \in \Gamma]\} = \int_A P(\eta, X_\tau, \Gamma)\mathbf{P}_x(d\omega), \tag{2}$$

$$A \subset \mathcal{F} \cap \{\omega : \tau(\omega) < \infty, \eta(\omega) < \infty\}.$$

Two examples of processes which are Markov but not strong Markov are now given. Case (i) is the first ever known example of such a process, proposed by A. A. Yushkevich (see Dynkin and Yushkevich 1956).

(i) Let $w = (w_t, t \geq 0, \mathbf{P}_x)$ be a Wiener process which can start from any point $x \in \mathbb{R}^1$. Define a new process $X = (X_t, t \geq 0)$ by

$$X_t = \begin{cases} w_t, & \text{if } w_0 \neq 0 \\ 0, & \text{if } w_0 = 0. \end{cases}$$

Then X is a homogeneous Markov process whose transition function is

$$P(t, x, \Gamma) = \begin{cases} (2\pi t)^{-1/2} \int_\Gamma \exp[-(u-x)^2/(2t)]\,\mathrm{d}u, & \text{if } x \neq 0 \\ \delta_0(\Gamma), & \text{if } x = 0. \end{cases}$$

(Here $\delta_0(\cdot)$ is a unit measure concentrated at the point 0.)

Let us check the usual Markov property for X. Clearly, P satisfies all the conditions for a transition function. We then need to establish the relation

$$\text{(3)} \qquad \mathbf{P}_x\{A \cap [X_{t+h} \in \Gamma]\} = \int_A P(h, X_t, \Gamma)\mathbf{P}_x(\mathrm{d}\omega)$$

for $t, h \geq 0$, $A \in \mathcal{F}_t$ and $\Gamma \in \mathcal{B}^1$. (Note that by equation (3) we express the Markov property of the family $(X_t, t \geq 0, \mathbf{P}_x)$, while the strong Markov property is given by (2).) If $x \neq 0$, then $X_t = w_t$ a.s. and (3) is reduced to the Markov property of the Wiener process. If $x = 0$, (3) is trivial since both sides are simultaneously either 1 or 0. Hence X is a Markov process.

Let us take $x \neq 0$, $\tau = \inf\{t : X_t = 0\}$, $\eta = (1-\tau) \bigvee 0$, $A = \{\tau \leq 1\}$ and $\Gamma = \mathbb{R}^1 \backslash \{0\}$. Then obviously $\tau < \infty$ a.s., $\eta < \infty$ a.s. Suppose X is strong Markov. Then the following relation would hold (see (2)):

$$\text{(4)} \qquad \mathbf{P}_x\{A \cap [X_{\tau+\eta} \in \Gamma]\} = \int_A P(\eta, X_\tau, \Gamma)\mathbf{P}_x(\mathrm{d}\omega).$$

However, the left-hand side is equal to

$$\mathbf{P}_x[\tau \leq 1, X_1 \neq 0] = \mathbf{P}_x[\tau \leq 1] = 2(1 - \Phi(|x|)) > 0$$

while the right-hand side is 0. Thus (4) is not valid and therefore the Markov process X is not strong Markov.

(ii) Let τ be a r.v. distributed exponentially with parameter 1, that is $\mathbf{P}[\tau > t] = \mathrm{e}^{-t}$. Define the process $X = (X_t, t \geq 0)$ by

$$X_t = X_t(\omega) = \max\{0, t - \tau(\omega)\}$$

and let $\mathcal{F}_t = \sigma\{X_s, s \leq t\}, t \geq 0$ be its own generated filtration.

It is easy to see that if $X_t = a > 0$ for some t, then $X_{t+s} = a + s$ for all $s \geq 0$. If for some t we have $X_t = 0$, then X_s must be zero for $s \leq t$, so it does not provide new information. Thus we conclude that X is a Markov process. Denote its transition function by $P(t, x, \Gamma)$. Let us show that X is not a strong Markov process. Indeed, the relation $[\omega : \tau(\omega) \geq t] = [\omega : X_t(\omega) = 0] \in \mathcal{F}_t$ shows that the r.v. τ is an $(\mathcal{F}_t)$-stopping time. For a given τ we have $\mathbf{P}[X_{\tau+s} = s] = 1$. Therefore

$$\mathbf{P}[X_{\tau+s} \leq x | \mathcal{F}_\tau] = \begin{cases} 1, & \text{if } x > s \\ 0, & \text{if } x \leq s. \end{cases} \tag{5}$$

If X were strong Markov, then the following relation would hold:

$$\mathbf{P}[X_{\tau+s} \leq x | \mathcal{F}_\tau] = \mathbf{P}[X_{\tau+s} \leq x | X_\tau] \quad \text{a.s.} \tag{6}$$

and the conditional probability on the right-hand side could be expressed according to (1) by the transition function $P(t, x, \Gamma)$, namely

$$\mathbf{P}[X_{\tau+s} \leq x | X_\tau] = P(s, X_\tau, \Gamma_x) \quad \text{where } \Gamma_x = (-\infty, x].$$

However,

$$\begin{aligned} P(s, X_\tau, \Gamma_x) &= P(s, 0, \Gamma_x) = \mathbf{P}[X_{t+s} \leq x | X_t = 0] \\ &= \begin{cases} 1, & \text{if } x > s \\ \mathrm{e}^{-(s-x)}, & \text{if } x \leq s. \end{cases} \end{aligned} \tag{7}$$

From (5) and (7) it follows that (6) is not satisfied. The process X is therefore Markov but not strong Markov.

20.9. Can a differential operator of order $k > 2$ be an infinitesimal operator of a Markov process?

Let $P(t, x, \Gamma)$, $t \geq 0$, $x \in \mathbb{R}^1$, $\Gamma \in \mathcal{B}^1$ be the transition function of a homogeneous Markov process $X = (X_t, t \geq 0)$ and $\{P^t, t \geq 0\}$ be the semigroup of operators associated with P: $P^t u(x) = \int_{\mathbb{R}^1} u(y) P(t, x, \mathrm{d}y)$. The *infinitesimal operator* corresponding to $\{P^t\}$ (and also to P and to X) is denoted by $\mathcal{A}$ and is defined by:

$$\mathcal{A}u(x) = \lim_{t \downarrow 0} \frac{1}{t} \left[\int_{\mathbb{R}^1} u(y) P(t, x, \mathrm{d}y) - u(x) \right]. \tag{1}$$

Let $D(\mathcal{A})$ be the *domain* of $\mathcal{A}$, that is $D(\mathcal{A})$ contains all functions $u(x)$, $x \in \mathbb{R}^1$ for which the limit in (1) exists in the sense of convergence in norm in the space where $\{P^t\}$ is considered.

Several important results concerning the infinitesimal operators of Markov processes and related topics can be found in the book by Dynkin (1965). In particular, Dynkin proves that under natural conditions the infinitesimal operator $\mathcal{A}$ is a differential operator of first or second order. In the latter case $D(\mathcal{A}) = \mathbb{C}^2(\mathbb{R}^1)$, the space of twice continuously differentiable functions. So we come to the following question: can a differential operator of order $k > 2$ be an infinitesimal operator?

Suppose the answer to this question is positive in the particular case $k = 3$ and let $\mathcal{A}u(x) = u'''(x)$ with $D(\mathcal{A}) = \mathbb{C}^3(\mathbb{R}^1)$, the space of three times continuously differentiable functions $u(x)$, $x \in \mathbb{R}^1$. However, if $\mathcal{A}$ is an infinitesimal operator, then according to the Hille–Yosida theorem (see Dynkin 1965; Wentzell 1981) $\mathcal{A}$ must satisfy the minimum principle: if $u \in D(\mathcal{A})$ is minimum at the point x_0, then $\mathcal{A}u(x_0) \geq 0$.

Take the function $u(x) = 2(\sin 2\pi x)^2 - (\sin 2\pi x)^3$. Obviously $u \in \mathbb{C}^3(\mathbb{R}^1)$ and it is a periodic function with period 1. It is easy to see that u takes its minimal value at $x = 0$ and, moreover, $u'''(0) < 0$. This implies that the minimum principle is violated. Thus in general a differential operator of order $k = 3$ cannot be an infinitesimal operator. Similar arguments can be used in the case $k > 3$.

SECTION 21. STATIONARY PROCESSES AND SOME RELATED TOPICS

Let $X = (X_t, t \in T \subset \mathbb{R}^1)$ be a real-valued stochastic process defined on the probability space $(\Omega, \mathcal{F}, \mathbf{P})$. We say that X is **strictly stationary** if for each $n \geq 1$ and $t_k, t_k + h \in T$, $k = 1, \dots, n$, the random vectors

$$(X_{t_1}, \dots, X_{t_n}) \quad \text{and} \quad (X_{t_1+h}, \dots, X_{t_n+h})$$

have the same distribution.

Suppose now that X is an L^2-process (or second-order process), that is $\mathbf{E}[X_t^2] < \infty$ for each $t \in T$. Such a process X is said to be **weakly stationary** if $\mathbf{E}X_t = c =$ constant for all $t \in T$ and the covariance function $C(s,t) = \mathbf{E}[(X_s - c)(X_t - c)]$ depends on s and t only through $t - s$. This means that there is a function $C(t)$ of one argument t, $t \in T$, such that $C(t) = \mathbf{E}[(X_s - c)(X_{s+t} - c)]$ for all s, $s + t \in T$.

On the same lines we can define weak and strict stationarity for multi-dimensional processes and for complex-valued processes. The notions of strict and weak stationarity were introduced by Khintchine (1934).

Let us note that the covariance function C of any weakly stationary process admits the so-called *spectral representation*. If $T = \mathbb{R}^1$ or $T = \mathbb{R}^+$ we have a *continuous-time weakly stationary process* and its covariance function C has the representation

$$C(t) = \int_{-\infty}^{\infty} \mathrm{e}^{it\lambda}\, \mathrm{d}F(\lambda)$$

where $F(\lambda)$, $\lambda \in \mathbb{R}^1$ is a non-decreasing, right-continuous and bounded function. F is called a *spectral d.f.*, while its derivative f, if it exists, is called a *spectral density function*. If $T = \mathbb{N}$ or $T = \overline{\mathbb{N}}$ we say that $X = (X_n)$ is a *discrete-time weakly stationary process* (or a *weakly stationary sequence*). In this case the covariance function C of X has the representation

$$C(n) = \int_{-\pi}^{\pi} e^{in\lambda} \, dF(\lambda)$$

where $F(\lambda)$, $\lambda \in [-\pi, \pi]$ possesses properties as in the continuous case.

Note that many useful properties of stationary processes and sequences can be derived under conditions in terms of C, F and f. It is important to note that stationary processes themselves also admit spectral representations in the form of integrals of the type $\int_{-\infty}^{\infty} e^{it\lambda} \, dZ_\lambda$ with respect to processes with orthogonal increments.

Let $X = (X_n, n \in \overline{\mathbb{N}})$ be a strictly stationary process. Denote by $\mathcal{M}_a^b$ the σ-field generated by the r.v.s. $X_a, X_{a+1}, \ldots, X_b$. Without going into details here we note that in terms of probabilities of events belonging to the σ-fields $\mathcal{M}_{-\infty}^k$ and $\mathcal{M}_{k+n}^\infty$ we can define some important conditions, such as φ-mixing strong mixing, regularity and absolute regularity, which are essential in studying stationary processes. In the examples below we give definitions of these conditions and analyse properties of the processes.

A complete presentation of the theory of stationary processes and several related topics can be found in the books by Parzen (1962), Cramér and Leadbetter (1967), Rozanov (1967), Gihman and Skorohod (1974/1979), Ibragimov and Linnik (1971), Ash and Gardner (1975), Ibragimov and Rozanov (1978) and Wentzell (1981).

In this section we consider only a few examples dealing with the stationarity property, as well as properties such as mixing and ergodicity.

21.1. On the weak and the strict stationary properties of stochastic processes

Since we shall be studying two classes of stationary processes, it is useful to clarify the relationship between them.

Firstly, if $X = (X_t, t \in \mathbb{R}^1)$ is a strictly stationary process, and moreover X is an L^2-process, then clearly X is also a weakly stationary process. However, X can be strictly stationary without being weakly stationary and this is the case when X is not an L^2-process. It is easy to construct examples of this.

Further, let θ be a r.v. with a uniform distribution on $[0, 2\pi]$ and let $Z_t = \sin \theta t$. Then the random sequence $(Z_n = \sin \theta n, n = 1, 2, \ldots)$ is weakly but not strictly stationary, while the process $(Z_t = \sin \theta t, t \in \mathbb{R}^1)$ is neither weakly nor strictly stationary. If we take another r.v., say ζ, which has an arbitrary distribution and does not depend on θ, then the process $Y = (Y_t, t \in \mathbb{R}^1)$ where $Y_t = \cos(\zeta t + \theta)$ is both weakly and strictly stationary.

Let us consider two other examples of weakly but not strictly stationary processes. Let ξ_1 and η_1 be r.v.s each distributed $\mathcal{N}(0, 1)$ and such that the distribution of (ξ_1, η_1) is not bivariate normal, and ξ_1 and η_1 are uncorrelated. Such examples exist and were described in Section 10. Now take an infinite sequence of independent copies of (ξ_1, η_1), that is

$$(\xi_1, \eta_1), (\xi_2, \eta_2), \ldots$$

which in this order are renamed $X_1, X_2, \ldots$, that is,

$$X_1 = \xi_1,\ X_2 = \eta_1,\ X_3 = \xi_2,\ X_4 = \eta_2, \ldots.$$

It is easy to check that the random sequence $(X_n, n = 1, 2, \ldots)$ is weakly stationary but not strictly stationary.

Finally, it is not difficult to construct a continuous-time process $X = (X_t, t \in \mathbb{R}^1)$ with similar properties. For $t \geq 0$ take X_t to be a r.v. distributed $\mathcal{N}(1, 1)$ and for $t < 0$ let X_t be exponentially distributed with parameter 1. Suppose also that for all $s \neq t$ the r.v.s. X_s and X_t are independent. Then X is a weakly but not strictly stationary process.

21.2. On the strict stationarity of a given order

Recall that the process $X = (X_t, t \in T \subset \mathbb{R}^1)$ is said to be *strictly stationary of order* m if for arbitrary $t_1, \ldots, t_m \in T$ and $t_1 + h, \ldots, t_m + h \in T$ the random vectors $(X_{t_1}, \ldots, X_{t_m})$ and $(X_{t_1+h}, \ldots, X_{t_m+h})$ have the same distribution. Clearly, the process X is strictly stationary if it is so of any order m, $m \geq 1$. It is easy to see that the m-order strictly stationary process X is also k-order strictly stationary for every k, $1 \leq k < m$. The following example determines if the converse is true.

Let ξ and η be independent r.v.s with the same non-degenerate d.f. $F(x)$, $x \in \mathbb{R}^1$. Define the sequence $(X_n, n = 1, 2, \ldots)$ as follows:

$$X_1 = \xi,\ X_2 = \xi,\ X_3 = \xi,\ X_4 = \eta,\ X_5 = \eta,\ X_6 = \xi,\ X_7 = \xi,\ X_8 = \xi,$$
$$X_9 = \eta,\ X_{10} = \eta,\ X_{11} = \xi, \ldots.$$

This means that

$$X_n = \begin{cases} \xi, & \text{if } n = 5k+1,\ 5k+2,\ 5k+3 \\ \eta, & \text{if } n = 5k+4,\ 5k+5, \text{ for } k = 0, 1, 2, \ldots. \end{cases}$$

It is obvious that the sequence $(X_n, n = 1, 2, \ldots)$ is strictly stationary of order 1. Let us check if it is strictly stationary of order 2. E.g. the random vectors (X_1, X_2), (X_2, X_3), (X_4, X_5), (X_6, X_7), ... are identically distributed. However, (X_3, X_4), that is, (X_{1+2}, X_{2+2}) has a distribution which differs from that of (X_1, X_2). Indeed, since $X_1 = \xi$, $X_2 = \xi$, $X_3 = \xi$ and $X_4 = \eta$ we have

$$\mathbf{P}[X_1 \leq x_1, X_2 \leq x_2] = \mathbf{P}[\xi \leq x_1, \xi \leq x_2] = \mathbf{P}[\xi \leq \min\{x_1, x_2\}]$$

$$= F(\min\{x_1, x_2\}) \neq F(x_1)F(x_2) = \mathbf{P}[\xi \leq x_1, \eta \leq x_2] = \mathbf{P}[X_3 \leq x_1, X_4 \leq x_2].$$

Therefore the sequence $(X_n, n = 1, 2, \ldots)$ is not strictly stationary of order 2. It is clear what conclusion can be drawn in the general case.

21.3. The strong mixing property can fail if we consider a functional of a strictly stationary strong mixing process

Suppose $X = (X_n, n \in \overline{\mathbb{N}})$ is a strictly stationary process satisfying the *strong mixing condition*. This means that there is a numerical sequence $\alpha(n) \downarrow 0$ as $n \to \infty$ such that

$$\sup_{A,B} |\mathbf{P}(AB) - \mathbf{P}(A)\mathbf{P}(B)| \leq \alpha(n)$$

where sup is taken over all events $A \in \mathcal{M}_{-\infty}^{k}$, $B \in \mathcal{M}_{k+n}^{\infty}$.

This condition is essential for establishing limit theorems for sequences of weakly dependent r.v.s (see Rosenblatt 1956, 1978; Ibragimov 1962; Ibragimov and Linnik 1971; Billingsley 1968).

Let $g(x)$, $x \in \mathbb{R}^1$, be a measurable function and $\xi = (\xi_n, n \in \overline{\mathbb{N}})$ a strictly stationary process. Then the process $(X_n, n \in \overline{\mathbb{N}})$ where $X_n = g(\xi_n)$ is again strictly stationary (see e.g. Breiman 1968). Suppose now $\xi = (\xi_n, n \in \overline{\mathbb{N}})$ is strongly mixing and $g(x)$, $x \in \mathbb{R}^\infty$ is a bounded and $\mathcal{B}^\infty$-measurable function. If we define the process $X = (X_n, n \in \overline{\mathbb{N}})$ by $X_n = g(\xi_n, \xi_{n+1}, \ldots)$, the question to consider is whether the functional g preserves the strong mixing property. In general the answer is negative and this is shown in the next example.

Let $\{\varepsilon_j, j \in \overline{\mathbb{N}}\}$ be a sequence of i.i.d. r.v.s. such that $\mathbf{P}[\varepsilon_j = 1] = \mathbf{P}[\varepsilon_j = 0] = \frac{1}{2}$. Define the random sequence $(X_j, j \in \overline{\mathbb{N}})$ where

$$X_j = 2^{-1}\varepsilon_j + 2^{-2}\varepsilon_{j+1} + \ldots + 2^{-k-1}\varepsilon_{j+k} + \ldots, \quad j \in \overline{\mathbb{N}}.$$

Since $\{\varepsilon_j\}$ consists of i.i d. r.v.s, then $\{\varepsilon_j\}$ is a strictly stationary sequence. This implies that the sequence $(X_j, j \in \overline{\mathbb{N}})$ is also strictly stationary. However, $\{\varepsilon_j\}$ satisfies the strong mixing condition and thus we could expect that $(X_j, j \in \overline{\mathbb{N}})$ satisfies this condition. Suppose this conjecture is right. Then according to Ibragimov (1962) the sequence of d.f.s

$$F_n(z) = \mathbf{P}[(X_1 + \ldots + X_n)/b_n - a_n \leq z], \quad z \in \mathbb{R}^1, \ n = 1, 2, \ldots$$

where a_n and b_n are norming constants, can converge as $n \to \infty$ only to a stable law. (For properties of stable distributions see Section 9.) Moreover, if the limit law of F_n has a parameter α, then necessarily $b_n = (\mathbf{V}[X_1 + \ldots + X_n])^{1/2} = n^{1/\alpha}h(n)$ where $h(n)$ is a slowly varying function.

Consider the random sequence

$$q_j = \sum_{k=1}^{\infty} r_k(X_j)k^{-3/4}$$

where $r_k(x)$, $k = 1, 2, \ldots$ are the Rademacher functions: $r_k(X_1) = \operatorname{sign}\sin(2^k\pi X_1)$ or $r_k = -1 + 2\varepsilon_k$ (ε_k as above). Since r_k, $k \geq 1$ are i.i.d. r.v.s,

$\mathbf{P}[r_k = \pm 1] = \frac{1}{2}$, we can easily see that

$$\mathbf{E}[g_1 g_{j+1}] = \sum_{k=1}^{\infty} k^{-3/4}(k+j)^{-3/4} > 2^{-3/4} j^{-3/4}$$

and

$$\sigma_n^2 = \mathbf{E}\left[\left(\sum_{j=1}^{n} g_j\right)^2\right] > n^{5/4}(1 + o(1)).$$

Moreover, as a consequence of our assumption that $\{X_j\}$ is strongly mixing, the sequence $\{g_j\}$ must satisfy the CLT, that is

$$\mathbf{P}[(g_1 + \ldots + g_n)/\sigma_n < z] \to \Phi(z), \; z \in \mathbb{R}^1 \quad \text{as} \quad n \to \infty.$$

However, as the variance σ_n^2 is greater than $n^{5/4}(1+o(1))$ it cannot be represented in the form $nh(n)$ with $h(n)$ a slowly varying function. This contradiction shows that the strictly stationary process $(X_j, j \in \overline{\mathbb{N}})$ defined above does not satisfy the strong mixing condition. This would be interesting even if we could only conclude that not every strictly stationary process is strongly mixing. Clearly, the example considered here provides a little more: the functional (X_j) of a strictly stationary and strong mixing process (ε_j) may not preserve the strong mixing property.

21.4. A strictly stationary process can be regular but not absolutely regular

Let $X = (X_t, t \in \mathbb{R}^1)$ be a strictly stationary process. We say that X is *regular* if the σ-field

$$\mathcal{M}_{-\infty} := \bigcap_{t \in \mathbb{R}^1} \mathcal{M}_{-\infty}^t$$

is trivial, that is if $\mathcal{M}_{-\infty}$ contains only events of probability 0 or 1. This condition can be expressed also in another form: for all $B \in \mathcal{M}_{-\infty}^{\infty}$ and $A \in \mathcal{M}_{-\infty}^{t}$ we have

$$\sup_A |\mathbf{P}(AB) - \mathbf{P}(A)\mathbf{P}(B)| \to 0 \quad \text{as} \quad t \to \infty.$$

Further, define $\rho(t) := \sup \mathbf{E}[\eta_1 \eta_2]$ where sup is taken over all r.v.s η_1 and η_2 such that η_1 is $\mathcal{M}_{-\infty}^s$-measurable, η_2 is $\mathcal{M}_{s+t}^{\infty}$-measurable, $\mathbf{E}\eta_1 = 0$, $\mathbf{E}\eta_2 = 0$, $\mathbf{E}[\eta_1^2] = 1$, $\mathbf{E}[\eta_2^2] = 1$. The quantity $\rho(t)$, $t \geq 0$ is called a *maximal correlation coefficient* between the σ-fields $\mathcal{M}_{-\infty}^s$ and $\mathcal{M}_{s+t}^{\infty}$. The process X is said to be *absolutely regular* (completely, strictly regular) if $\rho(t) \to 0$ as $t \to \infty$. Note that for stationary processes which are also Gaussian, the notion of absolute regularity coincides with the so-called strong mixing condition (see Ibragimov and Rozanov 1978).

It is obvious that any absolutely regular process is also regular. We now consider whether the converse is true.

Suppose X is a strictly stationary process and $f(\lambda)$, $\lambda \in \mathbb{R}^1$, is its spectral density function. Then X is regular iff

(1) $$\int_{-\infty}^{\infty} \frac{-\log f(\lambda)}{1+\lambda^2}\, d\lambda < \infty.$$

For the proof of this result and of many others we again refer the reader to the book by Ibragimov and Rozanov (1978).

Consider now a stationary process X whose spectral density is

(2) $$f(\lambda) = (\sin^2 \lambda^2 + 1)(\lambda^{-1} \sin \lambda)^{2p}$$

with p any positive integer. Then it is not difficult to check that f given by (2) satisfies (1). Hence X is a regular process. However, the process X and its spectral density f do not satisfy another condition which is necessary for a process to be absolutely regular (Ibragimov and Rozanov 1978, Th. 6.4.3). Thus we conclude that the stationary process X with spectral density f given by (2) is not absolutely regular even though it is regular.

21.5. Weak and strong ergodicity of stationary processes

Let $X = (X_n, n \geq 1)$ be a weakly stationary sequence with $\mathbf{E}X_n = 0$, $n \geq 1$. We say that X is *weakly ergodic* (or that X satisfies the WLLN) if

(1) $$\frac{1}{n}\sum_{k=1}^{n} X_k \xrightarrow{\text{P}} 0 \quad \text{as} \quad n \to \infty.$$

If (1) holds with probability 1, we say that X is *strongly ergodic* (or that X satisfies the SLLN).

If $X = (X_t, t \geq 0)$ is a weakly stationary (continuous-time) process with mean $\mathbf{E}X_t = 0$ and

(2) $$\frac{1}{T}\int_0^T X_t\, dt \xrightarrow{\text{P}} 0 \quad \text{as} \quad T \to \infty$$

then X is said to be *weakly ergodic* (to obey the WLLN); X is *strongly ergodic* if (2) is satisfied with probability 1 (now X obeys the SLLN).

There are many results concerning the ergodicity of stationary processes. The conditions guaranteeing a certain type of ergodicity can be expressed in different terms. Here we discuss two examples involving the covariance functions and the spectral d.f.

(i) Let $X = (X_n, n \geq 1)$ be a weakly stationary sequence such that $\mathbf{E}X_n = 0$, $\mathbf{E}[X_n^2] = 1$ and the covariance function is $C(n) = \mathbf{E}[X_k X_{k+n}]$. Then the condition

(3) $$\lim_{n\to\infty} C(n) = 0$$

is sufficient for the process X to be weakly ergodic (see Gihman and Skorohod 1974/1979; Gaposhkin 1973). Note that (3) also implies that $(1/n)\sum_{k=1}^{n} X_k \xrightarrow{\mathrm{L}^2} 0$ which means that (3) is a sufficient condition for X to be L^2-ergodic. Moreover, if we suppose additionally that X is strictly stationary, then it can be shown that condition (3) implies the strong ergodicity of X. Thus we come to the question: if X is only weakly stationary, can condition (3) ensure that X is strongly ergodic? It turns out that in general the answer is negative. It can be proved that there exists a weakly stationary sequence $X = (X_n, n \geq 1)$ such that its covariance function $C(n)$ satisfies the condition

$$C(n) = O[(\log\log n)^{-2}] \quad \text{as} \quad n \to \infty$$

(hence $C(n) \to 0$) so that X is weakly ergodic but $(1/n)\sum_{k=1}^{n} X_k$ diverges almost surely. Note that the construction of such a process as well as of a similar continuous-time process is given by Gaposhkin (1973).

(ii) We now consider a weakly stationary process $X = (X_t, t \in \mathbb{R}^+)$ with $\mathbf{E}X_t = 0$, $\mathbf{E}X_t = 1$ and covariance function $C(t) = \mathbf{E}[X_s X_{s+t}]$ and discuss the conditions which ensure the strong ergodicity of X.

Firstly let us formulate the following result (see Verbitskaya 1966). If the covariance function C satisfies the condition

$$\int_1^\infty \frac{1}{t} C(t)(\log t)^2 \,\mathrm{d}t < \infty$$

then the process X is strongly ergodic. Moreover, if the process X is bounded, then the condition $\int_1^\infty t^{-1}C(t)\,\mathrm{d}t < \infty$ is sufficient for the strong ergodicity of X.

Obviously this result contains conditions which are only sufficient for strong ergodicity. However, it is of general interest to look for necessary and sufficient conditions under which a stationary process will be strongly ergodic. The above result and other results in this area lead naturally to the conjecture that eventually such conditions can be expressed either as restrictions on the covariance function C at infinity, or as restrictions on the corresponding spectral d.f. around 0. The following example will show if this conjecture is true.

Consider two independent r.v.s, say ζ and θ, where ζ has an arbitrary d.f. $F(x)$, $x \in \mathbb{R}^1$, and θ is uniformly distributed on $[0, 2\pi]$. Let

$$X_t = \sqrt{2}\cos(\zeta t + \theta), \quad t \in \mathbb{R}^+.$$

Then the process $X = (X_t, t \in \mathbb{R}^+)$ is both weakly and strictly stationary,

$$\mathbf{E}X_t = 0, \quad \text{and} \quad C(t) = \int_{-\infty}^{\infty} \cos(tx)\,\mathrm{d}F(x).$$

In particular this explicit form of the covariance function of X shows that the d.f. F of the r.v. ζ is just the spectral d.f. of the process X. Obviously this fact is very convenient when studying the ergodicity of X.

Suppose F satisfies only one condition, namely it is continuous at 0:

$$F(0) - F(0-) = 0. \tag{4}$$

Recall that (4) is equivalent to the condition $\lim_{T\to\infty} \int_0^T C(t)\,\mathrm{d}t = 0$ which implies that X is weakly ergodic (see Gihman and Skorohod 1974/1979). Let us show that X is strongly ergodic. A direct calculation leads to:

$$\frac{1}{T}\int_0^T X_t\,\mathrm{d}t = \frac{\sqrt{2}}{T}\int_0^T \cos(\zeta t + \theta)\,\mathrm{d}t = O(1/T), \quad \text{if} \quad \zeta \neq 0 \tag{5}$$

and

$$\frac{1}{T}\int_0^T X_t\,\mathrm{d}t = \sqrt{2}\cos\theta, \quad \text{if} \quad \zeta = 0.$$

However, (4) implies that $\mathbf{P}[\zeta = 0] = 0$. From (5) we can then conclude that X is strongly ergodic.

Note especially that (4) is the only condition imposed on the spectral d.f. F of the process X. This example and other arguments given by Verbitskaya (1966) allow us to conclude that in general no restrictions on the spectral d.f. at a neighbourhood of 0 (excluding the continuity of F at 0) could be necessary conditions for the strong ergodicity of a stationary process.

21.6. A measure-preserving transformation which is ergodic but not mixing

Stationary processes possess many properties such as mixing and ergodicity which can be studied in a unified way as transformations of the probability space on which the processes are defined (see Ash and Gardner 1975; Rosenblatt 1979; Shiryaev 1995). We give some definitions first and then discuss an interesting example.

Let $(\Omega, \mathcal{F}, \mathbf{P})$ be a probability space and T a transformation of Ω into itself. T is called *measurable* if $T^{-1}(A) = \{\omega : T\omega \in A\} \in \mathcal{F}$ for all $A \in \mathcal{F}$. We say that $T : \Omega \mapsto \Omega$ is a *measure-preserving transformation* if $\mathbf{P}(T^{-1}A) = \mathbf{P}(A)$ for all $A \in \mathcal{F}$. If the event $A \in \mathcal{F}$ is such that $T^{-1}A = A$, A is called an *invariant event*. The class of all invariant events is a σ-field denoted by $\mathcal{J}$. If for every $A \in \mathcal{J}$ we have $\mathbf{P}(A) = 0$ or 1, the measure-preserving transformation T is said to be *ergodic*. The function $g : (\Omega, \mathcal{F}) \mapsto (\mathbb{R}^1, \mathcal{B}^1)$ is called invariant under T iff $g(T\omega) = g(\omega)$ for all ω. It can easily be shown that the measure-preserving transformation T is ergodic iff every T-invariant function is $\mathbf{P}$-a.s. constant. Finally, recall that T is said to be a *mixing transformation* on $(\Omega, \mathcal{F}, \mathbf{P})$ iff for all $A, B \in \mathcal{F}$,

$$\lim_{n\to\infty} \mathbf{P}(A \cap T^{-n}B) = \mathbf{P}(A)\mathbf{P}(B).$$

We now compare two of the notions introduced above—ergodicity and mixing.

Let T be a measure-preserving transformation on the probability space $(\Omega, \mathcal{F}, \mathbf{P})$. Then: (a) T is ergodic iff any T-invariant function is $\mathbf{P}$-a.s. constant; (b) T mixing implies that T is ergodic (see Ash and Gardner 1975; Rosenblatt 1979; Shiryaev 1995).

Let $\Omega = [0,1]$, $\mathcal{F} = \mathcal{B}_{[0,1]}$ and $\mathbf{P}$ be the Lebesgue measure. Consider the transformation $T\omega = (\omega + \lambda)(\text{mod } 1)$, $\omega \in \Omega$. It is easy to see that T is measure preserving. Thus we want to establish if T is ergodic and mixing.

Suppose first that λ is a rational number, $\lambda = k/m$ for some integer k and m. Then the set

$$A = \bigcup_{k=0}^{2m-2} \left\{\omega : \frac{k}{2m} \leq \omega < \frac{k+1}{2m}\right\}$$

is invariant and $\mathbf{P}(A) = \frac{1}{2}$. This implies that for λ rational, the transformation T cannot be ergodic.

Now let λ be an irrational number. Our goal is to show that in this case T is ergodic. Consider a r.v. $\xi = \xi(\omega)$ on $(\Omega, \mathcal{F}, \mathbf{P})$ with $\mathbf{E}[\xi^2] < \infty$. Then (see Ash 1972; Kolmogorov and Fomin 1970) the Fourier series $\sum_{n=-\infty}^{\infty} c_n \mathrm{e}^{2\pi i n \omega}$ of the function $\xi(\omega)$ is L^2-convergent and $\sum_{n=-\infty}^{\infty} |c_n|^2 < \infty$. Suppose that ξ is an invariant r.v. Since T is measure preserving we find for the Fourier coefficient c_n that

$$\begin{aligned} c_n &= \mathbf{E}[\xi(\omega)\mathrm{e}^{-2\pi i n \omega}] = \mathbf{E}[\xi(T\omega)\mathrm{e}^{-2\pi i n T\omega}] \\ &= \mathrm{e}^{-2\pi i n \lambda}\mathbf{E}[\xi(T\omega)\mathrm{e}^{-2\pi i n \omega}] \\ &= \mathrm{e}^{-2\pi i n \lambda}\mathbf{E}[\xi(\omega)\mathrm{e}^{-2\pi i n \omega}] = c_n \mathrm{e}^{-2\pi i n \lambda}. \end{aligned}$$

This implies that $c_n(1 - \mathrm{e}^{-2\pi i n \lambda}) = 0$. However, as we have assumed that λ is irrational, $e^{-2\pi i n \lambda} \neq 1$ for all $n \neq 0$. Hence $c_n = 0$ if $n \neq 0$ and $\xi(\omega) = c_0$ a.s., $c_0 = \text{constant}$. From statement (a) we can conclude that the transformation T is ergodic.

It remains for us to show that T is not mixing. Indeed, take the set $A = \{\omega : 0 \leq \omega \leq 1/2\}$ and let $B = A$. Since T is measure preserving and invertible, then for any n we get

$$\mathbf{P}(A \cap T^{-n}B) = \mathbf{P}(A \cap T^{-n}A) = \mathbf{P}(T^n A \cap A).$$

Let us fix a number $\varepsilon \in (0,1)$. Since λ is irrational, then for infinitely many n the difference between $\mathrm{e}^{2\pi i n \lambda}$ and $\mathrm{e}^{i.0} = 1$, in absolute value, does not exceed ε. The sets A and $T^n A$ overlap except for a set of measure less than ε. Thus

$$\mathbf{P}(A \cap T^{-n}B) \geq \mathbf{P}(A) - \varepsilon$$

and for $0 < \varepsilon < \frac{1}{8}$ we find

$$\mathbf{P}(A \cap T^{-n}B) > \tfrac{3}{8}.$$

If the transformation T were mixing, then

$$\mathbf{P}(A \cap T^{-n}B) \to \mathbf{P}(A)\mathbf{P}(B) \quad \text{as } n \to \infty$$

and it should be $\mathbf{P}(A)\mathbf{P}(B) \geq \frac{3}{8}$. On the other hand, since $\mathbf{P}(A) = \frac{1}{2}$,

$$\mathbf{P}(A)\mathbf{P}(B) = [\mathbf{P}(A)]^2 = \tfrac{1}{4}.$$

Thus we come to a contradiction, so the mixing property of T fails to hold. Therefore, for measure-preserving transformations, mixing is a stronger property than ergodicity.

21.7. On the convergence of sums of φ-mixing random variables

It is well known that in the case of independent r.v.s $\{X_n, n \geq 1\}$ the infinite series $\sum_{n=1}^{\infty} X_n$ is convergent simultaneously in distribution, in probability and with probability 1. This statement, called the Lévy equivalence theorem, can be found in a book by Ito (1984) and leads to the question: does a similar result hold for sequences of 'weakly' dependent r.v.s?

Let $\{X_n, n \geq 1\}$ be a stationary random sequence satisfying the so-called *φ-mixing condition*. This means that for some numerical sequence $\varphi(n) \downarrow 0$ as $n \to \infty$ we have

$$|\mathbf{P}(AB) - \mathbf{P}(A)\mathbf{P}(B)| \leq \varphi(n)\mathbf{P}(A)$$

where $A \in \mathcal{M}_1^m$, $B \in \mathcal{M}_{m+n}^{\infty}$, $m \geq 1$, $n \geq 1$ and $\mathbf{P}(A) > 0$.

Note that there are several results concerning the convergence of the partial sums $S_n = X_1 + \ldots + X_n$ as $n \to \infty$ of a φ-mixing sequence (see Stout 1974a, b). Let us formulate the following result from Stout (1974b).

The conditions (a) and (b) below are equivalent:

(a) $\sum_{n=1}^{\infty} X_n$ converges in distribution and $X_n \xrightarrow{\text{d}} 0$ as $n \to \infty$;
(b) $\sum_{n=1}^{\infty} X_n$ converges in probability.

Recall that for independent r.v.s a condition like $X_n \xrightarrow{\text{d}} 0$ is not involved. Since for φ-mixing sequences conditions (a) and (b) are equivalent it is clear that removal of the condition $X_n \xrightarrow{\text{d}} 0$ will surely make the series $\sum_{n=1}^{\infty} X_n$ not convergent in probability. An illustration follows.

Consider the sequence $\{\xi_n, n \geq 1\}$ of i.i.d. r.v.s with $\mathbf{P}[\xi_1 = \pm 1] = \frac{1}{2}$ and let $X_n = \xi_{n+1} - \xi_n$. It is easy to see that the new sequence $\{X_n, n \geq 1\}$ is φ-mixing with $\varphi(n) = 0$ for all $n \geq 2$. Clearly $S_n = \sum_{k=1}^{n} X_k = \xi_{n+1} - \xi_1$. It follows from here that S_n is convergent in distribution because S_n has the same distribution for all n, namely S_n takes the values 2, 0 and -2 with probabilities $\frac{1}{4}$, $\frac{1}{2}$ and $\frac{1}{4}$ respectively. Obviously $\{S_n\}$ is not convergent in probability as $n \to \infty$.

21.8. The central limit theorem for stationary random sequences

The classical CLT deals with independent r.v.s (see Section 17). Thus if we suppose that $\{X_n, n \geq 1\}$ is a sequence of 'weakly' dependent r.v.s we cannot expect that without additional assumptions the normed sums S_n/s_n will converge to the standard normal distribution. As usual, $S_n = X_1 + \ldots + X_n$, $s_n^2 = \mathbf{V}S_n$. There are works (see Rosenblatt 1956; Ibragimov and Linnik 1971; Davydov 1973; Bradley 1980) where under appropriate conditions the CLT is proved for some classes of stationary sequences (see Ibragimov and Linnik 1971; Bradley 1980) and for stationary random fields (see Bulinskii 1988).

We present below a few examples which show that for stationary sequences the normed sums S_n/s_n can behave differently as $n \to \infty$. In particular, the limit distribution, if it exists, need not be the normal distribution $\mathcal{N}(0,1)$.

(i) Let ξ be a r.v. distributed uniformly on [0, 1]. Consider the random sequence $\{X_n, n = 0, \pm 1, \ldots\}$ where $X_n = \cos(2\pi n\xi)$. It is easy to see that the variables X_n are uncorrelated (but not independent), so $\{X_n\}$ forms a weakly stationary sequence. If $S_n = X_1 + \ldots + X_n$ we can easily see that $\mathbf{E}S_n = 0$ and $\mathbf{V}S_n = \frac{1}{2}n$. Moreover,

$$S_n = \frac{1}{2} + \frac{1}{2}\frac{\sin(2\pi(n+\frac{1}{2})\xi)}{\sin(\pi\xi)}.$$

According to a result by Grenander and Rosenblatt (1957), we have

$$S_n \xrightarrow{\text{d}} Y := \frac{1}{2} + \frac{1}{2}\frac{\sin(\pi\eta)}{2\sin(\pi\xi)} \quad \text{as} \quad n \to \infty$$

where η is another r.v. uniformly distributed on [0, 1] and independent of ξ. Note especially that S_n itself, not the normed quantity S_n/s_n, has a limit distribution. Moreover, it is obvious that S_n/s_n does not converge to a r.v. distributed $\mathcal{N}(0,1)$.

(ii) Consider the sequence of r.v.s $\{X_n, n = 0, \pm 1, \ldots\}$ such that for an arbitrary integer n and non-negative integer m, the random vector $(X_n, X_{n+1}, \ldots, X_{n+m})$ has the following density:

$$\begin{aligned} f(x_n, x_{n+1}, \ldots, x_{n+m}) &= \frac{1}{2}(2\pi)^{-n/2}\sigma_1^{-n}\exp\left(-\frac{1}{2}\sigma_1^{-2}\sum_{k=0}^{m} x_{n+k}^2\right) \\ &+ \frac{1}{2}(2\pi)^{-n/2}\sigma_2^{-n}\exp\left(-\frac{1}{2}\sigma_2^{-2}\sum_{k=0}^{m} x_{n+k}^2\right). \end{aligned}$$

Here $\sigma_1 > 0$, $\sigma_2 > 0$ and we assume that $\sigma_1 \neq \sigma_2$. Obviously $\{X_n\}$ is a strictly stationary sequence. If $S_n = X_1 + \ldots + X_n$ it is not difficult to see that

$$\lim_{n\to\infty} \mathbf{P}[S_n/s_n \le x] := G(x) = \tfrac{1}{2}\Phi(\sigma_1 x) + \tfrac{1}{2}\Phi(\sigma_2 x).$$

Thus the limit distribution G of S_n/s_n is a mixture of two normal distributions and, since $\sigma_1 \neq \sigma_2$, G is not normal.

(iii) Let $\{X_n, n = 0, \pm 1, \ldots\}$ be a strictly stationary sequence with $\mathbf{E}[X_n^2] < \infty$ for all n. Denote by $\rho(n)$, $n \ge 1$, the maximal correlation coefficient associated with this sequence (see Example 21.4). Recall that in general

$$\rho(n) = \sup_{\eta_1, \eta_2}\{\mathbf{E}[(\eta_1 - \mathbf{E}\eta_1)(\eta_2 - \mathbf{E}\eta_2)]/(\mathbf{V}\eta_1\mathbf{V}\eta_2)^{1/2}\}$$

where η_1 is $\mathcal{M}_{-\infty}^{m}$-measurable, η_2 is $\mathcal{M}_{m+n}^{\infty}$-measurable, $0 < \mathbf{V}\eta_1, \mathbf{V}\eta_2 < \infty$, m is any integer and $n \geq 1$. Note that the condition

$$\rho(n) \to 0 \quad \text{as} \quad n \to \infty$$

plays an important role in the theory of stationary processes. In particular, the φ-mixing condition implies that $\rho(n) \to 0$ and, further, $\rho(\eta) \to 0$ implies the strong mixing condition (for details see Ibragimov and Linnik 1971).

Suppose $\{X_n, n = 0, \pm 1, \ldots\}$ is a strictly stationary sequence with $\mathbf{E}X_n = 0$ and $\mathbf{E}[X_n^2] < \infty$ for all n. Using the notation $S_n = X_1 + \ldots + X_n$ and $s_n^2 = \mathbf{E}[S_n^2]$ we formulate the following result of Ibragimov (1975). If $\rho(n) \to 0$ then either $\sup_n s_n^2 < \infty$ or $s_n^2 = nh(n)$ where $h(n)$ is a slowly varying function as $n \to \infty$. If $s_n^2 \to \infty$, $\rho(n) \to 0$ and for some $\delta > 0$, $\mathbf{E}[|X_0|^{2+\delta}] < \infty$, then $S_n/\sqrt{n} \xrightarrow{\text{d}} Y$ as $n \to \infty$ for a r.v. Y distributed $\mathcal{N}(0,1)$.

Our aim now is to see whether the conditions for this result can be weakened while preserving the asymptotic normality of $S_n/\sqrt{n}$. In particular, an example will be described in which instead of the condition $\mathbf{E}[|X_0|^{2+\delta}] < \infty$ we have $\mathbf{E}[|X_0|^{2+\delta}] = \infty$ for each $\delta > 0$ but $\mathbf{E}[|X_0|^2] < \infty$. This example is the main result in a paper by Bradley (1980) and is formulated as follows.

There exists a strictly stationary sequence $\{X_n, n = 0, \pm 1, \ldots\}$ of real-valued r.v.s such that: (a) $\mathbf{E}X_n = 0$ and $0 < \mathbf{V}X_n < \infty$; (b) $s_n^2 \to \infty$ as $n \to \infty$; (c) $\rho(n) \to 0$ as $n \to \infty$; (d) for each $\lambda > 0$ there is an increasing sequence of positive integers $\{n(k)\}$ such that $\lambda^{1/2} S_{n(k)}/s_{n(k)} \xrightarrow{\text{d}} \xi_\lambda$ as $k \to \infty$ where ξ_λ is a r.v. with a d.f. F_λ defined by

$$F_\lambda(x) = \mathrm{e}^{-\lambda} 1_{[0,\infty)}(x) + \sum_{k=1}^{\infty} \frac{\lambda^k}{k!} \mathrm{e}^{-\lambda} \frac{1}{\sqrt{2\pi k}} \int_{-\infty}^{x} \mathrm{e}^{-u^2/2k}\, du, \quad x \in \mathbb{R}^1.$$

Note that for each fixed $\lambda > 0$ the limit distribution F_λ is a Poisson mixture of normal distributions and has a point-mass at 0. Thus F_λ is not a normal distribution. Therefore the stationary sequence constructed above does not satisfy the CLT.

It is interesting to note that F_λ is an infinitely divisible but not a stable distribution. (Two other distributions with analogous properties were given in Example 9.7.)

Let us note finally that Herrndorf (1984) constructed an example of a stationary sequence (not m-dependent) of mutually uncorrelated r.v.s such that the strong mixing coefficient tends to zero 'very fast' but nevertheless the CLT fails to hold. For more recent results on this topic see the papers by Janson (1988) and Bradley (1989).

SECTION 22. DISCRETE-TIME MARTINGALES

Let $(X_n, n \geq 1)$ be a random sequence defined on the probability space $(\Omega, \mathcal{F}, \mathbf{P})$. We are also given the family $(\mathcal{F}_n, n \geq 1)$ of non-decreasing sub-σ-fields of $\mathcal{F}$, that is $\mathcal{F}_n \subset \mathcal{F}$ for each n and $\mathcal{F}_n \subset \mathcal{F}_{n+1}$. As usual, if we write $(X_n, \mathcal{F}_n, n \geq 1)$, this means that the sequence (X_n) is $(\mathcal{F}_n)$-adapted: X_n is $\mathcal{F}_n$-measurable for each n. The sequence $(X_n, n \geq 1)$ is integrable if $\mathbf{E}|X_n| < \infty$ for every $n \geq 1$. If $\sup_{n\geq 1} \mathbf{E}|X_n| < \infty$ we say that the given sequence is L^1-*bounded*, while if $\mathbf{E}[\sup_{n\geq 1} |X_n|] < \infty$ the sequence $(X_n, n \geq 1)$ is L^1-*dominated.*

The system $(X_n, \mathcal{F}_n, n \geq 1)$ is called a **martingale** if $\mathbf{E}|X_n| < \infty, n \geq 1$ and

$$\mathbf{E}[X_n|\mathcal{F}_m] = X_m \quad \text{a.s.} \tag{1}$$

for all $m \leq n$. If in (1) instead of equality we have $\mathbf{E}[X_n|\mathcal{F}_m] \leq X_m$ or $\mathbf{E}[X_n|\mathcal{F}_m] \geq X_m$, then we have a **supermartingale** or a **submartingale** respectively.

A *stopping time* with respect to $(\mathcal{F}_n)$ is a function $\tau : \Omega \mapsto \mathbb{N} \cup \{\infty\}$ such that $[\tau = n] \in \mathcal{F}_n$ for all $n \geq 1$. Denote by T the set of all bounded stopping times. Recall that the family $(a_\tau, \tau \in \mathrm{T})$ of real numbers (such a family is called a *net*) is said to converge to the real number b if for every $\varepsilon > 0$ there is $\tau_0 \in \mathrm{T}$ such that for all $\tau \in \mathrm{T}$ with $\tau \geq \tau_0$ we have $|a_\tau - b| < \varepsilon$.

Some definitions of systems whose properties are close to those of martingales but are in some sense generalizations of them are listed below. The random sequence $(X_n, \mathcal{F}_n, n \geq 1)$ is said to be:

(a) a **quasimartingale** if $\sum_{n=1}^{\infty} \mathbf{E}[|X_n - \mathbf{E}(X_{n+1}|\mathcal{F}_n)|] < \infty$;
(b) an **amart** if the net $(\mathbf{E}[X_\tau], \tau \in \mathrm{T})$ converges;
(c) a **martingale in the limit** if $\sup_{m\geq n} |\mathbf{E}(X_m|\mathcal{F}_n) - X_n| \xrightarrow{\text{a.s.}} 0$ as $n \to \infty$;
(d) a **game fairer with time** if $\sup_{m\geq n} |\mathbf{E}(X_m|\mathcal{F}_n) - X_n| \xrightarrow{\mathrm{P}} 0$ as $n \to \infty$;
(e) a **progressive martingale** if $A_n \subset A_{n+1}$ for $n \geq 1$ and $\mathbf{P}[\cup_{n=1}^{\infty} A_n] = 1$ where $A_n = [\mathbf{E}(X_{n+1}|\mathcal{F}_n) = X_n]$;
(f) an **eventual martingale** if $\mathbf{P}[\mathbf{E}(X_{n+1}|\mathcal{F}_n) \neq X_n \text{ i.o.}] = 0$.

Random sequences which possess the martingale, supermartingale or submartingale properties are of classic importance in the theory of stochastic processes. Complete presentations of them have been given by Doob (1953), Neveu (1975) and Chow and Teicher (1978).

The martingale generalizations (a)–(f) given above have appeared in recent years. Many results and references in this new area can be found in the works of Gut and Schmidt (1983) and Tomkins (1984a, b).

In this section we have included examples which illustrate the basic properties of martingales and martingale-like sequences (with discrete time) and reveal the relationships between them.

22.1. Martingales which are L^1-bounded but not L^1-dominated

Let $X = (X_n, \mathcal{F}_n, n \geq 1)$ be a martingale. The relation $\sup_{n\geq 1} \mathbf{E}|X_n| \leq \mathbf{E}[\sup_{n\geq 1} |X_n|]$ implies that every L^1-dominated martingale is also L^1-bounded.

This raises the question of whether the converse is true. The answer is negative and will be illustrated by a few examples.

(i) Consider the discrete space $\Omega = \{1, 2, \ldots\}$ with probability $\mathbf{P}$ on it defined by $\mathbf{P}(\{n\}) = \frac{1}{n} - \frac{1}{n+1}$, $n \in \mathbb{N}$. Let $(\mathcal{F}_n, n \geq 1)$ be the increasing sequence of σ-fields where $\mathcal{F}_n$ is generated by the partitions $\{\{1\}, \{2\}, \ldots, \{n\}, [n+1, \infty)\}$. Define the sequence $(X_n, n \geq 1)$ of r.v.s by

$$X_n = X_n(\omega) = (n+1) \times 1_{[n+1,\infty)}(\omega), \quad n \in \mathbb{N}.$$

Then $X = (X_n, \mathcal{F}_n, n \geq 1)$ is a positive martingale such that $\mathbf{E}X_n = 1$ for all $n \in \mathbb{N}$ and hence X is L^1-bounded. However, $\sup_{n \in \mathbb{N}} X_n(\omega) = \omega$ and clearly it is not integrable. Therefore the martingale X is not L^1-dominated.

(ii) Let $\Omega = [0, 1]$, $\mathcal{F} = \mathcal{B}_{[0,1]}$ and $\mathbf{P}$ be the Lebesgue measure. Define

$$X_n = X_\omega = \begin{cases} 0, & \text{if } 1/n < \omega \leq 1 \\ -n^2\omega + n, & \text{if } 0 \leq \omega \leq 1/n \end{cases}$$

and $\mathcal{F}_n = \sigma\{X_1, \ldots, X_n\}$. Then $(X_n, \mathcal{F}_n, n \geq 1)$ is a martingale. Since $\mathbf{E}X_n = \frac{1}{2}$ for each $n \in \mathbb{N}$ this martingale is L^1-bounded. However, its supremum, $\sup_{n \in \mathbb{N}} |X_n|$, is not integrable and the L^1-domination property fails to hold.

(iii) Let $w = (w(t), t \geq 0)$ be a standard Wiener process, $\mathcal{F}_t = \sigma\{w_s, s \leq t\}$. Take any numerical sequence $\{n_k, k \geq 1\}$ such that $0 < n_1 < n_2 < \cdots \to \infty$ as $k \to \infty$. Denote $M_k = \exp[w(n_k) - \frac{1}{2}n_k]$. Then it can be shown that $M = (M_k, \mathcal{F}_{n_k}, k \geq 1)$ is a non-negative martingale (and even that $M_k \xrightarrow{\text{a.s.}} 0$ as $k \to \infty$) which is integrable but $\mathbf{E}[\sup_{k \geq 1} M_k] = \infty$. Hence in this case the L^1-domination property again does not hold, despite the integrability of M. One additional example of an L^1-bounded but not L^1-dominated martingale will be given at the end of Example 22.2.

22.2. A property of a martingale which is not preserved under random stopping

Let $X = (X_n, \mathcal{F}_n, n \geq 1)$ be a martingale and $Y_n = \frac{1}{n}(X_1 + \cdots + X_n)$. Denote by T the set of all bounded $(\mathcal{F}_n)$-stopping times and introduce the following four conditions:

(1) $$\sup_{n \geq 1} \mathbf{E}|X_n| < \infty,$$

(2) $$\sup_{n \geq 1} \mathbf{E}|Y_n| < \infty,$$

(3) $$\sup_{\tau \in \mathrm{T}} \mathbf{E}|X_\tau| < \infty,$$

(4) $$\sup_{\tau \in \mathrm{T}} \mathbf{E}|Y_\tau| < \infty.$$

Obviously, conditions (3) and (4) can be considered as 'random stopped versions' of (1) and (2) respectively. It is well known (see Yamazaki 1972) that conditions (1) and (2) are equivalent; moreover, conditions (1) and (3) are also equivalent. Thus it is natural to assume that (3) and (4) are equivalent. However, as we shall now see, this conjecture is wrong.

Let $\tau \in \mathrm{T}$, that is τ is a positive integer-valued r.v. such that $\mathbf{P}[\tau < \infty] = 1$ and let $\mathbf{P}[\tau > n] > 0$ for every $n \geq 1$. Denote by $\mathcal{F}_n$ the σ-field generated by the events $[\tau = 1], [\tau = 2], \ldots, [\tau = n]$. Clearly, τ is an $(\mathcal{F}_n)$-stopping time. Let $\{b_n, n \geq 1\}$ be a non-increasing sequence of positive numbers such that $b_{k-1} - b_k = 0$ for those k for which $\mathbf{P}[\tau = k] = 0$, and in such cases we also put $(b_{k-1} - b_k)/\mathbf{P}[\tau = k] = 0$. Define the sequence $(X_n, n \geq 1)$ of r.v.s by

$$(5)\quad X_n(\omega) = \sum_{k=1}^{n}[(b_{k-1} - b_k)/\mathbf{P}[\tau = k]]1_{[\tau=k]}(\omega) + (b_n/\mathbf{P}[\tau > n])1_{[\tau>n]}(\omega).$$

Then it is not difficult to check that $X = (X_n, \mathcal{F}_n, n \geq 1)$ is a non-negative martingale. Indeed, taking into account that $[\tau = 1], \ldots, [\tau = n-1]$ and $[\tau > n-1]$ are atoms of $\mathcal{F}_n$, we can easily see that

$$\int_{[\tau=k]}(X_n - X_{n-1})\,\mathrm{d}\mathbf{P} = 0, \quad k = 1, \ldots, n-1,$$

$$\int_{[\tau>n-1]}(X_n - X_{n-1})\,\mathrm{d}\mathbf{P} = (b_{n-1} - b_n) + (b_n - b_{n-1}) = 0.$$

These relations imply the martingale property of X. We can check directly that condition (1) is satisfied and hence (2) and (3) hold. It then remains for us to clarify whether condition (4) is satisfied. To do this, consider the following variable $Y_\tau = (1/\tau)(X_1 + \ldots + X_\tau)$. Clearly,

$$Y_\tau \geq (1/\tau)\sum_{k=1}^{\tau-1} X_k = (1/\tau)\sum_{k=1}^{\tau-1}(b_k/\mathbf{P}[\tau > k])1_{[\tau>k]} := \eta.$$

Here η is a r.v. which takes the value $(1/n)\sum_{k=1}^{n-1}(b_k/\mathbf{P}[\tau > k])$ with probability equal to $\mathbf{P}[\tau = n]$. This implies that $\mathbf{E}Y_\tau \geq \mathbf{E}\eta$. So our aim is to estimate the expectation $\mathbf{E}\eta$. However, we need the following result from analysis (the proof is left to the reader): if $\{a_n, n \geq 1\}$ is a positive non-increasing sequence converging to zero and $\{b_n, n \geq 1\}$ is a non-negative and non-increasing sequence, then

$$(6)\quad \sum_{n=2}^{\infty}\left[\frac{1}{n}(a_{n-1} - a_n)\sum_{j=1}^{n-1}(b_j/a_j)\right] \geq \frac{1}{4}\sum_{n=1}^{\infty}[((a_{2^n} - a_{2^n-1})/a_{2^n})b_{2^n}].$$

Now let $a_n = \mathbf{P}[\tau > n]$, and take the sequence $\{b_n, n \geq 1\}$ used to define X by (5) to be non-increasing and bounded from below by some positive constant, that is $b_n \geq c = \text{constant} > 0$ for all $n \geq 1$. Then these two sequences, $\{a_n, n \geq 1\}$ and $\{b_n, n \geq 1\}$, satisfy the conditions required for the validity of (6). After some calculations we find that $\mathbf{E}\eta = \infty$ and hence

$$\mathbf{E}|Y_\tau| = \mathbf{E}Y_\tau \geq \mathbf{E}\eta = \infty.$$

Therefore condition (4) does not hold in spite of the fact that (1), (2) and (3) are satisfied.

Finally, let us look at the following possibility. It is easy to see that the martingale $(X_n, \mathcal{F}_n, n \geq 1)$ defined by (5) is uniformly integrable. If in particular we choose $b_n = 1/(n+1)$ and $\mathbf{P}[\tau = n] = 2^{-n}$, then we can check that $\mathbf{E}[\sup_{n\geq 1} X_n] = \infty$. Thus we obtain another example of a martingale which is L^1-bounded but not L^1-dominated (see also Example 22.1).

22.3. Martingales for which the Doob optional theorem fails to hold

Let $X = (X_n, \mathcal{F}_n, n \geq 0)$ be a martingale and τ be an $(\mathcal{F}_n)$-stopping time. Suppose the following two conditions are satisfied:

$$\text{(a)} \qquad \mathbf{E}[|X_\tau|] < \infty; \qquad\qquad \text{(b)} \qquad \lim_{n\to\infty} \int_{[\tau > n]} X_n \, \mathrm{d}\mathbf{P} = 0.$$

Then $\mathbf{E}X_\tau = \mathbf{E}X_0$.

This statement, called the *Doob optional theorem*, is considered in many books (see Doob 1953; Kemeny *et al* 1966; Neveu 1975). Conditions (a) and (b) together are sufficient conditions for the validity of the relation $\mathbf{E}X_\tau = \mathbf{E}X_0$. Our purpose now is to clarify whether both (a) and (b) are necessary.

(i) Let $\{\eta_n, n \geq 1\}$ be a sequence of i.i.d. r.v.s. Suppose η_1 takes only the values $-1, 0, 1$ and $\mathbf{E}\eta_1 = 0$. Define $X_n = \eta_1 + \cdots + \eta_n$ and $\mathcal{F}_n = \sigma\{\eta_1, \ldots, \eta_n\}$ for $n \geq 1$ and $X_0 = 0$, $\mathcal{F}_0 = \{\emptyset, \Omega\}$. Clearly $X = (X_n, \mathcal{F}_n, n \geq 0)$ is a martingale. If $\tau = \inf\{n : X_n = 1\}$, then τ is an $(\mathcal{F}_n)$-stopping time such that $\mathbf{P}[\tau < \infty] = 1$ and $X_\tau = 1$ a.s. Hence $\mathbf{E}X_0 = 0 \neq 1 = \mathbf{E}X_\tau$ which means that the Doob optional theorem does not hold for the martingale X and the stopping time τ. Let us check whether conditions (a) and (b) are satisfied.

It is easy to see that $\mathbf{E}|X_\tau| < \infty$ and thus condition (a) is satisfied. Furthermore,

$$0 = \int_\Omega X_n \, \mathrm{d}\mathbf{P} = \int_{[\tau \leq n]} X_n \, \mathrm{d}\mathbf{P} + \int_{[\tau > n]} X_n \, \mathrm{d}\mathbf{P} := J_1 + J_2.$$

The term J_1 is equal to the probability that level 1 has been reached by the martingale X in time n and this probability tends to 1 as $n \to \infty$. Since $J_1 + J_2 = 0$ we see that J_2 tends to -1, not to 0. Thus condition (b) is violated.

(ii) Let $\xi_1, \xi_2, \ldots$ be independent r.v.s where $\xi_n \sim \mathcal{N}(0, b_n)$. Here the variances $b_n, n \geq 1$, are chosen as follows. We take $b_1 = 1$ and $b_{n+1} = a_{n+1}^2 - a_n^2$ for $n \geq 1$

where $a_n = (n-1)^2/\log(3+n)$. The reason for this special choice will become clear later.

Define $X_n = \xi_1 + \cdots + \xi_n$ and $\mathcal{F}_n = \sigma\{\xi_1, \ldots, \xi_n\}$. Then $X = (X_n, \mathcal{F}_n, n \geq 0)$ is a martingale. Let g be a measurable function from $\mathbb{R}^1$ to $\mathbb{N}$ with $\mathbf{P}[g(\xi_1) = n] = p_n$ where $p_n = n^{-2} - (n+1)^{-2}, n \geq 1$. Thus $\tau := g(\xi_1)$ is a stopping time and moreover its expectation is finite. It can be shown that the relation $\mathbf{E}X_\tau = \mathbf{E}X_1$ does not hold. So let us check whether conditions (a) and (b) are satisfied.

Denote by F the d.f. of ξ_1 and let $S_1 = 0$, $S_n = \xi_2 + \cdots + \xi_n$ for $n \geq 2$. Thus ξ_1 is independent of $S_1, S_2, \ldots$ and $X_n = \xi_1 + S_n$ where $S_n \sim \mathcal{N}(0, a_n^2)$. Now we have to compute the quantities $\mathbf{E}|X_\tau|$ and $\int_{[\tau>n]} |X_n| \, \mathrm{d}\mathbf{P}$. We find

$$\begin{aligned}\int_{[\tau>n]} |X_n| \, \mathrm{d}\mathbf{P} &= \int_{[g>n]} \mathbf{E}[|y + S_n|] \, \mathrm{d}F(y) \\ &\leq \int_{[g>n]} \{|y| + \mathbf{E}|S_n|\} \, \mathrm{d}F(y) \\ &= \int_{[g>n]} |y| \, \mathrm{d}F(y) + c a_n \mathbf{P}[g > n]\end{aligned}$$

where $c = \mathbf{E}[|\xi_1|]$. It is easy to conclude that $\int_{[\tau>n]} |X_n| \, \mathrm{d}\mathbf{P} \to 0$ as $n \to \infty$ and hence condition (b) is satisfied. Furthermore,

$$\begin{aligned}\mathbf{E}|X_\tau| &= \int \mathbf{E}[|y + S_{g(y)}|] \, \mathrm{d}F(y) \geq \int \mathbf{E}[|S_{g(y)}|] \, \mathrm{d}F(y) \\ &= c \int a_{g(y)} \, \mathrm{d}F(y) = c \sum_{n=1}^{\infty} p_n a_n = \infty\end{aligned}$$

and condition (a) is not satisfied.

Examples (i) and (ii) show that both conditions (a) and (b) are essential for the validity of the Doob optional theorem.

22.4. Every quasimartingale is an amart, but not conversely

It is easy to show that every quasimartingale is also an amart (for details see Edgar and Sucheston 1976a; Gut and Schmidt 1983). However, the converse is not always true. This will be illustrated by two simple examples.

(i) Let $a_n = (-1)^n n^{-1}, n \geq 1$. Take $X_n = a_n$ a.s. and choose an arbitrary sequence $\{\tau_n, n \geq 1\}$ of bounded stopping times with the only condition that $\tau \uparrow \infty$ as $n \to \infty$. Since $a_n \to 0$ as $n \to \infty$, we have $X_{\tau_n} \xrightarrow{\text{a.s.}} 0$ as $n \to \infty$. Moreover, $|a_n| \leq 1$ implies that $\mathbf{E}X_{\tau_n} \to 0$ as $n \to \infty$. Hence for any increasing family of σ-fields $(\mathcal{F}_n, n \geq 1)$ to which (τ_n) are related, the system $(X_n, \mathcal{F}_n, n \geq 1)$ is an amart. However, $\sum_{n=1}^{\infty} \mathbf{E}|X_n - \mathbf{E}(X_{n+1}|\mathcal{F}_n)| = \sum_{n=1}^{\infty} |a_n - a_{n-1}| = \infty$ and the amart $(X_n, \mathcal{F}_n, n \geq 1)$ is not a quasimartingale.

(ii) Let $(X_n, n \geq 1)$ be a sequence of i.i.d. r.v.s such that $\mathbf{P}[X_n = 1] = \mathbf{P}[X_n = -1] = \frac{1}{2}$ and let $(c_n, n \geq 1)$ be positive real numbers, $c_n \downarrow 0$ as $n \to \infty$ and $\sum_{n=1}^{\infty} c_n = \infty$. Consider the sequence $(Y_n, n \geq 1)$ where $Y_n = c_n X_1 \dots X_n$ and the σ-fields $\mathcal{F}_n = \sigma\{X_1, \dots, X_n\}$. Clearly, Y_n is $\mathcal{F}_n$-measurable for every $n \geq 1$. Since a.s. $|Y_n| \leq c_n \downarrow 0$, $Y_{\tau_n} \xrightarrow{\text{a.s.}} 0$ as $n \to \infty$ for any sequence of bounded stopping times $(\tau_n, n \geq 1)$ such that $\tau_n \uparrow \infty$ as $n \to \infty$. Applying the dominated convergence theorem, we conclude that $\mathbf{E}Y_{\tau_n} \to 0$ as $n \to \infty$, so $Y = (Y_n, \mathcal{F}_n, n \geq 1)$ is an amart. However,

$$\sum_{n=1}^{\infty} \mathbf{E}|Y_n - \mathbf{E}(Y_{n+1}|\mathcal{F}_n)| = \sum_{n=1}^{\infty} \mathbf{E}|Y_n| = \sum_{n=1}^{\infty} c_n = \infty$$

and therefore the amart Y is not a quasimartingale.

22.5. Amarts, martingales in the limit, eventual martingales and relationships between them

(i) Let $\xi_1, \xi_2, \dots$ be a sequence of positive i.i.d. r.v.s such that $\mathbf{E}\xi_1 < \infty$ and $\mathbf{E}[\xi_1 \log^+ \xi_1] = \infty$. Consider the sequence $X_1, X_2, \dots$ where $X_n = \xi_n/n$ and the family $(\mathcal{F}_n, n \geq 1)$ with $\mathcal{F}_n = \sigma\{\xi_1, \dots, \xi_n\}$. It is easy to check that $X_n \xrightarrow{\text{a.s.}} 0$ as $n \to \infty$. Moreover, $\mathbf{E}X_n \to 0$ as $n \to \infty$ and $\mathbf{E}[\sup_{n \geq 1} X_n] = \infty$. It follows that $X = (X_n, \mathcal{F}_n, n \geq 1)$ is a martingale in the limit, but X is not an amart because the net $(\mathbf{E}X_\tau, \tau \in \mathrm{T})$ is unbounded where T is the set of all bounded $(\mathcal{F}_n)$-stopping times.

(ii) Consider the sequence $(\eta_n, n \geq 1)$ of independent r.v.s, $\mathbf{P}[\eta_n = 1] = n^{-2} = 1 - \mathbf{P}[\eta_n = 0]$. Let $\mathcal{F}_n = \sigma\{\eta_1, \dots, \eta_n\}$ and $X_n = \eta_1 + \dots + \eta_n$. Since

$$\mathbf{E}(X_n|\mathcal{F}_m) - X_m = \sum_{k=m+1}^{\infty} k^{-2} \Rightarrow \lim_{n \geq m \to \infty} (\mathbf{E}(X_n|\mathcal{F}_m) - X_m) = 0 \quad \text{a.s.}$$

we conclude that $X = (X_n, \mathcal{F}_n, n \geq 1)$ is a martingale in the limit. Moreover,

$$\mathbf{E}\left[\sum_{k=1}^{\infty} |\eta_k|\right] = \sum_{k=1}^{\infty} k^{-2} < \infty \quad \text{and} \quad |X_n| \leq \sum_{k=1}^{\infty} |\eta_k|, \quad n \geq 1$$

imply that X is even uniformly integrable. Despite these properties, X is not an eventual martingale. This follows from the relation

$$\mathbf{E}(X_n|\mathcal{F}_{n-1}) = X_{n-1} + n^{-2} \neq X_{n-1} \quad \text{for all } n \geq 2$$

and definition (f) (see the introductory notes in this section).

22.6. Relationships between amarts, progressive martingales and quasimartingales

(i) Let $(\xi_n, n \geq 1)$ be independent r.v.s such that $\mathbf{P}[\xi_n = 1] = n/(n+1) = 1 - \mathbf{P}[\xi_n = 0]$, $n \geq 1$. Define $\eta_1 = 1$ and for $n \geq 2$, $\eta_n = (-1)^{n-1}\xi_1\xi_2 \dots \xi_{n-1}$. Further, let $X_n = \eta_1 + \dots + \eta_n$ and $\mathcal{F}_n = \sigma\{\xi_1, \dots, \xi_n\}$. Obviously, for every n, X_n is either 0 or 1. Moreover, by the Borel–Cantelli lemma, $\mathbf{P}[\xi_n = 0 \text{ i.o.}] = 1$ which implies that $\mathbf{P}[\eta_n \neq 0 \text{ i.o.}] = 0$. However, $\mathbf{E}[\eta_n|\mathcal{F}_{n-1}] = \eta_{n-1}$ a.s. and $\eta_{n+1} = 0$ if $\eta_n = 0$. Hence $X = (X_n, \mathcal{F}_n, n \geq 1)$ is a progressive martingale. Let us check if X is a quasimartingale. We have

$$\mathbf{E}\eta_n = (-1)^{n-1} \prod_{k=1}^{n-1} \frac{k}{k+1} = \frac{(-1)^{n-1}}{n}$$

and

$$\sum_{n=1}^{\infty} \mathbf{E}|\mathbf{E}(\eta_n|\mathcal{F}_{n-1})| = \sum_{n=1}^{\infty} \frac{1}{n} = \infty.$$

Therefore the progressive martingale X is not a quasimartingale.

(ii) Let us now describe a random sequence which is a progressive martingale but not an amart.

Consider the sequence $(\xi_n, n \geq 1)$ of independent r.v.s where $\mathbf{P}[\xi_n = 1] = n/(n+1) = 1 - \mathbf{P}[\xi_n = 0]$ (case (i) above). Let $X_0 = 1$ and for $n \geq 1$, $X_n = n^2\xi_1\xi_2 \dots \xi_{n-1}$, and $\mathcal{F}_n = \sigma\{\xi_1, \dots, \xi_n\}$, $n \geq 1$. Clearly,

$$\mathbf{E}[X_n|\mathcal{F}_{n-1}] = X_n = \frac{n^2}{(n-1)^2}\xi_{n-1}X_{n-1} \quad \text{a.s.}$$

By the Borel–Cantelli lemma $\mathbf{P}[\xi_n = 0 \text{ i.o.}] = 1$ and since $X_{n-1} = 0$ implies that $X_n = 0$, we conclude that $\mathbf{P}[X_n \neq 0 \text{ i.o.}] = 0$. Consequently $X = (X_n, \mathcal{F}_n, n \geq 1)$ is a progressive martingale. However, $\mathbf{E}X_n = n \to \infty$ as $n \to \infty$ which shows that X cannot be an amart.

(iii) Recall that every quasimartingale is also an amart and a martingale in the limit. Let us illustrate that the converse is false.

Consider the sequence $(X_n, n \geq 1)$ given by $X_n = \sum_{k=1}^{n}(-1)^{k-1}k^{-1}$ and let $\mathcal{F}_n = \mathcal{F}_0$ for all $n \geq 1$. Then $X = (X_n, \mathcal{F}_n, n \geq 1)$ is an amart and also a martingale in the limit. Further, we have $0 < X_n < 1 + \sum_{k=1}^{\infty}(-1)^{k-1}k^{-1} < \infty$. However,

$$\sum_{n=1}^{\infty} \mathbf{E}|\mathbf{E}(X_n|\mathcal{F}_{n-1})| = \sum_{n=1}^{\infty} \frac{1}{n} = \infty$$

and therefore X is not a quasimartingale.

22.7. An eventual martingale need not be a game fairer with time

Let $(\xi_n, n \geq 1)$ be independent r.v.s such that $\mathbf{P}[\xi_n = -1] = 2^{-n} = 1 - \mathbf{P}[\xi_n = 1]$, $n \geq 1$. Let $\mathcal{F}_n = \sigma\{\xi_1, \ldots, \xi_n\}$, $\eta_1 = \xi_1$, $\eta_{n+1} = 2^n \xi_{n+1} I(\xi_n = -1)$ for $n \geq 1$ and $X_n = \eta_1 + \ldots + \eta_n$, $n \geq 1$. Then for $k > 1$ we find

$$\begin{aligned} \mathbf{E}[\eta_k|\mathcal{F}_{k-1}] &= 2^{k-1} I(\xi_{k-1} = -1)\mathbf{E}(\xi_k|\mathcal{F}_{k-1}) \\ &= (2^{k-1} - 1) I(\xi_{k-1} = -1). \end{aligned}$$

Hence

$$\textstyle\sum_{k=2}^{\infty} \mathbf{P}[\mathbf{E}(X_n|\mathcal{F}_{n-1}) \neq X_{n-1}] = \sum_{k=2}^{\infty} \mathbf{P}[\xi_{k-1} = -1] = \sum_{k=2}^{\infty} 2^{-k+1} < \infty.$$

Therefore $X = (X_n, \mathcal{F}_n, n \geq 1)$ is an eventual martingale.

Now take $m \geq 2$. Then

$$\begin{aligned} \mathbf{E}(X_{2m} - X_m|\mathcal{F}_m) &= \textstyle\sum_{k=m+1}^{2m} \mathbf{E}[\mathbf{E}(\eta_k|\mathcal{F}_{k-1})|\mathcal{F}_m] \\ &= \textstyle\sum_{k=m+2}^{2m} (2^{k-1} - 1)\mathbf{P}[\xi_{k-1} = -1] + (2^m - 1) I(\xi_m = -1) \\ &\geq \textstyle\sum_{k=m+2}^{2m} (1 - 2^{-k+1}) > (1 - 2^{-2m+1}) > \frac{1}{2}. \end{aligned}$$

Hence if $0 < \varepsilon < \frac{1}{2}$ we obtain

$$\mathbf{P}[|\mathbf{E}(X_{2m}|\mathcal{F}_m) - X_m| > \varepsilon] = 1 \quad \text{for all } m \geq 2.$$

This means that X is not a game fairer with time.

22.8. Not every martingale-like sequence admits a Riesz decomposition

Recall that the random sequence $(X_n, \mathcal{F}_n, n \geq 1)$ is said to admit the *Riesz decomposition* if $X_n = M_n + Z_n$, $n \geq 1$, where $(M_n, \mathcal{F}_n, n \geq 1)$ is a martingale and $\mathbf{E}[Z_n I_A] \to \infty$ as $n \to \infty$ for every $A \in \cup_{n=1}^{\infty} \mathcal{F}_n$. If this property holds then the sequence $(\mathbf{E}X_n)$ must converge since

$$\mathbf{E}X_n = \mathbf{E}M_n + \mathbf{E}Z_n = \mathbf{E}M_1 + \mathbf{E}[Z_n I_\Omega] \to \mathbf{E}M_1 \quad \text{as} \quad n \to \infty.$$

There are of course martingale-like sequences which admit the Riesz decomposition. However, this property does not always hold.

Consider the sequence $(\xi_n, n \geq 1)$ of i.i.d. r.v.s such that $\mathbf{P}[\xi_1 = 4] = \mathbf{P}[\xi_1 = 0] = \frac{1}{2}$. Let $X_n = \xi_1 \xi_2 \ldots \xi_n$ and $\mathcal{F}_n = \sigma\{\xi_1, \ldots, \xi_n\}$, $n \geq 1$. Since $\mathbf{E}X_n = 2^n \to \infty$, $(X_n, \mathcal{F}_n, n \geq 1)$ does not admit a Riesz decomposition. It remains for us to show that $(X_n, \mathcal{F}_n, n \geq 1)$ is a martingale-like sequence in the sense of at least one of the definitions (a)–(f) given in the introductory notes. In particular, it is easy to see that $\mathbf{E}[X_{n+1}|\mathcal{F}_n] = 2X_n$. Also we have $X_{n+1} = 0$ if $X_n = 0$. By the Borel–Cantelli lemma we conclude that $\mathbf{P}[X_n \neq 0 \text{ i.o.}] = 0$ and therefore $(X_n, \mathcal{F}_n, n \geq 1)$ is a progressive martingale.

22.9. On the validity of two inequalities for martingales

Here we shall consider two important inequalities for martingales and analyse the conditions under which they hold or fail to hold.

(i) Let $(X_n, \mathcal{F}_n, n \geq 1)$ be a martingale and $g : \mathbb{R}^1 \mapsto \mathbb{R}^1$ be a measurable function which is: (a) positive over $\mathbb{R}^+$; (b) even; and (c) convex; that is, for any $x, y \in \mathbb{R}^1$, $g(\frac{1}{2}(x+y)) \leq \frac{1}{2}g(x) + \frac{1}{2}g(y)$. Then for an arbitrary $\varepsilon > 0$,

$$(1) \qquad \mathbf{P}\left[\sup_{0 \leq k \leq n} |X_k| \geq \varepsilon\right] \leq \mathbf{E}[g(X_n)]/g(\varepsilon).$$

Note that this extension of the classical *Kolmogorov inequality* was obtained by Zolotarev (1961). Now we should like to show that the convexity of g is essential for the validity of (1).

Suppose g satisfies conditions (a) and (b) but not (c). Since g is not convex in this case, there exist a and h, $0 < h \leq a$, such that

$$(2) \qquad g(a) > \tfrac{1}{2}g(a-h) + \tfrac{1}{2}g(a+h).$$

Consider the r.v.s $X_1 = \xi_1$ and $X_2 = \xi_1 + \xi_2$ where ξ_1 and ξ_2 are independent, ξ_1 takes the values $\pm a$ with probability $\frac{1}{2}$ each, and ξ_2 takes the values $\pm h$ also with probability $\frac{1}{2}$ each. It is easy to check that $\mathbf{E}[X_2|X_1] = X_1$ a.s. Thus letting $\mathcal{F}_1 = \sigma\{\xi_1\}$, $\mathcal{F}_2 = \sigma\{\xi_1, \xi_2\}$ we find that the system $(X_k, \mathcal{F}_k, k = 1, 2)$ is a martingale. Since g is an even function, taking (2) into account we obtain

$$\begin{aligned} \mathbf{E}[g(X_2)] &= \tfrac{1}{4}[g(-a-h) + g(-a+h) + g(a-h) + g(a+h)] \\ &= \tfrac{1}{2}[g(a-h) + g(a+h)] < g(a) \end{aligned}$$

and

$$\mathbf{P}\left[\sup_{1 \leq k \leq 2} |X_k| \geq a\right] = 1 > \mathbf{E}[g(X_2)]/g(a).$$

Therefore inequality (1) does not hold for the martingale constructed above, taking $\varepsilon = a$.

(ii) Let $X = (X_n, \mathcal{F}_n, n \geq 1)$ be a martingale and $[X]_n = \sum_{j=1}^n (\Delta X_j)^2$, $\Delta X_j = X_j - X_{j-1}$, $X_0 = 0$ be its quadratic variation. Then for every $p > 1$ there are universal constants A_p and B_p (independent of X) such that

$$(3) \qquad A_p||\sqrt{[X]_n}||_p \leq ||X_n||_p \leq B_p||\sqrt{[X]_n}||_p$$

where $||X_n||_p = (\mathbf{E}(|X_n|^p))^{1/p}$.

Note that inequalities (3), called *Burkholder inequalities*, are often used in the theory of martingales (for details see Burkholder and Gundy 1970; Shiryaev 1995).

We shall now check that the condition on p, namely $p > 1$, is essential. By a simple example we can illustrate that (3) fails to hold if $p = 1$.

Let $\xi_1, \xi_2, \dots$ be independent Bernoulli r.v.s with $\mathbf{P}[\xi_i = 1] = \mathbf{P}[\xi_i = -1] = \frac{1}{2}$ and let $X_n = \sum_{j=1}^{n\wedge\tau} \xi_j$ where $\tau = \inf\{n \geq 1 : \sum_{j=1}^{n} \xi_j = 1\}$. If $\mathcal{F}_n = \sigma\{\xi_1, \dots, \xi_n\}$ then it is easy to see that the sequence $X = (X_n, \mathcal{F}_n, n \geq 1)$ is a martingale with the property

$$||X_n||_1 = \mathbf{E}|X_n| = 2\mathbf{E}[X_n^+] \to 2 \quad \text{as} \quad n \to \infty.$$

However,

$$||\sqrt{[X]_n}||_1 = \mathbf{E}(\sqrt{[X]_n}) = \mathbf{E}\left\{\left(\sum_{j=1}^{n\wedge\tau} 1\right)^{1/2}\right\} = \mathbf{E}\{\sqrt{\tau \wedge n}\} \to \infty \text{ as } n \to \infty.$$

Therefore in general inequalities (3) cannot hold for $p = 1$.

22.10. On the convergence of submartingales almost surely and in L^1-sense

Let $(X_n, \mathcal{F}_n, n \geq 1)$ be a submartingale satisfying the condition

$$\sup_{n\geq 1} \mathbf{E}|X_n| < \infty. \tag{1}$$

Then, according to the classical Doob theorem, the limit $X_\infty := \lim_{n\to\infty} X_n$ exists a.s. and $\mathbf{E}|X_\infty| < \infty$. Moreover, if $(X_n, \mathcal{F}_n, n \geq 1)$ is a uniformly integrable submartingale, then there is a r.v. X_∞ with $\mathbf{E}|X_\infty| < \infty$ such that $X_n \xrightarrow{\text{a.s.}} X_\infty$ and $X_n \xrightarrow{\mathrm{L}^1} X_\infty$ as $n \to \infty$. The proof of these and of many other close results can be found in the books by Doob (1953), Neveu (1975), Chow and Teicher (1978) and Shiryaev (1995).

Let us now consider a few examples with the aim of illustrating the importance of the conditions under which the above results hold.

(i) Let $\{\xi_n, n \geq 1\}$ be i.i.d. r.v.s with $\mathbf{P}[\xi_1 = 0] = \mathbf{P}[\xi_1 = 2] = \frac{1}{2}$. Define $X_n = \xi_1 \cdots \xi_n$ and $\mathcal{F}_n = \sigma\{\xi_1, \dots, \xi_n\}$, $n \geq 1$. Then $(X_n, \mathcal{F}_n, n \geq 1)$ is a martingale with $\mathbf{E}X_n = 1$ for all $n \geq 1$. Hence condition (1) implies that $X_n \xrightarrow{\text{a.s.}} X_\infty$ as $n \to \infty$ where X_∞ is a r.v. with $\mathbf{E}|X_\infty| < \infty$. Clearly we have $\mathbf{P}[X_n = 2^n] = 2^{-n}$, $\mathbf{P}[X_n = 0] = 1 - 2^{-n}$ and $X_\infty = 0$ a.s. However, $\mathbf{E}|X_n - X_\infty| = \mathbf{E}X_n = 1$. Therefore $X_n \overset{\mathrm{L}^1}{\not\to} X_\infty$ despite the a.s. convergence of X_n to X_∞.

(ii) Let $(\Omega, \mathcal{F}, \mathbf{P})$ be a probability space defined by $\Omega = [0, 1]$, $\mathcal{F} = \mathcal{B}_{[0,1]}$ and let $\mathbf{P}$ be the Lebesgue measure. On this space we consider the random sequence $(X_n, n \geq 1)$ where $X_n = X_n(\omega) = 2^n$ if $\omega \in [0, 2^{-n}]$ and $X_n = X_n(\omega) = 0$ if $\omega \in (2^{-n}, 1]$ and let $\mathcal{F}_n = \sigma\{X_1, \dots, X_n\}$. Then $(X_n, \mathcal{F}_n, n \geq 1)$ is a martingale with $\mathbf{E}X_n = 1$

for all $n \geq 1$. Hence by (1), $X_n \xrightarrow{\text{a.s.}} X_\infty$ as $n \to \infty$ with $X_\infty = 0$. Again, as above, $X_n \stackrel{L^1}{\not\to} 0$ as $n \to \infty$.

So, having examples (i) and (ii), we conclude that the Doob condition (1) guarantees a.s. convergence but not convergence in the L^1-sense. In both cases we have $\mathbf{E}|X_n| = 1$, $n \geq 1$, which means that the corresponding martingales are not uniformly integrable.

(iii) Let us consider this further. Recall that the martingale $X = (X_n, \mathcal{F}_n, n \geq 1)$ is said to be *regular* if there exists an integrable r.v. ξ such that $X_n = \mathbf{E}[\xi|\mathcal{F}_n]$ a.s. for each $n \geq 1$. Clearly, if the parameter n takes only a finite number of values, say $n = 1, \ldots, N$, then such a martingale is regular since $X_n = \mathbf{E}[X_N|\mathcal{F}_n]$. However, if $n \in \mathbb{N}$, the martingale need not be regular.

Note first the following result (see Shiryaev 1995): the martingale X is regular iff X is uniformly integrable. In this case $X_n = \mathbf{E}[X_\infty|\mathcal{F}_n]$ where $X_\infty =: \lim_{n\to\infty} X_n$.

Consider the sequence $(\xi_k, k \geq 1)$ of i.i.d. r.v.s each distributed $\mathcal{N}(0,1)$ and let $S_n = \xi_1 + \ldots + \xi_n$, $X_n = \exp(S_n - \frac{1}{2}n)$, $\mathcal{F}_n = \sigma\{\xi_1, \ldots, \xi_n\}$. Then we can easily check that $X = (X_n, \mathcal{F}_n, n \geq 1)$ is a martingale. Applying the SLLN to the sequence $(\xi_k, k \geq 1)$ we find that

$$X_\infty := \lim_{n\to\infty} X_n = \lim_{n\to\infty} \exp\left[n\left(\frac{1}{n}S_n - \frac{1}{2}\right)\right] = 0 \quad \text{a.s.}$$

Therefore a.s.

$$X_n \neq \mathbf{E}[X_\infty|\mathcal{F}_n] = 0.$$

Thus we have shown that the martingale X is not regular and it can be verified that it is not uniformly integrable.

22.11. A martingale may converge in probability but not almost surely

Recall that for series of independent r.v.s the two kinds of convergence, in probability and with probability 1, are equivalent (see e.g. Loève 1978, Ito 1984, Rao 1984). This result leads to the following question for the martingale $M = (M_n, \mathcal{F}_n, n \geq 1)$. If we know that M_n converges in probability as $n \to \infty$, does this imply its convergence with probability 1? (The converse, of course, is always true.)

(i) Let $(\xi_n, n \geq 1)$ be a sequence of independent r.v.s where

$$\mathbf{P}[\xi_n = \pm 1] = (2n)^{-1}, \quad \mathbf{P}[\xi_n = 0] = 1 - n^{-1}, \quad n \geq 1.$$

Consider a new sequence $(X_n, n \geq 0)$ given by $X_0 = 0$ and

$$X_n = \begin{cases} \xi_n, & \text{if } X_{n-1} = 0 \\ nX_{n-1}|\xi_n|, & \text{if } X_{n-1} \neq 0, \quad n \geq 1. \end{cases}$$

Let $\mathcal{F}_n = \sigma\{\xi_1, \ldots, \xi_n\}$. We can easily verify that the following four statements hold:

(a) $X = (X_n, \mathcal{F}_n, n \geq 1)$ is a martingale;
(b) for each $n \geq 1$, $X_n = 0$ iff $\xi_n = 0$;
(c) $\mathbf{P}[X_n = 0] = \mathbf{P}[\xi_n = 0] = 1 - n^{-1}$;
(d) $\mathbf{P}[X_n \neq 0 \text{ i.o.}] = \mathbf{P}[\xi_n \neq 0 \text{ i.o.}] = 1$.

Note that statement (d) follows from the relation $\sum_{n=1}^{\infty} \mathbf{P}[|\xi_n| = 1] = \infty$.

We are interested in the behaviour of X_n as $n \to \infty$. Obviously, (c) implies that $X_n \xrightarrow{\text{P}} 0$ as $n \to \infty$. However, (d) shows that $\mathbf{P}[\omega : X_n(\omega) \text{ converges}] = 0$. Thus the martingale X converges in probability but not with probability 1.

(ii) Let $(\xi_n, n \geq 1)$ be a sequence of i.i.d. r.v.s each taking the values ± 1 with probability $\frac{1}{2}$. Define $\mathcal{F}_n = \sigma\{\xi_1, \dots, \xi_n\}$ and let $(B_n, n \geq 1)$ be a sequence of events adapted to the family $(\mathcal{F}_n)$, that is $B_n \in \mathcal{F}_n$ for each $n \geq 1$ and such that $\lim_{n\to\infty} \mathbf{P}(B_n) = 0$ and $\mathbf{P}(\limsup_{n\to\infty} B_n) = 1$. Consider the random sequence $(X_n, n \geq 1)$ where $X_1 = 0$ and

$$X_{n+1} = X_n(1 + \xi_{n+1}) + 1_{B_n}\xi_{n+1}, \quad n \geq 1.$$

It is easy to check that $X = (X_n, \mathcal{F}_n, n \geq 1)$ is a martingale. Since

$$\mathbf{P}[X_{n+1} \neq 0] \leq \tfrac{1}{2}\mathbf{P}[X_n \neq 0] + \mathbf{P}(B_n)$$

we conclude that

$$\lim_{n\to\infty} \mathbf{P}(X_n = 0) = 1, \quad \mathbf{P}[\omega : X_n(\omega) \text{ converges}] = 0.$$

Therefore the martingale X is a.s. divergent despite the fact that it converges in probability.

(iii) The existence of martingales obeying some special properties can be proved by using the following result (see Bojdecki 1977). Let the probability space $(\Omega, \mathcal{F}, \mathbf{P})$ consist of $\Omega = [0, 1]$, $\mathcal{F} = \mathcal{B}_{[0,1]}$ and $\mathbf{P}$ the Lebesgue measure. For any sequence $(\xi_n, n \geq 1)$ of simple r.v.s (ξ_j is *simple* if it takes a finite number of values) there exists a martingale $(X_n, \mathcal{F}_n, n \geq 1)$ such that

$$\mathbf{P}[\xi_n = X_n \text{ for all sufficiently large } n] = 1.$$

Recall that there are sequences of simple r.v.s converging in probability but not a.s., and other sequences which are bounded but not converging. Thus in these particular cases we come to the following two statements.

(a) There exists a martingale $(X_n, \mathcal{F}_n, n \geq 1)$ such that

$$X_n \xrightarrow{\text{P}} 0 \text{ as } n \to \infty \quad \text{but} \quad \mathbf{P}[\omega : X_n(\omega) \text{ converges to } 0] = 0.$$

(b) There exists a martingale $(X_n, \mathcal{F}_n, n \geq 1)$ such that

$$\mathbf{P}[\omega : (X_n, n \geq 1) \text{ is bounded}] = 1 \quad \text{but} \quad \mathbf{P}[\omega : X_n(\omega) \text{ converges}] = 0.$$

22.12. Zero-mean martingales which are divergent with a given probability

(i) Let $(\xi_n, n \geq 1)$ be a sequence of i.i.d. r.v.s with $\mathbf{E}\xi_1 = 0$ and $\mathbf{E}|\xi_1| > 0$. Take another sequence $(\eta_n, n \geq 1)$ of independent r.v.s with $\mathbf{E}\eta_n = 0$, $\mathbf{E}[\eta_n^2] = n^{-2}$, $n \geq 1$, and consider the two series, $\sum_{n=1}^{\infty} \xi_n$ and $\sum_{n=1}^{\infty} \eta_n$. According to Chung and Fuchs (1951), the series $\sum_{n=1}^{\infty} \xi_n$ diverges a.s. On the other hand, the series $\sum_{n=1}^{\infty} \eta_n$ converges by the Kolmogorov three-series theorem.

Assume that $(\xi_n, n \geq 1)$ and $(\eta_n, n \geq 1)$ are independent of each other and take another r.v. X_0 which is independent of both sequences (ξ_n) and (η_n) and is such that $\mathbf{P}[X_0 = 1] = p = 1 - \mathbf{P}[X_0 = -1]$ where p is any fixed number in the interval $[0, 1]$. Define the new sequence $(X_n, n \geq 1)$ as:

$$X_n = \xi_n I(X_0 = 1) + \eta_n I(X_0 = -1).$$

Let $\mathcal{F}_n = \sigma\{X_1, \ldots, X_n\}$ and put $S_n = \sum_{k=1}^{n} X_k$. Then $(S_n, \mathcal{F}_n, n \geq 1)$ is a martingale with $\mathbf{E}S_n = 0$, $n \geq 1$. The question of obvious interest is what happens to the sequence (S_n) when $n \to \infty$. Since

$$S_n = I(X_0 = 1) \sum_{k=1}^{n} \xi_k + I(X_0 = -1) \sum_{k=1}^{n} \eta_k$$

it follows that

$$\mathbf{P}[S_n \text{ converges}] = \mathbf{P}[X_0 = -1] = 1 - p, \quad \mathbf{P}[S_n \text{ diverges}] = \mathbf{P}[X_0 = 1] = p.$$

(ii) Let $(w_t, t \geq 0)$ be a standard Wiener process on $(\Omega, \mathcal{F}, \mathbf{P})$ which is adapted to the given filtration $(\mathcal{F}_t, t \geq 0)$ where $\mathcal{F}_0 = \{\emptyset, \Omega\}$ and $\mathcal{F} = \bigvee_{t \geq 0} \mathcal{F}_t$. Let us take an event $A \in \mathcal{F}$ with $0 < \mathbf{P}(A) < 1$. Define the random sequence

$$X_n = X_n(\omega) = \begin{cases} 0, & \text{if } \omega \in A \\ w_n(\omega), & \text{if } \omega \in A^c, \quad n \geq 1. \end{cases}$$

Then $(X_n, n \geq 1)$ is a martingale with respect to the filtration $(\mathcal{F}_n, n \geq 1)$. This is a simple consequence of the martingale property of the Wiener process. Indeed, for any $n > m$ we have a.s.

$$\mathbf{E}[X_n|\mathcal{F}_m] = \mathbf{E}[w_n I(A^c)|\mathcal{F}_m] = I(A^c)\mathbf{E}[w_n|\mathcal{F}_m] = I(A^c)w_m = X_m.$$

Furthermore, it is well known (see Freedman 1971) that

$$\mathbf{P}\left[\limsup_{n\to\infty} w_n = \infty\right] = 1, \quad \mathbf{P}\left[\liminf_{n\to\infty} w_n = -\infty\right] = 1.$$

From these relations we conclude that

$$\mathbf{P}[\omega : X_n(\omega) \text{ converges as } n \to \infty] = \mathbf{P}(A)$$

where, to repeat, $\mathbf{P}(A)$ is a fixed number between 0 and 1.

22.13. More on the convergence of martingales

Here we present three examples of martingales $X = (X_n, \mathcal{F}_n, n \geq 1)$ which satisfy the condition $\sup_{n\geq 1} |X_n| < \infty$ a.s. but have quite different behaviour as $n \to \infty$. It will be shown that X may not be convergent, or convergent with a given probability (as in Example 22.12), or a.s. divergent.

(i) Let $(\xi_k, k \geq 1)$ be independent r.v.s with $\mathbf{P}[\xi_k = 2^k - 1] = 2^{-k}$ and $\mathbf{P}[\xi_k = -1] = 1 - 2^{-k}$. Defining $\tau = \inf\{k : \xi_k \neq -1\}$ we find that

$$\mathbf{P}[\tau = \infty] = \prod_{k=1}^{\infty} (1 - 2^{-k}) > 0.$$

Consider the sequence $(X_n, n \geq 1)$ and the family $(\mathcal{F}_n, n \geq 1)$ defined by

$$X_n = \sum_{k=1}^{n} (-1)^k 1_{[\tau \geq k]} \xi_k, \quad \mathcal{F}_n = \sigma\{\xi_1, \dots, \xi_n\}, \quad n \geq 1.$$

Then $(X_n, \mathcal{F}_n, n \geq 1)$ is a martingale and for $X^* = \sup_{n\geq 1} |X_n|$ we have

$$\mathbf{P}[X^* > 2^n] \leq \sum_{k=n+1}^{\infty} (1/2^k) = 2^{-n}.$$

Hence for all $n \geq 1$, we have $2^n \mathbf{P}[X^* > 2^n] \leq 1$ and $\lambda \mathbf{P}[X^* > \lambda] \leq 2$ for arbitrary $\lambda > 0$.

Thus we have shown that $X^* < \infty$ a.s. However, on the set $[\tau = \infty]$ which has positive probability, X_n alternates between 1 and 0, and hence (X_n) does not converge as $n \to \infty$.

(ii) Let $(\xi_n, n \geq 1)$ be independent r.v.s such that

$$\begin{aligned} \mathbf{P}[\xi_{2n} = 1] &= 1 - n^{-2} = 1 - \mathbf{P}[\xi_{2n} = -(n^2 - 1)], \\ \mathbf{P}[\xi_{2n-1} = -1] &= 1 - n^{-2} = 1 - \mathbf{P}[\xi_{2n-1} = n^2 - 1], \quad n \geq 1. \end{aligned}$$

Obviously $\mathbf{E}\xi_n = 0$ for all $n \geq 1$. By the Borel–Cantelli lemma

$$\mathbf{P}[\xi_{2n+1} + \xi_{2n} \neq 0 \text{ i.o.}] = 0 \quad \text{and} \quad \mathbf{P}[|\xi_n| \neq 1 \text{ i.o.}] = 0.$$

Let $S_n = \xi_1 + \dots + \xi_n$ and $\mathcal{F}_n = \sigma\{\xi_1, \dots, \xi_n\}$. Define the stopping time

$$\tau = \inf\{n \geq m : |\xi_n| \neq 1\}$$

where $m = m(p)$ is chosen so that $\mathbf{P}[\cup_{n=m}^{\infty}\{|\xi_n| \neq 1\}] < 1 - p$ for some fixed p, $0 < p < 1$. Finally, let $X_n = S_{\tau\wedge n}$. Then it is easy to check that $(X_n, \mathcal{F}_n, n \geq 1)$ is a martingale and

$$X_n = S_n I(\tau > n) + S_\tau I(\tau \leq n).$$

Let us note that $S_n I(\tau > n)$ is either 0 or $+1$ or -1, and so for each $n \geq m$

$$|X_n| \leq 1 + |S_\tau| I(\tau < \infty).$$

However, $X_n = S_n$ on the set $[\tau = \infty]$ and thus S_n diverges a.s. since its summands ξ_k alternate between 1 and -1 for all large n. Therefore

$$\mathbf{P}[X_n \text{ diverges as } n \to \infty] \geq \mathbf{P}[\tau = \infty] > p.$$

(iii) What is the limit behavior of a martingale $\{X_n, \mathcal{F}_n, n = 1, 2, \ldots\}$ whose differences are bounded, i.e. $|X_{n+1} - X_n| \leq M < \infty$ a.s.? Define two events $C = \{\lim_n X_n \text{ exists and is finite}\}$, $D = \{\overline{\lim}_n X_n = +\infty \text{ and } \underline{\lim}_n X_n = -\infty\}$. Then, as shown in Durrett (1991), we have $\mathbf{P}(C \cup D) = 1$.

Is this kind of property valid if we replace the boundedness condition above by a weaker one, e.g. by $\sup |X_n| < \infty$? To answer, consider a sequence $U_1, U_2, \ldots$ of i.i.d. uniform r.v.s on $(0, 1)$ and define a Markov chain $(X_n,\ n = 1, 2, \ldots)$ starting from position $X_1 = 0$ and evolving for $n = 1, 2, \ldots$ as follows:

$$X_{n+1} = \begin{cases} 1, & \text{if } X_n = 0, U_{n+1} \geq \frac{1}{2} \\ -1, & \text{if } X_n = 0, U_{n+1} < \frac{1}{2} \\ 0, & \text{if } X_n \neq 0, U_{n+1} > n^{-2} \\ n^2 X_n, & \text{if } X_n \neq 0, U_{n+1} < n^{-2}. \end{cases}$$

With $\mathcal{F}_n = \sigma(X_1, \ldots, X_n)$, we have

$$\mathbf{E}(X_{n+1}|\mathcal{F}_n) = \frac{1}{2}\mathbf{1}_{\{X_n=0\}} - \frac{1}{2}\mathbf{1}_{\{X_n=0\}} + n^2 X_n \frac{1}{n^2} = X_n$$

and hence $(X_n, \mathcal{F}_n, n = 1, 2, \ldots)$ is a martingale. Since $\sum_{n=1}^{\infty}(1/n^2) < \infty$, the Borel–Cantelli lemma implies that $\mathbf{P}(X_n = a \text{ i.o.}) = 1$ for $a = -1, 0$ and 1 and that $\sup |X_n| < \infty$, i.e. the martingale will eventually 'oscillate' from the initial position 0 to ± 1 and back to 0.

22.14. A uniformly integrable martingale with a nonintegrable quadratic variation

Suppose $M = (M_n, n = 0, 1, \ldots)$ is a uniformly integrable martingale. Then the series $\sum_{n \geq 1} \Delta_n M$ of the successive differences $\Delta_n M = M_n - M_{n-1}$ ($M_0 = 0$) is L^1-convergent. A natural question is if L^1-convergence also holds for all subseries $\sum_{n \geq 1} v_n \Delta_n M$ called Burkholder *martingale transforms*. Here $v_n \in \{0, 1\}$, $n \geq 1$.

Dozzi and Imkeller (1990) have shown that the integrability of the quadratic variation $S(M) := \{\sum_{n \geq 1}(\Delta_n M)^2\}^{1/2}$ implies that all series $\sum_{n \geq 1} v_n \Delta_n M$ are L^1-convergent. Moreover, if $S(M)$ is not integrable, then there is a sequence $\{v_n, n \geq 1\}$ such that $\sum_{n \geq 1} v_n \Delta_n M$ is not integrable.

Let us describe an explicit example of a uniformly integrable martingale M with a nonintegrable quadratic variation $S(M)$ and construct a nonintegrable martingale transform $\sum_{n\geq 1} v_n \Delta_n M$.

Consider the probability space $(\Omega, \mathcal{F}, \mathbf{P})$ where $\Omega = [1, \infty)$, $\mathcal{F}$ is the σ-field of the Lebesgue-measurable sets in Ω and $\mathbf{P}(\mathrm{d}\omega) = c\mathrm{e}^{-\omega}\mathrm{d}\omega$. Here $c = \mathrm{e}$ is the norming constant and $\mathbf{P}$ corresponds to a shifted exponential distribution $\mathcal{E}xp(1)$. Introduce the r.v. M_∞ and the filtration $(\mathcal{F}_k, k = 1, 2, \ldots)$ as follows:

$$M_\infty = \mathrm{e}^\omega \omega^{-2},\ \omega \in \Omega;\ \mathcal{F}_k = \sigma([1, k]) \vee \{[k, \infty)\},\ k \geq 1.$$

Since M_∞ is integrable, the conditional expectation $\mathbf{E}[M_\infty | \mathcal{F}_k]$ is well defined and is $\mathcal{F}_k$-measurable for each $k \geq 1$. Hence with $M_k := \mathbf{E}[M_\infty | \mathcal{F}_k]$ we obtain the martingale $M = (M_k, \mathcal{F}_k, k \geq 1)$. Let us derive some properties of M. For this we use the following representation of M:

$$M_k(\omega) = \mathrm{e}^\omega \omega^{-2} 1_{[1,k)}(\omega) + \mathrm{e}^k k^{-1} 1_{[k,\infty)}(\omega), \quad k \geq 1.$$

For $A \in \sigma([1, k])$ this is trivial, and

$$\int_{[k,\infty)} M_\infty \,\mathrm{d}\mathbf{P} = c \int_{[k,\infty)} \mathrm{e}^\omega \omega^{-2} \mathrm{e}^{-\omega} \,\mathrm{d}\omega = \mathrm{e}^k k^{-1} \mathbf{P}([k, \infty)) = \int_{[k,\infty)} M_k \,\mathrm{d}\mathbf{P}.$$

Similar reasoning shows that M is uniformly integrable. The next property of M is based on the variable ($[\omega]$ is the integer part of ω)

$$M^*(\omega) := \sup_{k\geq 1} |M_k(\omega)| = \mathrm{e}^{[\omega]} [\omega]^{-1}.$$

Obviously M^* is not integrable, that is $M^* \notin \mathrm{L}^1(\Omega, \mathcal{F}, \mathbf{P})$, and the Davis inequality (see e.g. Dellacherie and Meyer (1982) or Liptser and Shiryaev (1989)) implies that $S(M) \notin \mathrm{L}^1(\Omega, \mathcal{F}, \mathbf{P})$.

Thus we have described a uniformly integrable martingale whose quadratic variation is not integrable.

It now remains for us to construct a sequence $(v_k, k \geq 1)$, $v_k \in \{0, 1\}$, such that the partial sums $N_n := \sum_{k=1}^n v_k \Delta_k M$ are a.s. convergent as $n \to \infty$ but not L^1-convergent.

Since $M_0 := 0$, then $\Delta_1 M(\omega) = M_1(\omega) = \mathrm{e}$ and choosing $v_k = \frac{1}{2}(1 + (-1)^k)$, $k \geq 1$, and using the above representation of M we easily find that

$$N_{2n}(\omega) = \sum_{k=1}^n \left\{ \left(\frac{\mathrm{e}^\omega}{\omega^2} - \frac{\mathrm{e}^{2k-1}}{2k-1} \right) 1_{[2k-1,2k)}(\omega) + \left(\frac{\mathrm{e}^{2k}}{2k} - \frac{\mathrm{e}^{2k-1}}{2k-1} \right) 1_{[2k,\infty)}(\omega) \right\}.$$

This shows in particular that $N_\infty := \lim_{n\to\infty} N_n$ exists a.s. If we write explicitly $N_{2n}(\omega) 1_{[2l,2l+1)}(\omega)$ for $l \leq n$ and denote $B = \cup_{l\geq 1} [2l, 2l+1)$, then we see by a direct calculation that

$$\int_B N_\infty \,\mathrm{d}\mathbf{P} = \sum_{l=1}^\infty \sum_{k=1}^l \left(\frac{\mathrm{e}^{2k}}{2k} - \frac{\mathrm{e}^{2k-1}}{2k-1} \right) c \int_{2l}^{2l+1} \mathrm{e}^{-\omega} \,\mathrm{d}\omega = \infty.$$

Therefore the martingale transform $(N_n, n \geq 1)$ is not L^1-convergent.

It is interesting to note the case when $M_n = \sum_{k=1}^n X_k$, $n \geq 1$, with X_k independent r.v.s, $\mathbf{E}X_k = 0$, $k \geq 1$. Here uniform integrability of $(M_n, n \geq 1)$ implies integrability of the quadratic variation $S(M)$.

SECTION 23. CONTINUOUS-TIME MARTINGALES

Suppose we have given a complete probability space $(\Omega, \mathcal{F}, \mathbf{P})$ and a filtration $(\mathcal{F}_t, t \geq 0)$ which satisfies the usual conditions: $\mathcal{F}_t \subset \mathcal{F}$ for each t; if $s < t$, then $\mathcal{F}_s \subset \mathcal{F}_t$, $(\mathcal{F}_t)$ is right-continuous; each $\mathcal{F}_t$ contains all $\mathbf{P}$-null sets of $\mathcal{F}$. As usual, the notation $(X_t, \mathcal{F}_t, t \geq 0)$ means that the stochastic process $(X_t, t \geq 0)$ is adapted with respect to $(\mathcal{F}_t)$, that is for each t, X_t is $\mathcal{F}_t$-measurable.

The process $X = (X_t, \mathcal{F}_t, t \geq 0)$ with $\mathbf{E}|X_t| < \infty$ for all $t \geq 0$ is called a *martingale*, *submartingale* or *supermartingale*, if $s \leq t$ implies respectively that $\mathbf{E}[X_t|\mathcal{F}_s] = X_s$ a.s., $\mathbf{E}[X_t|\mathcal{F}_s] \geq X_s$ a.s., or $\mathbf{E}[X_t|\mathcal{F}_s] \leq X_s$ a.s.

We say that the martingale $M = (M_t, \mathcal{F}_t, t \geq 0)$ is an L^p*-martingale*, $p \geq 1$, if $\mathbf{E}[|X_t|^p] < \infty$ for all $t \geq 0$. If $p = 2$ we use the term *square integrable martingale*.

A r.v. T on Ω with values in $\mathbb{R}^+ \cup \{\infty\}$ is called a *stopping time* with respect to $(\mathcal{F}_t)$ (or that T is an $(\mathcal{F}_t)$-stopping time) if for all $t \in \mathbb{R}^+$, $[T \leq t] \in \mathcal{F}_t$.

Let $X = (X_t, \mathcal{F}_t, t \geq 0)$ be a right-continuous process. X is said to be a *local martingale* if there exists an increasing sequence $(T_n, n \geq 1)$ of $(\mathcal{F}_t)$-stopping times with $T_n \xrightarrow{\text{a.s.}} \infty$ as $n \to \infty$ such that for each n the process $(X_{t \wedge T_n}, \mathcal{F}_t, t \geq 0)$ is a uniformly integrable martingale. Further, X is called *locally square integrable* if $(X_{t \wedge T_n}, \mathcal{F}_t, t \geq 0)$ are square integrable martingales, that is if for each n, $\mathbf{E}[X^2_{t \wedge T_n}] < \infty$.

If $M = (M_t, \mathcal{F}_t, t \geq 0)$ is a square integrable martingale, then there exists a unique predictable increasing process denoted by $\langle M \rangle = (\langle M_t \rangle, \mathcal{F}_t, t \geq 0)$ and called a *quadratic variation* of M, such that $(M_t^2 - \langle M \rangle_t, \mathcal{F}_t, t \geq 0)$ is a martingale.

Suppose $X = (X_t, \mathcal{F}_t, t \geq 0)$ is a càdlàg process (that is, X is right-continuous with left-hand limits) where the filtration $(\mathcal{F}_t)$ satisfies the usual conditions, and assume for simplicity that $\mathcal{F}_{0-} = \mathcal{F}_0$, $\mathcal{F}_{\infty-} = \mathcal{F}$. The process X is said to be a *semimartingale* if it has the following decomposition:

$$X_t = X_0 + M_t + A_t, \quad t \geq 0$$

where $M = (M_t, \mathcal{F}_t, t \geq 0)$ is a local martingale with $M_0 = 0$, and $A = (A_t, \mathcal{F}_t, t \geq 0)$ is a right-continuous process, $A_0 = 0$, with paths of locally finite variation.

A few other notions will be introduced and analysed in the examples below.

A great number of papers and books devoted to the theory of martingales and its various applications have been published recently. For an intensive and complete presentation of the theory of martingales we refer the reader to books by Dellacherie and Meyer (1978, 1982), Jacod (1979), Métivier (1982), Elliott (1982), Durrett

(1984), Kopp (1984), Jacod and Shiryaev (1987), Liptser and Shiryaev (1989), Revuz and Yor (1991) and Karatzas and Shreve (1991).

For the present section we have chosen a few examples which illustrate the relationship between different but close classes of processes obeying one or another martingale-type property. In general, the examples in this section can be considered jointly with the examples in Section 22.

23.1. Martingales which are not locally square integrable

We now introduce and study close subclasses of martingale-like processes. This makes it necessary to compare these subclasses and clarify the relationships between them. In particular, the examples below show that in general a process can be a martingale without being locally square integrable. We shall suppose that the probability space $(\Omega, \mathcal{F}, \mathbf{P})$ is complete and the filtration $(\mathcal{F}_t, t \geq 0)$ satisfies the usual conditions.

(i) Let us construct a uniformly integrable martingale $X = (X_t, \mathcal{F}_t, t \geq 0)$ such that for every $(\mathcal{F}_t)$-stopping time T, T is not identically zero, we have $\mathbf{E}[X_T^2] = \infty$. Obviously such an X cannot be locally square integrable.

Let $\Omega = \mathbb{R}^+$, $\mathcal{F} = \mathcal{B}^+$ and $\mathcal{F}_t$ be the σ-field generated by $\tau \wedge t$ where τ is a r.v. distributed exponentially with parameter $1 : \mathbf{P}[\tau > x] = \mathrm{e}^{-x}$, $x \geq 0$. Moreover, $\mathcal{F}$ and $\mathcal{F}_t$ are assumed to be completed by all $\mathbf{P}$-null sets of Ω. According to Dellacherie (1970) the following two statements hold.

(a) $(\mathcal{F}_t)$ is an increasing right-continuous sequence of σ-fields without points of discontinuity.

(b) The r.v. T is a stopping time with respect to $(\mathcal{F}_t)$ iff there exists a number $u \in \mathbb{R}^+ \cup \{\infty\}$ such that $T \geq \tau$ a.s. on the set $[\tau \leq u]$ and $T = u$ a.s. on the set $[\tau > u]$.

Thus for each stopping time T with $\mathbf{P}[T = 0] < 1$ there exists $u \in \mathbb{R}^+ \cup \{0\}$ such that $\tau \wedge u = T$ a.s.

Consider now the r.v. $Z = \tau^{-1/2}\mathrm{e}^{\tau/2} I_{[0<\tau\leq 1]}$. Obviously we have

$$\mathbf{E}Z = \int_0^1 x^{-1/2}\mathrm{e}^{x/2}\mathrm{e}^{-x}\,\mathrm{d}x = \int_0^1 x^{-1/2}\mathrm{e}^{-x/2}\,\mathrm{d}x < \infty.$$

So Z is an integrable r.v. Take the process $X = (X_t, t \geq 0)$ where

$$X_t = \mathbf{E}[Z|\mathcal{F}_t].$$

Then X is a right-continuous martingale which is uniformly integrable.

The next step is to check whether X is locally square integrable. To see this, we use the following representation found by Doleans-Dade (1971):

$$X_t = Z I_{[\tau \leq t]} + \frac{\mathbf{E}[Z I_{[\tau > t]}]}{\mathbf{P}[\tau > t]} I_{[\tau > t]}.$$

Further, for every $a \in (0,1)$ we have

$$X^2_{\tau\wedge a} \geq Z^2 I_{[\tau\leq\tau\wedge a]} = \tau^{-1}\mathrm{e}^{\tau} I_{[0<\tau\leq a]} \Rightarrow \mathbf{E}[X^2_{\tau\wedge a}] \geq \int_0^a x^{-1}\mathrm{e}^{x}\mathrm{e}^{-x}\,\mathrm{d}x = \infty.$$

Now let T be a stopping time such that $\mathbf{P}[T = 0] < 1$ and $a \in (0,1)$ so that $\tau \wedge a \leq T$ a.s. Then the inequality $\mathbf{E}[X^2_T] < \infty$, which is necessary for square integrability, is not possible because this would imply that $\mathbf{E}[X^2_{\tau\wedge a}] \leq \mathbf{E}[X^2_T] < \infty$ which leads to a contradiction.

Therefore the martingale X is not locally square integrable.

(ii) Let the r.v. τ be the moment of the first jump of a homogeneous Poisson process $N = (N_t, t \geq 0)$ with parameter 1. Define the filtration $(\mathcal{F}_t, t \geq 0)$ where $\mathcal{F}_t = \sigma\{N_s, s \leq t\}$ and the process $m = (m_t, t \geq 0)$ by

$$m_t = \tau^{-1/2} I_{[\tau\leq t]} - 2\sqrt{\tau \wedge t}.$$

According to Kabanov (1974), the process m has the following representation as a Stieltjes integral:

$$m_t = \int_0^t s^{-1/2} I_{[\tau\geq s]}(\mathrm{d}N_s - \mathrm{d}s).$$

It can be derived from here that $(m_t, \mathcal{F}_t, t \geq 0)$ is a martingale. It also obeys other properties but the question to ask is whether m is locally square integrable. To answer this we again use the result of Dellacherie (1970) cited above. So, take any $(\mathcal{F}_t)$-stopping time T. Then $T \wedge \tau = c \wedge \tau$ for some constant c and for any $c > 0$ we have $\mathbf{E}[\tau^{-1} I_{[\tau\leq c]}] = \infty$. Hence $\mathbf{E}[m^2_T] = \infty$ and the martingale m is not locally square integrable.

(iii) Let ξ be a r.v. defined on $(\Omega, \mathcal{F}, \mathbf{P})$. Consider the process $M = (M_t, t \geq 0)$ and the filtration $(\mathcal{F}_t, t \geq 0)$ given by

$$M_t = \begin{cases} 0, & \text{if } 0 \leq t < 1 \\ \xi - \mathbf{E}\xi, & \text{if } t \geq 1, \end{cases} \qquad \mathcal{F}_t = \begin{cases} \{\emptyset, \Omega\}, & \text{if } 0 \leq t < 1 \\ \mathcal{F}^{\xi} = \sigma\{\xi\}, & \text{if } t \geq 1. \end{cases}$$

In addition, suppose that $\mathbf{E}|\xi| < \infty$ but $\mathbf{E}[\xi^2] = \infty$. Then it is easy to verify that $(M_t, \mathcal{F}_t, t \geq 0)$ is a martingale. Following the definition we see that this martingale, which is also a local martingale, is not locally square integrable.

23.2. Every martingale is a weak martingale but the converse is not always true

Let $M = (M_t, \mathcal{F}_t, t \geq 0)$ be a stochastic process. We say that M is a *weak martingale* if for each n there exists a right-continuous and uniformly integrable martingale $M^n = (M^n_t, \mathcal{F}_t, t \geq 0)$ such that $M_t = M^n_t$ for $0 \leq t < T_n$, where $(T_n, n \geq 1)$ is an increasing sequence of $(\mathcal{F}_t)$-stopping times with $T_n \xrightarrow{\text{a.s.}} \infty$ as

$n \to \infty$. It is convenient to say that a stopping time T *reduces* a right-continuous process $M = (M_t, \mathcal{F}_t, t \geq 0)$ if there exists a uniformly integrable martingale $H = (H_t, \mathcal{F}_t, t \geq 0)$ such that $M_t = H_t$ for $0 \leq t < T$.

It is obvious from the above definition that every martingale and every local martingale are also weak martingales. This observation leads naturally to the question of whether or not the converse statement is correct. The answer is contained in the next example.

Let $\pi = (\pi_t, t \geq 0)$ be a Poisson process with parameter $\lambda > 0$, $\pi_0 = 0$ and $(\mathcal{F}_t, t \geq 0)$ be its own generated filtration: $\mathcal{F}_t = \mathcal{F}_t^{\pi} = \sigma\{\pi_s, s \leq t\}$. Let τ be the first jump time of π so τ is an exponential r.v. with parameter λ. An easy computation shows that

$$\mathbf{E}[\tau - \lambda^{-1} | \mathcal{F}_\tau] = \begin{cases} t, & \text{if } t < \tau \\ \tau - \lambda^{-1}, & \text{if } t \geq \tau. \end{cases}$$

This relation will help us to construct the example we require. Indeed, for a suitable probability space, consider a sequence of such independent Poisson processes $\pi^n = (\pi_t^n, t \geq 0), n \geq 1$, where π^n has parameter λ_n and suppose that $\lambda_n \to 0$ as $n \to \infty$. Let τ_n be the first jump time of the process π_n. Denote by $\tilde{\mathcal{F}}_t$ the σ-field generated by the r.v.s π_s^n for all n and $s \leq t$ and including all sets of measure zero. Thus the family $(\tilde{\mathcal{F}}_t, t \geq 0)$ is right-continuous. Consider the process $M = (M_t, \tilde{\mathcal{F}}_t, t \geq 0)$ where $M_t = t$. Using the independence of the processes π^n we obtain analogously that

$$\mathbf{E}[\tau_n - \lambda_n^{-1} | \mathcal{F}_t] = \begin{cases} t, & \text{if } t < \tau_n \\ \tau_n - \lambda_n^{-1}, & \text{if } t \geq \tau_n. \end{cases}$$

This relation shows that τ_n reduces M. If we take, for instance, $\lambda_n = n^{-3}$ then the series $\sum_n \mathbf{P}[\tau_n \leq n] = \sum_n (1 - \mathrm{e}^{-n\lambda_n})$ converges and the Borel–Cantelli lemma says that $\tau_n \xrightarrow{\text{a.s.}} \infty$ as $n \to \infty$. This and a result of Kazamaki (1972a) imply that the process M is a weak martingale. However, M is not a martingale, which is seen immediately if we stop M at a fixed time u.

Therefore we have described an example of a continuous and bounded weak martingale which is not a martingale.

23.3. The local martingale property is not always preserved under change of time

Again, let $(\Omega, \mathcal{F}, \mathbf{P})$ be a complete probability space and $(\mathcal{F}_t, t \geq 0)$ a filtration satisfying the usual conditions. All martingales considered here are assumed to be $(\mathcal{F}_t)$-adapted and right-continuous.

By a *change of time* $(\tau_t, \mathcal{F}_t, t \geq 0)$ we mean a family of $(\mathcal{F}_t)$-stopping times (τ_t) such that for all $\omega \in \Omega$ the mapping $\tau_\cdot(\omega)$ is increasing and right-continuous.

If $X = (X_t, \mathcal{F}_t, t \geq 0)$ is a stochastic process, denote by $\tilde{X} = (X_{\tau_t}, \mathcal{F}_{\tau_t}, t \geq 0)$ the new process obtained from X by a change of time. So if X obeys some useful property, it is of general interest to know whether the new process $\tilde{X}$ obeys the same property. In particular, if X is a martingale or a weak martingale we want to know whether under some mild conditions the process $\tilde{X}$ is a martingale or a weak martingale respectively (see Kazamaki 1972a, b). Thus we come to the question: does a change of time preserve the local martingale property?

Let $M = (M_t, \mathcal{F}_t, t \geq 0)$, $M_0 = 0$ be a continuous martingale with

$$\mathbf{P}[\limsup_{t\to\infty} M_t = \infty] = 1.$$

In particular, we can choose M to be a standard Wiener process w. The r.v. τ_t defined by $\tau_t = \inf\{u : M_u > t\}$ is a finite $(\mathcal{F}_t)$-stopping time. Clearly, $\tau_0 = 0$ and $\tau_\infty = \infty$ a.s. It is easy to see that the change of time $(\tau_t, t \geq 0)$ satisfies the relation $M_{\tau_t} = t$ which is a consequence of the continuity of M. However, the process $\tilde{M} = (t, \mathcal{F}_{\tau_t}, t \geq 0)$ is not a local martingale.

Therefore in general the local martingale property is not invariant under a change of time. Dellacherie and Meyer (1982) give very general results on semimartingales when the semimartingale property is preserved under a change of time.

23.4. A uniformly integrable supermartingale which does not belong to class (D)

Let $X = (X_t, t \in \mathbb{R}^+)$ be a measurable process. We say that X is bounded in L^1 with respect to a given filtration $(\mathcal{F}_t, t \in \mathbb{R}^+)$ if the number

$$\|X\|_1 = \sup_\tau \mathbf{E}[|X_\tau| I_{[t<\infty]}]$$

where sup is taken aver all $(\mathcal{F}_t)$-stopping times τ, is finite. If, moreover, all the r.v.s $X_\tau I_{[\tau<\infty]}$ are uniformly integrable, X is said to belong to *class* (D). Several results characterizing this class can be found in the book by Dellacherie and Meyer (1982). In particular, it is shown there that every discrete-time uniformly integrable supermartingale belongs to class (D). This leads naturally to the question of the validity of a similar result for continuous-time supermartingales. The example below shows that in the continuous case such a result does not hold.

Let $w = (w_t, t \in \mathbb{R}^+)$ be a standard Wiener process in $\mathbb{R}^3$ starting at $t = 0$ at a point x different from the origin. Take the superharmonic function $h(y) = 1/|y|$, $y \in \mathbb{R}^3$ (this is just the so-called Newtonian potential) and consider the stochastic process $X = (X_t, t \in \mathbb{R}^+)$ where $X_t = h(w_t)$. Our purpose now is to study the properties of the process X. Since h is a superharmonic function and the process w is a martingale, we conclude that X is a positive supermartingale with respect to the filtration $(\mathcal{F}_t, t \in \mathbb{R}^+)$ with $\mathcal{F}_t = \sigma\{w_s, s \leq t\}$. Moreover, X has continuous trajectories. As the trajectories of w in $\mathbb{R}^3$ diverge to infinity as $t \to \infty$ (see Freedman 1971), $X_t \xrightarrow{\text{a.s.}} 0$ as $t \to \infty$ and we get $X_\infty = 0$. Using the explicit form of the distribution of w, we find that the expectation $\mathbf{E}[X_t]$ is a continuous function of t on $[0, \infty]$. Moreover, for every sequence (t_n) of elements of $[0, \infty]$ converging to $t \in [0, \infty]$ we have $X_{t_n} \xrightarrow{L^1} X_t$. So the mapping $t \mapsto X_t$ of $[0, \infty]$ into the space L^1 is continuous and since $[0, \infty]$ is compact, the r.v.s $X_t, t \in [0, \infty]$ are uniformly integrable (see Dellacherie and Meyer 1978).

Therefore the process X is a uniformly integrable supermartingale which is even continuous. It remains for us to check if X belongs to class (D). For this purpose we use the following result (see Johnson and Helms 1963). Let Z be a positive right-continuous supermartingale and let

$$\tau_n = \tau_n(\omega) = \inf\{t : Z_t(\omega) \geq n\}.$$

Then Z belongs to class (D) iff $\lim_{n \to \infty} \mathbf{E}[Z_{\tau_n} I_{[\tau_n < \infty]}] = 0$.

In our case the process X is continuous, $X_{\tau_n} = n$ on the set $[\tau_n < \infty]$ where $\tau_n = \inf\{t : X_t \geq n\}$ and obviously $\int_{[\tau_n < \infty]} X_{\tau_n} \, d\mathbf{P} = n\mathbf{P}[\tau_n < \infty]$. On the other hand, $\tau_n = \inf\{t : |w_t| \leq 1/n\}$ and we have

$$\mathbf{P}[\tau_n < \infty] = \begin{cases} 1, & \text{if } |x| \leq 1/n \\ 1/(n|x|), & \text{if } |x| \geq 1/n. \end{cases}$$

Hence $n\mathbf{P}[\tau_n < \infty] = 1/|x|$ for sufficiently large n, $n\mathbf{P}[\tau_n < \infty]$ does not tend to 0 as $n \to \infty$ and according to the result of Johnson and Helms (1963) quoted above, the process X does not belong to class (D).

23.5. L^p-bounded local martingale which is not a true martingale

Recall that the process $M = (M_t, \mathcal{F}_t, t \geq 0)$ is called an L^p-*martingale*, $p \geq 1$, iff it is a martingale and $M_t \in L^p$ for each $t \geq 0$. If $\sup_t \mathbf{E}[|M_t|^p] < \infty$ we say that M is L^p-*bounded*. For simplicity, let $M_0 = 0$. For $p \in [0, \infty)$, M is called a *local* L^p-*martingale* if there is a sequence $\{\tau_n, n \geq 1\}$ of $(\mathcal{F}_t)$-stopping times such that $t_n \uparrow \infty$ as $n \to \infty$ and for each n the process $M^n = (M_{t \wedge \tau_n}, \mathcal{F}_t, t \geq 0)$ is an L^p-martingale.

In Example 23.1 we established that there are martingales and local martingales which are not locally square integrable. Similarly, we shall show below that an L^p-bounded local martingale need not be a true martingale.

(i) Let $w = (w_t, t \geq 0)$ be a standard Wiener process in $\mathbb{R}^3$. Let $h : \mathbb{R}^3 \setminus \{0\} \mapsto \mathbb{R}^1$ be a function defined by $h(x) = |x|^{-1}$ for $x \in \mathbb{R}^3 \setminus \{0\}$ and let $\tau_n = \inf\{t > 0 : |w_t| \leq n^{-1}\}$. Then $\{\tau_n, n \geq 1\}$ is an increasing sequence of $(\mathcal{F}_t)$-stopping times, $\mathcal{F}_t = \mathcal{F}_t^w$, with $\tau_n \to \infty$ a.s. as $n \to \infty$. The function h is harmonic in the domain $\mathbb{R}^3 \setminus \{0\}$ which obviously contains the domain $D_n = \{x : |x| > n^{-1}\}$ for each $n \geq 1$. Define a function g_n on the closure $\overline{D}_n$ of D_n by $g_n(x) = \mathbf{E}_x[h(w_{\tau_n})]$, $x \in \overline{D}_n$ where $\mathbf{E}_x$ denotes the expectation given $w_0 = x$ a.s. Since w is a strong Markov process (see Dynkin 1965; Freedman 1971; Wentzell 1981) with spherical symmetry, g_n possesses the mean-value property that its average value over the surface of any sufficiently small ball about $x \in D_n$ equals its value at x (see Dynkin 1965). This implies that g_n is a harmonic function in D_n and it can be shown that g_n is continuous in $\overline{D}_n$ with boundary values equal to those of the function h. By the maximum principle for harmonic functions we conclude that $g_n = h$ in $\overline{D}_n$ for all n. Moreover, for $n \geq 1$, $x \in D_n$ and each fixed t we have the following relation:

$$\mathbf{E}_x[h(w_{\tau_n})|\mathcal{F}_t] = 1_{[\tau_n \leq t]} h(w_{t\wedge\tau_n}) + 1_{[\tau_n > t]} \mathbf{E}_x[h(w_{\tau_n})|\mathcal{F}_t] \text{ a.s.}$$

The strong Markov property of w gives

$$\mathbf{E}_x[h(w_{\tau_n})|\mathcal{F}_t] = \mathbf{E}_{w_t}[h(w_{\tau_n})] = g_n(w_t) \text{ a.s.}$$

on the set $[\tau_n > t]$. So if we combine these two relations and take into account that $g_n = h$ in $\overline{D}_n$ we have the equality

$$\mathbf{E}_x[h(w_{\tau_n})|\mathcal{F}_t] = h(w_{t\wedge\tau_n}) \text{ a.s.}$$

Recall now that the initial state of the Wiener process is $w_0 \neq (0,0,0)$ and let $w_0 = x_0$. Then for all sufficiently large n we have $x_0 \in D_n$. Thus we conclude that the process $(h(w_{t\wedge\tau_n}), t \geq 0)$ is a bounded martingale. This implies that $(h(w_t), t \geq 0)$ is a local martingale. So it remains for us to clarify whether this local martingale is a true martingale. We have

$$\mathbf{E}_{x_0}[h(w_0)] = x_0$$

and we want to find $\mathbf{E}_{x_0}[h(w_t)]$. If $t > 0$ and $c > 2|x_0|$ then

$$\begin{aligned}
\mathbf{E}_{x_0}[h(w_t)] &= (2\pi t)^{-3/2} \int_{\mathbb{R}^3} |y|^{-1} \exp(-|y - x_0|^2/(2t))\, dy \\
&\leq (2\pi t)^{-3/2} \left\{ \int_{|y|\leq c} |y|^{-1}\, dy + \int_{|y|>c} |y|^{-1} \exp(-|y|^2/(8t))\, dy \right\} \\
&\leq (2\pi t)^{-3/2} c_1 c^2 + c_2 c^{-1}.
\end{aligned}$$

Here $c_1 > 0$ and $c_2 > 0$ are constants not depending on c and t. Obviously, if $t \to \infty$, $c \to \infty$ and $ct^{-3/4} \to 0$ then $\mathbf{E}_{x_0}[h(w_t)] \to 0$. Hence for all sufficiently large t we obtain

$$\mathbf{E}_{x_0}[h(w_t)] \neq \mathbf{E}_{x_0}[h(w_0)].$$

This relation means that the process $(h(w_t), t \geq 0)$ is not a true martingale despite the fact that it is a local martingale.

A calculation similar to the one above shows that $h(w_t) \in \mathrm{L}^2$ for each t and also $\sup_t \mathbf{E}[h^2(w_t)] < \infty$. Therefore $(h(w_t), t \geq 0)$ is an L^2-bounded local martingale although, let us repeat, it is not a true martingale. It would be useful for the reader to compare this case with Example 23.4.

(ii) Let us briefly consider another interesting example. Let $X = (X_t, t \geq 0)$ be a *Bessel process* of order l, $l \geq 2$. Recall that X is a continuous Markov process whose infinitesimal operator on the space of twice differentiable functions has the form

$$\frac{1}{2}\frac{\mathrm{d}^2}{\mathrm{d}x^2} + \frac{l-1}{2x}\frac{\mathrm{d}}{\mathrm{d}x}.$$

Note that if l is integer, X is identical in law with the process $(|w(t)|, t \geq 0)$ where $|w(t)| = (w_1^2(t) + \cdots + w_l^2(t))^{1/2}$ and $((w_1(t), \ldots, w_l(t)), t \geq 0)$ is a standard Wiener process in $\mathbb{R}^l$ (see Dynkin 1965; Rogers and Williams 1994).

Suppose X starts from a point $x > 0$, that is $X_0 = x$ a.s., and consider the process $M = (M_t, t \geq 0)$ where

$$M_t = 1/X_t^{l-2}.$$

If $\mathcal{F}_t = \sigma\{X_s, s \leq t\}$ then it can be shown that $(M_t, \mathcal{F}_t, t \geq 0)$ is a local continuous martingale which, however, is not a martingale because $\mathbf{E}M_t$ vanishes when $t \to \infty$ (compare with case (i)). On the other hand, $\mathbf{E}[M_t^p] < \infty$ for any p such that $p < l/(l-2)$. Thus, if l is close to 2, p is 'big enough' and we have a continuous local martingale which is 'sufficiently' integrable in the sense that M belongs to the space L^p for 'sufficiently'large p; despite this fact, the process M is not a true martingale.

23.6. A sufficient but not necessary condition for a process to be a local martingale

We shall start by considering the following. Let $X = (X_t, \mathcal{F}_t, t \geq 0)$ be a càdlàg process with $X_0 = 0$ and $A = (A_t, \mathcal{F}_t, t \geq 0)$ be a continuous increasing process such that $A_0 = 0$. Assume that for $\lambda \in \mathbb{R}^1$ the process $Z^\lambda = (Z_t^\lambda, \mathcal{F}_t, t \geq 0)$ defined by

$$Z_t^\lambda = \exp(\lambda X_t - \tfrac{1}{2}\lambda^2 A_t)$$

is a local martingale. Then X is a continuous local martingale and $A = \langle X \rangle$. Here $A = \langle X \rangle$ is the unique predictable process of finite variation such that $X^2 - \langle X \rangle$ is a martingale. (For details see Dellacherie and Meyer (1982) or Métivier (1982).)

This result is due to M. Yor and is presented here in a form suggested by C. Stricker. It can also be found in a paper by Meyer and Zheng (1984).

Now we shall show that the continuity of A and the condition $A_0 = 0$ are essential for the validity of this result.

Let $X^n = (X_t^n, \mathcal{F}_t, t \geq 0)$ be a sequence of centred Gaussian martingales such that X^n has the following increasing process $A^n = (A_t^n, \mathcal{F}_t, t \geq 0)$:

$$A_t^n = \begin{cases} 0, & \text{if } t \leq c \\ 1, & \text{if } t \geq c + 1/n \\ \text{linear in between}, & c = \text{constant}. \end{cases}$$

We now consider the limiting case as $n \to \infty$. Referring the reader to a paper by Meyer and Zheng (1984) for details, we get $A_t^n \to A_t$ and $X_t^n \to X_t$ weakly in the space $\mathbb{D}$ where $A_t = Z_{[t\geq c]}$ and $X_t = \xi 1_{[t\geq c]}$ with ξ a r.v. distributed $\mathcal{N}(0,1)$. It is not difficult to check that for each λ the process $Z^\lambda = (Z_t^\lambda, \mathcal{F}_t, t \geq 0)$, where $Z_t^\lambda = \exp(\lambda X_t - \frac{1}{2}\lambda^2 A_t)$, is a martingale. However, neither A nor X is continuous. Moreover, if $c = 0$, the property $X_0 = 0$ a.s. no longer holds.

23.7. A square integrable martingale with a non-random characteristic need not be a process with independent increments

Let $V = (X_t, \mathcal{F}_t, t \geq 0)$ be a square integrable martingale defined on the complete probability space $(\Omega, \mathcal{F}, \mathbf{P})$ where the filtration $(\mathcal{F}_t, t \geq 0)$ satisfies the usual conditions. The well known Lévy theorem asserts that if X is continuous and its characteristic $\langle X\rangle$ is deterministic, then X is a Gaussian process with independent increments (see Grigelionis 1977; Jacod 1979).

Our purpose now is to answer the following question. Is it true that any square integrable martingale X with a non-random characteristic $\langle X\rangle$ is a process with independent increments?

Let $\Omega = [0,1]$, $\mathbf{P}$ be the Lebesgue measure and the σ-field $\mathcal{F}$ be generated by the following three r.v.s η_0, η_1 and η_2, where

$$\eta_0 = 0 \quad \text{for all } \omega \in \Omega,$$

$$\eta_1 = \begin{cases} -1, & \text{if } \omega \in [0, \frac{1}{2}) \\ 1, & \text{if } \omega \in [\frac{1}{2}, 1], \end{cases}$$

$$\eta_2 = \begin{cases} -2, & \text{if } \omega \in [0, \frac{1}{4}) \\ 0, & \text{if } \omega \in [\frac{1}{4}, \frac{1}{2}) \\ 1 - \sqrt{3/2}, & \text{if } \omega \in [\frac{1}{2}, \frac{2}{3}) \\ 1, & \text{if } \omega \in [\frac{2}{3}, \frac{5}{6}) \\ 1 + \sqrt{3/2}, & \text{if } \omega \in [\frac{5}{6}, 1]. \end{cases}$$

Denote by $\tilde{\mathcal{F}}_i$ the σ-field generated by the r.v. η_i, $i = 0, 1, 2$, and fix the points $s_0 = 0$, $s_1 = 1$, $s_2 = 2$, $s_3 = \infty$. Consider the stochastic process $X = (X_t, t \geq 0)$ defined by

$$X_t = \sum_{k=0}^{2} \eta_k I_{[s_k, s_{k+1})}(t), \quad t \geq 0$$

and introduce the family $(\mathcal{F}_t, t \geq 0)$ of increasing and right-continuous sub-σ-fields of $\mathcal{F}$ where $\mathcal{F}_t = \tilde{\mathcal{F}}_k$ for $t \in [s_k, s_{k+1})$, $k = 0, 1, 2$. It is easy to check that $X = (X_t, \mathcal{F}_t, t \geq 0)$ is a martingale (and is bounded). Moreover, its characteristic $\langle X\rangle$ can be found explicitly, namely:

$$\langle X\rangle_t = \begin{cases} 0, & \text{if } 0 \leq t < 1 \\ 1, & \text{if } 1 \leq t < 2 \\ 2, & \text{if } t \geq 2. \end{cases}$$

Obviously the characteristic $\langle X\rangle$ is non-random. Further, the relations

$$\mathbf{P}[X_1 - X_0 = 1, X_2 - X_1 = 1] = 0 \neq \tfrac{1}{8} = \mathbf{P}[X_1 - X_0 = 1]\mathbf{P}[X_2 - X_1 = 1]$$

imply that the increments of the process X are not independent.

Therefore we have constructed a square integrable martingale whose characteristic $\langle X\rangle$ is non-random, but this does not imply that X has independent increments. It may be noted that here the process X varies only by jumps while in the Lévy theorem X is supposed to be continuous. Thus the continuity condition is essential for this result.

A correct generalization of the Lévy theorem to arbitrary square integrable martingales (not necessarily continuous) was given by Grigelionis (1977). (See also the books of Liptser and Shiryaev 1989 or Jacod and Shiryaev 1987.)

23.8. The time-reversal of a semimartingale can fail to be a semimartingale

Let $w = (w_t, t \geq 0)$ be a standard Wiener process in $\mathbb{R}^1$. Take some measurable function h which maps the space $\mathbb{C}[0,1]$ one–one to the interval [0,1]. Define the r.v. $\tau = \tau(\omega) = h(\{w_s(\omega), 0 \leq s \leq 1\})$ and the process $X = (X_t, t \geq 0)$ where

$$X_t = \begin{cases} w_t, & \text{if } 0 \leq t \leq 1 \\ w_1, & \text{if } 1 < t \leq 1 + \tau \\ w_{t-\tau}, & \text{if } t \geq 1 + \tau. \end{cases}$$

Thus X is a Wiener process with a flat spot of length $\tau \leq 1$ interpolated from $t = 1$ to $t = 1 + \tau$. Since τ is measurable with respect to the σ-field $\sigma\{X_s, s \leq t\}$, it is easy to see that X is a martingale (and hence a semimartingale).

Now we shall reverse the process X from the time $t = 2$. Let

$$\tilde{X}_t = X_{2-t} \text{ for } 0 \leq t \leq 2.$$

Denote by $(\tilde{\mathcal{F}}_t)$ the natural filtration of X. Note that the variable τ is $\tilde{\mathcal{F}}_1$-measurable, hence so is $\{X_t, 0 \leq t \leq 1\}$, since it is the time-reversal $h^{-1}(\tau)$. Thus $\tilde{\mathcal{F}}_t = \tilde{\mathcal{F}}_1$ for $1 < t < 2$. This means that any martingale with respect to the filtration $(\tilde{\mathcal{F}}_t)$ will be constant on the interval $(1,2)$ and any semimartingale will have a finite variation there. However, the Wiener process w has an infinite variation on each interval and therefore $\tilde{X}$ has an infinite variation on the interval $(1,2)$. Hence $\tilde{X}$, which was defined as the time-reversal of X, is not a semimartingale relative to its own generated filtration $(\tilde{\mathcal{F}}_t)$. According to a result by Stricker (1977), the process $\tilde{X}$ cannot be a semimartingale with respect to any other filtration.

23.9. Functions of semimartingales which are not semimartingales

Let $X = (X_t, \mathcal{F}_t, t \geq 0)$ be a semimartingale on the complete probability space $(\Omega, \mathcal{F}, \mathbf{P})$ and the family of σ-fields $(\mathcal{F}_t, t \geq 0)$ satisfies the usual conditions. The following result is well known and often used. If $f(x), x \in \mathbb{R}^1$, is a function of the space $\mathbb{C}^2(\mathbb{R}^1)$ or f is a difference of two convex functions, then the process $Y = (Y_t, \mathcal{F}_t, t \geq 0)$ where $Y_t = f(X_t)$ is again a semimartingale (see Dellacherie and Meyer 1982).

In general, it is not surprising that for some 'bad' functions f the process $Y = f(X)$ fails to be a semimartingale. However, it would be useful to have at least one particular example of this kind.

Take the function $f(x) = |x|^\alpha, 1 < \alpha < 2$. Consider the process $Y = f(X)$, that is $Y = |X|^\alpha$ and try to clarify whether Y is a semimartingale. In order to do this we need the following result (see Yor 1978): if X is a continuous local martingale, $X_0 = 0$ a.s., then statements (a) and (b) below are equivalent:

(a) $X \equiv 0$; (b) the local time of X at 0 is $L^0 = 0$.

Let us suppose that the process $Y = |X|^\alpha$ is a semimartingale. Then applying the Itô formula (see Dellacherie and Meyer 1982; Elliott 1982; Métivier 1982; Chung and Williams 1990) for $\beta = 1/\alpha > 1$ we obtain

$$|X_t| = Y_t^\beta = \beta \int_0^t Y_s^{\beta-1} \, \mathrm{d}Y_s + \tfrac{1}{2}\beta(\beta - 1) \int_0^t Y_s^{\beta-2} \, \mathrm{d}\langle Y^{\mathrm{c}}, Y^{\mathrm{c}} \rangle_s$$

and

$$L_t^0 = \int_0^t 1_{[X_s=0]} \, \mathrm{d}|X_s| = \int_0^t 1_{[Y_s=0]} \, \mathrm{d}(Y^\beta)_s = 0, \quad t \geq 0.$$

Thus by the above result we can conclude that $X \equiv 0$. This contradiction shows that the process $Y = |X|^\alpha$ is not a semimartingale.

The following particular case of this example is fairly well known. If w is the standard Wiener process, then $|w|^\alpha, 0 < \alpha < \frac{1}{2}$, is not a semimartingale (see Protter 1990). Other useful facts concerning the semimartingale properties of functions of semimartingales can be found in the books by Yor (1978), Liptser and Shiryaev (1989), Protter (1990), Revuz and Yor (1991), Karatzas and Shreve (1991) and Yor (1992, 1996).

23.10. Gaussian processes which are not semimartingales

One of the 'best' representatives of Gaussian processes is the Wiener process which is also a martingale, and hence a semimartingale. Since any Gaussian process is square integrable, it seems natural to ask the following questions. What is the relationship between the Gaussian and semimartingale properties of a stochastic process? In particular, is any Gaussian process a semimartingale?

Our aim now is to construct a family $\{X^{(\alpha)}\}$ of Gaussian processes depending on a parameter α such that for some α, $X^{(\alpha)}$ is a semimartingale, while for other α, it is not. Indeed, consider the function

$$K^{(\alpha)}(s,t) = \tfrac{1}{2}(s^\alpha + t^\alpha - |s - t|^\alpha), \quad s,t \in \mathbb{R}^+, \quad \alpha \in [1,2].$$

It can be shown that for each $\alpha \in [1, 2]$ the function $K^{(\alpha)}$ is positive definite. This implies (see Doob 1953; Ash and Gardner 1975) that for each $\alpha \in [1, 2]$ there exists a Gaussian process, say $X^{(\alpha)} = (X_t^{(\alpha)}, t \in \mathbb{R}^+)$, such that $\mathbf{E}X_t^{(\alpha)} = 0$ and its covariance function is $\mathbf{E}[X_s^{(\alpha)} X_t^{(\alpha)}] = K^{(\alpha)}(s, t)$, $s, t \in \mathbb{R}^+$.

The next step is to verify whether or not the process $X^{(\alpha)}$ is a semimartingale (with respect to its natural filtration). It is easy to see that for $\alpha = 1$ we have $K^{(1)}(s, t) = \min\{s, t\}$. This fact and the continuity of any of the processes $X^{(\alpha)}$ imply that $X^{(1)}$ is the standard Wiener process. Further, if $\alpha = 2$ we obtain that $X_t^{(2)} = t\xi$ where ξ is a r.v. distributed $\mathcal{N}(0, 1)$. Therefore in these two particular cases, $\alpha = 1$ and $\alpha = 2$, the corresponding Gaussian processes $X^{(1)}$ and $X^{(2)}$ are semimartingales. To determine what happens if $1 < \alpha < 2$ we need the following result of A. Butov (see Liptser and Shiryaev 1989). Suppose $X = (X_t, t \geq 0)$ is a Gaussian process with zero mean and covariance function $\Gamma(s, t)$, $s, t \geq 0$ and conditions (a) and (b) below are satisfied.

(a) There does not exist a non-negative and non-decreasing function F of bounded variation such that $(\Gamma(t, t) + \Gamma(s, s) - 2\Gamma(s, t))^{1/2} \leq F(t) - F(s)$, $s < t$.

(b) For any interval $[0, T] \subset \mathbb{R}^+$ and any partition $0 = t_0 < t_1 < \ldots < t_n = T$ with $\max_k (t_{k+1} - t_k) \to 0$ we have $\sum_{k=0}^{n-1} (X_{t_{k+1}} - X_{t_k})^2 \xrightarrow{\mathrm{P}} 0$ as $n \to \infty$.

Then the process X is not a semimartingale.

Now let us check conditions (a) and (b) for the process $X^{(\alpha)}$. We have

$$(K^{(\alpha)}(t, t) + K^{(\alpha)}(s, s) - 2K^{(\alpha)}(s, t))^{1/2} = |t - s|^{\alpha/2}.$$

However, the function $|t - s|^{\alpha/2}$ with $1 < \alpha < 2$ is not representable in the form $F(t) - F(s)$ for some non-negative and non-decreasing F of bounded variation. So condition (a) is satisfied. Furthermore, for $t > s$ we can easily calculate that $\mathbf{E}[(X_t^{(\alpha)} - X_s^{(\alpha)})^2] = |t - s|^\alpha$. It follows that

$$\sum_{k=0}^{n-1} \mathbf{E}[(X_{t_{k+1}}^{(\alpha)} - X_{t_k}^{(\alpha)})^2] \leq T \max_{0 \leq k \leq n-1} (t_{k+1} - t_k)^{\alpha - 1} \to 0 \text{ as } n \to \infty$$

which implies the validity of condition (b).

Thus the Gaussian process $X^{(\alpha)}$ is not a semimartingale if $1 < \alpha < 2$.

Therefore we have constructed the family $\{X^{(\alpha)}, 1 \leq \alpha \leq 2\}$ of Gaussian (indeed, continuous) processes such that some members of this family, those for $\alpha = 1$ and $\alpha = 2$, are semimartingales, while others, when $1 < \alpha < 2$, are not semimartingales.

Consider another interpretation of the above case. Recall that a *fractional standard Brownian motion* $B_H = (B_H(t), t \in \mathbb{R}^1)$ with scaling parameter H, $0 < H < 1$, is a Gaussian process with zero mean, $B_H(0) = 0$ a.s. and covariance function

$$r(s, t) = \mathbf{E}[B_H(s) B_H(t)] = \tfrac{1}{2}[|s|^{2H} + |t|^{2H} - |s - t|^{2H}], \quad s, t \in \mathbb{R}^1$$

(compare $r(s,t)$ with $K^{(\alpha)}(s,t)$ above) (see Mandelbrot and Van Ness 1968).

Hence for any H, $\frac{1}{2} < H < 1$, the fractional Brownian motion B_H is not a semimartingale.

A very interesting general problem (posed as far as we know by A. N. Shiryaev) is to characterize the class of Gaussian processes which are also semimartingales. Useful results on this topic can be found in papers by Emery (1982), Jain and Monrad (1982), Stricker (1983), Enchev (1984, 1988) and Galchuk (1985) and the book by Liptser and Shiryaev (1989).

23.11. On the possibility of representing a martingale as a stochastic integral with respect to another martingale

(i) Let the process $X = (X_t, t \in [0,T])$ be a martingale relative to its own generated filtration $(\mathcal{F}_t^X, t \in [0,T])$. Suppose $M = (M_t, t \in [0,T])$ is another process which is a martingale with respect to $(\mathcal{F}_t^X)$. The question is whether M can be represented as a stochastic integral with respect to X, that is whether there exists a 'suitable' function $\varphi_s, s \in [0,T])$ such that $M_t = \int_0^t \varphi_s \, \mathrm{d}X_s$. One reason for asking this question is that there is an important case when the answer is positive, e.g. when X is a standard Wiener process (see Clark 1970; Liptser and Shiryaev 1977/78; Dudley 1977).

The following example shows that in some cases the answer to the above question is negative.

Consider two independent Wiener processes, say $w = (w_t, t \geq 0)$ and $v = (v_t, t \geq 0)$. Let $X_t = \int_0^t W_s \, \mathrm{d}v_s$ and $\mathcal{F}_t^X = \sigma\{X_s, s \leq t\}$. Then $\langle X \rangle_t$ is $\mathcal{F}_t^X$-measurable and since $\langle X \rangle_t = \int_0^t w_s^2 \, \mathrm{d}s$ it follows that w_t^2 is $\mathcal{F}_t^X$-measurable. Hence the process $M = (M_t, t \geq 0)$ where $M_t = w_t^2 - t$ is an L^2-martingale with respect to the filtration $(\mathcal{F}_t^X, t \geq 0)$. Suppose now that the martingale M can be represented as a stochastic integral with respect to X: that is, for some predictable function $(H_s(\omega), s \geq 0)$ with $\mathbf{E}[\int_0^\infty H_s^2 \, \mathrm{d}\langle X \rangle_s] < \infty$ we have $M_t = \int_0^t H_s \, \mathrm{d}X_s$, $t \geq 0$. Since by the Ito formula we have $M_t = 2\int_0^t w_s \, \mathrm{d}w_s$, it follows that

$$M_t = 2\int_0^t w_s \, \mathrm{d}w_s = \int_0^t H_s \, \mathrm{d}X_s = \int_0^t H_s w_s \, \mathrm{d}v_s.$$

These relations imply that

$$\begin{aligned} 0 &= \mathbf{E}\left\{\left[2\int_0^t w_s \, \mathrm{d}w_s - \int_0^t H_s w_s \, \mathrm{d}v_s\right]^2\right\} \\ &= 4\mathbf{E}\left[\int_0^t w_s^2 \, \mathrm{d}s\right] + \mathbf{E}\left[\int_0^t H_s^2 w_s^2 \, \mathrm{d}s\right] \end{aligned}$$

which of course is not possible.

Therefore the martingale $M = (M_t, \mathcal{F}_t^X, t \geq 0)$ cannot be represented as a stochastic integral with respect to the martingale X.

(ii) Let X be a r.v. which is measurable with respect to the σ-field $\mathcal{F}_1^w$ generated by the Wiener process w in the interval [0,1]. Clearly in this case X is a functional of the Wiener process and it is natural to expect that X has some representation through w. The following useful result can be found in the book by Liptser and Shiryaev (1977/78). Let the r.v. X be square integrable, that is $\mathbf{E}[X^2] < \infty$. Suppose additionally that the r.v. X and the Wiener process $w = (w(t), t \in [0,1])$ form a Gaussian system. Then there exists a deterministic measurable function $g(t)$, $t \in [0,1]$, with $\int_0^1 g^2(t)\, dt < \infty$ such that

$$X = \mathbf{E}X + \int_0^1 g(t) dw(t). \tag{1}$$

We now want to show that the conditions ensuring the validity of this result cannot be weakened. In particular, the condition that the pair (X, w) is a Gaussian system, cannot be removed. Indeed, consider the process

$$X_t = \int_0^t h(w(s))\, dw(s), \quad t \in [0,1]$$

where $h(x) = 1$ if $x \geq 0$ and $h(x) = -1$ if $x < 0$. It is easy to check that $(X_t, t \in [0,1])$ is a Wiener process. Therefore the r.v. $X = X_1$ is a Gaussian and $\mathcal{F}_1^w$-measurable r.v. However, X cannot be represented in the form given by (1) with a deterministic function g.

SECTION 24. POISSON PROCESS AND WIENER PROCESS

The Poisson process and the Wiener process play important roles in the theory of stochastic processes, similar to the roles of the Poisson and the normal distributions in the theory of probability. In previous sections we considered the Poisson and the Wiener processes in order to illustrate some basic properties of stochastic processes. Here we shall analyse other properties of these two processes, but for convenience let us give the corresponding definitions again.

We say that $w = (w_t, t \geq 0)$ is a standard *Wiener process* if: (i) $w_0 = 0$ a.s.; (ii) any increment $w_t - w_s$ where $s < t$ is distributed normally, $\mathcal{N}(0, t-s)$; (iii) for each $n \geq 3$ and any $0 \leq t_1 < t_2 < \ldots < t_n$ the increments $w_{t_2} - w_{t_1}, w_{t_3} - w_{t_2}, \ldots, w_{t_n} - w_{t_{n-1}}$ are independent.

The process $N = (N_t, t \geq 0)$ is said to be a (homogeneous) *Poisson process* with parameter λ, $\lambda > 0$, if: (i) $N_0 = 0$ a.s.; (ii) any increment $N_t - N_s$ where $s < t$ has a Poisson distribution with parameter $\lambda(t-s)$; (iii) for each $n \geq 3$ and any $0 \leq t_1 < t_2 < \ldots < t_n$ the increments $N_{t_2} - N_{t_1}, N_{t_3} - N_{t_2}, \ldots, N_{t_n} - N_{t_{n-1}}$ are independent.

Note that the processes w and N can also be defined in different but equivalent ways. In particular we can consider the non-standard Wiener process, the Wiener process with drift, the non-homogeneous Poisson process, etc. Another possibility is to give the martingale characterization of each of these processes. The reader can find numerous important and interesting results concerning the Wiener and Poisson processes in the books by Freedman (1971), Yeh (1973), Cinlar (1975), Liptser and Shiryaev (1977/78), Wentzell (1981), Chung (1982), Durrett (1984), Kopp (1984), Protter (1990), Karatzas and Shreve (1991), Revuz and Yor (1991), Yor (1992, 1996) and Rogers and Williams (1994).

24.1. On some elementary properties of the Poisson process and the Wiener process

(i) Take the standard Wiener process w and the Poisson process N with parameter 1 and let $\tilde{N} = (\tilde{N}_t, t \geq 0)$ where $\tilde{N}_t = N_t - t$ is the so-called centred Poisson process. It is easy to calculate their covariance functions:

$$C_w(s,t) = \min\{s,t\}, \quad C_{\tilde{N}}(s,t) = \min\{s,t\}, \quad s,t \geq 0.$$

Therefore these two quite different processes have the same covariance functions.

Further, if we denote $\mathcal{F}_t^w = \sigma\{w_s, s \leq t\}$ and $\mathcal{F}_t^N = \sigma\{N_s, s \leq t\}$ then each of the processes $(w_t, \mathcal{F}_t^w, t \geq 0)$ and $(\tilde{N}_t, \mathcal{F}_t^N, t \geq 0)$ is a square integrable martingale. Recall that for every square integrable martingale $M = (M_t, \mathcal{F}_t, t \geq 0)$ we can find a unique process denoted by $\langle M \rangle = (\langle M \rangle_t, t \geq 0)$ and called a quadratic variation process, such that $M^2 - \langle M \rangle$ is a martingale with respect to $(\mathcal{F}_t)$ (see Dellacherie and Meyer 1982; Elliott 1982; Métivier 1982; Liptser and Shiryaev 1977/78, 1989). In our case we easily see that

$$\langle w \rangle_t = t \quad \text{and} \quad \langle \tilde{N} \rangle_t = t.$$

Again, two very different square integrable martingales have the same quadratic processes. Obviously, in both cases $\langle w \rangle$ and $\langle \tilde{N} \rangle$ are deterministic functions (indeed, continuous), the processes w and N have independent increments, w is a.s. continuous, while almost all trajectories of N are discontinuous (increasing stepwise functions, left- or right-continuous, with unit jumps only).

Therefore, neither the covariance function nor the quadratic variation characterize the processes w and N uniquely.

(ii) The above reasoning can be extended. Take the function

$$C(s,t) = \mathrm{e}^{-2\lambda|s-t|}, \quad s,t \geq 0,\ \lambda = \text{constant} > 0.$$

It can be checked that $C(s,t)$ is positive definite and hence there exists a Gaussian stationary process with zero-mean function and covariance function equal to C. We shall now construct two stationary processes, say X and Y, each with a covariance function C; moreover, X will be defined by the Wiener process w and Y by the Poisson process N with parameter λ.

Consider the process $X = (X_t, t \geq 0)$ where

$$X_t = \mathrm{e}^{-\beta t} w(\alpha \mathrm{e}^{2\beta t}).$$

Here $\alpha > 0$ and $\beta > 0$ are fixed constants. This process X is called the *Ornstein–Uhlenbeck process* with parameters α and β. It is easy to conclude that X is a continuous stationary Gaussian and Markov process with $\mathbf{E}X_t = 0, t \geq 0$ and covariance function $C_X(s,t) = \alpha \mathrm{e}^{-\beta|s-t|}$. So, if we take $\alpha = 1$ and $\beta = 2\lambda$, we obtain $C_X(s,t) = \mathrm{e}^{-2\lambda|s-t|}$ (the function given at the beginning).

Further, let $Y = (Y_t, t \geq 0)$ be a process defined by

$$Y_t = Y_0(-1)^{N_t}$$

where Y_0 is a r.v. taking two values, 1 and -1, with probability $\frac{1}{2}$ each and Y_0 does not depend on N. The process Y, called a *random telegraph signal*, is a stationary process with $\mathbf{E}Y_t = 0, t \geq 0$ and covariance function $C_Y(s,t) = \mathrm{e}^{-2\lambda|s-t|}$. Obviously, Y takes only two values, 1 and -1; it is not continuous and is not Gaussian.

Thus using the processes w and N we have constructed in very different ways two new processes, X and Y, which have the same covariance functions.

(iii) Here we look at other functionals of the processes w and N. Consider the processes $U = (U_t, t \geq 0)$ and $V = (V_t, t \geq 0)$ defined by

$$U_t = \int_0^t X_s \, \mathrm{d}s, \quad V_t = v(N_t)$$

where X is the Ornstein–Uhlenbeck process introduced above and the function v is such that $v(2n) = v(2n+1) = n, n = 0, 1, \ldots$.

What can we say about the processes U and V? Obviously, U is Gaussian because it is derived from the Gaussian process X by a linear operation. Direct calculation shows that $\mathbf{E}U_t = 0, t \geq 0$ and

$$C_U(s,t) = \frac{2}{\beta}\min\{s,t\} + \frac{1}{\beta^2}[\mathrm{e}^{-\beta\min\{s,t\}} + \mathrm{e}^{-\beta\max\{s,t\}} - \mathrm{e}^{-\beta|s-t|-1}].$$

Clearly, if we take $\beta = 2$, then for large s, t and $|s-t|$ we have $C_U(s,t) \approx \min\{s,t\}$. So we can say that, asymptotically, the process U has the same covariance structure as the Wiener process. Both processes are continuous but some of their other properties are very different. In particular, U is not Markov and is not a process with stationary increments.

Consider now the process V. Does this process obey the properties of the original process N? From the definition it follows that V is a counting process which, however, only counts the arrivals $t_2, t_4, t_6, \ldots$ from N. Further, for $0 < t - h < t < t + h$ we have

$$\mathbf{P}[V_{t+h} - V_t = 0] = \mathrm{e}^{-\lambda h} + \tfrac{1}{2}\lambda h \mathrm{e}^{-\lambda h}(1 - \mathrm{e}^{-2\lambda h})$$

which means that V is not a process with stationary increments. Moreover, it is easy to establish that the increments of V are not independent. Finally, from the relations

$$\lim_{r\downarrow 0} \mathbf{P}[V_{t+h} = 1 | V_t = 1, V_{t-r} = 0] = \mathbf{P}[N_h = 0 \text{ or } 1]$$

and

$$\mathbf{P}[N_h = 0 \text{ or } 1] > \mathbf{P}[V_{t+h} = 1 | V_t = 1]$$

we conclude that V is not a Markov process.

Thus the process V, obtained as a function of the Poisson process N, does not obey at least three of the essential properties of N. Actually, this is not so surprising, since v as defined above is not a one–one function.

24.2. Can the Poisson process be characterized by only one of its properties?

Recall that we can construct a Poisson process in the interval [0,1] by choosing the number of points according to a Poisson distribution and then distributing them independently of each other and uniformly on [0,1] (see Doob 1953).

We now consider an example of a point process, say S, on the interval [0,1] such that the number of points in any subinterval has a Poisson distribution with given parameter λ, but the numbers of points in disjoint subintervals are not independent. Obviously such a process cannot be a Poisson process.

Fix $\lambda > 0$, choose a number n with probability $\mathrm{e}^{-\lambda}\lambda^n/n!$, $n = 0, 1, \ldots$ and define the d.f. F_n of the n points $t_1, \ldots, t_n$ of S as follows.

If $n \neq 3$ let

$$F_n(x_1, \ldots, x_n) = x_1 \ldots x_n$$

and if $n = 3$ let

$$\begin{aligned} F_3(x_1, x_2, x_3) &= x_1 x_2 x_3 \\ &+ \varepsilon x_1 x_2 x_3 (x_1 - x_2)^2 (x_1 - x_3)^2 (x_2 - x_3)^2 (1 - x_1)(1 - x_2)(1 - x_3). \end{aligned} \tag{1}$$

It is easy to see that for sufficiently small $\varepsilon > 0$, F_3 is a d.f. Moreover, it is obvious that the process S described by the family of d.f.s $\{F_n\}$ is not a Poisson process. Thus it remains for us to show that the number of points of S in any subinterval of [0,1] has a Poisson distribution. For positive integers $m < n$ and $(a, b) \subset [0, 1]$ we have

$$\begin{aligned} G_{m,n}(a, b) &= \mathbf{P}_n[\text{exactly } m \text{ of } t_1, \ldots, t_n \in (a, b)] \\ &= \binom{n}{m} \mathbf{P}_n[t_1, \ldots, t_m \in (a, b), t_{m+1}, \ldots, t_n \notin (a, b)] \\ &= \binom{n}{m} \mathbf{E}_n \left[\prod_{j=1}^{m} (X_b(t_j) - X_a(t_j)) \prod_{j=m+1}^{n} (X_a(t_j) + X_1(t_j) - X_b(t_j)) \right] \end{aligned} \tag{2}$$

where $X_a(t) = 1$ if $t < a$ and $X_a(t) = 0$ if $t \geq a$. Moreover, since

$$\mathbf{E}[X_{a_1}(t_1) \cdots X_{a_n}(t_n)] = F_n(a_1, \ldots, a_n)$$

then in (2) only terms of the form $F_n(a_1, \ldots, a_n)$ appear where for each i, a_i is equal either to a, or to b, or to 1. Hence if

$$(3) \qquad F_n(a_1, \ldots, a_n) = a_1 \cdots a_n$$

for all such values of $a_1, \ldots, a_n$, then $G_{m,n}(a, b)$ in (2) will be the same as in the Poisson case. For $n \neq 3$ this follows from the choice of F_n as a uniform d.f. For $n = 3$, relation (3) follows from (1) and the remark before (3).

The final conclusion is that the Poisson process cannot be characterized by only one of its properties even if this is the most important property.

24.3. The conditions under which a process is a Poisson process cannot be weakened

Let ν be a point process on $\mathbb{R}^1$, $\nu(I)$ denote the number of points which fall into the interval I and $|I|$ be the length of I. Recall that the stationary Poisson process can be characterized by the following two properties.

(A) For any interval I, $\mathbf{P}[\nu(I) = k] = \dfrac{(\lambda|I|)^k}{k!} \mathrm{e}^{-\lambda|I|}$, $k = 0, 1, 2, \ldots$.

(B) For any number n of disjoint intervals $I_1, \ldots, I_n$, the r.v.s $\nu(I_1), \ldots, \nu(I_n)$ are independent, $n = 2, 3, \ldots$.

In Example 24.2 we have seen that condition (B) cannot be removed if the Poisson property of the process is to be preserved. Suppose now that condition (A) is satisfied but (B) is replaced by another condition which is weaker, namely:

(B_2) for any disjoint intervals I_1 and I_2 the r.v.s $\nu(I_1)$ and $\nu(I_2)$ are independent.

Thus we come to the following question (posed in a similar form by A. Rényi): do conditions (A) and (B_2) imply that ν is a Poisson process? The construction below shows that in general the answer is negative.

For our purpose it is sufficient to construct the process ν in the interval [0,1]. Let ν be a Poisson process with parameter λ with respect to a given probability measure $\tilde{\mathbf{P}}$. The idea is to introduce another measure, denoted by $\mathbf{P}$, with respect to which the process ν will not be Poisson.

Define the unconditional and conditional probabilities of ν with respect to $\mathbf{P}$ by the relations

$$(1) \qquad \mathbf{P}[\nu([0,1]) = k] = \tilde{\mathbf{P}}[\nu([0,1]) = k] = \frac{\lambda^k}{k!} \mathrm{e}^{-\lambda}, \quad k = 0, 1, 2, \ldots,$$

$$(2) \qquad \mathbf{P}[\cdot \,|\nu([0,1]) = k] = \tilde{\mathbf{P}}[\cdot \,|\nu([0,1]) = k], \quad \text{if } k \neq 5.$$

If $\nu([0,1]) = k$ and we take a random permutation of the k points of a Poisson process which fall into [0,1], then the distribution of the k-dimensional vector obtained is the same as the distribution of a vector whose components are independent and uniformly distributed in [0,1], that is its conditional d.f. $\tilde{F}_k$ given $\nu([0,1]) = k$ has the form

$$\tilde{F}_k(x_1, \ldots, x_k) = x_1 \cdots x_k \text{ where } 0 \leq x_j \leq 1,\ j = 1, \ldots, k.$$

From (2) it follows that for $k \neq 5$ the d.f. F_k of ν about **P** satisfies the relations

$$F_k = \tilde{F}_k. \tag{3}$$

For $k = 5$ and $0 \leq x_j \leq 1$, $j = 1, \ldots, 5$ we define F_5 as follows:

(4)

$$\begin{aligned} F_5(x_1, \ldots, x_5) &= x_1 \cdots x_5 + \varepsilon x_1 \cdots x_5 (1 - x_1) \cdots (1 - x_5) \prod_{1 \leq i < j \leq 5} (x_j - x_i) \\ &= \tilde{F}_5(x_1, \ldots, x_5) + H(x_1, \ldots, x_5). \end{aligned}$$

It is easy to check that for ε positive and sufficiently small the mixed partial derivative $(\partial^5/\partial x_1 \cdots \partial x_5)F_5(x_1, \ldots, x_5)$ is a probability density function and thus F_5 is a d.f. It is obvious that our process ν, and also the measure **P**, are determined by (1), (2), (3) and (4). Moreover, (4) means that ν is not a Poisson process.

Clearly it remains for us to verify that the probability measure **P** satisfies conditions (A) and (B_2). These conditions are satisfied for the measure $\tilde{\mathbf{P}}$, so it is sufficient to prove that for disjoint intervals I_1 and I_2 we have

$$\mathbf{P}[\nu(I_1) = k_1, \nu(I_2) = k_2] = \tilde{\mathbf{P}}[\nu(I_1) = k_1, \nu(I_2) = k_2].$$

By the definition of **P** we see that we need to establish the relation

$$\mathbf{P}[\nu(I_1){=}k_1, \nu(I_2){=}k_2 | \nu([0,1]){=}5] = \tilde{\mathbf{P}}[\nu(I_1){=}k_1, \nu(I_2){=}k_2 | \nu([0,1]){=}5]. \tag{5}$$

The probability in the left-hand side of (5) is a finite sum of the form $\sum(\pm F_5(\alpha_1, \ldots, \alpha_5))$ where each α_j is either 0, or 1, or the endpoint of one of the intervals I_1 or I_2. So the difference between the two sides of (5) is equal to $\sum(\pm H(\alpha_1, \ldots, \alpha_5))$. Obviously, each term in this sum is 0. This is clear if 0 or 1 occurs among the αs; if not, then at least two of the αs are the same, so H vanishes again.

Therefore the measure **P** satisfies conditions (A) and (B_2). This means that we have described a process ν which obeys the properties (A) and (B_2), but nevertheless ν is not a Poisson process.

Condition (B_2) can be replaced by a slightly stronger condition of the same type, (B_M), which includes the mutual independence of the r.v.s $\nu(I_1), \ldots, \nu(I_M)$ for any M disjoint intervals $I_1, \ldots, I_M$ where, let us emphasise, M is finite. The conclusion in this case is the same as for $M = 2$.

24.4. Two dependent Poisson processes whose sum is still a Poisson process

Let $X = (X(t), t \geq 0)$ and $Y = (Y(t), t \geq 0)$ be Poisson processes with given parameters. It is well known and easy to check that if X and Y are independent then their sum $X + Y = (X(t) + Y(t), t \geq 0)$ is also a Poisson process. Let us now consider the converse question. X and Y are Poisson processes and we know that their sum $X + Y$ is also a Poisson process. Does it follow that X and Y are independent? The example below shows that in general the answer is negative.

Let $g(x, y) = \mathrm{e}^{-x-y}$ for $x \geq 0$ and $y \geq 0$, and $g(x, y) = 0$ otherwise. So g is the density of a pair of independent exponentially distributed r.v.s each of rate 1. We introduce the function

$$f_1(x,y) = \begin{cases} \alpha, & \text{if } (0 \le x < 1, 3 \le y < 4) \text{ or } (1 \le x < 2, 2 \le y < 3) \\ & \text{or } (2 \le x < 3, 0 \le y < 1) \text{ or } (3 \le x < 4, 1 \le y < 2) \\ -\alpha, & \text{if } (0 \le x < 1, 2 \le y < 3) \text{ or } (1 \le x < 2, 3 \le y < 4) \\ & \text{or } (2 \le x < 3, 1 \le y < 2) \text{ or } (3 \le x < 4, 0 \le y < 1) \\ 0, & \text{otherwise} \end{cases}$$

where $\alpha =$ constant, $0 < \alpha < \mathrm{e}^{-6}$ and define

$$f(x,y) = g(x,y) + f_1(x,y), \quad (x,y) \in \mathbb{R}^2.$$

It is easy to check that: (a) f is a density of some d.f. on $\mathbb{R}^2$; (b) the marginals of f are exponential of rate 1; (c) for each non-negative measurable function h on $\mathbb{R}^2$ such that $h(x, y) = h(y, x)$ the following equality holds:

$$(1) \qquad \int_{\mathbb{R}^2} f(x,y)h(x,y)\,\mathrm{d}x\,\mathrm{d}y = \int_{\mathbb{R}^2} g(x,y)h(x,y)\,\mathrm{d}x\,\mathrm{d}y.$$

Now let $\Omega = \mathbb{R}^2 \times \mathbb{R}^2 \times \cdots$ be the infinite and countable product of the space $\mathbb{R}^2$ with itself such that $\Omega = (\mathbb{R}^2)^{\mathbb{N}}$. Define $W_n(\omega) = (U_n(\omega), V_n(\omega))$, $n \geq 1$, as the nth coordinate of $\omega \in \Omega$. Let $\mathcal{A}$ be the σ-field generated by the coordinates. We shall provide $(\Omega, \mathcal{A})$ with two different probability measures, say **P** and **Q**, as follows:

(a) **P** is a measure for which $\{W_n, n \geq 1\}$ is a sequence of independent r.v.s, W_1 has density f and each W_n, $n \geq 2$, has density g;

(b) **Q** is a measure for which $\{W_n, n \geq 1\}$ is a sequence of i.i.d. r.v.s each having the same density g.

Put $U_0 = V_0 = 0$ and define the processes X, Y and $Z = X + Y$ where:

$$X(t) = n, \text{ if } \sum_{k=0}^{n} U_k \le t < \sum_{k=0}^{n+1} U_k,$$

$$Y(t) = n, \text{ if } \sum_{k=0}^{n} V_k \le t < \sum_{k=0}^{n+1} V_k,$$

$$Z(t) = X(t) + Y(t).$$

$\{U_n, n \geq 1\}$ is a sequence of independent exponential r.v.s of rate 1 with respect to each of the measures **P** and **Q**. The same holds for the sequence $\{V_n, n \geq 1\}$. Hence X and Y are Poisson processes with respect to both **P** and **Q**. Moreover, X and Y are independent for **Q** which implies that Z is a Poisson process for **Q**.

The next step is to show that X and Y are not independent for **P**. This will follow from the relation

$$\mathbf{P}[X(2) = 0, X(3) \geq 1, Y(1) \geq 1] = \mathbf{P}[X(2) = 0, X(3) \geq 1]\mathbf{P}[Y(1) \geq 1] + \alpha.$$

Now let $B \subset \Omega$ and $\tilde{B}$ be the set of all points $((x_1, y_1), \ldots, (x_n, y_n), \ldots)$ such that $((y_1, x_1), \ldots, (y_n, x_n), \ldots) \in B$. Using relation (1) we can prove that $\mathbf{P}(B) = \mathbf{Q}(B)$ for any measurable subset $B \subset \Omega$ such that $\tilde{B} = B$ (for details see Jacod 1975).

It remains for us to show that $Z = (Z(t),\ t \geq 0)$ is a Poisson process for the measure **P**. Note first that each event B which depends only on the process Z (this means that B belongs to the σ-field generated by Z) satisfies the equality $\tilde{B} = B$. Since Z is a Poisson process for the measure **Q**, Z must also be a Poisson process for the measure **P**. More precisely, if $s_1 \leq t_1 \leq \ldots \leq s_n \leq t_n$ we see from the above reasoning that

$$\mathbf{P}\left[\bigcap_{k=1}^{n}\{Z(t_k) - Z(s_k) = n_k\}\right] = \mathbf{Q}\left[\bigcap_{k=1}^{n}\{Z(t_k) - Z(s_k) = n_k\}\right]$$
$$= \prod_{k=1}^{n} \exp\{-2(t_k - s_k)\}[2(t_k - s_k)]^{n_k}/n_k!\,.$$

Obviously this relation illustrates the fact that Z is a Poisson process with respect to the probability measure **P**.

Note that the present example is in some sense an analogue to Example 12.3 where we considered an interesting property of dependent Poisson r.v.s.

24.5. Multidimensional Gaussian processes which are close to the Wiener process

Recall that $w = ((w_1(t), \ldots, w_n(t)), t \geq 0)$ is said to be an n-dimensional standard Wiener process if each component $w_j = (w_j(t), t \geq 0)$, $j = 1, \ldots, n$, is a onedimensional standard Wiener process and $w_1, \ldots, w_n$ are independent. Further, the linear combinations

$$Y(t) = \sum_{j=1}^{n} \lambda_j w_j(t), \ \ t \geq 0, \ \ \lambda_j \in \mathbb{R}^1$$

are often called the *projections* of w and it is very easy to see that

$$\text{(1)} \qquad \mathbf{E}[Y(s)Y(t)] = \left(\sum_{j=1}^{n} \lambda_j^2\right) \min\{s, t\}.$$

Suppose now $X = ((X_1(t), \ldots, X_n(t)), t \geq 0)$ is a Gaussian process whose projections $Z(t) = \sum_{j=1}^n \lambda_j X_j(t)$, $\lambda_j \in \mathbb{R}^1$, satisfy the relation

(2) $$\mathbf{E}[Z(s)Z(t)] = \left(\sum_{j=1}^n \lambda_j^2\right) \min\{s, t\}.$$

Comparing (2) and (1) we see that in some sense the projections of the process X behave like those of the Wiener process w. Since in (2) and (1), $\lambda_1, \ldots, \lambda_n$ are arbitrary numbers in $\mathbb{R}^1$, and s and t are also arbitrary in $\mathbb{R}^+$, we could conjecture that X is a standard Wiener process in $\mathbb{R}^n$. However, the example below shows that in general this is not the case. To see this, consider for simplicity the case $n = 2$. Take two independent Wiener processes, $w_1 = (w_1(t), t \geq 0)$ and $w_2 = (w_2(t), t \geq 0)$, and define the process $X = ((X_1(t), X_2(t)), t \geq 0)$ by

$$X_1(t) = w_1(\tfrac{2}{3}t) + w_2(\tfrac{1}{3}t), \quad X_2(t) = w_1(\tfrac{1}{3}t) - w_2(\tfrac{2}{3}t).$$

Then $w = ((w_1(t), w_2(t)), t \geq 0)$ is a standard Wiener process in $\mathbb{R}^2$ and for any $\lambda_1, \lambda_2 \in \mathbb{R}^1$ the projections $Y(t) = \lambda_1 w_1(t) + \lambda_2 w_2(t)$ satisfy (1). Further, if we take the same λ_1, λ_2 we can easily show that the projections $Z(t) = \lambda_1 X_1(t) + \lambda_2 X_2(t)$ of the Gaussian process X satisfy (2). However, this coincidence of the covariances of the projections of X and w does not imply that X is a standard Wiener process in $\mathbb{R}^2$. It is enough to note that the components X_1 and X_2 of X are not independent.

Note that the Gaussian process X with property (2) will be a standard Wiener process in $\mathbb{R}^n$ if we impose some other conditions (see Hardin 1985).

24.6. On the Wald identities for the Wiener process

Let $w = (w(t), t \geq 0)$ be a standard Wiener process and τ be an $(\mathcal{F}_t^w)$-stopping time. The following three relations

(1) $$\mathbf{E}w(\tau) = 0,$$

(2) $$\mathbf{E}[w^2(\tau)] = \mathbf{E}\tau,$$

(3) $$\mathbf{E}[\exp(w(\tau) - \tfrac{1}{2}\tau)] = 1$$

are called the *Wald identities for the Wiener process*. Let us introduce three conditions, namely

(1*) $$\mathbf{E}\sqrt{\tau} < \infty,$$

(2*) $$\mathbf{E}\tau < \infty,$$

(3*) $$\mathbf{E}[\exp(\tfrac{1}{2}\tau)] < \infty.$$

Note that (1*), (2*) and (3*) are sufficient conditions for the validity of (1), (2) and (3) respectively (see Burkholder and Gundy 1970; Novikov 1972; Liptser and Shiryaev 1977/78).

Our purpose here is to analyse these conditions. In particular, to clarify what happens to (1), (2) and (3) when changing (1*), (2*) and (3*).

Firstly, take the stopping time $\tau_1 = \inf\{t \geq 0 : w(t) \geq 1\}$. By the continuity of the Wiener process w we have $w(\tau_1) = 1$ and hence $\mathbf{E}w(\tau_1) = 1$ but not 0 as in (1). However, the r.v. τ_1 has density $(2\pi t^3)^{-1/2} \exp(-1/(2t))$, $t > 0$, and it is easy to check that $\mathbf{E}[\tau_1^\delta] < \infty$ for all $\delta < \frac{1}{2}$ but $\mathbf{E}[\tau_1^{1/2}] = \infty$, so (1*) is violated. Obviously, identity (2) is also not satisfied because $\mathbf{E}[w^2(\tau_1)] = 1 \neq \mathbf{E}\tau_1 = \infty$.

Regarding the identity (1) we can go further. Among many other results Novikov (1983) proved the following statement. Let $f(t), t \geq 0$, be a positive, continuous and non-decreasing function such that

$$\int_1^\infty t^{-3/2} f(t)\,\mathrm{d}t = \infty.$$

Then for any $(\mathcal{F}_t^w)$-stopping time τ with $\mathbf{E}[f(\tau)] < \infty$ and $\mathbf{E}[|w(\tau)|] < \infty$ we have $\mathbf{E}w(\tau) = 0$. Let us show that the integrability condition for f cannot be weakened.

Suppose that $\tilde{f}$ is positive, continuous and non-decreasing, $\tilde{f}(0) > 0$ and $\int_1^\infty t^{-3/2}\tilde{f}(t)\,\mathrm{d}t < \infty$. Consider the stopping time $\tau_2 = \inf\{t \geq 0 : w(t) \geq 1 - \tilde{f}(t)\}$. It can be shown (for details see Novikov (1983)) that

$$\mathbf{E}[|w(\tau_2)|] < \infty, \quad \mathbf{E}[\tilde{f}(\tau_2)] < \infty \ \text{ but } \ \mathbf{E}w(\tau_2) > 0.$$

Now consider condition (3*) and the identity (3). It is not difficult to show that (3*) cannot be essentially weakened. Indeed, define the stopping time

$$\tau_a = \inf\{t \geq 0 : w(t) \leq -1 + at\}$$

where a is an arbitrary real number. Since τ_a has the density

$$(2\pi t^3)^{-1/2} \exp[-\tfrac{1}{2}(-1 + at)^2/t], \quad t > 0$$

it is easy to verify that $\mathbf{E}[\exp(\frac{1}{2}a^2\tau_a)] < \infty$ for each a. Furthermore,

$$\mathbf{E}[\exp(w(\tau_a) - \tfrac{1}{2}\tau_a)] = \begin{cases} 1, & \text{if } a \geq 1 \\ c_a < 1, & \text{if } a < 1. \end{cases}$$

Here c_a is a constant depending on a. Its exact value is not important but it is essential that $c_a < 1$. Therefore the coefficient $\frac{1}{2}$ in the exponent in condition (3*) is the 'best possible' case for which the Wald identity (3) still holds.

The Wald identity (3) is closely connected with a more general problem of characterization of the uniform integrability of the class of exponential continuous local martingales (see Liptser and Shiryaev 1977/78; Novikov 1979; Kazamaki and Sekiguchi 1983; Liptser and Shiryaev 1989; Kazamaki 1994). (It is useful to compare (3) and (3*) with the description in Example 24.7.)

24.7. Wald identity and a non-uniformly integrable martingale based on the Wiener process

Let us formulate first the following very recent and general result (see Novikov 1996). Suppose $X = (X_t, \mathcal{F}_t, t \geq 0)$ is a square integrable local martingale with bounded jumps ($|\Delta X_t = |X_t - X_{t-}| \leq$ constant a.s. for each t) and such that $\langle X\rangle_\infty < \infty$ a.s. and $\mathbf{E}[X_\infty]$ exists. Then

$$\liminf_{t\to\infty}(\sqrt{t}\,\mathbf{P}[\langle X\rangle_\infty > t]) \geq \sqrt{2/\pi}|\mathbf{E}[X_\infty]| \tag{1}$$

and in particular $\liminf_{t\to\infty}(\sqrt{t}\,\mathbf{P}[\langle X\rangle_\infty > t]) = 0$ implies $\mathbf{E}[X_\infty] = 0$.

From this result we can easily derive an elegant corollary for the Wiener process $w = (w_t, t \geq 0)$. Let τ be a $(\mathcal{F}_t^w)$-stopping time such that $\tau < \infty$ a.s. and $\mathbf{E}[w_\tau]$ exists. Then the process $X_t := w_{t\wedge\tau}$, $\tau \geq 0$ is a square integrable local martingale (even continuous) with $\langle X\rangle_\infty = \tau$ and $X_\infty = w_\tau$ and in this case (1) takes the form

$$\liminf_{t\to\infty}(\sqrt{t}\,\mathbf{P}[\tau > t]) \geq \sqrt{2/\pi}|\mathbf{E}[w_\tau]|. \tag{2}$$

In particular, $\liminf_{t\to\infty}(\sqrt{t}\,\mathbf{P}[\tau > t]) = 0 \Rightarrow \mathbf{E}[w_\tau] = 0$. Example 24.6 shows that the Wald identity $\mathbf{E}[w_\tau] = 0$ does not hold for the stopping times $\tau_A = \inf\{t : w_t = A\}$, A is a real number. Note however that $\mathbf{P}[\tau_A > t] \approx \sqrt{2/\pi}|A|t^{-1/2}$ for large t, implying that $\liminf_{t\to\infty}(\sqrt{t}\,\mathbf{P}[\tau_A > t]) > 0$.

Thus we arrive at the question: is there a more general martingale X satisfying the conditions in the above result of Novikov and such that

$$\liminf_{t\to\infty}(\sqrt{t}\,\mathbf{P}[\langle X\rangle_\infty > t]) = 0 \quad\text{but}\quad \limsup_{t\to\infty}(\sqrt{t}\,\mathbf{P}[\langle X\rangle_\infty > t]) > 0 \tag{3}$$

and, if so, what additional conclusion can be derived?

Let us show by a specific example that both relations (3) are possible. Indeed, take the increasing sequence $1 = t_1 < t_2 < t_3 < \ldots$ and define the function $g(s)$, $s \geq 0$, where

$$g(s) = \begin{cases} 1, & \text{if } 0 \leq s < 1 = t_1 \\ 1/s, & \text{if } t_i \leq s < t_{i+1} \text{ for odd } i \\ 2, & \text{if } t_i \leq s < t_{i+1} \text{ for even } i,\ i = 1, 2, \ldots. \end{cases}$$

Introduce the following two stopping times:

$$\tau_1 = \inf\{t \geq 0 : w_t = 1\} \quad\text{and}\quad \tau = \inf\{t \geq \tau_1 : w_t = 0\}$$

and define the process $m = (m_t, t \geq 0)$ by

$$m_t = \int_0^{t\wedge\tau} g(s)\,dw_s.$$

Then m is a square integrable local martingale which is continuous and such that $\langle m\rangle_t = G(t) = \int_0^t g^2(s)\,ds$ and $\langle m\rangle_\infty = G(\tau) < \infty$ a.s. Moreover, the relations

$$m_\infty = \int_0^\tau g(s)\,dw_s = 2w_\tau + \int_0^\tau (2 - g(s))\,dw_s,\ w_\tau = 0 \text{ a.s.},$$

$$\int_0^\infty (2 - g(s))^2 ds \leq 1 + \int_1^\infty (1/s)^2\, ds \leq 2$$

imply that m_∞ is integrable. The next step is to check that for large t we have $\mathbf{P}[\tau > t] \approx c{\cdot}t^{-1/2}$ and, since $G(t), t \geq 0$ is strictly monotone (due to the special choice of g above), there is an inverse function G^{-1} and

$$\mathbf{P}[\langle m\rangle_\infty > t] = \mathbf{P}[\tau > G^{-1}(t)] \approx c{\cdot}(G^{-1}(t))^{-1/2}.$$

Thus we conclude that

$$\liminf_{t\to\infty}(\sqrt{t}\,\mathbf{P}[\langle m\rangle_\infty > t]) = 0 \quad (\Rightarrow \mathbf{E}[m_\infty] = 0)$$

while

(4) $$\limsup_{t\to\infty}(\sqrt{t}\,\mathbf{P}[\langle m\rangle_\infty > t]) > 0.$$

It should be noted that (4) is a sufficient and necessary condition for the process m to be non-uniformly integrable (see Azema *et al* 1980). Therefore we have described a continuous square integrable local martingale $m = (m_t, t \geq 0)$ with $\mathbf{E}[m_\infty] = 0$ but despite these properties, m is not uniformly integrable.

24.8. On some properties of the variation of the Wiener process

(i) Let us consider the Wiener process w in the unit interval [0,1]. For any fixed $p \geq 1$ let

$$V_p(w) = \sup_{\pi_n} \sum_{k=0}^{n-1} |w(t_{k+1} - w(t_k)|^p$$

where sup is taken over all finite partitions $\pi_n = \{0 = t_0 < t_1 < \ldots < t_n = 1\}$ of [0,1]. The quantity $V_p(w)$ is called a *p-variation* (or maximal p-variation) of the Wiener process in [0,1]. Let us also introduce the so-called *expected p-variation* of w as $\mathbf{E}[V_p(w)]$.

We are interested in the conditions ensuring that $V_p(w)$ and $\mathbf{E}[V_p(w)]$ take finite values. It is better to consider an even more general situation.

Suppose $X = (X(t), t \in [0,1])$ is a separable Gaussian process with $\mathbf{E}X(t) = 0$, $t \in [0,1]$ and let $e_X(s,t) = \mathbf{E}|X(s) - X(t)|$. Firstly, according to the 0–1 law for Gaussian processes, the probability $\mathbf{P}[V_p(X) < \infty]$ is either 1 or 0 (see Jain and Monrad 1983). Further, it can be shown that if $\mathbf{P}[V_p(X) < \infty] = 1$, then it is also true that $\mathbf{E}[V_p(X)] < \infty$ (see Fernique 1974). Since

$$\mathbf{E}\left[\sup_{\pi_n} \sum_{k=0}^{n-1} |X(t_{k+1}) - X(t_k)|^p\right] \geq \sup_{\pi_n} \mathbf{E}\left[\sum_{k=0}^{n-1} |X(t_{k+1}) - X(t_k)|^p\right]$$

$$\geq c_p \sup_{\pi_n} \sum_{k=0}^{n-1} e_X^p(t_k, t_{k+1})$$

we conclude that the condition

$$\text{(1)} \qquad \sup_{\pi_n} \sum_{k=0}^{n-1} e_X^p(t_k, t_{k+1}) < \infty$$

is necessary for the Gaussian process X to have trajectories of finite p-variation with probability 1.

Take the particular case $p = 1$. The equality

$$\mathbf{E}\left[\sup_{\pi_n} \sum_{k=0}^{n-1} |X(t_{k+1}) - X(t_k)|\right] = \sup_{\pi_n} \sum_{k=0}^{n-1} e_X(t_k, t_{k+1})$$

shows that if $p = 1$, then condition (1) is also sufficient to ensure that the variation $V_1(X)$ of order 1 is finite. If $p > 1$, condition (1) is not sufficient to ensure that $V_p(X) < \infty$ a.s. To demonstrate this, consider the Wiener process w again. In particular, for $p = 2$ we have

$$\sup_{\pi_n} \sum_{k=0}^{n-1} e_w^2(t_k, t_{k+1}) < \infty$$

that is, (1) is satisfied. However, the Wiener process w has an infinite variation on every interval. Therefore the finiteness of the expected p-variation, $p > 1$, does not in general imply that the trajectories of the process have a.s. finite p-variation for $p = 1$.

(ii) Let us now consider some properties of the *quadratic variation* $V_2(w, \pi_n)$ of the Wiener process w, which is defined by

$$\text{(2)} \qquad V_2(w, \pi_n) = \sum_{k=0}^{n-1} [w(t_{k+1}) - w(t_k)]^2.$$

It is useful to recall the following classical result (see Lévy 1940). If the partition π_n is defined by $\{k2^{-n}, k = 0, \dots, 2^n\}$ then with probability 1

$$V_2(w, \pi_n) = \sum_{k=0}^{2^n-1} \left[w\left(\frac{k+1}{2^n}\right) - w\left(\frac{k}{2^n}\right)\right]^2 \to 1 \quad \text{as } n \to \infty.$$

(Note that the limit value 1 is simply the length of the interval [0,1].) Obviously, in this particular case the diameter of π_n is $d_n = 2^{-n}$ which tends to 0 'very quickly' as $n \to \infty$. Thus we come to the question of the limit behaviour of $V_2(w, \pi_n)$ as $n \to \infty$ and $d_n \to 0$. Dudley (1973) proved that the condition $d_n = o(1/\log n)$ implies that $V_2(w, \pi_n) \xrightarrow{\text{a.s.}} 1$ as $n \to \infty$. Even in more general situations he has shown that $o(1/\log n)$ is the 'best possible' order of d_n. More precisely, there exists a sequence $\{\pi_n\}$ of partitions of the interval [0,1] with $d_n = O(1/\log n)$ and such that $V_2(w, \pi_n)$ does not converge a.s. to 1 as $n \to \infty$; $V_2(w, \pi_n)$ will converge a.s. to a number which is (strictly) greater than 1. A paper by Fernandez De La Vega (1974) gives details concerning the construction of $\{\pi_n\}$ with $d_n = O(1/\log n)$ and proof that the quadratic variation $V_2(w, \pi_n)$ converges a.s. to a number $1 + \delta$ where $\delta > 0$.

(iii) Finally, let us mention another interesting result. It can be shown that if the diameter d_n of the partition π_n of the interval [0,1] is of order less than $(1/\log n)^\alpha$ for any $0 < \alpha < 1$, then the quadratic variation $V_2(w, \pi_n)$ of the Wiener process w diverges a.s. as $n \to \infty$. For details we refer the reader to a paper by Wrobel (1982).

24.9. A Wiener process with respect to different filtrations

The Wiener process $w = (w_t, t \geq 0)$ obeys several useful properties. One of them is that w is a martingale which, moreover, is square integrable (see Liptser and Shiryaev 1977/78; Kallianpur 1980; Durrett 1984; Protter 1990).

Recall, however, that for some martingale $M = (M_t, t \geq 0)$ we mean that M is adapted with respect to a suitable filtration $(\mathcal{F}_t, t \geq 0)$, that is for each $t \geq 0$, M_t is $\mathcal{F}_t$-measurable. In the case of the Wiener process w we can start with some of its definitions and establish that w is a martingale with respect to its own generated filtration $(\mathcal{F}_t^w, t \geq 0)$: $\mathcal{F}_t^w = \sigma\{w_s, s \leq t\}$. Note that in general a process can be adapted with different filtrations; in particular, a process can be a martingale about different filtrations. Hence it is interesting to consider the following question. What is the role of the filtration and what happens if we replace one filtration by another? One possible answer will be given in the example below.

Let $(\Omega, \mathcal{F}, \mathbf{P})$ be a probability space and let $(\mathcal{X}_t, t \geq 0)$ and $(\mathcal{Y}_t, t \geq 0)$ be two filtrations on this space. Suppose $w = (w_t, t \geq 0)$ is a Wiener process with respect to each of the filtrations $(\mathcal{X}_t, t \geq 0)$ and $(\mathcal{Y}_t, t \geq 0)$. Now let us define a new filtration, say $(\mathcal{F}_t, t \geq 0)$, where $\mathcal{F}_t = \mathcal{X}_t \vee \mathcal{Y}_t$ is the σ-field generated by the union of $\mathcal{X}_t$ with $\mathcal{Y}_t$. How is the process $w = (w_t, t \geq 0)$ connected with the new filtration $(\mathcal{F}_t, t \geq 0)$? In particular, is it true that w is a Wiener process with respect to $(\mathcal{F}_t, t \geq 0)$? Intuitively we could expect that the answer to the last question is positive. However, the example below shows that such a conjecture is false.

Suppose we have found two r.v.s, say X and Y, which satisfy the following three conditions:

(a) X does not depend on the process $w = (w_t, t \geq 0)$;
(b) Y does not depend on the process $w = (w_t, t \geq 0)$;
(c) the process $w = (w_t, t \geq 0)$ and the σ-field $\sigma(X, Y)$ generated by the r.v.s X and Y are dependent.

Now, denote $\mathcal{F}_t^w = \sigma\{w_s, s \leq t\}$ and define $\mathcal{X}_t$ and $\mathcal{Y}_t$ as:

$$\mathcal{X}_t = \mathcal{F}_t^w \vee \sigma(X), \quad \mathcal{Y}_t = \mathcal{F}_t^w \vee \sigma(Y).$$

It is easy to see that the new filtration $(\mathcal{F}_t, t \geq 0)$ where $\mathcal{F}_t = \mathcal{X}_t \vee \mathcal{Y}_t$ is such that $\mathcal{F}_t = \mathcal{F}_t^w \vee \sigma(X, Y)$.

Clearly $w = (w_t, t \geq 0)$ is a Wiener process with respect to each of the filtrations $(\mathcal{X}_t, t \geq 0)$ and $(\mathcal{Y}_t, t \geq 0)$. However, $w = (w_t, t \geq 0)$ is not a Wiener process with respect to $(\mathcal{F}_t, t \geq 0)$ which follows from condition (c).

Hence it only remains for us to construct r.v.s X and Y satisfying conditions (a), (b) and (c). For simplicity we consider the Wiener process w in the interval [0,1].

Let X be a r.v. with $\mathbf{P}[X = 1] = P[X = -1] = \frac{1}{2}$ and suppose X is independent of $w = (w_t, 0 \leq t \leq 1)$. Define the r.v. Y by $Y = \frac{1}{2}|X + \text{sign}(w_1)|$. Obviously the r.v. $\text{sign}(w_1)$ is $\sigma(X, Y)$-measurable and thus condition (c) is satisfied. Condition (a) is satisfied by construction. Let us check the validity of (b). For this, let $0 \leq t_1 < t_2 < \ldots < t_n \leq 1$ be any subdivision of [0,1]. Then for arbitrary continuous and bounded functions $f(x)$, $x \in \mathbb{R}^1$, and $g(x_1, \ldots, x_n)$, $x_1, \ldots, x_n \in \mathbb{R}^1$, we have

$$\begin{aligned} \mathbf{E}[f(Y)g(w_{t_1}, \ldots, w_{t_n})] &= \mathbf{E}\{\mathbf{E}[f(Y)g(w_{t_1}, \ldots, w_{t_n})|X, w_1]\} \\ &= \mathbf{E}\{f(Y)E[g(w_{t_1}, \ldots, w_{t_n})|w_1]\}. \end{aligned}$$

Since $\mathbf{E}[g(w_{t_1}, \ldots, w_{t_n})|w_1]$ is a (measurable) function of w_1 only, it is sufficient to show that the variables Y and w_1 are independent. Obviously, 1 and 0 are the possible values of Y, and the event $[Y = 0]$ can occur only if $w_1 \in B$ where $B \subset (-\infty, 0)$, while $[Y = 1]$ is possible only if $w_1 \in B \subset (0, \infty)$. Further, the relation $[Y = 0] \cap [w_1 \in B] = [X = -1] \cap [w_1 \in B]$ holds for any set $B \in \mathcal{B}^1$ and hence we have

$$\mathbf{P}\{[Y = 0] \cap [w_1 \in B]\} = \tfrac{1}{2}\mathbf{P}[w_1 \in B].$$

Analogously,

$$\mathbf{P}\{[Y = 1] \cap [w_1 \in B]\} = \tfrac{1}{2}\mathbf{P}[w_1 \in B].$$

Therefore the variables Y and w_1 are independent and so condition (b) is also satisfied.

24.10. How to enlarge the filtration and preserve the Markov property of the Brownian bridge

Let $X = (X_t, t \geq 0)$ be a real-valued Markov process on the probability space $(\Omega, \mathcal{F}, \mathbf{P})$ and let $\mathbf{E}|X_t| < \infty$ for all $t \geq 0$. For s, t and u with $0 \leq s < t < u$, let us call t the 'present' time. Then, with respect to the 'present' time t, the σ-field $\mathcal{H}_s = \sigma\{X_v, v \leq s\}$ is the 'past' (the 'history') of the process X up to time s, while the σ-field $\mathcal{F}_u = \sigma\{X_v, v \geq u\}$ is called the 'future' of X from time u on.

Denote by $\mathcal{H}_s \vee \mathcal{F}_u$, $s \leq u$, the minimal σ-field generated by the union of $\mathcal{H}_s$ and $\mathcal{F}_u$. The Markov property of X, usually written as $\mathbf{P}[X_t \in \Gamma|\mathcal{H}_s] = \mathbf{P}[X_t \in \Gamma|X_s]$ a.s., can be expressed in the following equivalent form involving the 'past' and the 'future' in a symmetric way (see e.g. Al-Hussaini and Elliott 1989):

$$\mathbf{E}[X_t|\mathcal{H}_s \vee \mathcal{F}_u] = \mathbf{E}[X_t|X_s, X_u] \text{ a.s.} \tag{1}$$

It is not difficult to derive from (1) the corollary:

$$\mathbf{E}[X_t|\mathcal{H}_s \vee \sigma(X_u)] = \mathbf{E}[X_t|X_s, X_u] \text{ a.s.} \tag{2}$$

Our goal now is to determine if the 'information' $\mathcal{H}_s \vee \sigma(X_u)$ can be enlarged whilst still keeping the Markov property (2). Let us consider a standard Wiener process $w = (w_t, t \geq 0)$ and let $1 < s < t < 2$. By $\mathcal{H}_s^w = \sigma\{w_v, v \leq s\}$ we denote the 'past' of w about the 'present' time t and $\sigma\{w_1 + w_2\}$ is the σ-field generated by the r.v. $w_1 + w_2$. Note that the value w_2 plays the role of the fixed 'future' of the process w_t at time $t = 2$. In such a case we speak about a *Brownian bridge process*.

We want to compare two conditional expectations, $\mathbf{E}[w_t|\mathcal{H}_s^w \vee \sigma(w_1 + w_2)]$ and $\mathbf{E}[w_t|w_s, w_1 + w_2]$. In view of (2) we could suggest that these two quantities coincide. Let us check if this is true. The Markov property of w implies

$$\begin{aligned}\mathbf{E}[w_t|\mathcal{H}_s^w \vee \sigma(w_1 + w_2)] &= \mathbf{E}[w_t|\mathcal{H}_s^w \vee \sigma(w_2)] \\ &= \mathbf{E}[w_t|w_s, w_2] \text{ a.s.}\end{aligned}$$

Since w is a Gaussian process with independent increments, we can easily derive the following two relations:

$$\mathbf{E}[w_t|w_s, w_2] = [(2-t)w_s + (t-s)w_2]/(2-s) \text{ a.s.}$$

and

$$\mathbf{E}[w_t|w_s, w_1 + w_2] = [(1+s)(1+t)w_s + s(1+t)(w_1 + w_2)]/[(1+s)^2 + 5s] \text{ a.s.}$$

Thus we have shown that

$$\mathbf{E}[w_t|\mathcal{H}_s^w \vee \sigma(w_1 + w_2)] \neq \mathbf{E}[w_t|w_s, w_1 + w_2].$$

Hence the Markov property expressed by (2) will not be preserved if the 'past' $\mathcal{H}_s^w$ is enlarged by 'new information' taken strictly from the 'future'.

SECTION 25. DIVERSE PROPERTIES OF STOCHASTIC PROCESSES

This section covers only a few counterexamples concerning different properties of stochastic processes. All new notions are defined in the examples themselves. Obviously, far more counterexamples could be considered here, but for various reasons we have restricted ourselves to indicating additional sources of interesting but rather complicated counterexamples in the Supplementary Remarks.

25.1. How can we find the probabilistic characteristics of a function of a stationary Gaussian process?

Let $X = (X_t, t \in \mathbb{R}^1)$ be a real-valued stationary Gaussian process with $\mathbf{E}X_t = 0$, $t \in \mathbb{R}^1$, and covariance function $C(t) = \mathbf{E}[X_s X_{s+t}]$, $s, t \in \mathbb{R}^1$. Then the finite-dimensional distributions of X, and hence any other probabilistic characteristics, are completely determined by $C(t), t \in \mathbb{R}^1$. In other words, if X is any Gaussian process and we know its moments of order 1 and 2 (that is, we know the mean function $\mathbf{E}X_t$ and the covariance function $\mathbf{E}[X_s X_t]$), then we can find all probabilistic characteristics of X. It is interesting to clarify whether a similar fact holds for the process $Y = (Y_t, t \in \mathbb{R}^1)$ which is a function of X. We consider the following particular case

$$Y_t = X_t^2, \quad t \in \mathbb{R}^1.$$

Does there exist a universal constant m such that the moments of Y of order not greater than m are enough to determine the distributions of Y? As was mentioned above, for Gaussian processes the answer is positive and $m = 2$.

For fixed $\varepsilon \in \mathbb{R}^1$ with $|\varepsilon| < 1$ and integer $n \geq 2$, introduce the function

$$f(\lambda) = \lambda^2(1 - \cos\lambda)(1 + \varepsilon \cos n\lambda), \quad \lambda \in \mathbb{R}^1.$$

It can be shown that the Fourier transform $C_{\varepsilon,n}$ of f has the form:

$$(1) \qquad C_{\varepsilon,n}(t) = \begin{cases} \frac{1}{2}\varepsilon(1 - |t - n|), & \text{if } |t - n| \leq 1 \\ 1 - |t|, & \text{if } |t| \leq 1 \\ \frac{1}{2}\varepsilon(1 - |t + n|), & \text{if } |t + n| \leq 1 \\ 0, & \text{otherwise.} \end{cases}$$

Moreover, for $|\varepsilon| < 1$ and $n \geq 2$ the function $C_{\varepsilon,n}(t)$, $t \in \mathbb{R}^1$, is positive definite. Therefore (see Doob 1953; Ash and Gardner 1975) there is a centred stationary Gaussian process, say X, with covariance function equal to $C_{\varepsilon,n}$.

Now take $Y_t = X_t^2$, $t \in \mathbb{R}^1$, and suppose that we know all the moments of Y of order not greater than m where m is a fixed natural number. This means that we know the quantities $\mathbf{E}[Y_{t_1} Y_{t_2} \cdots Y_{t_k}]$ for all $k \leq m$ and arbitrary $t_1, t_2, \ldots, t_k \in \mathbb{R}^1$. However,

$$(2) \qquad \mathbf{E}[Y_{t_1} Y_{t_2} \cdots Y_{t_k}] = \mathbf{E}[X_{t_1}^2 X_{t_2}^2 \cdots X_{t_k}^2]$$

and since $X_{t_1}^2, X_{t_2}^2, \ldots, X_{t_k}^2$ is a product of $2k$ Gaussian r.v.s, this includes the repetitions, applying the well known Wick lemma we obtain

$$\mathbf{E}[Y_{t_1} Y_{t_2} \cdots Y_{t_k}] = \sum_{\pi} C_{\varepsilon,n}(t_{\pi 1} - t_{\pi k}) C_{\varepsilon,n}(t_{\pi 2} - t_{\pi 1}) \cdots C_{\varepsilon,n}(t_{\pi k} - t_{\pi(k-1)})$$

where this summation is taken over the group of all permutations π of the k elements $t_1, t_2, \ldots, t_k$.

Now we show (by an appropriate choice of k) that the information contained in (2) is not sufficient to determine the sign of the parameter ε. Indeed, let us first clarify which of the terms in (2) give a non-zero contribution. It is easy to see that non-zero terms are those in which the difference $|t_{\pi i} - t_{\pi(i-1)}|$ is either smaller than 1 or is between $k - 1$ and $k + 1$. This observation is based on the explicit form (1) of the covariance function $C_{\varepsilon,n}$; together with the equality $(t_{\pi 1} - t_{\pi k}) + \cdots + (t_{\pi k} - t_{\pi(k-1)}) = 0$, it implies that if we choose k such that $n > 2m \geq 2k$ then the number of terms in (2) whose arguments are close to n is the same as the number of terms with arguments close to $(-n)$. Obviously this means that the parameter ε in (2) has an even power and thus its sign is lost.

We have shown that for an arbitrary positive integer m, there exists a centred stationary Gaussian process X such that the moments of order not greater than m of $Y_t = X_t^2, t \in \mathbb{R}^1$, are not sufficient to determine the distributions of Y.

25.2. Cramér representation, multiplicity and spectral type of stochastic processes

Let $X = (X(t), t \geq 0)$ be a real-valued L^2-stochastic process defined on a given probability space $(\Omega, \mathcal{F}, \mathbf{P})$. Denote by $\mathcal{H}_t(X)$ the closed (in L^2-sense) linear manifold generated by the r.v.s $\{X(s), 0 \leq s \leq t\}$ and let

$$\mathcal{H}(X) = \bigcup_{t \in \mathbb{R}^+} \mathcal{H}_t(X).$$

Suppose now that $Y = (Y(t), t \geq 0)$ is an L^2-process with orthogonal increments. Then $\mathcal{H}(Y)$ coincides with the set of all integrals $\int_0^\infty g(u)\,\mathrm{d}Y(u)$ where $\int_0^\infty g^2(u)\,\mathrm{d}F(u) < \infty$ and $\mathrm{d}F(u) = \mathbf{E}[\mathrm{d}Y^2(u)]$. Thus we come to the following interesting and important problem. For a given process X, find a process Y with orthogonal increments such that

$$\text{(1)} \qquad \mathcal{H}_t(X) = \mathcal{H}_t(Y) \ \text{ for each } t > 0.$$

Regarding this problem and other related topics we refer the reader to works by Cramér (1960, 1964), Hida (1960), Ivkovic *et al* (1974) and Rozanov (1977). In particular, Hida (1960) suggested the first example of a process X for which relation (1) is not possible. Take two independent standard Wiener processes, $w_1 = (w_1(t), t \geq 0)$ and $w_2 = (w_2(t), t \geq 0)$, and define the process $X = (X(t), t \geq 0)$ by

$$X(t) = \begin{cases} w_1(t), & \text{if } t \text{ is rational} \\ w_2(t), & \text{if } t \text{ is irrational.} \end{cases}$$

Obviously, X is discontinuous at each t and we have the representation

$$\mathcal{H}_t(X) = \mathcal{H}_t(w_1) \oplus \mathcal{H}_t(w_2), \quad t > 0$$

where as usual the symbol $\oplus$ denotes the sum of orthogonal subspaces.

The general solution of the coincidence problem (1) was found by Cramér (1964) and can be described as follows. Let $F_1, F_2, \ldots, F_N$ be an arbitrary sequence of measures on $\mathbb{R}^+$ ordered by absolute continuity, namely:

$$(2) \qquad F_1 \succ F_2 \succ \cdots \succ F_N.$$

Here N is either a fixed natural number or infinity. Then there exists a continuous process X and N mutually orthogonal processes $Y_1, \ldots, Y_N$ each with orthogonal increments and $\mathrm{d}F_n(t) = \mathbf{E}[(\mathrm{d}Y_n(t))^2]$, $n = 1, \ldots, N$, such that

$$(3) \qquad \mathcal{H}_t(X) = \sum_{n=1}^{N} \mathcal{H}_t(Y_n), \quad t > 0.$$

This general result implies in particular that

$$(4) \qquad X(t) = \sum_{n=1}^{N} \int_0^t g_n(t,u)\,\mathrm{d}Y_n(u)$$

where the functions g_n, $n = 1, \ldots, N$, satisfy the condition

$$\sum_{n=1}^{N} \int_0^t g_n^2(t,u)\,\mathrm{d}F_n(u) < \infty.$$

The equality (4) is called the *Cramér representation* for the process X while the sequence (2) is called the *spectral type* of X. Finally, the number N in (2) (also in (3) and (4)) is called the *multiplicity* of X.

Our purpose now is to illustrate the relationships between the notions introduced above by suitable examples.

(i) Suppose $Y = (Y(t), t \geq 0)$ is an arbitrary L^2-process with orthogonal increments and $\mathbf{E}[(\mathrm{d}Y(t))^2] = \mathrm{d}t$. Consider the process $X = (X(t), t \geq 0)$,

$$(5) \qquad X(t) = \int_0^t h(t,u)\,\mathrm{d}Y(u)$$

where h is some (deterministic) function. Comparing (5) and (4) we see that X has a Cramér-type representation and it is natural to expect that the multiplicity of X is equal to 1. However, the kernel h can be chosen such that the multiplicity of X is greater than 1. Indeed, take

$$h(t,u) = 0, \text{ if } 0 \leq t \leq t_0 \quad \text{and} \quad a \leq u \leq b < t_0$$

where $0 < a < b < t_0$ are fixed numbers. Since for $t > 0$ any increment $Y(d) - Y(c)$ with $a \leq c < d \leq b$ belongs to $\mathcal{H}_t(Y)$ and is orthogonal to $\mathcal{H}_t(X)$, we conclude that $\mathcal{H}_t(X) \subset \mathcal{H}_t(Y)$ (the inclusion is strong). Further, the function h can be chosen arbitrarily for $0 \leq t \leq t_0$ and $u \notin [a, b]$. Take, for example,

$$h(t,u) = \begin{cases} \sin u, & \text{if } u \text{ is rational} \\ \cos u, & \text{if } u \text{ is irrational} \end{cases}$$

and suppose that $b - a = 2\pi k$ for some natural number k. Then for any $t > t_0$, $X(t)$ is equal either to Z_1 or to Z_2 where Z_1 and Z_2 are r.v.s defined by

$$Z_1 = \int_a^b \sin u \, \mathrm{d}Y(u), \quad Z_2 = \int_a^b \cos u \, \mathrm{d}Y(u).$$

Obviously,

$$\mathbf{E}[Z_1 Z_2] = \int_0^{2\pi k} \sin u \cos u \, \mathrm{d}u = 0$$

which means that Z_1 and Z_2 are orthogonal. Moreover, both Z_1 and Z_2 are orthogonal to the space $\mathcal{H}_{t_0}(X)$. Thus for any $t > t_0$, $\mathcal{H}_t(X)$ consists of $\mathcal{H}_{t_0}(X)$, Z_1 and Z_2. Consequently

$$\bigcap_{\delta > 0} \mathcal{H}_{t_0+\delta}(X) = \mathcal{H}_{t_0}(X) \oplus Z_1 \oplus Z_2.$$

According to a result by Cramér (1960), the point t_0 is a point of increase of the space $\mathcal{H}_{t_0}(X)$ with dimension equal to 2. Therefore the multiplicity of the process X at time t_0 cannot be less than 2.

(ii) Let $X_1 = (X_1(t), t \geq 0)$ and $X_2 = (X_2(t), t \geq 0)$ be Gaussian processes. Denote by $\mathbf{P}_1$ and $\mathbf{P}_2$ the probability measures induced by these processes in the sample function space. It is well known that $\mathbf{P}_1$ and $\mathbf{P}_2$ are either equivalent or singular (see Ibragimov and Rozanov 1978). The question of whether $\mathbf{P}_1$ and $\mathbf{P}_2$ are equivalent is closely related to some property of the corresponding spectral types of the processes X_1 and X_2. The following result can be found in the book by Rozanov (1977). If the measures $\mathbf{P}_1$ and $\mathbf{P}_2$ of the Gaussian processes X_1 and X_2 are equivalent, then X_1 and X_2 have the same spectral type. The next example shows whether the converse is true.

Consider two processes, the standard Wiener process $w = (w(t), t \geq 0)$ and the process $\xi = (\xi(t), t \geq 0)$ defined by

$$\xi(t) = h(t) w(t), \quad t \geq 0.$$

Here h is a function which will be chosen in a special way: h is a non-random continuous function which is not absolutely continuous in any interval. Additionally, let $0 < c_1 \leq h(t) \leq c_2 < \infty$ for some constants c_1, c_2 and all t. It is obvious that $\mathcal{H}_t(\xi) = \mathcal{H}_t(w)$ for each $t > 0$. This implies that the processes ξ and w have the same spectral type.

Denote by $\mathbf{P}_w$ and $\mathbf{P}_\xi$ the measures in the space $\mathbb{C}(\mathbb{R}^+)$ induced by the processes w and ξ respectively. Clearly, it remains for us to see whether these measures are equivalent. Indeed, if C_w and C_ξ are the covariance functions of w and ξ, then the difference between them is

$$\Delta(s,t) = C_w(s,t) - C_\xi(s,t) = [1 - h(s)h(t)] \min\{s,t\}.$$

Since the function $\Delta(s,t)$, $(s,t) \in \mathbb{R}^2$, is not absolutely continuous, using a well known criterion (see Ibragimov and Rozanov 1978) we conclude that the measures $\mathbf{P}_w$ and $\mathbf{P}_\xi$ are not equivalent despite the coincidence of the spectral types of the processes w and ξ.

25.3. Weak and strong solutions of stochastic differential equations

A large class of stochastic processes can be obtained as solutions of stochastic differential equations of the type

$$(1) \qquad X(t) = X_0 + \int_0^t a(s, X(s))\, ds + \int_0^t \sigma(s, X(s))\, dw(s), \quad t \geq 0$$

where a and σ^2, the drift and the diffusion coefficients respectively, satisfy appropriate conditions, and $\int_0^t \sigma(\cdot)\, dw(s)$ is a stochastic integral (in the sense of K. Ito) with respect to the standard Wiener process.

Let us define two kinds of solutions of (1), weak and strong, and then analyse the relationship between them.

Let $w = (w(t), t \geq 0)$ be a standard Wiener process on the probability space $(\Omega, \mathcal{F}, \mathbf{P})$. Suppose that w is adapted to the family $(\mathcal{F}_t, t \geq 0)$ of non-decreasing sub-σ-fields of $\mathcal{F}$. If there exists an $(\mathcal{F}_t)$-adapted process $X = (X(t), t \geq 0)$ satisfying (1) a.s., we say that (1) has a *strong solution*. If (1) has at most one $(\mathcal{F}_t)$-strong solution, we say that *strong uniqueness* (pathwise uniqueness) holds for (1), or that (1) has a unique strong solution. Further, if there exist a probability space $(\Omega, \mathcal{F}, \mathbf{P})$, a family $(\mathcal{F}'_t, t \geq 0)$ of non-decreasing sub-σ-fields of $\mathcal{F}'$, and two $(\mathcal{F}'_t)$-adapted processes, $X' = (X'(t), t \geq 0)$ and $w' = (w'(t), t \geq 0)$ such that w' is a standard Wiener process and (X', w') satisfy (1) a.s., we say that a *weak solution* exists. If the law of the process X' is unique (that is, the measure in the space $\mathbb{C}$ generated by X' is unique), we say that *weak uniqueness* holds for (1), or that (1) has a unique weak solution.

There are many books dealing entirely or partially with the theory of stochastic differential equations (see Doob 1953; McKean 1969; Gihman and Skorohod 1972, 1979; Liptser and Shiryaev 1977/78; Krylov 1980; Jacod 1979; Kallianpur 1980; Ikeda and Watanabe 1981; Durrett 1984).

The purpose of the two examples below is to clarify the relationship between the weak and strong solutions of (1), looking at both aspects, existence and uniqueness. The survey paper by Zvonkin and Krylov (1981) provides a very useful and detailed analysis of these two concepts (see also Barlow 1982; Barlow and Perkins 1984; Protter 1990; Karatzas and Shreve 1991).

Let us briefly describe the first interesting example in this field. Consider the stochastic differential equation

$$x(t) = \int_0^t |x(s)|^\alpha \, dw(s), \quad t \geq 0$$

where $\alpha > 0$ is a fixed parameter. It can be shown that for $\alpha \geq \frac{1}{2}$ this equation has only one strong solution (with respect to the family $(\mathcal{F}_t^w)$), namely $x \equiv 0$. However, for $0 < \alpha < \frac{1}{2}$ it has infinitely many solutions. For the proof of this result we refer the reader to the original paper of Girsanov (1962) (see also McKean 1969). Thus the above stochastic equation has a strong solution for any $\alpha > 0$, but this strong solution need not be unique.

Among a variety of results concerning the properties of the solutions of stochastic differential equations (1), we quote the following (see Yamada and Watanabe 1971): strong uniqueness of the solution of (1) implies its weak uniqueness.

Of course, this result is not surprising. However, it can happen that a weak solution exists and is unique, but no strong solution exists. For details of such an example we refer the reader to a book by Stroock and Varadhan (1979).

Let us now consider an example of a stochastic differential equation which has a unique weak solution but several (at least two) strong solutions.

Take the function $\sigma(x) = 1$ if $x \geq 0$ and $\sigma(x) = -1$ if $x < 1$ and consider the stochastic equation

$$x(t) = x_0 + \int_0^t \sigma(x(s)) \, dw(s), \quad t \geq 0. \tag{2}$$

Firstly let us check that (2) has a solution. Suppose for simplicity that $x_0 = 0$ and let

$$x(t) = w(t) \quad \text{and} \quad \tilde{w}(t) = \int_0^t \sigma(x(s)) \, dw(s), \quad t \geq 0.$$

Then $\tilde{w}$ is a continuous local martingale with $\langle \tilde{w} \rangle_t = t$ and so $\tilde{w}$ is a Wiener process. Moreover, the pair $(x(t), \tilde{w}(t), t \geq 0)$ is a solution of (2). Hence the stochastic equation (2) has a weak solution. Weak uniqueness of (2) follows from the fact that for any solution x, the stochastic integral $\int_0^t \sigma(x(s)) \, dw(s)$, with the function σ defined above, is again a Wiener process.

It remains for us to show that strong uniqueness does not hold for the stochastic equation (2). Obviously, $\sigma(-x) = -\sigma(x)$ for $x \neq 0$ and if $x_0 = 0$ and $(x(t), t \geq 0)$ is a solution of (2), then the process $(-x(t), t \geq 0)$ is also its solution.

Moreover, it is not only strong uniqueness which cannot hold for equation (2)—the stochastic equation (2) does not have a strong solution at all. This can be shown by using the local time technique (for details see Karatzas and Shreve 1991).

25.4. A stochastic differential equation which does not have a strong solution but for which a weak solution exists and is unique

Let $(\Omega, \mathcal{F}, \mathbf{P})$ be a complete probability space on which a standard Wiener process $w = (w(t), t \geq 0)$ is given. Suppose that $a(t, x)$ is a real-valued function on $[0, 1] \times \mathbb{C}([0, 1])$ defined as follows. Let $(t_k, k = 0, -1, -2, \ldots)$ be a sequence contained in the interval [0,1] and such that $t_0 = 1 > t_{-1} > t_{-2} > \ldots \to 0$ as $k \to \infty$. If for $x \in \mathbb{C}([0, 1])$ we have $a(0, x) = 0$ and if $t > 0$, let

$$a(t, x) = \left\{ \frac{x_{t_k} - x_{t_{k-1}}}{t_k - t_{k-1}} \right\} \text{ for } t_k \leq t \leq t_{k+1}, \quad k = -1, -2, \ldots$$

where $\{\alpha\}$ denotes the fractional part of the real number α and x_t denotes the value of the continuous function x at the point t. Clearly, a satisfies the usual measurability conditions, a is $(\mathcal{C}_t)$-adapted where $\mathcal{C}_t = \sigma\{x_s, s \leq t\}$ and $\int_0^1 a^2(t, x)\, dt < \infty$ for each $x \in \mathbb{C}([0, 1])$.

Consider the following stochastic differential equation:

$$(1) \qquad \xi_t = \int_0^t a(s, \xi)\, ds + w_t, \quad t \in [0, 1].$$

Firstly, according to general results given by Liptser and Shiryaev (1977/78), Stroock and Varadhan (1979) and Kallianpur (1980), equation (1) has a weak solution and this solution is unique.

Let us now determine whether equation (1) has a strong solution. Suppose the answer is positive, that is (1) has a strong solution $(\xi_t, 0 \leq t \leq 1)$ which is $(\mathcal{F}_t^w)$-adapted where $\mathcal{F}_t^w = \sigma\{w_s, s \leq t\}$. Then if $t_k \leq t < t_{k+1}$ we obtain from (1) that

$$\xi_t - \xi_{t_k} = \int_{t_k}^t \left\{ \frac{\xi_{t_k} - \xi_{t_{k-1}}}{t_k - t_{k-1}} \right\} ds + w_t - w_{t_k}.$$

Using the notations

$$\eta_k = \left\{ \frac{\xi_{t_k} - \xi_{t_{k-1}}}{t_k - t_{k-1}} \right\} \quad \text{and} \quad \varepsilon_{k+1} = \frac{w_{t_{k+1}} - w_{t_k}}{t_{k+1} - t_k}$$

we arrive at the relation

$$\eta_{k+1} = \{\eta_k\} + \varepsilon_{k+1}, \quad k = 0, -1, -2, \ldots.$$

Since we supposed that a strong solution of (1) exists, η_k must be $\mathcal{F}_{t_k}^w$-measurable and, moreover, the family of r.v.s $\{\eta_m, m = k, k-1, \ldots\}$ is independent of ε_{k+1}. This independence and the equality

$$(2) \qquad e^{2\pi i \eta_{k+1}} = e^{2\pi i \{\eta_k\}} e^{2\pi i \varepsilon_{k+1}}$$

easily lead to the relation

$$d_{k+1} = d_k \mathbf{E}[e^{2\pi i \varepsilon_{k+1}}] = d_k e^{-2\pi^2/(t_{k+1}-t_k)}$$

where we have introduced the notation $d_k = \mathbf{E}[e^{2\pi i \eta_k}]$. Thus, for any $n = 0, 1, \ldots$, we get inductively that

$$d_{k+1} = d_{k-n} \exp\left[-2\pi^2 \left(\frac{1}{t_{k+1} - t_k} + \cdots + \frac{1}{t_{k+1-n} - t_{k-n}}\right)\right].$$

It follows that $|d_{k+1}| \leq e^{-2\pi^2 n}$ for any n and so $d_{k+1} \to 0$ as $n \to \infty$. Hence

$$d_k = 0 \text{ for } k = 0, -1, -2, \ldots.$$

From (2) and the relation for η_{k+1} we find that

$$e^{2\pi i \eta_{k+1}} = e^{2\pi i \eta_{k-n}} e^{2\pi i(\varepsilon_{k+1} + \cdots + \varepsilon_{k+1-n})}$$

and also

$$\mathbf{E}[e^{2\pi i \eta_{k+1}} | \mathcal{F}^w_{[t_{k-n}, t_{k+1}]}] = d_{k-n} e^{2\pi i(\varepsilon_{k+1} + \cdots + \varepsilon_{k+1-n})}$$

where $\mathcal{F}^w_{[t_{k-n}, t_{k+1}]} = \sigma\{w_t - w_s, t_{k-n} \leq s \leq t \leq t_{k+1}\}$. Since $d_{k-n} = 0$ we have

$$\mathbf{E}[e^{2\pi i \eta_{k+1}} | \mathcal{F}^w_{[t_{k-n}, t_{k+1}]}] = 0.$$

Now, if $n \to \infty$, then $\mathcal{F}^w_{[t_{n-k}, t_{k+1}]} \uparrow \mathcal{F}^w_{t_{k+1}}$ and since η_{k+1} is $\mathcal{F}^w_{t_{k+1}}$-measurable for each k, we come to the equality

$$0 = \mathbf{E}[e^{2\pi i \eta_{k+1}} | \mathcal{F}^w_{t_{k+1}}] = e^{2\pi i \eta_{k+1}}.$$

It is obvious, however, that this is not possible and this contradiction is a direct result of our assumption that (1) has a strong solution. Therefore, despite the fact that the stochastic differential equation (1) has a unique weak solution, it has no strong solution.

In Examples 25.3 and 25.4 we analysed a few stochastic differential equations and have seen that the properties of their solutions (existence, non-existence, uniqueness, non-uniqueness) in the weak and strong sense depend completely on either the drift coefficient or the diffusion coefficient.

More details on stochastic differential equations, not only theory but also examples and intricate counterexamples, can be found in many books (e.g. Liptser and Shiryaev 1977/78 and 1989; Jacod 1979; Strook and Varadhan 1979; Kallianpur 1980; Ikeda and Watanabe 1981; Jacod and Shiryaev 1987; Protter 1990, Karatzas and Shreve 1991; Revuz and Yor 1991; Rogers and Williams 1994; Nualart 1995).

Supplementary Remarks

Section 1. Classes of random events

Examples 1.1, 1.2, 1.3, 1.4 and 1.7 or their modifications can be found in many books. These examples are part of so-called probabilistic folklore. The idea of Example 1.5 is taken from Bauer (1996). Example 1.6 is based on arguments by Neveu (1965) and Kingman and Taylor (1966). Other interesting counterexamples and ideas for constructing counterexamples can be found in works by Chung (1974), Broughton and Huff (1977), Williams (1991) and Billingsley (1995).

Section 2. Probabilities

Example 2.1 could be classified as folklore. Example 2.2 belongs to Breiman (1968). The presentation of Example 2.3 follows that in Neveu (1965) and Shiryaev (1995). Example 2.4 was originally suggested by Doob (1953) and has since been included in many books; see Halmos (1974), Loève (1978), Laha and Rohatgi (1979), Rao (1979) and Billingsley (1995). We refer the reader also to works by Pfanzagl (1969), Blake (1973), Rogers and Williams (1994) and Billingsley (1995) where other interesting counterexamples concerning conditional probabilities can be found.

Section 3. Independence of random events

Since the concept of independence plays a central role in probability theory, it is no wonder that we find it treated in almost all textbooks and lecture notes. Many examples concerning the independence properties of collections of random events could be qualified as probabilistic folklore. For Example 3.1 see Feller (1968) or Bissinger (1980). Example 3.2(i), suggested by Bohlmann (1908), and 3.2(ii), suggested by Bernstein (1928), seem to be the oldest among all examples included into this book. Example 3.2(iii) is due to Feller (1968) and 3.2(v) to Roussas (1973). Examples 3.2(iv) and 3.3(ii) were suggested by an anonymous referee. Example 3.3(i) is given by Ash (1970) and Shiryaev (1995). The idea of Examples 3.3(iii) and 3.7 was taken from Neuts (1973). Example 3.4(i) belongs to Wong (1972) and case (ii) of the same example was suggested by Ambartzumian (1982). Example 3.5 is based on the papers of Wang *et al* (1993) and Stoyanov (1995). Example 3.6 is considered by Papoulis (1965). Example 3.7 is given by Sevastyanov *et al* (1985). For other counterexamples the reader is referred to works by Lancaster (1965), Kingman and Taylor (1966), Crow (1967), Moran (1968), Ramachandran (1975), Chow and Teicher (1978), Grimmett and Stirzaker (1982), Lopez and Moser (1980), Falk and Bar-Hillel (1983), Krewski and Bickis (1984), Wang *et al* (1993), Stoyanov (1995), Shiryaev (1995), Billingsley (1995) and Mori and Stoyanov (1995/1996).

Section 4. Diverse properties of random events and their probabilities

The idea of Example 4.1 came from Gelbaum (1976) and, as the author noted, case (ii) was originally suggested by E. O. Thorp. Example 4.2 is folklore. Example 4.3 belongs to Krewski and Bickis (1984). Example 4.5 is from Rényi (1970). Several other counterexamples can be found in works by Lehmann (1966), Hawkes (1973), Ramachandran (1974), Lee (1985) and Billingsley (1995).

Section 5. Distribution functions of random variables

Different versions of Examples 5.1, 5.3, 5.6 and 5.7 can be found in many sources and definitely belong to folklore. Example 5.2 was suggested by Zubkov (1986). Examples like 5.5 are noted by Gnedenko (1962), Cramér (1970) and Laha and Rohatgi (1979). Case (ii) of Example 5.8 is described by Ash (1970) and case (iii) is given by Olkin *et al* (1980). Cases (iv) and (v) of the same example are considered by Gumbel (1958) and Fréchet (1951). A paper by Clarke (1975) covers Example 5.9. Example 5.10(i) is treated by Chung (1953), while case (ii) is presented by Dharmadhikari and Jogdeo (1974). Example 5.12 follows the presentation in Dharmadhikari and Joag-Dev (1988). The last example, 5.13, is described by Hengartner and Theodorescu (1973). Other counterexamples concerning properties of one-dimensional and multi-dimensional d.f.s can be found in the works of Thomasian (1969), Feller (1971), Dall'Aglio (1960, 1972, 1990), Barndorff-Nielsen (1978), Rüschendorf (1991), Rachev (1991), Mikusinski *et al* (1992) and Kalashnikov (1994).

Section 6. Expectations and conditional expectations

Example 6.1 belongs to Simons (1977). Example 6.2 is due to Takacs (1985) and is the answer to a problem proposed by Emmanuele and Villani (1984). Example 6.4 and other related topics can be found in Piegorsch and Casella (1985). Example 6.5, suggested by Churchill (1946), is probably the first example to be found of a non-symmetric distribution with vanishing odd-order moments. Example 6.6 is indicated by Bauer (1996). Examples 6.7 and 6.8 can be classified as folklore. Example 6.9 belongs to Enis (1973) (see also Rao (1993)) while Example 6.10 was taken from Laha and Rohatgi (1979). The idea of Example 6.11 is taken from Dharmadhikari and Joag-Dev (1988). Finally, Example 6.12 belongs to Tomkins (1975a). Several other counterexamples concerning the integrability properties of r.v.s, conditional expectations and some related topics can be found in works by Robertson (1968), B. Johnson (1974), Witsenhausen (1975), Rao (1979), Leon and Massé (1992), Bryc and Smolenski (1992), Zieba (1993) and Rao (1993).

Section 7. Independence of random variables

Examples 7.1(i), 7.8, 7.9(i) and (ii), 7.10(i), (ii) and (iii), and 7.12 can be described as folklore. Examples 7.1(ii), (iii), and 7.8 follow some ideas by Feller (1968, 1971). Example 7.2 is due to Pitman and Williams (1967), who assert that this is the first example of three pairwise independent but not mutually independent r.v.s in the absolutely continuous case. Example 7.3(i) is based on a paper by Wang (1979), case (ii) is considered by Han (1971), while case (iii) is outlined by Ying (1988). Example 7.4 is based on a paper by Wang (1990). Runnenburg (1984) is the originator of Example 7.5. Drossos (1984) suggested Example 7.6(i) to me and attributed it to E. Lukacs and R. Laha. Case (ii) of the same example was suggested by Falin (1985) and case (iii) is indicated by Rohatgi (1976). The description of Example 7.7(i) follows an idea of Fisz (1963) and Laha and Rohatgi (1979). Examples 7.7(ii) and 7.12 are indicated by Rényi (1970). The idea of Example 7.7(ii) belongs to Flury (1986). Case (iii) of Example 7.10 was suggested by an anonymous referee. Example 7.11(iii) is taken from Ash and Gardner (1975). Examples 7.14(i) and (ii) belong to Chow and Teicher (1978) while case (iii) of the same example is considered by Cinlar (1975). Case (i) of Example 7.15 follows an idea of Billingsley (1995) and case (ii) is indicated by Johnson and Kotz (1977). Finally, Example 7.16 is based on a paper by Kimeldorf and Sampson (1978). Note that a great number of additional counterexamples concerning the independence and dependence properties of r.v.s can be found in works by Geisser and Mantel (1962), Tsokos (1972), Roussas (1973), Coleman (1974), Chung (1974), Joffe (1974), Fortet (1977), Gänssler and Stute (1977), Loève (1978), Wang (1979), O'Brien (1980), Grimmett and Stirzaker (1982), Galambos (1984), Gelbaum (1985, 1990), Heilmann and Schröter (1987), Ahmed (1990), Dall'Aglio (1990), Dall'Aglio *et al* (1991), Durrett (1991), Whittaker (1991), Liu and Diaconis (1993) and Mori (1995).

Section 8. Characteristic and generating functions

Example 8.1 belongs to Gnedenko (1937) and can be classified as one of the most significant classical counterexamples in probability theory. Example 8.2 is contained in many books; see those by Fisz (1963), Moran (1968) and Ash (1972). Examples 8.3, 8.4, 8.5 and 8.6, or versions of them, can be found in the book by Lukacs (1970) and in later books by other authors. Example 8.7 was suggested by Zygmund (1947) and our presentation follows that in Rényi (1970) and Lamperti (1966). Example 8.8 is described by Wintner (1947) and also by Sevastyanov *et al* (1985). Example 8.9 is given by Linnik and Ostrovskii (1977). Finally, Example 8.10 is presented in a form close to that given by Lukacs (1970) and Laha and Rohatgi (1979). Note that other counterexamples on the topics in this section can be found in works by Ramachandran (1967), Thomasian (1969), Feller (1971), Loève (1977/1978), Chow and Teicher (1978), Rao (1984), Rohatgi (1984), Dudley (1989) and Shiryaev (1995).

Section 9. Infinitely divisible and stable distributions

Example 9.1 and other versions of it can be classified as folklore. Example 9.2 belongs to Gnedenko and Kolmogorov (1954) (see also Laha and Rohatgi (1979)). Example 9.3(i) is based on a paper by Shanbhag *et al* (1977) and answers a question proposed by Steutel (1973). Case (ii) of Example 9.3 as well as Example 9.4 are considered by Rohatgi *et al* (1990). Example 9.5 is described by Linnik and Ostrovskii (1977). Example 9.6 belongs to Lévy (1948), but some arguments from Griffiths (1970) are also needed (also see Rao (1984)). Ibragimov (1972) proposed Example 9.7. Example 9.8 could also be considered as probabilistic folklore. The last example, 9.9, belongs to Lukacs (1970). Let us note that several other counterexamples which are interesting but rather complicated can be found in works by Ramachandran (1967), Steutel (1970), Kanter (1975), O'Connor (1979), Marcus (1983), Hansen (1988), Evans (1991), Jurek and Mason (1993), Rutkowski (1995) and Bondesson *et al* (1996).

Section 10. Normal distribution

Some of the examples in this section are popular among probabilists and statisticians and can be found in different sources. In particular, cases (ii), (iii) and (iv) of Example 10.1 are noted respectively by Roussas (1973), Morgenstern (1956) and Olkin *et al* (1980). The idea of Example 10.2 is indicated by Papoulis (1965). Example 10.3(i) is based on papers by Pierce and Dykstra (1969) and Han (1971). Case (ii) of the same example is considered by Bühler and Mieshke (1981). Example 10.4(i) in this form belongs to Ash and Gardner (1975) and case (iii) is treated by Ijzeren (1972). Hamedani and Tata (1975) describe Examples 10.5 and 10.7, while Example 10.6 is considered by Hamedani (1984). Moran (1968) proposed the problem of finding a non-normal density such that both conditional densities are normal. Example 10.8 presents one of the possible answers. Case (i) is a result of my discussions with N. V. Krylov and A. M. Zubkov, while case (ii) is due to Ahsanullah and Sinha (1986). Example 10.9 is given by Breiman (1969). Finally, Example 10.10 was suggested by Kovatchev (1996). Many useful facts, including counterexamples, concerning the normal distribution can be found in the works of Anderson (1958), Steck (1959), Geisser and Mantel (1962), Grenander (1963), Thomasian (1969), Feller (1971), Kowalski (1973), Vitale (1978), Hahn and Klass (1981), Melnick and Tenenbein (1982), Ahsanullah (1985), Devroye (1986), Janson (1988), Castillo and Galambos (1989), Whittaker (1991) and Hamedani (1992).

Section 11. The moment problem

Example 11.1 follows the presentation of Berg (1988). Example 11.2(i) was originally suggested by Heyde (1963a) and has since been included in many textbooks; see Feller (1971), Rao (1973), Billingsley (1995), Laha and Rohatgi (1979). Case (ii) of this example belongs to Leipnik (1981). Example 11.3 is considered in a recent paper by Targhetta (1990). Example 11.4 is mentioned by Hoffmann-Jorgensen (1994), but also see Lukacz (1970) and Berg (1988). Example 11.5 follows an idea from Carnal and Dozzi (1989). Our presentation of Example 11.6 follows that in Kendall and Stuart (1958) and Shiryaev (1995). Examples 11.7 and 11.8 belong to Jagers (1983) and Fu (1984) respectively. As far as we know these are the first examples of this kind in the discrete case (also see Schoenberg (1983)). Example 11.9(i) is based on a paper by Dharmadhikari (1965). Case (ii) of the same example is considered by Chow and Teicher (1978). Both cases of Example 11.10 belong to Heyde (1963b). Example 11.12 is based on papers by Heyde (1963b) and Hall (1981). Example 11.13 is treated by Heyde (1975). Note that other counterexamples concerning the moment problem as well as related topics can be found in works by Fisz (1963), Neuts (1973), Prohorov and Rozanov (1969), Lukacs (1970), Schoenberg (1983), Devroye (1986), Berg and Thill (1991), Slud (1993), Hoffmann-Jorgensen (1994) and Shiryaev (1995). Readers interested in the history of progress in the moment problem are referred to works by Shohat and Tamarkin (1943), Kendall and Stuart (1958), Heyde (1963b), Akhiezer (1965) and Berg (1995).

Section 12. Characterization properties of some probability distributions

Example 12.1 was suggested by Zubkov (1986). General characterization theorems for the binomial distribution can be found in Ramachandran (1967) and Chow and Teicher (1978). Example 12.2 is given by Klimov and Kuzmin (1985). Example

12.3 belongs to Steutel (1984) but, according to Jacod (1975), a similar result was proved by R. Serfling and included in a preprint which unfortunately I have never seen. Example 12.4 is a natural continuation of the reasoning in Example 12.1. Example 12.5 belongs to Philippou and Hadjichristos (1985). Example 12.6 is given by Rossberg *et al* (1985). Example 12.7 is based on an idea of Robertson *et al* (1988). Laha (1958) is the author of Example 12.8(i), while case (iv) of this example uses an idea from Mauldon (1956). Case (v) of Example 12.8 is discussed by Letac (1995). Baringhaus and Henze (1989) invented Example 12.9. Example 12.10 is based on a paper by Blank (1981). The idea of Example 12.11 can be found in the book by Syski (1991). Example 12.12 is outlined by Rohatgi (1976) and Example 12.14 was suggested to me by Seshadri (1986). Note that other counterexamples and useful facts concerning the characterization-type properties of different classes of probability distributions can be found in works by Mauldon (1956), Dykstra and Hewett (1972), Kagan *et al* (1973), Gani and Shanhag (1974), Huang (1975), Galambos and Kotz (1978), Ahlo (1982), Azlarov and Volodin (1983), Hwang and Lin (1984), Rossberg *et al* (1985), Too and Lin (1989), Balasubrahmanyan and Lau (1991), Letac (1991, 1995), Prakasa Rao (1992), Yellott and Iverson (1992), Braverman (1993) and Huang and Li (1993).

Section 13. Diverse properties of random variables

Example 13.1(i) is folklore while case (ii) is due to Behboodian (1989). Example 13.2 is indicated by Feller (1971), but we have followed the presentation given by Kelker (1971). Example 13.3 is outlined by Barlow and Proshan (1966). Example 13.4 is based on a paper by Pavlov (1978). The notion of exchangeability is intensively treated by Feller (1971), Chow and Teicher (1978), Laha and Rohatgi (1979) and Aldous (1985). Example 13.5 is based on these sources and on discussions with Rohatgi (1986). Diaconis and Dubins (1980) suggested Example 13.6, but a similar statement can also be found in the book by Feller (1971). The idea of Example 13.7 is indicated by Galambos (1987). Example 13.8 belongs to Taylor *et al* (1985). Example 13.9 is considered by Gut and Janson (1986). Other counterexamples classified as 'diverse' can be found in the works of Bhattacharjee and Sengupta (1966), Ord (1968), Fisher and Walkup (1969), Brown (1972), Burdick (1972), Dykstra and Hewett (1972), Klass (1973), Gleser (1975), Cambanis *et al* (1976), Freedman (1980), Tong (1980), Franken and Lisek (1982), Laue (1983), Chen and Shepp (1983), Galambos (1984), Aldous (1985), Taylor *et al* (1985), Hüsler (1989) and Metry and Sampson (1993). For more abstract topics, see Laha and Rohatgi (1979), Rao (1979), Tjur (1980), Vahaniya *et al* (1989), Gelbaum and Olmsted (1990), Dall'Aglio *et al* (1991), Ledoux and Talagrand (1991), Kalashnikov (1994) and Rao (1995).

Section 14. Various kinds of convergence of sequences of random variables

Examples 14.1, 14.2, 14.4, 14.5, 14.6, 14.8(i), 14.10(ii), 14.12(i) or their modifications can be found in many publications. These examples can be classified as belonging to probabilistic folklore. Examples 14.3(i) and 14.7(i) are based on arguments by Roussas (1973), Laha and Rohatgi (1979) and Bauer (1996). Example 14.3(ii) is due to Grimmett and Stirzaker (1982). Examples 14.7(ii) and 14.8(ii) are considered by Thomas (1971). Fortet (1977) has described Example 14.7(iii). Example 14.7(iv) is treated by Chung (1974). The idea of Example 14.9 is indicated by Feller (1971), Lukacs (1975) and Billingsley (1995). Cases (i) and (ii) of example 14.10 were suggested by Grimmett (1986) and Zubkov (1986) respectively. Example 14.11 is due to Rohatgi (1986) and a similar example is given in Serfling (1980). Case (ii) of Example 14.12 is briefly discussed by Cuevas (1987). Example 14.13 is presented in a form which is close to that of Ash and Gardner (1975). In Example 14.14 we follow Hsu and Robbins (1947) and Chow and Teicher (1978). Lukacs (1975) considers Examples 14.15 and 14.18. Example 14.17 was suggested by Zubkov (1986). Cases (i) and (ii) of Example 14.16 are described following Lukacs (1975) and Billingsley (1995) respectively. Note that other useful counterexamples can be found in works by Neveu (1965), Kingman and Taylor (1966), Hettmansperger and Klimko (1974), Stout (1974a), Dudley (1976), Gänssler and Stute (1977), Bartlett (1978), Rao (1984), Ledoux and Talagrand (1991), Lessi (1993) and Shiryaev (1995).

Section 15. Laws of large numbers

Example 15.1(i) and its modifications can be classified as folklore. Examples 15.1(ii), 15.3 and 15.4 belong to Geller (1978). In Example 15.2 we follow the presentations of Lukacs (1975) and Bauer (1996). The statement in Example 15.5 is contained in many books: see those by Fisz (1963) or Lukacs (1975). Révész (1967) is the author of Example 15.6. Example 15.7 is based on papers by Prohorov (1950) and Fisz (1959). Example 15.8 is due to Hsu and Robbins (1947). Taylor and Wei (1979) describe Example 15.9. For a presentation of Example 15.10 see Stoyanov *et al* (1988). For the presentation of Example 15.11 we used papers by Jamison *et al* (1965), Chow and Teicher (1971) and Wright *et al* (1977). The classical Example 15.12 is described by Feller (1968). Finally, let us note that other counterexamples about the laws of large numbers and related topics can be found in works by Prohorov (1959), Jamison *et al* (1965), Lamperti (1966), Révész (1967), Chow and Teicher (1971), Feller (1971), Stout (1974a), Wright *et al* (1977), Asmussen and Kurtz (1980), Hall and Heyde (1980), Csörgő *et al* (1983), Dobric (1987), Ramakrishnan (1988) and Chandra (1989).

Section 16. Weak convergence of probability measures and distributions

Example 16.1 and other similar examples were originally described by Billingsley (1968) and have since appeared in many books and lecture notes. Chung (1974) considered Example 16.2 and its variations can be classified as folklore. Example 16.3, suggested by Robbins (1948), is presented in a form similar to that in Fisz (1963). Clearly, Example 16.4 belongs to probabilistic folklore. The idea of Example 16.5 was suggested by Zolotarev (1989). Takahasi (1971/72) is the originator of Example 16.6. The idea of Example 16.7 is indicated by Feller (1971). Example 16.8 is considered by Kendall and Rao (1950). Example 16.9 is outlined by Rohatgi (1976). Example 16.10 is described by Laube (1973). Other counterexamples devoted to weak convergence and related topics can be found in works by Billingsley (1968, 1995), Breiman (1968), Sibley (1971), Borovkov (1972), Roussas (1972), Chung (1974), Lukacs (1975) and Eisenberg and Shixin (1983).

Section 17. Central limit theorem

Example 17.1(i) is based on arguments given by Ash (1972) and Chow and Teicher (1978). Cases (ii) and (iii) of the same example are considered by Thomasian (1969). Obviously Examples 17.2 and 17.5 can be classified as folklore. Example 17.3 is considered by Ash (1972). The idea of Example 17.4 is to be found in Feller (1971). Zubkov (1986) suggested Example 17.6. Case (i) of Example 17.7 is considered by Gnedenko and Kolmogorov (1954), while case (ii) is taken from Malisic (1970) and is presented as it is given by Stoyanov *et al* (1988). Additional counterexamples concerning the CLT can be found in works by Gnedenko and Kolmogorov (1954), Fisz (1963), Rényi (1970), Feller (1971), Chung (1974), Landers and Rogge (1977), Laha and Rohatgi (1979), Rao (1984), Shevarshidze (1984), Janson (1988) and Berkes *et al* (1991).

Section 18. Diverse limit theorems

Example 18.1(i) is considered by Billingsley (1995). Case (ii) of this example and Example 18.3 are considered by Chow and Teicher (1978). Examples 18.2 and 18.4 are covered in many sources. Tomkins (1975a) is the author of Example 18.5 and 18.6 belongs to Neveu (1975). Basterfield (1972) considered Example 18.7 and noted that this example was suggested by Williams. Examples 18.8 and 18.9 are considered by Lukacs (1975). Example 18.10 belongs to Arnold (1966), but also see Lukacs (1975). Example 18.11 is based on a paper by Stout (1979). Example 18.12 is given by Breiman (1967). Vasudeva (1984) treated Example 18.13. Example

18.14 is due to Resnik (1973). A great number of other counterexamples concerning the limit behaviour of random sequences can be found in the literature. However, some of these counterexamples are either very specialized or very complicated. The interested reader is referred to works by Spitzer (1964), Kendall (1967), Feller (1968, 1971), Moran (1968), Sudderth (1971), Roussas (1972), Greenwood (1973), Berkes (1974), Chung (1974), Stout (1974a), Kuelbs (1976), Hall and Heyde (1980), Serfling (1980), Tomkins (1980), Rosalsky and Teicher (1981), Prohorov (1983), Daley and Hall (1984), Boss (1985), Kahane (1985), Wittmann (1985), Sato (1987), Alonso (1988), Barbour *et al* (1988), Hüsler (1989), Adler (1990), Jensen (1990), Tomkins (1990, 1992, 1996), Hu (1991), Ledoux and Talagrand (1991), Rachev (1991), Williams (1991), Adler *et al* (1992), Fu (1993), Rosalsky (1993), Klesov (1995) and Rao (1995).

Section 19. Basic notions on stochastic processes

Example 19.1 is based on remarks by Ash and Gardner (1975) and Billingsley (1995). Examples 19.2, 19.3, 19.4(i), 19.6(i), 19.7, 19.8 and 19.10(i) or modifications of them can be found in many textbooks and can be classified as probabilistic folklore. Case (ii) of Example 19.4 is considered by Yeh (1973). Example 19.5(i) is described by Kallianpur (1980), case (ii) belongs to Cambanis (1975), while case (iii) is given by Dellacherie (1972) and Elliott (1982). Example 19.6(ii) is due to Masry and Cambanis (1973). Example 19.9 is based entirely on a paper by Wang (1982). Cases (ii) and (iii) of Example 19.10 are given in a form similar to that of Morrison and Wise (1987). For other counterexamples concerning the basic characteristics and properties of stochastic processes we refer the reader to works by Dudley (1973), Kallenberg (1973), Wentzell (1981), Dellacherie and Meyer (1982), Elliott (1982), Métivier (1982), Doob (1984), Hooper and Thorisson (1988), Edgar and Sucheston (1992), Rogers and Williams (1994), Billingsley (1995) and Rao (1995).

Section 20. Markov processes

Examples 20.1(i) and 20.2(iii) are probabilistic folklore. Example 20.1, cases (ii) and (iii), are due to Feller (1968, 1959). Case (iv) of Example 20.1 as well as Example 20.2(i) and (ii) are considered by Rosenblatt (1971, 1974). Example 20.2(iv) is discussed by Freedman (1971). Arguments which are essentially from Isaacson and Madsen (1976) are used to describe Example 20.3. According to Holewijn and Hordijk (1975), Example 20.4 was suggested by Runnenburg. Example 20.5 is due to Tanny (1974) and O'Brien (1982). Speakman (1967) considered Example 20.6. Example 20.7 is considered by Dynkin (1965) and Wentzell (1981). Example 20.8(i) is due to A. A. Yushkevich (see Dynkin and Yushkevich 1956; and also Dynkin 1961; Wentzell 1981). Case (ii) of the same example is based on an idea from Wong (1971).

Example 20.9 is considered by Ito (1963). A great number of other counterexamples (some of them very complicated) can be found in the works of Chung (1960, 1982), Dynkin (1961, 1965), Breiman (1968), Chung and Walsh (1969), Kurtz (1969), Feller (1971), Freedman (1971), Rosenblatt (1971, 1974), Gnedenko and Solovyev (1973), D. P. Johnson (1974), Tweedie (1975), Monrad (1976), Lamperti (1977), Iosifescu (1980), Wentzell (1981), Portenko (1982), Salisbury (1986, 1987), Ethier and Kurtz (1986), Grey (1989), Liu and Neuts (1991), Revuz and Yor (1991), Alabert and Nualart (1992), Ihara (1993), Meyn and Tweedie (1993), Courbage and Hamdan (1994), Rogers and Williams (1994), Pakes (1995) and Eisenbaum and Kaspi (1995).

Section 21. Stationary processes and some related topics

Examples 21.1 and 21.2 and other versions of them are probabilistic folklore. Example 21.3 is considered by Ibragimov (1962). Example 21.4 is based on arguments by Ibragimov and Rozanov (1978). Case (i) of Example 21.5 is discussed by Gaposhkin (1973), while case (ii) of the same example can be found in the paper by Verbitskaya (1966). Example 21.6 can be found in more than one source: we follow the presentation given by Shiryaev (1995); see also Ash and Gardner (1975). Example 21.7 is due to Stout (1974b). Cases (i) and (ii) of Example 21.8 are found in the work of Grenander and Rosenblatt (1957), while case (iii) is discussed by Bradley (1980). For other counterexamples we refer the reader to works by Krickeberg (1967), Billingsley (1968), Breiman (1969), Ibragimov and Linnik (1971), Davydov (1973), Rosenblatt (1979), Bradley (1982, 1989), Herrndorf (1984), Robertson and Womak (1985), Eberlein and Taqqu (1986), Cambanis *et al* (1987), Dehay (1987a, 1987b), Janson (1988), Rieders (1993), Doukhan (1994) and Rosalsky *et al* (1995).

Section 22. Discrete-time martingales

Examples 22.1(iii), 22.4(i) and 22.10 can be classified as probabilistic folklore. Example 22.1(i) is given by Neveu (1975), while case (ii) of the same example was proposed by Küchler (1986). Example 22.2 is based on arguments by Yamazaki (1972). Case (i) and case (ii) of Example 22.3 are considered respectively by Kemeny *et al* (1965) and Freedman (1971). Examples 22.4(ii) and 22.5(i) were suggested by Melnikov (1983). Tomkins (1975b) described Examples 22.5(ii) and 22.7. Examples 22.6 and 22.8 can be found in Tomkins (1984b) and (1984a) respectively. Zolotarev (1961) is the author of Example 22.9, case (i), while case (ii) can be found in Shiryaev (1984). Example 22.11(i) is given by Stout (1974a) with an indication that it belongs to G. Simons. Case (ii) of the same example is treated by Neveu (1975), while the general possibility presented by case (iii) was suggested by Bojdecki (1985). Example 22.12(ii) was suggested by Marinescu (1985) and is given here in the form proposed by Iosifescu (1985). Example 22.13(i) is considered by Edgar and

Sucheston (1976a). Example 22.13(iii) is based on Durrett (1991) and suggested by P. Chigansky. The last example, 22.14, is based on a paper by Dozzi and Imkeller (1990). Other counterexamples concerning the properties of discrete-time martingales can be found in works by Cuculescu (1970), Nelson (1970), Baez-Duarte (1971), Ash (1972), Gilat (1972), Mucci (1973), Austin *et al* (1974), Stout (1974a), Edgar and Sucheston (1976a, 1976b, 1977), Blake (1978, 1983), Janson (1979), Rao (1979), Alvo *et al* (1981), Gut and Schmidt (1983), Tomkins (1984a, b), Alsmeyer (1990) and Durrett (1991).

Section 23. Continuous-time martingales

Example 23.1(i) belongs to Doleans-Dade (1971). Case (ii) and case (iii) of the same example are described by Kabanov (1974) and Stricker (1986) respectively. According to Kazamaki (1972a), Example 23.2 was suggested by P. A. Meyer. Example 23.3 is given by Kazamaki (1972b). Johnson and Helms (1963) have given Example 23.4, but here we follow the presentation given by Dellacherie and Meyer (1982) and Rao (1979). Case (i) of Example 23.5 is treated by Chung and Williams (1990) (see also Revuz and Yor (1991)) while case (ii) was suggested to me by Yor (1986) (see Karatzas and Shreve (1991)). Example 23.6 is presented by Meyer and Zheng (1984). Example 23.7, considered by Radavicius (1980), is an answer to a question posed by B. Grigelionis. Example 23.8 belongs to Walsh (1982). Yor (1978) has treated topics covering Example 23.9. Example 23.10 was communicated to me by Liptser (1985) (see also Liptser and Shiryaev (1989)). According to Kallianpur (1980), Example 23.11(i) belongs to H. Kunita, and the presentation given here is due to Yor. Case (ii) of the same example is considered by Liptser and Shiryaev (1977). Several other counterexamples (some very complicated) can be found in works by Dellacherie and Meyer (1982), Métivier (1982), Kopp (1984), Liptser and Shiryaev (1989), Isaacson (1971), Kazamaki (1974, 1985a), Surgailis (1974), Edgar and Sucheston (1976b), Monroe (1976), Sekiguchi (1976), Stricker (1977, 1984), Janson (1979), Jeulin and Yor (1979), Azema *et al* (1980), Kurtz (1980), Enchev (1984, 1988), Bouleau (1985), Merzbach and Nualart (1985), Williams (1985), Ethier and Kurtz (1986), Jacod and Shiryaev (1987), Dudley (1989), Revuz and Yor (1991), Yor (1992, 1996), Kazamaki (1994) and Pratelli (1994).

Section 24. Poisson process and Wiener process

Example 24.1 and its versions can be found in many sources and so can be classified as probabilistic folklore. According to Goldman (1967), Example 24.2 is due to L. Shepp. We present Example 24.3 following the paper of Szasz (1970). Example 24.4 belongs to Jacod (1975). Hardin (1985) described Example 24.5. Example 24.6, cases (i), (ii) and (iii), was treated by Novikov (1972, 1979, 1983) (but see also

Liptser and Shiryaev (1977)). Example 24.7 was created recently by Novikov (1996). Case (i) of Example 24.8 is considered by Jain and Monrad (1983); for case (ii) see Dudley (1973) and Fernandez De La Vega (1974). Case (iii) of the same example is the main result of Wrobel's work (1982). An anonymous enthusiast from Marseille wrote a letter describing the idea behind Example 24.9. Example 24.10 belongs to Al-Hussaini and Elliott (1989). Several other counterexamples can be found in the works of Moran (1967), Thomasian (1969), Wang (1977), Novikov (1979), Jain and Monrad (1983), Kazamaki and Sekiguchi (1983), Panaretos (1983), Williams (1985), Daley and Vere-Jones (1988), Mueller (1988), Huang and Li (1993) and Yor (1992, 1996).

Finally, let us pose one interesting question concerning the Wiener process. Suppose $X = (X_t, t \geq 0)$ is a process such that: (i) $X_0 = 0$ a.s.; (ii) any increment $X_t - X_s$ with $s < t$ is distributed $\mathcal{N}(0, t-s)$; (iii) any two increments, $X_{t_2} - X_{t_1}$ and $X_{t_4} - X_{t_3}$, where $0 \leq t_1 < t_2 \leq t_3 < t_4$, are independent. **Question**: Do these conditions imply that X is a Wiener process? **Conjecture**: No. (This was confirmed. See the paper by Föllmer, Wu and Yor (2000) cited in the Appendix.)

Section 25. Diverse properties of stochastic processes

Example 25.1 belongs to Grünbaum (1972). Case (i) of Example 25.2 is indicated in the work of Ephremides and Thomas (1974), while case (ii) of the same example was suggested to me by Ivkovic (1985). Example 25.3 is due to H. Tanaka (see Yamada and Watanabe 1971; Zvonkin and Krylov 1981; Durrett 1984). Example 25.4 was originally considered by Tsirelson (1975), but the proof of the non-existence of the strong solution given here belongs to N. V. Krylov (see also Liptser and Shiryaev 1977; Kallianpur 1980). For a variety of further counterexamples we refer the reader to the following sources: Kadota and Shepp (1970), Borovkov (1972), Davies (1973), Cairoli and Walsh (1975), Azema and Yor (1978), Rao (1979), Hasminskii (1980), Hill *et al* (1980), Kallianpur (1980), Krylov (1980), Métivier and Pellaumail (1980), Chitashvili and Toronjadze (1981), Csörgő and Révész (1981), Föllmer (1981), Liptser and Shiryaev (1981, 1982), Washburn and Willsky (1981), Kabanov *et al* (1983), Le Gall and Yor (1983), Melnikov (1983), Van der Hoeven (1983), Zaremba (1983), Barlow and Perkins (1984), Hoover and Keisler (1984), Engelbert and Schmidt (1985), Ethier and Kurtz (1986), Rogers and Williams (1987, 1994), Rutkowski (1987), Küchler and Sorensen (1989), Maejima (1989), Anulova (1990), Ihara (1993), Schachermayer (1993), Assing and Manthey (1995), Hu and Pérez-Abreu (1995) and Rao (1995).

References

AAP	=	*Advances in Applied Probability*
AMM	=	*American Mathematical Monthly*
AmS	=	*American Statistician*
AMS	=	*Annals of Mathematical Statistics*
AP	=	*Annals of Probability*
AS	=	*Annals of Statistics*
JAP	=	*Journal of Applied Probability*
JMVA	=	*Journal of Multivariate Analysis*
JSPI	=	*Journal of Statistical Planning and Inference*
JTP	=	*Journal of Theoretical Probability*
LNM	=	*Lecture Notes in Mathematics*
LNS	=	*Lecture Notes in Statistics*
PTRF	=	*Probability Theory and Related Fields (formerly ZW)*
SAA	=	*Stochastic Analysis and Applications*
SPA	=	*Stochastic Processes and Their Applications*
SPL	=	*Statistics and Probability Letters*
TPA	=	*Theory of Probability and Its Applications (transl. of: Teoriya Veroyatnostey i Primeneniya)*
ZW	=	*Zeitschrift für Wahrscheinlichkeitstheorie und verwandte Gebiete (new title PTRF since 1986)*

Adell, J. A. (1996) Personal communication.

Adler, A. (1990) On the nonexistence of the LIL for weighted sums of identically distributed r.v.s. *J. Appl. Math. Stoch. Anal.* **3**, 135–140.

Adler, A., Rosalsky, A. and Taylor, R. L. (1992) Some SLLNs for sums of random elements. *Bull. Inst. Math. Acad. Sinica* **20**, 335–357.

Ahlo, J. (1982) A class of random variables which are not continuous functions of several independent random variables. *ZW* **60**, 497–500.

Ahmed, A. H. N. (1990) Negative dependence structures through stochastic ordering. *Trab. Estadistica* **5**, 15–26.

Ahsanullah, M. (1985) Some characterizations of the bivariate normal distribution. *Metrika* **32**, 215–217.

Ahsanullah, M. and Sinha, B. K. (1986) On normality via conditional normality. *Calcutta Statist. Assoc. Bulletin* **35**, 199–202.

Akhiezer, N. I. (1965) *The Classical Moment Problem and Some Related Questions in Analysis*. Hafner, New York. (Russian edn 1961)

Al-Hussaini, A. N. and Elliott, R. (1989) Markov bridges and enlarged filtrations. *Canad. J. Statist.* **17**, 329–332.

Alabert, A. and Nualart, D. (1992) Some remarks on the conditional independence and the Markov property. In: *Stochastic Analysis and Related Topics*. Eds H. Koreslioglu and A. Ustunel. Birkhäuser, Basel. 343–364.

Aldous, D. J. (1985) Exchangeability and related topics. *LNM* **1117**, 1–186.

Alonso, A. (1988) A counterexample on the continuity of conditional expectations. *J. Math. Analysis Appl.* **129**, 1–5.

Alsmeyer, G. (1990) Convergence rates in LLNs for martingales. *SPA* **36**, 181–194.

Alvo, M., Cabilio, P. and Feigin, P. D. (1981) A class of martingales with non-symmetric distributions. *ZW* **58**, 87–93.

Ambartzumian, R. A. (1982) Personal communication.

Anderson, T. W. (1958) *An Introduction to Multivariate Statistical Analysis*. John Wiley & Sons, New York.

Anulova, S. V. (1990) Counterexamples: SDE with linearly increasing coefficients may have an explosive solution within a domain. *TPA* **35**, 336–338.

Arnold, L. (1966) Über die Konvergenz einer zufälligen Potenzreihe. *J. Reine Angew. Math.* **222**, 79–112.

Arnold, L. (1967) Convergence in probability of random power series and a related problem in linear topological spaces. *Israel J. Math.* **5**, 127–134.

Ash, R. (1970) *Basic Probability Theory*. John Wiley & Sons, New York.

Ash, R. (1972) *Real Analysis and Probability*. Acad. Press, New York.

Ash, R. B. and Gardner, M. F. (1975) *Topics in Stochastic Processes*. Acad. Press, New York.

Asmussen, S. and Kurtz, T. (1980) Necessary and sufficient conditions for complete convergence in the law of large numbers. *AP* **8**, 176–182.

Assing, S. and Manthey, R. (1995) The behaviour of solutions of stochastic differential inequalities. *PTRF* **103**, 493–514.

Austin, D. G., Edgar, G. A. and Ionescu Tulcea, A. (1974) Pointwise convergence in terms of expectations. *ZW* **30**, 17–26.

Azema, J. and Yor, M. (eds) (1978) Temps locaux. *Asterisque* **52–53**.

Azema, J., Gundy, R. F. and Yor, M. (1980) Sur l'integrabilite uniforme des martingales continues. *LNM* **784**, 53–61.

Azlarov, T. A. and Volodin, N. A. (1983) On the discrete analog of the Marshall–Olkin distribution. *LNM* **982**, 17–23.

Baez-Duarte, L. (1971) An a.e. divergent martingale that converges in probability. *J. Math. Analysis Appl.* **36**, 149–150.

Bagchi, A. (1989) Personal communication.

Balasanov, Yu. G. and Zhurbenko, I. G. (1985) Comments on the local properties of the sample functions of random processes. *Math. Notes* **37**, 506–509.

Balasubrahmanyan, R. and Lau, K. S. (1991) *Functional Equations in Probability Theory*. Acad. Press, Boston.

Baringhaus, L., Henze, N. and Morgenstern, D. (1988) Some elementary proofs of the normality of $XY/(X^2+Y^2)^{1/2}$ when X and Y are normal. *Comp. Math. Appl.* **15**, 943–944.

Baringhaus, L. and Henze, N. (1989) An example of normal $XY/(X^2+Y^2)^{1/2}$ with non-normal X, Y. Preprint, Univ. Hannover.

Barbour, A. D., Holst, L. and Janson, S. (1988) Poisson approximation with the Stein–Chen method and coupling. Preprint, Uppsala Univ.

Barlow, M. T. (1982) One-dimensional stochastic differential equation with no strong solution. *J. London Math. Soc.* (2) **26**, 335–347.

Barlow, R. E. and Proshan, F. (1966) Tolerance and confidence limits for classes of distributions based on failure rate. *AMS* **37**, 1593–1601.

Barlow, M. T. and Perkins, E. (1984) One-dimensional stochastic differential equations involving a singular increasing process. *Stochastics* **12**, 229–249.

Barndorff-Nielsen, O. (1978) *Information and Exponential Families*. John Wiley & Sons, Chichester.

Bartlett, M. (1978) *An Introduction to Stochastic Processes* (3rd edn). Cambr. Univ. Press., Cambridge.

Basterfield, J. G. (1972) Independent conditional expectations and Llog L. *ZW* **21**, 233–240.

Bauer, H. (1996) *Probability Theory*. Walter de Gruyter, Berlin.

Behboodian, J. (1989) Symmetric sum and symmetric product of two independent r.v.s. *J. Theoret. Probab.* **2**, 267–270.

Belyaev, Yu. K. (1985) Personal communication.

Berg, C. (1988) The cube of a normal distribution is indeterminate. *AP* **16**, 910–913.

Berg, C. (1995) Indeterminate moment problems and the theory of entire functions. *J. Comput. Appl. Math.* **65**, 27–55.

Berg, C. and Thill, M. (1991) Rotation invariant moment problem. *Acta Math.* **167**, 207–227.

Berkes, I. (1974) The LIL for subsequences of random variables. *ZW* **30**, 209–215.

Berkes, I., Dehking, H. and Mori, T. (1991) Counterexamples related to the a.s. CLT. *Studia Sci. Math. Hungarica* **26**, 153–164.

Bernstein, S. N. (1928) *Theory of Probability*. Gostechizdat, Moscow, Leningrad. (In Russian; preliminary edition 1916)

Bhattacharjee, A. and Sengupta, D. (1966) On the coefficient of variation of the classes L and $\overline{L}$. *SPL* **27**, 177–180.

Bhattacharya, R. and Waymire, E. (1990) *Stochastic Processes and Applications*. John Wiley & Sons, New York.

Billingsley, P. (1968) *Convergence of Probability Measures*. John Wiley & Sons, New York.

Billingsley, P. (1995) *Probability and Measure* (3rd edn). John Wiley & Sons, New York.

Bischoff, W. and Fieger, W. (1991) Characterization of the multivariate normal distribution by conditional normal distributions. *Metrika* **38**, 239–248.

Bissinger, B. (1980) Stochastic independence versus intuitive independence. *Two-Year College Math. J.* **11**, 122–123.

Blackwell, D. and Dubins, L.-E. (1975) On existence and non-existence of proper regular conditional probabilities. *AP* **3**, 741–752.

Blake, L. H. (1973) Simple extensions of measures and the preservation of regularity of conditional probabilities. *Pacific J. Math.* **46**, 355–359.

Blake, L. H. (1978) Every amart is a martingale in the limit. *J. London Math. Soc.* (2), **18**, 381–384.

Blake, L. H. (1983) Some further results concerning equiconvergence of martingales. *Rev. Roum. Math. Pure Appl.* **28**, 927–932.

Blank, N. M. (1981) On the definiteness of functions of bounded variation and of d.f.s. by the asymptotic behavior as $x \to \infty$. In: *Problems of Stability of Stochastic Models*. Eds V. M. Zolotarev and V. V. Kalashnikov. Inst. Systems Sci., Moscow, 10–15. (In Russian)

Block, H. W., Sampson, A. R. and Savits, T. H. (eds) (1991) *Topics in Statistical Dependence*. (IMS Series, vol. **16**). Inst. Math. Statist., Hayward (CA).

Blumenthal, R. M. and Getoor, R. K. (1968) *Markov Processes and Potential Theory*. Acad. Press, New York.

Blyth, C. (1986) Personal communication.

Bohlmann, G. (1908) Die Grundbergiffe der Wahrscheinlichkeitsrechnung in Ihrer Anwendung auf die Lebensversicherung. In: *Atti dei 4. Congresso Internationale del Matematici*, (Roma 1908), vol. **3**. Ed G. Castelnuovo. 244–278.

Bojdecki, T. (1977) *Discrete-time Martingales*. Warsaw Univ. Press, Warsaw. (In Polish)

Bojdecki, T. (1985) Personal communication.

Bondesson, L., Kristiansen, G. K. and Steutel, F. W. (1996) Infinite divisibility of r.v.s and their integer parts. *SPL* **28**, 271–278.

Borovkov, A. A. (1972) Convergence of distributions of functionals of stochastic processes. *Russian Math. Surveys* **27**, 1–42.

Boss, D. D. (1985) A converse to Scheffé's theorem. *AS* **13**, 423–427.

Bouleau, N. (1985) About stochastic integrals with respect to processes which are not semimartingales. *Osaka J. Math.* **22**, 31–34.

Bradley, R. C. (1980) A remark on the central limit question for dependent random variables. *JAP* **17**, 94–101.

Bradley, R. C. (1982) Counterexamples to the CLT under strong mixing conditions, I and II. *Colloq. Math. Soc. Janos Bolyai* **36**, 153–171 and **57**, 59–67.

Bradley, R. (1982) Personal communication.

Bradley, R. (1989) A stationary, pairwise independent, absolutely regular sequences for which the CLT fails. *PTRF* **81**, 1–10.

Braverman, M. S. (1993) Remarks on characterization of normal and stable distributions. *J. Theoret. Probab.* **6**, 407–415.

Breiman, L. (1967) On the tail behavior of sums of independent random variables. *ZW* **9**, 20–25.

Breiman, L. (1968) *Probability*. Addison-Wesley, Reading (MA).

Breiman, L. (1969) *Probability and Stochastic Processes*. Houghton Mifflin, Boston.

Broughton, A. and Huff, B. W. (1977) A comment on unions of σ-fields. *AMM* **84**, 553–554.

Brown, J. B. (1972) Stochastic metrics. *ZW* **24**, 49–62.

Bryc, W. and Smolenski, W. (1992) On the stability problem for conditional expectation. *SPL* **15**, 41–46.

Bühler, W. J. and Mieshke, K. L. (1981) On $(n-1)$-wise and joint independence and normality of n random variables. *Commun. Statist. Theor. Meth.* **10**, 927–930.

Bulinskii, A. (1988) On different mixing conditions and asymptotic normality. *Soviet Math. Doklady* **37**, 443–447.

Bulinskii, A. (1989) Personal communications.

Burdick, D. L. (1972) A note on symmetric random variables. *AMS* **43**, 2039–2040.

Burkholder, D. L. and Gundy, R. F. (1970) Extrapolation and interpolation of quasi-linear operators of martingales. *Acta Math.* **124**, 249–304.

Cacoullos, T. (1985) Personal communication.

Cairoli, R. and Walsh, J. B. (1975) Stochastic integrals in the plane. *Acta Math.* **134**, 111–183.

Cambanis, S. (1975) The measurability of a stochastic process of second order and its linear space. *Proc. Amer. Math. Soc.* **47**, 467–475.

Cambanis, S., Simons, G. and Stout, W. (1976) Inequalities for $\mathbf{E}[k(X,Y)]$ when the marginals are fixed. *ZW* **36**, 285–294.

Cambanis, S., Hardin, C. D. and Weron, A. (1987) Ergodic properties of stationary stable processes. *SPA* **24**, 1–18.

Candiloro, S. (1993) Personal communication.

Capobianco, M. and Molluzzo, J. C. (1978) *Examples and Counterexamples in Graph Theory*. North-Holland, Amsterdam.

Carnal, H. and Dozzi, M. (1989) On a decomposition problem for multivariate probability measures. *JMVA* **31**, 165–177.

Castillo, E. and Galambos, J. (1987) Bivariate distributions with normal conditionals. In: *Proc. Intern. Assoc. Sci.-Techn. Development* (Cairo'87). Acta Press, Anaheim (CA). 59–62.

Castillo, E. and Galambos, J. (1989) Conditional distributions and the bivariate normal distribution. *Metrika* **36**, 209–214.

Chandra, T. K. (1989) Uniform integrability in the Cesàro sense and the weak LLNs. *Sankhya* **A-51**, 309–317.

Chen, R. and Shepp, L. A. (1983) On the sum of symmetric r.v.s. *AmS* **7**, 236.
Chernogorov, V. G. (1996) Personal communication.
Chitashvili, R. J. and Toronjadze, T. A. (1981) On one-dimensional SDEs with unit diffusion coefficient. Structure of solutions. *Stochastics* **4**, 281–315.
Chow, Y. and Teicher, H. (1971) Almost certain summability of independent identically distributed random variables. *AMS* **42**, 401–404.
Chow, Y. S. and Teicher, H. (1978) *Probability Theory: Independence, Interchangeability, Martingales*. Springer, New York.
Chung, K. L. (1953) Sur les lois de probabilite unimodales. *C. R. Acad. Sci. Paris* **236**, 583–584.
Chung, K. L. (1960) *Markov Chains with Stationary Transition Probabilities*. Springer, Berlin.
Chung, K. L. (1974) *A Course in Probability Theory* (2nd edn). Acad. Press, New York.
Chung, K. L. (1982) *Lectures from Markov Processes to Brownian Motion*. Springer, New York.
Chung, K. L. (1984) Personal communication.
Chung, K. L. and Fuchs, W. H. J. (1951) On the distribution of values of sums of random variables. *Memoirs Amer. Math. Soc.* **6**.
Chung, K. L. and Walsh, J. B. (1969) To reverse a Markov process. *Acta Math.* **123**, 225–251.
Chung, K. L. and Williams, R. J. (1990) *Introduction to Stochastic Integration* (2nd edn). Birkhäuser, Boston.
Churchill, E. (1946) Information given by odd moments. *AMS* **17**, 244–246.
Cinlar, E. (1975) *Introduction to Stochastic Processes*. Prentice-Hall, Englewood Cliffs (NJ).
Clark, J. M. C. (1970) The representation of functionals of Brownian motion by stochastic integrals. *AMS* **41**, 1282–1295.
Clarke, L. E. (1975) On marginal density functions of continuous densities. *AMM* **82**, 845–846.
Coleman, R. (1974) *Stochastic Processes. Problem Solvers*. Allen & Unwin, London.
Courbage, M. and Hamdan, D. (1994) Chapman–Kolmogorov equation for non-Markovian shift-invariant measures. *AP* **22**, 1662–1677.
Cramér, H. (1936) Über eine Eigenschaft der normalen Verteilungfunktion. *Math. Z* **41**, 405–414.
Cramér, H. (1960) On some classes of non-stationary stochastic processes. In: *Proc. 4th Berkeley Symp. Math. Statist. Probab.* **2**. Univ. California Press, Berkeley. 57–78.
Cramér, H. (1964) Stochastic processes as curves in Hilbert space. *TPA* **9**, 169–179.
Cramér, H. (1970) *Random Variables and Probability Distributions*. Cambr. Univ. Press, Cambridge.
Cramér, H. and Leadbetter, M. R. (1967) *Stationary and Related Stochastic Processes*. John Wiley & Sons, New York.
Crow, E. L. (1967) A counterexample on independent events. *AMM* **74**, 716–717.
Csörgő, M. and Révész, P. (1981) *Strong Approximations in Probability and Statistics*. Akad. Kiadó, Budapest, and Acad. Press, New York.
Csörgő, S., Tandori, K. and Totik, V. (1983) On the strong law of large numbers for pairwise independent random variables. *Acta Math. Hungar.* **42**, 319–330.
Cuculescu, I. (1970) Nonuniformly integrable non-negative martingales. *Rev. Roum. Math. Pure Appl.* **15**, 327–337.
Cuevas, A. (1987) Density estimation: robustness versus consistency. In: *New Perspectives in Theoretical and Applied Statistics*. Ed M. L. Puri. John Wiley & Sons, New York. 259–264.
Cuevas, A. (1989) Personal communications.
Daley, D. J. and Hall, P. (1984) Limit laws for the maximum of weighted and shifted i.i.d. r.v.s. *AP* **12**, 571–587.

Daley, D. J. and Vere-Jones, D. (1988) *An Introduction to the Theory of Point Processes*. Springer, New York.

Dall'Aglio, G. (1960) Les fonctions extrêmes de la classe de Fréchet a 3 dimensions. *Publ. Inst. Statist. Univ. Paris* **9**, 175–188.

Dall'Aglio, G. (1972) Fréchet classes and compatibility of distribution functions. *Symp. Math.* **9**, 131–150.

Dall'Aglio, G. (1990) Somme di variabili aleatorie e convoluzioni. Preprint # 6/1990, Dip. Statist., Univ. Roma "La Sapienza".

Dall'Aglio, G. (1995) Personal communication.

Dall'Aglio, G., Kotz, S. and Salinetti, G. (eds) (1991) *Advances in Probability Distributions with Given Marginals*. (Proc. Symp., Roma'90). Kluwer Acad. Publ., Dordrecht.

Davies, P. L. (1973) A class of almost nowhere differentiable stationary Gaussian processes which are somewhere differentiable. *JAP* **10**, 682–684.

Davis, M. H. A. (1990) Personal communication.

Davydov, Yu. A. (1973) Mixing conditions for Markov chains. *TPA* **18**, 312–328.

De La Cal, J. (1996) Personal communication.

Dehay, D. (1987a) SLLNs for weakly harmonizable processes. *SPA* **24**, 259–267.

Dehay, D. (1987b) On a class of asymptotically stationary harmonizable processes. *JMVA* **22**, 251–257.

Dellacherie, C. (1970) Un example de la théorie generale des processus. *LNM* **124**, 60–70.

Dellacherie, C. (1972) *Capacities et Processus Stochastiques*. Springer, Berlin.

Dellacherie, C. and Meyer, P.-A. (1978) *Probabilities and Potential*. **A**. North-Holland, Amsterdam.

Dellacherie, C. and Meyer, P.-A. (1982) *Probabilities and Potential*. **B**. North-Holland, Amsterdam.

Devroye, L. (1986) *Non-uniform Random Variate Generation*. Springer, New York.

Devroye, L. (1988) Personal communication.

Dharmadhikari, S. W. (1965) An example in the problem of moments. *AMM* **72**, 302–303.

Dharmadhikari, S. W. and Jogdeo, K. (1974) Convolutions of α-modal distributions. *ZW* **30**, 203–208.

Dharmadhikari, S. and Joag-Dev, K. (1988) *Unimodality, Convexity and Applications*. Acad. Press, New York.

Diaconis. P. and Dubins, L. (1980) Finite exchangeable sequences. *AP* **8**, 745–764.

Dilcher, K. (1992) Personal communication.

Dobric, V. (1987) The law of large numbers: examples and counterexamples. *Math. Scand.* **60**, 273–291.

Dodunekova, R. D. (1985) Personal communication.

Doléans-Dade, C. (1971) Une martingale uniformément intégrable mais non localement de carré intégrable. *LNM* **191**, 138–140.

Doob, J. L. (1953) *Stochastic Processes*. John Wiley & Sons, New York.

Doob, J. L. (1984) *Classical Potential Theory and its Probabilistic Counterpart*. Springer, New York.

Doukhan, P. (1994) *Mixing: Properties and Examples*. (Lecture Notes in Statist. **85**.) Springer, New York.

Dozzi, M. (1985) Personal communication.

Dozzi, M. and Imkeller, P. (1990) On the integrability of martingale transforms. Preprint, Univ. Bern.

Drossos, C. (1984) Personal communication.

Dryginoff, M. B. L. (1996) Personal communication.

Dudley, R. M. (1972) A counterexample on measurable processes. In: *Proc. Sixth Berkeley Symp. Math. Statist. Probab.* **II**. Univ. California Press, Berkeley. 57–66.

Dudley, R. M. (1973) Sample functions of the Gaussian processes. *AP* **1**, 66–103.

Dudley, R. M. (1976) *Probabilities and Metrics*. Lecture Notes Ser. no. 45. Aarhus Univ., Aarhus.

Dudley, R. M. (1977) Wiener functionals as Ito integrals. *AP* **5**, 140–141.

Dudley, R. M. (1989) *Real Analysis and Probability*. Wadsworth & Brooks, Pacific Grove (CA).

Durrett, R. (1984) *Brownian Motion and Martingales in Analysis*. Wadsworth & Brooks, Monterey (CA).

Durrett, R. (1991) *Probability: Theory and Examples*. Wadsworth, Belmont (CA).

Dykstra, R. L. and Hewett, J. E. (1972) Examples of decompositions chi-squared variables. *AmS* **26**(4), 42–43.

Dynkin, E. B. (1961) *Theory of Markov Processes*. Prentice-Hall, Englewood Cliffs (NJ). (Russian edn 1959)

Dynkin, E. B. (1965) *Markov Processes*. Vols **1** and **2**. Springer, Berlin. (Russian edn 1963)

Dynkin, E. B. and Yushkevich, A. A. (1956) Strong Markov processes. *TPA* **1**, 134–139.

Eberlein, E. and Taqqu, M. (eds) (1986) *Dependence in Probability and Statistics*. Birkhäuser, Boston.

Edgar, G. A. and Sucheston, L. (1976a) Amarts: A class of asymptotic martingales. A. Discrete parameter. *JMVA* **6**, 193–221.

Edgar, G. A. and Sucheston, L. (1976b) Amarts: A class of asymptotic martingales. B. Continuous parameter. *JMVA* **6**, 572–591.

Edgar, G. A. and Sucheston, L. (1977) Martingales in the limit and amarts. *Proc. Am. Math. Soc.* **67**, 315–320.

Edgar, G. A. and Sucheston, L. (1992) *Stopping Times and Directed Processes*. Cambr. Univ. Press, New York.

Eisenbaum, N. and Kapsi, H. (1995) A counterexample for the Markov property of local time for diffusions on graphs. *LNM* **1613**, 260–265.

Eisenberg, B. and Shixin, G. (1983) Uniform convergence of distribution functions. *Proc. Am. Math. Soc.* **88**, 145–146.

Elliott, R. J. (1982) *Stochastic Calculus and Applications*. Springer, New York.

Emery, M. (1982) Covariance des semimartingales Gaussienes. *C. R. Acad. Sci. Paris Ser. I* **295**, 703–705.

Emmanuele, G. and Villani, A. (1984) Problem 6452. *AMM* **91**, 144.

Enchev, O. B. (1984) *Gaussian random functionals*. Math. Research Report. Techn. Univ. Rousse, Rousse (BG).

Enchev, O. B. (1985) Personal communication.

Enchev, O. (1988) Hilbert-space-valued semimartingales. *Boll. Unione Mat. Italiana* B **2**(7), 19–39.

Engelbert, H. J. and Schmidt, W. (1985) On solutions of one-dimensional stochastic differential equations without drift. *ZW* **68**, 287–314.

Enis, P. (1973) On the relation $\mathbf{E}[\mathbf{E}(X|Y)] = \mathbf{E}X$. *Biometrika* **60**, 432–433.

Ephremides, A. and Thomas, J. B. (1974) On random processes linearly equivalent to white noise. *Inform. Sci.* **7**, 133–156.

Ethier, S. N. and Kurtz, T. G. (1986) *Markov Processes. Characterization and Convergence*. John Wiley & Sons, New York.

Evans, S. N. (1991) Association and infinite divisibility for the Wishart distribution and its diagonal marginals. *J. Multivar. Analysis* **36**, 199–203.

Faden, A. M. (1985) The existence of regular conditional probabilities: necessary and sufficient conditions. *AP* **13**, 288–298.

Falk, R. and Bar-Hillel, M. (1983) Probabilistic dependence between events. *Two-Year College Math. J.* **14**, 240–247.

Falin, G. I. (1985) Personal communication.

Feller, W. (1946) A limit theorem for random variables with infinite moments. *Am. J. Math.* **68**, 257–262.

Feller, W. (1959) Non-Markovian processes with the semigroup property. *AMS* **30**, 1252–1253.

Feller, W. (1968) *An Introduction to Probability Theory and its Applications* **1** (3rd edn). John Wiley & Sons, New York.

Feller, W. (1971) *An Introduction to Probability Theory and its Applications* **2** (2nd edn). John Wiley & Sons, New York.

Fernandez De La Vega, W. (1974) On almost sure convergence of quadratic Brownian variation. *AP* **2**, 551–552.

Fernique, X. (1974) Régularité des trajectoires des fonctions aléatoires Gaussiennes. *LNM* **480**, 1–96.

Fisher, L. and Walkup, D. W. (1969) An example of the difference between the Lévy and Lévy–Prohorov metrics. *AMS* **40**, 322–324.

Fisz, M. (1959) On necessary and sufficient conditions for the validity of the SLLN expressed in terms of moments. *Bull. Acad. Polon. Sci. Ser. Math.* **7**, 221–225.

Fisz, M. (1963) *Probability Theory and Mathematical Statistics* (3rd edn). John Wiley & Sons, New York.

Flury, B. K. (1986) On sums of random variables and independence. *AmS* **40**, 214–215.

Föllmer, H. (1981) Dirichlet processes. *LNM* **851**, 476–478.

Föllmer, H. (1986) Personal communication.

Fortet, R. (1977) *Elements of Probability Theory*. Gordon and Breach, London.

Franken, P. and Lisek, B. (1982) On Wald's identity for dependent variables. *ZW* **60**, 143–150.

Fréchet, M. (1951) Sur les tableaux de corrélation dont les marges sont donées. *Ann. Univ. Lyon* **14**, 53–77.

Freedman, D. (1971) *Brownian Motion and Diffusion*. Holden-Day, San Francisco.

Freedman, D. A. (1980) A mixture of i.i.d. r.v.s. need not admit a regular conditional probability given the exchangeable σ-field. *ZW* **51**, 239–248.

Fu, J. C. (1984) The moments do not determine a distribution. *AmS* **38**, 294.

Fu, J. C. (1993) Poisson convergence in reliability of a large linearly connected systems as related to coin tossing. *Statistica Sinica* **3**, 261–275.

Gaidov, S. (1986) Personal communication.

Galambos, J. (1984) *Introductory Probability Theory*. Marcel Dekker, New York.

Galambos, J. (1987) *The Asymptotic Theory of Extreme Order Statistics* (2nd edn). Krieger, Malabar (FL).

Galambos, J. (1988) *Advanced Probability Theory*. Marcel Dekker, New York.

Galambos, J. and Kotz, S. (1978) *Characterizations of Probability Distributions*. (*LNM* **675**). Springer, Berlin.

Galchuk, L. I. (1985) Gaussian semimartingales. In: *Statistics and control of stochastic processes. Proc. Steklov Seminar* 1984. Eds N. Krylov, R. Liptser and A. Novikov. Optimization Software, New York. 102–121.

Gani, J. and Shanbhag, D. N. (1974) An extension of Raikov's theorem derivable from a result in epidemic theory. *ZW* **29**, 33–37.

Gänssler, P. and Stute, W. (1977) *Wahrscheinlichkeitstheorie*. Springer, Berlin.

Gaposhkin, V. F. (1973) On the SLLN for second-order stationary processes and sequences. *TPA* **18**, 372–375.

Geisser, S. and Mantel, N. (1962) Pairwise independence of jointly dependent variables. *AMS* **33**, 290–291.

Gelbaum, B. R. (1976) Independence of events and of random variables. *ZW* **36**, 333–343.

Gelbaum, B. R. (1985) Some theorems in probability theory. *Pacific J. Math.* **118**, 383–391.

Gelbaum, B. R. and Olmsted, J. M. H. (1964) *Counterexamples in Analysis*. Holden-Day, San Francisco.

Gelbaum, B. R. and Olmsted, J. M. H. (1990) *Theorems and Counterexamples in Mathematics*. Springer, New York.

Geller, N. L. (1978) Some examples of the WLLN and SLLN for averages of mutually independent random variables. *AmS* **32**, 34–36.

Gihman, I. I. and Skorohod, A. V. (1972) *Stochastic Differential Equations*. Springer, Berlin. (Russian edn 1968)

Gihman, I. I. and Skorohod, A. V. (1974/79) *Theory of Stochastic Processes*. Vols **1**, **2** and **3**. Springer, New York. (Russian edns 1971/75)

Gilat, D. (1972) Convergence in distribution, convergence in probability and almost sure convergence of discrete martingales. *AMS* **43**, 1374–1379.

Girsanov, I. V. (1962) An example of nonuniqueness of the solution of Ito stochastic integral equation. *TPA* **7**, 325–331.

Gleser, L. J. (1975) On the distribution of the number of successes in independent trials. *AP* **3**, 182–188.

Gnedenko, B. V. (1937) Sur les fonctions caractéristiques. *Bull. Univ. Moscou, Ser. Internat., Sect. A* **1**, 16–17.

Gnedenko, B. V. (1943) Sur la distribution limite du terme maximum d'une serie aléatoire. *Ann. Math.* **44**, 423–453.

Gnedenko, B. V. (1962) *The Theory of Probability*. Chelsea, New York. (Russian edn 1960)

Gnedenko, B. V. (1985) Personal communication.

Gnedenko, B. V. and Kolmogorov, A. N. (1954) *Limit Distributions for Sums of Independent Random Variables*. Addison-Wesley, Cambridge (MA). (Russian edn 1949)

Gnedenko, B. V. and Solovyev, A. D. (1973) On the conditions for existence of final probabilities for a Markov process. *Math. Operationasforsch. Statist.* **4**, 379–390.

Goldman, J. R. (1967) Stochastic point processes: Limit theorems. *AMS* **38**, 771–779.

Golec, J. (1994) Personal communication.

Goode, J. M. (1995) Personal communication.

Greenwood, P. (1973) Asymptotics of randomly stopped sequences with independent increments. *AP* **1**, 317–321.

Grenander, U. (1963) *Probabilities on Algebraic Structures*. Almqvist & Wiksell, Stockholm and John Wiley & Sons, New York.

Grenander, U. and Rosenblatt, M. (1957) *Statistical Analysis of Stationary Time Series*. John Wiley & Sons, New York.

Grey, D. R. (1989) A note on explosiveness of Markov branching processes. *AAP* **21**, 226–228.

Griffiths, R. C. (1970) Infinitely divisible multivariate gamma distributions. *Sankhya* **A32**, 393–404.

Grigelionis, B. (1977) On martingale characterization of stochastic processes with independent increments. *Lithuanian Math. J.* **17**(1), 52–60.

Grigelionis, B. (1986) Personal communication.

Grimmett, G. (1986) Personal communication.

Grimmett, G. R. and Stirzaker, D. R. (1982) *Probability and Stochastic Processes*. Clarendon Press, Oxford.

Groeneboom, P. and Klaassen, C. A. J. (1982) Solution to Problem 121. *Statist. Neerlandica* **36**, 160–161.

Grünbaum, F. A. (1972) An inverse problem for Gaussian processes. *Bull. Am. Math. Soc.* **78**, 615–616.

Gumbel, E. (1958) Distributions à plusieurs variables dont les marges sont données. *C. R. Acad. Sci. Paris* **246**, 2717–2720.

Gut, A. and Schmidt, K. D. (1983) *Amarts and Set Function Processes*. (*LNM* **1042**). Springer, Berlin.

Gut, A. and Janson, S. (1986) Converse results for existence of moments and uniform integrability for stopped random walks. *AP* **14**, 1296–1317.

Gyöngy, I. (1985) Personal communication.

Hahn, M. G. and Klass, M. J. (1981) The multidimensional CLT for arrays normed by affine transformations. *AP* **9**, 611–623.

Hall, P. (1981) A converse to the Spitzer–Rosen theorem. *AP* **9**, 633–641.

Hall, P. and Heyde, C. C. (1980) *Martingale Limit Theory and its Application.* Acad. Press, New York.

Halmos, P. R. (1974) *Measure Theory*. Springer, New York.

Hamedani, G. G. (1984) Nonnormality of linear combinations of normal random variables. *AmS* **38**, 295–296.

Hamedani, G. G. (1992) Bivariate and multivariate normal characterizations. *Commun. Statist. Theory Methods* **21**, 2665–2688.

Hamedani, G. G. and Tata, M. N. (1975) On the determination of the bivariate normal distribution from distributions of linear combinations of the variables. *AMM* **82**, 913–915.

Han, C. P. (1971) Dependence of random variables. *AmS* **25**(4), 35.

Hansen, B. G. (1988) On the log-concave and log-convex infinitely divisible sequences and densities. *AP* **16**, 1832–1839.

Hardin, C., Jr. (1985) A spurious Brownian motion. *Proc. Am. Math. Soc.* **93**, 350.

Hasminskii, R. Z. (1980) *Stochastic Stability of Differential Equations.* Sijthoff & Nordhoff, Alphen aan den Rijn. (Russian edn 1969)

Hawkes, J. (1973) On the covering of small sets by random intervals. *Quart. J. Math.* **24**, 427–432.

Heilmann, W.-R. and Schröter, K. (1987) Eine Bemerkung über bedingte Wahrscheinlichkeiten, bedingte Erwartungswerte und bedingte Unabhängigkeit. *Blätter* **28**, 119–126.

Hengartner, W. and Theodorescu, R. (1973) *Concentration Functions.* Acad. Press, New York.

Herrndorf, N. (1984) A functional central limit theorem for weakly dependent sequences of random variables. *AP* **12**, 141–153.

Hettmansperger, T. P. and Klimko, L. A. (1974) A note on the strong convergence of distributions. *AS* **2**, 597–598.

Heyde, C. C. (1963a) On a property of the lognormal distribution. *J. Royal Statist. Soc.* **B29**, 392–393.

Heyde, C. C. (1963b) Some remarks on the moment problem. I and II. *Quart. J. Math.* (2) **14**, 91–96, 97–105.

Heyde, C. C. (1975) Kurtosis and departure from normality. In: *Statistical Distributions in Scientific Work.* **1**. Eds G. P. Patil *et al.* Reidel, Dordrecht. 193–221.

Heyde, C. C. (1986) Personal communication.

Hida, T. (1960) Canonical representations of Gaussian processes and their applications. *Memoirs Coll. Sci. Univ. Kyoto* **23**, 109–155.

Hill, B. M., Lane, D. and Sudderth, W. (1980) A strong law for some generalized urn processes. *AP* **8**, 214–226.

Hoffmann-Jorgensen, J. (1994) *Probability with a View Toward Statistics* vols **1** and **2**. Chapman & Hall, London.

Holewijn, P. J. and Hordijk, A. (1975) On the convergence of moments in stationary Markov chains. *SPA* **3**, 55–64.

Hooper, P. M. and Thorisson, H. (1988) On killed processes and stopped filtrations. *Stoch. Analysis Appl.* **6**, 389–395.

Hoover, D. N. and Keisler, H. J. (1984) Adapted probability distributions. *Trans. Am. Math. Soc.* **286**, 159–201.

Houdré, C. (1993) Personal communication.

Hsu, P. L. and Robbins, H. (1947) Complete convergence and the law of large numbers. *Proc. Nat. Acad. Sci. USA* **33**(2), 25–31.

Hu, T. C. (1991) On the law of the iterated logarithm for arrays of random variables. *Commun. Statist. Theory Methods* **20**, 1989–1994.

Hu, Y. and Pérez-Abreu, V. (1995) On the continuity of Wiener chaos. *Bol. Soc. Mat. Mexicana* **1**, 127–135.

Huang, J. S. (1975) A note on order statistics from Pareto distribution. *Skand. Aktuarietidskr.* **3**, 187–190.

Huang, W. J. and Li, S. H. (1993) Characterization of the Poisson process using the variance. *Commun. Statist. Theory Methods* **22**, 1371–1382.

Hüsler, J. (1989) A note on the independence and total dependence of max i.d. distributions. *AAP* **21**, 231–232.

Hwang, J. S. and Lin, G. D. (1984) Characterizations of distributions by linear combinations of moments of order statistics. *Bull. Inst. Math. Acad. Sinica* **12**, 179–202.

Ibragimov, I. A. (1962) Some limit theorems for stationary processes. *TPA* **7**, 361–392.

Ibragimov, I. A. (1972) On a problem of C. R. Rao on infinitely divisible laws. *Sankhya* **A34**, 447–448.

Ibragimov, I. A. (1975) Note on the CLT for dependent random variables. *TPA*, **20**, 135–141.

Ibragimov, I. A. (1983) On the conditions for the smoothness of trajectories of random functions. *TPA* **28**, 229–250.

Ibragimov, I. A. and Linnik, Yu. V. (1971) *Independent and Stationary Sequences of Random Variables*. Wolters-Noordhoff, Gröningen. (Russian edn 1965)

Ibragimov, I. A. and Rozanov, Yu. A. (1978) *Gaussian Random Processes*. Springer, Berlin. (Russian edn 1970)

Ihara, S. (1993) *Information Theory for Continuous Systems*. World Scientific, Singapore.

Ihara, S. (1995) Personal communication.

Ijzeren, J. van (1972) A bivariate distribution with instructive properties as to normality, correlation and dependence. *Statist. Neerland.* **26**, 55–56.

Ikeda, N. and Watanabe, S. (1981) *Stochastic Differential Equations and Diffusion Processes*. North-Holland, Amsterdam.

Iosifescu, M. (1980) *Finite Markov Processes and their Applications*. John Wiley & Sons, Chichester and Tehnica, Bucharest.

Iosifescu, M. (1985) Personal communication.

Isaacson, D. (1971) Continuous martingales with discontinuous marginal distributions. *AMS* **42**, 2139–2142.

Isaacson, D. L. and Madsen, R. W. (1976) *Markov Chains. Theory and Applications*. John Wiley & Sons, New York.

Ito, K. (1963) *Stochastic Processes*. Inostr. Liter., Moscow. (In Russian; transl. from Japanese)

Ito, K. (1984) *Introduction to Probability Theory*. Cambr. Univ. Press, Cambridge.

Ivkovic, Z. (1985) Personal communication.

Ivkovic, Z., Bulatovic, J., Vukmirovic, J. and Zivanovic, S. (1974) *Applications of Spectral Multiplicity in Separable Hilbert Space to Stochastic Processes*. Matem. Inst., Belgrade.

Jacod, J. (1975) Two dependent Poisson processes whose sum is still a Poisson process. *JAP* **12**, 170–172.

Jacod, J. (1979) *Calcul Stochastique et Problème de Martingales*. (*LNM* **714**). Springer, Berlin.

Jacod, J. and Shiryaev, A. (1987) *Limit Theorems for Stochastic Processes*. Springer, Berlin.

Jagers, A. A. (1983) Solution to Problem 650. *Nieuw Archief voor Wiskunde*. Ser. 4, **1**, 377–378.

Jagers, A. A. (1988) Personal communication.

Jain, N. C. and Monrad, D. (1982) Gaussian quasimartingales. *ZW* **59**, 139–159.

Jain, N. C. and Monrad, D. (1983) Gaussian measures in B_p. *AP* **11**, 46–57.

Jamison, B., Orey, S. and Pruitt, W. (1965) Convergence of weighted averages of independent random variables. *ZW* **4**, 40–44.

Janković, S. (1988) Personal communication.

Janson, S. (1979) *A two-dimensional martingale counterexample*. Report no. 8. Inst. Mittag-Leffler, Djursholm.

Janson, S. (1988) Some pairwise independent sequences for which the central limit theorem fails. *Stochastics* **23**, 439–448.

Jensen, U. (1990) An example concerning the convergence of conditional expectations. *Statistics* **21**, 609–611.

Jeulin, T. and Yor, M. (1979) Inegalité de Hardy, semimartingales et faux-amis. *LNM* **721**, 332–359.

Joffe, A. (1974) On a set of almost deterministic k-dependent r.v.s. *AP* **2**, 161–162.

Joffe, A. (1988) Personal communication.

Johnson, B. R. (1974) An inequality for conditional distributions. *Math. Mag.* **47**, 281–283.

Johnson, D. P. (1974) Representations and classifications of stochastic processes. *Trans. Am. Math. Soc.* **188**, 179–197.

Johnson, G. and Helms, L. L. (1963) Class (D) supermartingales. *Bull. Am. Math. Soc.* **69**, 59–62.

Johnson, N. S. and Kotz, S. (1977) *Urn Models and their Application*. John Wiley & Sons, New York.

Jurek, Z. J. and Mason, J. D. (1993) *Operator-Limit Distributions in Probability Theory*. John Wiley & Sons, New York.

Kabanov, Yu. M. (1974) Integral representations of functionals of processes with independent increments. *TPA* **19**, 853–857.

Kabanov, Yu. M. (1985) Personal communication.

Kabanov, Yu. M., Liptser, R. Sh. and Shiryaev, A. N. (1983) Weak and strong convergence of the distributions of counting processes. *TPA* **28**, 303–306.

Kadota, T. T. and Shepp, L. A. (1970) Conditions for absolute continuity between a certain pair of probability measures. *ZW* **16**, 250–260.

Kagan, A. M., Linnik, Yu. V. and Rao, C. R. (1973) *Characterization Problems in Mathematical Statistics*. John Wiley & Sons, New York. (Russian edn 1972)

Kahane, J.-P. (1985) *Some Random Series of Functions* (2nd edn). Cambr. Univ. Press, Cambridge.

Kaishev, V. (1985) Personal communication.

Kalashnikov, V. (1994) *Topics on Regenerative Processes*. CRC Press, Boca Raton (FL).

Kalashnikov, V. (1996) Personal communication.

Kallenberg, O. (1973) Conditions for continuity of random processes without discontinuities of second kind. *AP* **1**, 519–526.

Kallianpur, G. (1980) *Stochastic Filtering Theory*. Springer, New York.

Kallianpur, G. (1989) Personal communication.

Kalpazidou, S. (1985) Personal communication.

Kanter, M. (1975) Stable densities under change of scale and total variation inequalities. *AP* **3**, 697–707.

Karatzas, I. and Shreve, S. E. (1991) *Brownian Motion and Stochastic Calculus* (2nd edn). Springer, New York.

Katti, S. K. (1967) Infinite divisibility of integer-valued r.v.s. *AMS* **38**, 1306–1308.

Kazamaki, N. (1972a) Changes of time, stochastic integrals and weak martingales. *ZW* **22**, 25–32.

Kazamaki, N. (1972b) Examples of local martingales. *LNM* **258**, 98–100.

Kazamaki, N. (1974) On a stochastic integral equation with respect to a weak martingale. *Tohoku Math. J.* **26**, 53–63.

Kazamaki, N. (1985a) A counterexample related to A_p-weights in martingale theory. *LNM* **1123**, 275–277.
Kazamaki, N. (1985b) Personal communication.
Kazamaki, N. (1994) *Continuous Exponential Martingales and BMO*. (*LNM* **1579**). Springer, Berlin.
Kazamaki, N. and Sekiguchi, T. (1983) Uniform integrability of continuous exponential martingales. *Tohoku Math. J.* **35**, 289–301.
Kelker, D. (1971) Infinite divisibility and variance mixture of the normal distribution. *AMS* **42**, 802–808.
Kemeny, J. G., Snell, J. L. and Knapp, A. W. (1966) *Denumerable Markov Chains*. Van Nostrand, Princeton (NJ).
Kendall, D. G. (1967) On finite and infinite sequences of exchangeable events. *Studia Sci. Math. Hung.* **2**, 319–327.
Kendall, D. G. (1985) Personal communication.
Kendall, D. G. and Rao, K. S.(1950) On the generalized second limit theorem in the theory of probability. *Biometrika* **37**, 224–230.
Kendall, M. G. and Stuart, A. (1958) *The Advanced Theory of Statistics*. **1**. Griffin, London.
Kenderov, P. S. (1992) Personal communication.
Khintchine, A. Ya. (1934) Korrelationstheorie der stationären stochastischen Prozesse. *Math. Ann.* **109**, 604–615.
Kimeldorf, D. and Sampson, A. (1978) Monotone dependence. *AS* **6**, 895–903.
Kingman, J. F. C. and Taylor, S. J. (1966) *Introduction to Measure and Probability*. Cambr. Univ. Press, Cambridge.
Klass, M. J. (1973) Properties of optimal extended-valued stopping rules for S_n/n. *AP* **1**, 719–757.
Klesov, O. I. (1995) Convergence a.s. of multiple sums of independent r.v.s. *TPA* **40**, 52–65.
Klimov, G. P. and Kuzmin, A. D. (1985) *Probability, Processes, Statistics. Exercises with Solutions*. Moscow Univ. Press, Moscow. (In Russian)
Klopotowski, A. (1996) Personal communication.
Kolmogorov, A. N. (1956) *Foundations of the Theory of Probability*. Chelsea, New York. (German edn 1933; Russian edns 1936 and 1973)
Kolmogorov, A. N. and Fomin, S. V. (1970) *Introductory Real Analysis*. Prentice-Hall, Englewood Cliffs (NJ). (Russian edn 1968)
Kopp, P. E. (1984) *Martingales and Stochastic Integrals*. Cambr. Univ. Press, Cambridge.
Kordzahia, N. (1996) Personal communication.
Kotz, S. (1996) Personal communication.
Kovatchev, B. (1996) Personal communication.
Kowalski, C. J. (1973) Nonnormal bivariate distributions with normal marginals. *AmS* **27**(3), 103–106.
Krein, M. (1944) On one extrapolation problem of A. N. Kolmogorov. *Doklady Akad. Nauk SSSR* **46**(8), 339–342. (In Russian)
Krengel, U. (1989) Personal communication.
Krewski, D. and Bickis, M. (1984) A note on independent and exhaustive events. *AmS* **38**, 290–291.
Krickeberg, K. (1967) Strong mixing properties of Markov chains with infinite invariant measure. In: *Proc. 5th Berkeley Symp. Math. Statist. Probab.* **2**, part **II**. Univ. California Press, Berkeley. 431–446.
Kronfeld, B. (1982) Personal communication.
Krylov, N. V. (1980) *Controlled Diffusion Processes*. Springer, New York. (Russian edn 1977)
Krylov, N. V. (1985) Personal communication.
Küchler, U. (1986) Personal communication.

Küchler, U. and Sorensen, M. (1989) Exponential families of stochastic processes: A unified semimartingale approach. *Int. Statist. Rev.* **57**, 123–144.

Kuelbs, J. (1976) A counterexample for Banach space valued random variables. *AP* **4**, 684–689.

Kurtz, T. G. (1969) A note on sequences of continuous parameter Markov chains. *AMS* **40**, 1078–1082.

Kurtz, T. G. (1980) The optional sampling theorem for martingales indexed by directed sets. *AP* **8**, 675–681.

Kuznetsov, S. (1990) Personal communication.

Kwapien, S. (1985) Personal communication.

Laha, R. G. (1958) An example of a nonnormal distribution where the quotient follows the Cauchy law. *Proc. Nat. Acad. Sci. USA* **44**, 222–223.

Laha, R. and Rohatgi, V. (1979) *Probability Theory*. John Wiley & Sons, New York.

Lamperti, J. (1966) *Probability*. Benjamin, New York.

Lamperti, J. (1977) *Stochastic Processes*. Springer, New York.

Lancaster, H. O. (1965) Pairwise statistical independence. *AMS* **36**, 1313–1317.

Landers, D. and Rogge, L. (1977) A counterexample in the approximation theory of random summation. *AP* **5**, 1018–1023.

Laube, G. (1973) Weak convergence and convergence in the mean of distribution functions. *Metrika* **20**, 103–105.

Laue, G. (1983) Existence and representation of density functions. *Math. Nachricht.* **114**, 7–21.

Le Breton, A. (1989, 1996) Personal communications.

Le Gall, J. and Yor, M. (1983) Sur l'équation stochastique de Tsirelson. *LNM* **986**, 81–88.

Lebedev, V. (1985) Personal communication.

Ledoux, M. and Talagrand, M. (1991) *Probability in Banach Space*. Springer, New York.

Lee, M.-L. T. (1985) Dependence by total positivity. *AP* **13**, 572–582.

Lehmann, E. L. (1966) Some concepts of dependence. *AMS* **37**, 1137–1153.

Leipnik, R. (1981) The lognormal distribution and strong non-uniqueness of the moment problem. *TPA* **26**, 850–852.

Leon, A. and Massé, J.-C. (1992) A counterexample on the existence of the L_1-median. *SPL* **13**, 117–120.

Lessi, O. (1993) *Corso di Probabilità*. Metria, Padova.

Letac, G. (1991) Counterexamples to P. C. Consul's theorem about the factorization of the GPD. *Canad. J. Statist.* **19**, 229–232.

Letac, G. (1995) *Integration and Probability: Exercises and Solutions*. Springer, New York.

Lévy, P. (1940) Le mouvement Brownien plan. *Am. J. Math.* **62**, 487–550.

Lévy, P. (1948) The arithmetic character of Wishart's distribution. *Proc. Cambr. Phil. Soc.* **44**, 295–297.

Lévy, P. (1965) *Processus Stochastique et Mouvement Brownien* (2nd edn). Gauthier-Villars, Paris.

Lindemann, I. (1995) Personal communication.

Linnik, Yu. V. and Ostrovskii, I. V. (1977) *Decomposition of Random Variables and Vectors*. Am. Math. Soc., Providence (RI). (Russian edn 1972)

Liptser, R. Sh. (1985) Personal communication.

Liptser, R. Sh. and Shiryaev, A. N. (1977/78) *Statistics of Random Processes*. **1** & **2**. Springer, New York. (Russian edn 1974)

Liptser, R. Sh. and Shiryaev, A. N. (1981) On necessary and sufficient conditions in the functional CLT for semimartingales. *TPA* **26**, 130–135.

Liptser, R. Sh. and Shiryaev, A. N. (1982) On a problem of necessary and sufficient conditions in the functional CLT for local martingales. *ZW* **59**, 311–318.

Liptser, R. Sh. and Shiryaev, A. N. (1989) *Theory of Martingales*. Kluwer Acad. Publ., Dordrecht. (Russian edn 1986)
Liu, D. and Neuts, M. F. (1991) Counterexamples involving Markovian arrival processes. *Stoch. Models* **7**, 499–509.
Liu, J. S. and Diaconis, P. (1993) Positive dependence and conditional independence for bivariate exchangeable random variables. Techn. Rep. 430, Dept. Statist., Harvard Univ.
Loève, M. (1977/78) *Probability Theory*. **1** & **2** (4th edn). Springer, New York.
Lopez, G. and Moser, J. (1980) Dependent events. *Pi-Mu-Epsilon* **7**, 117–118.
Lukacs, E. (1970) *Characteristic Functions* (2nd edn). Griffin, London.
Lukacs, E. (1975) *Stochastic Convergence* (2nd edn). Acad. Press, New York.
McKean, H. P. (1969) *Stochastic Integrals*. Acad. Press, New York.
Maejima, M. (1989) Self-similar processes and limit theorems. *Sugaku Expos.* **2**, 103–123.
Malisic, J. (1970) *Collection of Exercises in Probability Theory with Applications*. Gradjevinska Kniga, Belgrade. (In Serbo-Croatian)
Mandelbrot, B. B. and Van Ness, J. W. (1968) Fractional Brownian motions, fractional noises and applications. *SIAM Rev.* **10**, 422–437.
Marcus, D. J. (1983) Non-stable laws with all projections stable. *ZW* **64**, 139–156.
Marinescu, E. (1985) Personal communication.
Masry, E. and Cambanis, S. (1973) The representation of stochastic processes without loss of information. *SIAM J. Appl. Math.* **25**, 628–633.
Mauldon, J. G. (1956) Characterizing properties of statistical distributions. *Quart. J. Math. Oxford* **7**(2), 155–160.
Melnick, E. L. and Tenenbein, A. (1982) Misspecification of the normal distribution. *AmS* **36**, 372–373.
Melnikov, A. V. (1983) Personal communication.
Merzbach, E. and Nualart, D. (1985) Different kinds of two-parameter martingales. *Israel J. Math.* **52**, 193–208.
Métivier, M. (1982) *Semimartingales. A Course on Stochastic Processes*. Walter de Gruyter, Berlin.
Métivier, M. and Pellaumail, J. (1980) *Stochastic Integration*. Acad. Press, New York.
Metry, M. H. and Sampson, A. R. (1993) Ordering for positive dependence on multivariate empirical distribution. *Ann. Appl. Probab.* **3**, 1241–1251.
Meyer, P.-A. and Zheng, W. A. (1984) Tightness criteria for laws of semimartingales. *Ann. Inst. H. Poincaré* **B20**, 353–372.
Meyn, S. P. and Tweedie, R. L. (1993) *Markov Chains and Stochastic Stability*. Springer, London.
Mikusinski, P., Sherwood, H. and Taylor, M. D. (1992) Shuffles of min. *Stochastica* **13**, 61–74.
Molchanov, S. A. (1986) Personal communication.
Monrad, D. (1976) Lévy processes: absolute continuity of hitting times for points. *ZW* **37**, 43–49.
Monroe, I. (1976) Almost sure convergence of the quadratic variation of martingales: a counterexample. *AP* **4**, 133–138.
Moran, P. A. P. (1967) A non-Markovian quasi-Poisson process. *Studia Sci. Math. Hungar.* **2**, 425–429.
Moran, P. A. P. (1968) *An Introduction to Probability Theory*. Oxford Univ. Press, New York.
Morgenstern, D. (1956) Einfache Beispiele zweidimensionaler Verteilungen. *Mitt. Math. Statist.* **8**, 234–245.
Mori, T. F. (1995) Personal communication.
Morrison, J. M. and Wise, G. L. (1987) Continuity of filtrations of σ-algebras. *SPL* **6**, 55–60.
Mucci, A. G. (1973) Limits for martingale-like sequences. *Pacific J. Math.* **48**, 197–202.
Mueller, C. (1988) A counterexample for Brownian motion on manifolds. *Contemporary Math.* **73**, 217–221.

Mutafchiev, L. (1986) Personal communication.

Negri, I. (1995) Personal communication.

Nelsen, R. B. (1992) Personal communication.

Nelson, P. I. (1970) A class of orthogonal series related to martingales. *AMS* **41**, 1684-1694.

Neuts, M. (1973) *Probability*. Allyn & Bacon, Boston.

Neveu, J. (1965) *Mathematical Foundations of the Calculus of Probability*. Holden-Day, San Francisco.

Neveu, J. (1975) *Discrete Parameter Martingales*. North-Holland, Amsterdam.

Novikov, A. A. (1972) On an identity for stochastic integrals. *TPA* **17**, 717–720.

Novikov, A. A. (1979) On the conditions of the uniform integrability of continuous nonnegative martingales. *TPA* **24**, 820–824.

Novikov, A. A. (1983) A martingale approach in problems of first crossing time of nonlinear boundaries. *Proc. Steklov Inst. Math.* **158**, 141–163.

Novikov, A. A. (1985) Personal communication.

Novikov, A. A. (1996) Martingales, Tauberian theorems and gambling systems. Preprint.

Nualart, D. (1995) *The Malliavin Calculus and Related Topics*. Springer, New York.

O'Brien, G. L. (1980) Pairwise independent random variables. *AP* **8**, 170–175.

O'Brien, G. L. (1982) The occurrence of large values in stationary sequences. *ZW* **61**, 347–353.

O'Connor, T. A. (1979) Infinitely divisible distributions with unimodal Lévy spectral functions. *AP* **7**, 494–499.

Olkin, I., Gleser, L. and Derman, C. (1980) *Probability Models and Applications*. Macmillan, New York.

Ord, J. K. (1968) The discrete Student's distribution. *AMS* **39**, 1513–1516.

Pakes, A. G. (1995) Quasi-stationary laws for Markov processes: examples of an always proximate absorbing state. *AAP* **27**, 120–145.

Pakes, A. G. and Khattree, R. (1992) Length-biasing, characterizations of laws and the moment problem. *Austral. J. Statist.* **34**, 307–322.

Panaretos, J. (1983) On Moran's property of the Poisson distribution. *Biometr. J.* **25**, 69–76.

Papageorgiou, H. (1985) Personal communication.

Papoulis, A. (1965) *Probability, Random Variables and Stochastic Processes*. McGraw-Hill, New York.

Parzen, E. (1960) *Modern Probability Theory & Applications*. John Wiley & Sons, New York.

Parzen, E. (1962) *Stochastic Processes*. Holden-Day, San Francisco.

Parzen, E. (1993) Personal communication.

Pavlov, H. V. (1978) Some properties of the distributions of the class NBU. In: *Math. and Math. Education*, **4** (Proc. Spring Conf. UBM). Academia, Sofia. 283–285. (In Russian)

Peligrad, M. (1993) Personal communication.

Pesarin, F. (1990) Personal communication.

Petkova, E. (1994) Personal communications.

Petrov, V. V. (1975) *Sums of Independent Random Variables*. Springer, Berlin. (Russian edn 1972)

Pfanzagl, J. (1969) On the existence of regular conditional probabilities. *ZW* **11**, 244–256.

Pflug, G. (1991) Personal communication.

Philippou, A. N. (1983) Poisson and compound Poisson distributions of order k and some of their properties. *Zap. Nauchn. Semin. LOMI AN SSSR* (Leningrad) **130**, 175–180.

Philippou, A. N. and Hadjichristos, J. H. (1985) *A note on the Poisson distribution of order k and a result of Raikov*. Preprint, Univ. Patras, Patras, Greece.

Piegorsch, W. W. and Casella, G. (1985) The existence of the first negative moments. *AmS* **39**, 60–62.

Pierce, D. A. and Dykstra, R. L. (1969) Independence and the normal distribution. *AmS* **23**(4), 39.
Pitman, E. J. G. and Williams, E. G. (1967) Cauchy-distributed functions of Cauchy variates. *AMS* **38**, 916–918.
Pirinsky, Ch. (1995) Personal communication.
Portenko, N. I. (1982) *Generalized Diffusion Processes*. Naukova Dumka, Kiev. (In Russian)
Portenko, N. I. (1986) Personal communication.
Prakasa Rao, B. L. S. (1992) *Identifiability in Stochastic Models*. Acad. Press, Boston.
Pratelli, L. (1994) Deux contre-exemples sur la convergence d'intégrales anticipative. *LNM* **1583**, 110–112.
Prohorov, Yu. V. (1950) The strong law of large numbers. *Izv. Akad. Nauk SSSR, Ser. Mat.* **14**, 523–536. (In Russian)
Prohorov, Yu. V. (1956) Convergence of random processes and limit theorems in probability theory. *TPA* **1**, 157–214.
Prohorov, Yu. V. (1959) Some remarks on the strong law of large numbers. *TPA* **4**, 204–208.
Prohorov, Yu. V. (1983) On sums of random vectors with values in Hilbert space. *TPA* **28**, 375–379.
Prohorov, Yu. V. and Rozanov, Yu. A. (1969) *Probability Theory*. Springer, Berlin. (Russian edn 1967)
Protter, P. (1990) *Stochastic Integration and Differential Equations. A New Approach*. Springer, New York.
Puri, M. L. (1993) Personal communication.
Rachev, S. T. (1991) *Probability Metrics and the Stability of Stochastic Models*. John Wiley & Sons, Chichester.
Radavicius, M. (1980) On the question of the P. Lévy theorem generalization. *Litovsk. Mat. Sbornik* **20**(4), 129–131. (In Russian)
Raikov, D. A. (1938) On the decomposition of Gauss and Poisson laws. *Izv. Akad. Nauk SSSR, Ser. Mat.* **2**, 91–124. (In Russian)
Ramachandran, B. (1967) *Advanced Theory of Characteristic Functions*. Statist. Society, Calcutta.
Ramachandran, D. (1974) Mixtures of perfect probability measures. *AP* **2**, 495–500.
Ramachandran, D. (1975) On the two definitions of independence. *Colloq. Math.* **32**, 227–231.
Ramakrishnan, S. (1988) A sequence of coin toss variables for which the strong law fails. *AMM* **95**, 939–941.
Rao, C. R. (1973) *Linear Statistical Inference and its Applications* (2nd edn). John Wiley & Sons, New York.
Rao, M. M. (1979) *Stochastic Processes and Integration*. Sijthoff & Noordhoff, Alphen.
Rao, M. M. (1984) *Probability Theory with Applications*. Acad. Press, Orlando.
Rao, M. M. (1993) *Conditional Measures and Applications*. Marcel Dekker, New York.
Rao, M. M. (1995) *Stochastic Processes: General Theory*. Kluwer Acad. Publ., Dordrecht.
Regazzini, E. (1992) Personal communication.
Rényi, A. (1970) *Probability Theory*. Akad. Kiadó, Budapest, and North-Holland, Amsterdam.
Resnik, S. I. (1973) Record values and maxima. *AP* **1**, 650–662.
Révész, P. (1967) *The Laws of Large Numbers*. Akad. Kiadó, Budapest and Acad. Press, New York.
Revuz, D. and Yor, M. (1991) *Continuous Martingales and Brownian Motion*. Springer, Berlin.
Riedel, M. (1975) On the one-sided tails of infinitely divisible distributions. *Math. Nachricht.* **70**, 115–163.
Rieders, E. (1993) The size of the averages of strongly mixing r.v.s. *SPL* **18**, 57–64.

Robbins, H. (1948) Convergence of distributions. *AMS* **19**, 72–76.
Robertson, J. B. and Womak, J. M. (1985) A pairwise independent stationary stochastic process. *SPL* **3**, 195–199.
Robertson, L. C., Shortt, R. M. and Landry, S. S. (1988) Dice with fair sums. *AMM* **95**, 316–328.
Robertson, T. (1968) A smoothing property for conditional expectations given σ-lattices. *AMM* **75**, 515–518.
Robinson, P. M. (1990) Personal communication.
Rogers, L. C. G. and Williams, D. (1987) *Diffusions, Markov Processes and Martingales.* Vol. **2**: *Ito calculus*. John Wiley & Sons, Chichester.
Rogers, L. C. G. and Williams, D. (1994) *Diffusions, Markov Processes and Martingales.* Vol. **1**: *Foundations* (2nd edn). John Wiley & Sons, Chichester.
Rohatgi, V. (1976) *Introduction to Probability Theory*. John Wiley & Sons, New York.
Rohatgi, V. (1984) *Statistical Interference*. John Wiley & Sons, New York.
Rohatgi, V. (1986) Personal communication.
Rohatgi, V. K., Steutel, F. W. and Székely, G. J. (1990) Infinite divisibility of products and quotients of iid random variables. *Math. Sci.* **15**, 53–59.
Rosalsky, A. (1993) On the almost certain limiting behaviour of normed sums of identically distributed positive random variables. *SPL* **16**, 65–70.
Rosalsky, A. and Teicher, H. (1981) A limit theorem for double arrays. *AP* **9**, 460–467.
Rosalsky, A., Stoyanov, J. and Presnell, B. (1995) An ergodic-type theorem à la Feller for nonintegrable strictly stationary continuous time process. *Stoch. Anal. Appl.* **13**, 555–572.
Rosenblatt, M. (1956) A central limit theorem and a strong mixing condition. *Proc. Nat. Acad. Sci. USA* **42**, 43–47.
Rosenblatt, M. (1971) *Markov Processes. Structure and Asymptotic Behavior*. Springer, Berlin.
Rosenblatt, M. (1974) *Random Processes*. Springer, New York.
Rosenblatt, M. (1979) Dependence and asymptotic independence for random processes. In: *Studies in Probability Theory*. **18**. Math. Assoc. of America, Washington (DC). 24–45.
Rossberg, H.-J., Jesiak, B. and Siegal, G. (1985) *Analytic Methods of Probability Theory*. Akademie, Berlin.
Rotar, V. (1985) Personal communication.
Roussas, G. (1972) *Contiguity of Probability Measures*. Cambr. Univ. Press, Cambridge.
Roussas, G. (1973) *A First Course in Mathematical Statistics*. Addison-Wesley, Reading (MA).
Royden, H. L. (1968) *Real Analysis* (2nd edn). Macmillan, New York.
Rozanov, Yu. A. (1967) *Stationary Random Processes*. Holden-Day, San Francisco. (Russian edn 1963)
Rozanov, Yu. A. (1977) *Innovation Processes*. Winston & Sons, Washington (DC). (Russian edn 1974)
Rozovskii, B. L. (1988) Personal communication.
Rudin, W. (1966) *Real and Complex Analysis*. McGraw-Hill, New York.
Rudin, W. (1973) *Functional Analysis*. McGraw-Hill, New York.
Rudin, W. (1994) Personal communication.
Runnenburg, J. Th. (1984) Problem 142 with the solution. *Statist. Neerland.* **39**, 48–49.
Rüschendorf, L. (1991) On conditional stochastic ordering of distributions. *AAP* **23**, 46–63.
Rutkowski, M. (1987) Strong solutions of SDEs involving local times. *Stochastics* **22**, 201–218.
Rutkowski, M. (1995) Left and right linear innovation for a multivariate SαS random variables. *SPL* **22**, 175–184.
Salisbury, T. S. (1986) An increasing diffusion. In: *Seminar in Stochastic Processes* 1984. Eds E. Cinlar *et al.* Birkhäuser, Basel. 173–194.

Salisbury, T. S. (1987) Three problems from the theory of right processes. *AP* **15**, 263–267.
Sato, H. (1987) On the convergence of the product of independent random variables. *J. Math. Kyoto Univ.* **27**, 381–385.
Schachermayer, W. (1993) A counterexample to several problems in the theory of asset pricing. *Math. Finance* **3**, 217–229.
Schoenberg, I. J. (1983) Solution to Problem 650. *Nieuw Archiff Vor Wiskunde, Ser.* 4, **1**, 377–378.
Sekiguchi, T. (1976) Note on the Krickeberg decomposition. *Tohoku Math. J.* **28**, 95–97.
Serfling, R. (1980) *Approximation Theorems of Mathematical Statistics*. John Wiley & Sons, New York.
Seshadri, V. (1986) Personal communication.
Sevastyanov, B. A., Chistyakov, V. P. and Zubkov, A. M. (1985) *Problems in the Theory of Probability*. Mir, Moscow. (Russian edn 1980)
Shanbhag, D. N., Pestana, D. and Sreehari, M. (1977) Some further results in infinite divisibility. *Math. Proc. Cambr. Phil. Soc.* **82**, 289–295.
Shevarshidze, T. (1984) On the multidimensional local limit theorem for densities. In: *Limit Theorems and Stochastic Equations*. Ed. G. M. Manya. Metsniereba, Tbilisi. 12–53.
Shiryaev, A. N. (1985) Personal communication.
Shiryaev, A. (1995) *Probability* (2nd edn). Springer, New York. (Russian edn 1980)
Shohat, J. and Tamarkin, J. (1943) *The Problem of Moments*. Am. Math. Soc., New York.
Shur, M. G. (1985) Personal communication.
Sibley, D. (1971) A metric for weak convergence of distribution functions. *Rocky Mountain J. Math.* **1**, 437–440.
Simons, G. (1977) An unexpected expectation. *AP* **5**, 157–158.
Slud, E. V. (1993) The moment problem for polynomial forms in normal random variables. *AP* **21**, 2200–2214.
Solovay, R. M. (1970) A model of set theory in which every set of reals is Lebesgue measurable. *Ann. Math.* **92**, 1–56.
Solovyev, A. D. (1985) Personal communication.
Speakman, J. M. O. (1967) Two Markov chains with common skeleton. *ZW* **7**, 224.
Spitzer, F. (1964) *Principles of Random Walk*. Van Nostrand, Princeton, (NJ).
Steck, G. P. (1959) A uniqueness property not enjoyed by the normal distribution. *AMS* **29**, 604–606.
Steen, L. A. and Seebach, J. A. (1978) *Counterexamples in Topology* (2nd edn). Springer, New York.
Steutel, F. W. (1970) *Preservation of Infinite Divisibility Under Mixing*. **33**. Math. Centre Tracts, Amsterdam.
Steutel, F. W. (1973) Some recent results in infinite divisibility. *SPA* **1**, 125–143.
Steutel, F. W. (1984) Problem 153 and its solution. *Statist. Neerland.* **38**, 215.
Steutel, F. W. (1989) Personal communications.
Stout, W. (1974a) *Almost Sure Convergence*. Acad. Press, New York.
Stout, W. (1974b) On convergence of φ-mixing sequences of random variables. *ZW* **31**, 69–70.
Stout, W. (1979) Almost sure invariance principle when $\mathbf{E}X_1^2 = \infty$. *ZW* **49**, 23–32.
Stoyanov, J. (1995) Dependency measure for sets of random events or random variables. *SPL* **23**, 13–20.
Stoyanov, J., Mirazchiiski, I., lgnatov, Zv. and Tanushev, M. (1988) *Exercise Manual in Probability Theory*. Kluwer Acad. Publ., Dordrecht. (Bulgarian edn 1985; Polish edn 1991)
Strassen, V. (1964) An invariance principle for the law of the iterated logarithm. *ZW* **3**, 211–226.
Stricker, C. (1977) Quasimartingales, martingales locales, semimartingales et filtrations naturelles. *ZW* **39**, 55–64.

Stricker, C. (1983) Semimartingales Gaussiennes—application au problème de l'innovation. *ZW* **64**, 303–312.

Stricker, C. (1984) Integral representation in the theory of continuous trading. *Stochastics* **13**, 249–265.

Stricker, C. (1986) Personal communication.

Stroock, D. W. and Varadhan, S. R. S. (1979) *Multidimensional Diffusion Processes*. Springer, New York.

Stroud, T. F. (1992) Personal communication.

Sudderth, W. D. (1971) A 'Fatou equation' for randomly stopped variables. *AMS* **42**, 2143–2146.

Surgailis, D. (1974) Characterization of a supermartingale by some stopping times. *Lithuanian Math. J.* **14**(1), 147–150.

Syski, R. (1991) *Introduction to Random Processes* (2nd edn). Marcel Dekker, New York.

Szasz, D. O. H. (1970) Once more on the Poisson process. *Studia Sci. Math. Hungar.* **5**, 441–444.

Székely, G. J. (1986) *Paradoxes in Probability Theory and Mathematical Statistics*. Akad. Kiadó, Budapest and Kluwer Acad. Publ., Dordrecht.

Takacs, L. (1985) Solution to Problem 6452. *AMM* **92**, 515.

Takahasi, K. (1971/72) An example of a sequence of frequency function which converges to a frequency function in the mean of order 1 but nowhere. *J. Japan Statist. Soc.* **2**, 33–34.

Tanny, D. (1974) A zero–one law for stationary sequences. *ZW* **30**, 139–148.

Targhetta, M. L. (1990) On a family of indeterminate distributions. *J. Math. Anal. Appl.* **147**, 477–479.

Taylor, R. L. and Wei, D. (1979) Laws of large numbers for tight random elements in normed linear spaces. *AP* **7**, 150–155.

Taylor, R. L., Daffer, P. Z. and Patterson, R. F. (1985) *Limit Theorems for Sums of Exchangeable Random Variables*. Rowman & Allanheld, Totowa (NJ).

Thomas, J. (1971) *An Introduction to Applied Probability and Random Processes*. John Wiley & Sons, New York.

Thomasian, A. J. (1957) Metrics and norms on spaces of random variables. *AMS* **28**, 512–514.

Thomasian, A. (1969) *The Structure of Probability Theory with Applications*. McGraw-Hill, New York.

Tjur, T. (1980) *Probability Based on Radon Measures*. John Wiley & Sons, Chichester.

Tjur, T. (1986) Personal communication.

Tomkins, R. J. (1975a) On the equivalence of modes of convergence. *Canad. Math. Bull.* **10**, 571–575.

Tomkins, R. J. (1975b) Properties of martingale-like sequences. *Pacific J. Math.* **61**, 521–525.

Tomkins, R. J. (1980) Limit theorems without moment hypotheses for sums of independent random variables. *AP* **8**, 314–324.

Tomkins, R. J. (1984a) Martingale generalizations and preservation of martingale properties. *Canad. J. Statist.* **12**, 99–106.

Tomkins, R. J. (1984b) Martingale generalizations. In: *Topics in Applied Statistics*. Eds Y. P. Chaubey and T. D. Dviwedi. Concordia Univ., Montreal. 537–548.

Tomkins, R. J. (1986) Personal communication.

Tomkins, R. J. (1990) A generalized LIL. *SPL* **10**, 9–15.

Tomkins, R. J. (1992) Refinements of Kolmogorov's LIL. *SPL* **14**, 321–325.

Tomkins, R. J. (1996) Refinement of a 0–1 law for maxima. *SPL* **27**, 67–69.

Tong, Y. L. (1980) *Probability Inequalities in Multivariate Distributions*. Acad. Press, New York.

Too, Y. H. and Lin, G. D. (1989) Characterizations of uniform and exponential distributions. *SPL* **7**, 357–359.

Tsirelson, B. S. (1975) An example of a stochastic equation having no strong solution. *TPA* **20**, 416–418.
Tsokos, C. (1972) *Probability Distributions: An Introduction to Probability Theory with Applications*. Duxbury Press, Belmont (CA).
Twardowska, K. (1991) Personal communication.
Tweedie, R. L. (1975) Sufficient conditions for ergodicity and recurrence of Markov chains on a general state space. *SPA* **3**, 385–403.
Tzokov, V. S. (1996) Personal communication.
Vahaniya, N. N., Tarieladze, V. I. and Chobanyan, S. A. (1989) *Probability Distributions in Banach Spaces*. Kluwer Acad. Publ., Dordrecht. (Russian edn 1985)
Van der Hoeven, P. C. T. (1983) *On Point Processes*. **165**. Math. Centre Tracts, Amsterdam.
Van Eeden, C. (1989) Personal communication.
Vandev, D. L. (1986) Personal communication.
Vasudeva, R. (1984) Chover's law of the iterated logarithm and weak convergence. *Acta Math. Hung.* **44**, 215–221.
Verbitskaya, I. N. (1966) On conditions for the applicability of the SLLN to wide sense stationary processes. *TPA* **11**, 632–636.
Vitale, R. A. (1978) Joint vs individual normality. *Math Magazine* **51**, 123.
Walsh, J. B. (1982) A non-reversible semimartingale. *LNM* **920**, 212.
Wang, A. (1977) Quadratic variation of functionals of Brownian motion. *AP* **5**, 756–769.
Wang, Y. H. (1979) Dependent random variables with independent subsets. *AMM* **86**, 290–292.
Wang, Y. H. (1990) Dependent random variables with independent subsets—II. *Canad. Math. Bull.* **33**, 24–28.
Wang, Y. H., Stoyanov, J. and Shao, Q.-M. (1993) On independence and dependence properties of sets of random events. *AmS* **47**, 112–115.
Wang, Zh. (1982) A remark on the condition of integrability in quadratic mean for second order random processes. *Chinese Ann. Math.* **3**, 349–352. (In Chinese)
Washburn, R. B. and Willsky, A. S. (1981) Optional sampling of submartingales indexed by partially observed sets. *AP* **9**, 957–970.
Wentzell, A. D. (1981) *A Course in the Theory of Stochastic Processes*. McGraw-Hill, New York. (Russian edn 1975)
Whittaker, J. (1991) *Graphical Models in Applied Multivariate Statistics*. John Wiley & Sons, Chichester.
Williams, D. (1991) *Probability with Martingales*. Cambr. Univ. Press, Cambridge.
Williams, R. J. (1984) Personal communication.
Williams, R. J. (1985) Reflected Brownian motion in a wedge: semimartingale property. *ZW* **69**, 161–176.
Wintner, A. (1947) *The Fourier Transforms of Probability Distributions*. Baltimore (MD). (Published by the author.)
Witsenhausen, H. S. (1975) On policy independence of conditional expectations. *Inform. Control* **28**, 65–75.
Wittmann, R. (1985) A general law of iterated logarithm. *ZW* **68**, 521–543.
Wong, C. K. (1972) A note on mutually independent events. *AmS* **26**, April, 27–28.
Wong, E. (1971) *Stochastic Processes in Information and Dynamical Systems*. McGraw-Hill, New York.
Wright, F. T., Platt, R. D. and Robertson, T. (1977) A strong law for weighted averages of i.i.d. r.v.s. with arbitrarily heavy tails. *AP* **5**, 586–590.
Wrobel, A. (1982) On the almost sure convergence of the square variation of the Brownian motion. *Probab. Math. Statist.* **3**, 97–101.

Yamada, T. and Watanabe, S. (1971) On the uniqueness of solutions of SDEs. I and II. *J. Math. Kyoto Univ.* **11**, 115–167, 553–563.
Yamazaki, M. (1972) Note on stopped average of martingales. *Tohoku Math. J.* **24**, 41–44.
Yanev, G. P. (1993) Personal communication.
Yeh, J. (1973) *Stochastic Processes and the Wiener integrals.* Marcel Dekker, New York.
Yellott, J. and Iverson, G. J. (1992) Uniqueness properties of higher-order autocorrelation functions. *J. Optical Soc. Am.* **9**, 388–404.
Ying, P. (1988) A note on independence of random events and random variables. *Natural Sci. J. Hunan Normal Univ.* **11**, 19–21. (In Chinese)
Yor, M. (1978) Un exemple de processus qui n'est pas une semi-martingale. *Asterisque* **52–53**, 219–221.
Yor, M. (1986, 1996) Personal communications.
Yor, M. (1989) De convex resultats sur l'équation de Tsirelson. *C. R. Acad. Sci. Paris, Ser. I* **309**, 511–514.
Yor, M. (1992) *Some Aspects of Brownian Motion. Part I: Some Special Functionals.* Birkhäuser, Basel.
Yor, M. (1996) *Some Aspects of Brownian Motion. Part II: Recent Martingale Problems.* Birkhäuser, Basel.
Zabczyk, J. (1986) Personal communication.
Zanella, A. (1990) Personal communication.
Zaremba, P. (1983) Embedding of semimartingales and Brownian motion. *Litovsk. Mat. Sbornik* **23**(1), 96–100.
Zbaganu, G. (1985) Personal communication.
Zieba, W. (1993) Some special properties of conditional expectations. *Acta Math. Hungar.* **62**, 385–393.
Zolotarev, V. M. (1961) Generalization of the Kolmogorov inequality. *Issled. Mech. Prikl. Matem. (MFTI)* **7**, 162–166. (In Russian)
Zolotarev, V. M. (1986) *One-dimensional Stable Distributions.* Am. Math. Soc., Providence (RI). (Russian edn 1983)
Zolotarev, V. M. (1989) Personal communication.
Zolotarev, V. M. and Korolyuk, V. S. (1961) On a hypothesis proposed by B. V. Gnedenko. *TPA* **6**, 431–434.
Zubkov, A. M. (1986) Personal communication.
Zvonkin, A. K. and Krylov, N. V. (1981) On strong solutions of SDEs. *Selecta Math. Sovietica* **1**, 19–61. (Russian publication 1975)
Zygmund, A. (1947) A remark on characteristic functions. *AMS* **18**, 272–276.
Zygmund, A. (1968) *Trigonometric Series* Vols **1** and **2** (2nd edn). Cambr. Univ. Press, Cambridge.

APPENDIX

We use here the same standard terminology, abbreviations and notations as in the main body of the book and as generally accepted. The Appendix consists of two parts.

First we give alphabetically **Key Words** followed by reference citations in a chronological order. These are sources where the reader can find a counterexample on that topic.

Then we provide complete bibliographic data of all references. Most of the papers and books, with a few exceptions, included in the **New References**, are published after 1996. All references are carefully selected from a huge pile of several thousands of 'candidates'. Chosen are references containing clearly formulated statements or facts which perfectly fall into the category *counterexamples in probability and stochastic processes*.

There is a huge amount of counterexamples in this area, some are available in the literature, others are 'private possession' of professional researchers and/or teachers in stochastics. Anybody is welcome to contact me (`stoyanovj@gmail.com`) with specific suggestions, comments, sources, etc. There are ways to make all these available to the readers.

Key Words and citations

backward SDE: Hu and Peng (1995), Jia (2008), Crepey and Matoussi (2008), Delong and Imkeller (2010), Fan and Jiang (2010), Ma and Zhang (2011)

Borel-Cantelli lemma: Chandra (2012)

bounded potential: Kartashov and Mishura (2004)

Brownian bridge: Gnedin and Pitman (2005)

Brownian motion: Salisbury (1986), Tsirelson (1997), Föllmer *et al* (2000), Levental and Erickson (2003), Banuelos *et al* (2004), Bass *et al* (2005), Mansuy and Yor

(2006), Kuwada and Sturm (2007), Morters and Peres (2010), Chaumont and Yor (2012), Shiryaev (2012)

canonical correlation: Cuadras (2005)

characteristic functions: Bingham *et al* (1989), Rozovskii (1999), Ushakov (1999), Bisgaard and Sasvári (2000), Shiryaev (2012), Sasvári (2013), Shiryaev *et al* (2013)

characterization of distributions: Van Eeden (1991), Wesolowski (1992), Gilat (1994), Sapatinas (1995), Remapla and Nguyen (1996), Wesolowski (1996), Arnold and Wesolowski (1997), Ilinskii and Chistyakov (1997), Cuesta-Albertos *et al* (1998), Arnold *et al* (1999), Klaassen *et al* (2000), Székely and Rao (2000), Feldman and Graczyk (2000), Lin and Hu (2000), Christoph *et al* (2001), Hamedani *et al* (2007), Jones (2008), Chaumont and Yor (2012), Misiewicz and Wesolowski (2012), Cacoullos and Papadatos (2013)

CLT: Gamkrelidze (1968, 1995), Hall and Wightwick (1986), Bingham *et al* (1989), Cuesta-Albertos and Matrán (1991), Song (1999), Shorack (2000), Dudley (2002), Häggström (2005), Koralov and Sinai (2007), Bradley (2007, 2009), Grzenda and Zieba (2008), Björklund (2010), Conze and Le Borgne (2011)

CLT and LLN in abstract spaces: Giné and Zinn (1983), Zalesskii *et al* (1991), Guan and Yang (2001), Shirikyan (2006), Rosalsky and Thanh (2007, 2009)

coagulation and fragmentation processes: Bertoin (2006), Pitman (2006)

coalescent: Kingman (1982), Bertoin and Le Gall (2000, 2003), Pitman (2006), Bertoin (2006)

complete convergence: Szynal (1993), Sung (1997)

completeness of SDEs: Elworthy (1978, 1982), Grigoryan (2009), Li and Scheutzow (2011)

compound sums: Schmidli (1999), Watanabe (2008), Albin (2008b), Embrechts *et al* (2012), Foss *et al* (2013)

conditional CLT: Ouchti and Volný (2008)

conditional distribution: Bai and Shepp (1994)

convergence of random sequences: Szynal (1993), Isaac (1999), Dudley (2002), Galambos and Simonelli (2003), Roussas (2005), Dupuis and Wang (2005), Athreya and Lahiri (2006), Koshnevisan (2007), Li *et al* (2010), DasGupta (2010), Shiryaev (2012), Shiryaev *et al* (2013)

convolution of distributions: DasGupta (1994), Shimura (1997), Watanabe (2008), Albin (2008b), Shiryaev (2012), Foss *et al* (2013)

correlation: Arnold and Villasenor (2013), Cuadras (2013)

dependency measure: Stoyanov (1998), Drouet Mari and Kotz (2001), Cuadras (2005), Székely *et al* (2007), Székely and Rizzo (2009)

diffusions: Engländer (2004), Veretennikov (2006), Del Tenno (2009), Bass and Chen (2010), Elworthy *et al* (2010), Collet *et al* (2012)

Dirichlet forms: Fukushima *et al* (2012), Chen and Fukushima (2012)

distributions: Aslmeyer (1996), Letac (1996), Wesolowski (1996), Bertin *et al* (1997), Shimura (1997), Bogachev (1998), Majumdar (1999), Dewan and Prakasa Rao (2000), Shorack (2000), Bose *et al* (2002), Khuri and Khuri (2003), Prokhorov and Ushakov (2003), Roussas (2004), Suhov and Kelbert (2005), Hu *et al* (2006), Hult (2006), Koshnevisan (2007), Denisov and Zwart (2007), Bélisle and Melfi (2008), Michel (2008), Watanabe (2008), Boman and Lindskog (2009), Kundu and Nanda (2010), DasGupta (2010), Backhausz (2011), Kochar and Xu (2011), Păltănea and Zbăganu (2011), Cacoullos and Papadatos (2013), Shiryaev *et al* (2013), Foss *et al* (2013)

empirical probability measures: Devroye and Györfi (1990)

equivalent martingale measures: Delbaen and Schachermayer (1997, 1998), Schweizer (1999), Rásonyi and Stettner (2005), Rokhlin (2008, 2010), Černý and Kallsen (2008), Barndorff-Nielsen and Shiryaev (2011)

ergodicity: Kifer (1986), Boyarsky and Gora (1997), Norris (1997), Athreya and Lahiri (2006), Suhov and Kelbert (2008), Conze and Le Borgne (2011)

exchangeability: Hu (1997), Pitman (2006), Rao and Swift (2006), Bogachev *et al* (2008), Schweinberg (2010), Aldous (2010), Gnedin *et al* (2010)

excursion process: Salisbury (1986)

explosion times: Ichihara (1982), Grigoryan (2009)

extreme values: Embrechts *et al* (2009), Embrechts *et al* (2012)

exponential integrability: Krylov (2002), Revuz and Yor (2003), Jeanblanc *et al* (2009), Chen (2012), Ruf (2012), Sokol (2012, 2013)

filtering problem: Baxendale *et al* (2004), Elworthy *et al* (2010), Van Handel (2012)

filtrations: Feldman (1996), Tsirelson (1997), Coquet *et al* (2000), Mansuy and Yor (2006)

fractional Brownian motion: Pavlov (1980), Cheridito (2001), Pipiras and Taqqu (2001), Embrechts and Maejima (2002), Bojdecki *et al* (2007), Mishura (2008), Biagini *et al* (2008)

fragmentation processes: Bertoin (2006), Gnedin and Pitman (2007)

free chaos: Junge *et al* (2007)

free probability: Voiculescu (2000), Hiai and Petz (2000), Biane (2003), Rao (2005), Nica and Speicher (2006)

free random variables: Bercovici and Voiculescu (1995), Benaych-Georges (2005), Nica and Speicher (2006), Kargin (2007a, 2007b)

functional limit theorems: Jacod and Shiryaev (2003)

Gaussian functional: Dorogovtsev (2007), Chaumont and Yor (2012)

Gaussian measures: Remapla and Nguyen (1996), Bogachev (1998), Rao (2005), Chaumont and Yor (2012)

Gibbs measures: Maes *et al* (1999)

Hardy's condition: Stoyanov and Lin (2012)

heavy tailed distributions: Bingham *et al* (1989), Samorodnitsky and Taqqu (1994), Asmussen *et al* (1999), Embrechts *et al* (2012), Foss *et al* (2013)

high-dimensional approximations: Fujikoshi *et al* (2010)

independence: Stoyanov (1998), Ostrovska (2002, 2006), Roussas (2004), Athreya and Lahiri (2006), Bisgaard and Sasvári (2006), Stepniak and Owsiany (2009), Doukhan *et al* (2010), Chaumont and Yor (2012), Shiryaev (2012), Nelsen and Úbeda-Flores (2012), Arnold and Villasenor (2013)

infinite divisibility: Sapatinas (1995), Kristiansen (1995), Bondesson *et al* (1996), Rosinski and Zak (1997), Sato (1999), Berg (2000), Zempléni (2001), Babu and Manstavichyus (2002), Bose *et al* (2002), Iksanov and Jurek (2003), Steutel and van Horn (2004), Wulfsohn (2005), Hu *et al* (2006), Sapatinas and Shanbhag (2010), Barndorff-Nielsen and Shiryaev (2011), Maejima and Ueda (2013)

invariance principle: Peligrad and Utev (2006)

large deviations: Dinwoodie (1991), Glasserman and Wang (1997), Asmussen (2002), Asmussen *et al* (2002), Nagaev (2002), Garsia (2004), Ridder and Schwartz (2005), Liu and Wu (2009), Móri (2009), Dembo and Zeitouni (2010)

law of iterated logarithm (LIL): Giné and Mason (1998), Guan and Yang (2001)

Lévy measure: Luczak (1997), Sato (1999), Kyprianou (2006), Bertoin (2007), Doney (2007), Applebaum (2009), Sapatinas and Shanbhag (2010)

Lévy processes: Sato (1999), Kyprianou (2006), Doney (2007), Bertoin (2007), Appelbaum (2009), Stroock (2011)

LLN: Houdré and Lacey (1996), Isaac (1999)

Markov chains: Norris (1997), Fulman and Wilmer (1999), Diaconis and Freedman (1999), Stenflo (2001), Kallenberg (2002), Hamdan (2002), Kendall and Montana (2002), Häggström (2005), Roberts and Rosenthal (2006), Jerrum (2006), Suhov and Kelbert (2008), Saloff-Coste and Züniga (2009), Alsmeyer (2010), Goldberg and Jerrum (2012), Collet *et al* (2012), Shiryaev (2012), Bielecki *et al* (2013)

Markov jump linear system: Fragoso and Baczynski (2002), Fragoso and Rocha (2005)

Markov processes: Carmichel *et al* (1984), Salisbury (1986), Koole (1993), Burdzy and Kendall (2000), Bertoin and Yor (2002), Daduna (2006), Tian (2006), Markus and Rosen (2006), Lowther (2009), Van Handel (2009), Baral *et al* (2010), Dai Pra *et al* (2010), Bjöttcher (2011), Fukushima *et al* (2012), Chen and Fukushima (2012), Jaroszewska (2013)

Markov semigroup: Eberle (1999)

Martin boundary: Salisbury (1986)

martingales: Baldi (1986), Imkeller (1986), Choulli and Stricker (1998), Harenbrock and Schmitz (1992), Heinkel (1996), Feldman and Smorodinsky (1997), Majer and Mancino (1997), Delbaen and Schachermayer (1998), Elworthy *et al* (1999), Isaac (1999), Krylov (2002), Lejay (2002), De La Rue (2002), Strasser (2003), Revuz and Yor (2003), Kuo (2006), Mansuy and Yor (2006), Cherny (2007), Iksanov and Marynich (2008), Stoica (2008), Albin (2008a), Jeanblanc *et al* (2009), Barndorff-Nielsen and Shiryaev (2011), Chaumont and Yor (2012), Shiryaev (2012)

mixing conditions: Rosinski and Zak (1997), Veretennikov (2006), Bradley (2007), Dedecker *et al* (2007), Doukhan *et al* (2009), Doukhan *et al* (2010)

moment problem: Berg and Valent (1994), Lin (1997), Pedersen (1998), Stoyanov (2000), Bisgaard and Sasvári (2000), Berg (2000), Bryc (2001), Pakes (2001), Pakes *et al* (2001), Bertoin and Yor (2002), Christiansen (2003), De Jeu (2003), Wulfsohn (2005), Mahmoud (2008), DasGupta (2008), Kleiber (2012, 2013), Berschneider and Sasvári (2012), Stoyanov and Lin (2012), Sasvári (2013)

moment sequences: Bisgaard and Sasvári (2000), Christiansen (2003), Berg and Durán (2004), Berg (2005), Chalender and Partington (2007)

normal distribution: Wesolowski (1992), DasGupta (1994), Remapla and Nguyen (1996), Kagan *et al* (1997), Arnold *et al* (1999), Hamedani *et al* (2007), Cuesta-Albertos *et al* (2007)

Novikov condition: Krylov (2002), Jeanblanc *et al* (2009), Sokol (2012, 2013), Ruf (2012)

occupancy problems: Bogachev *et al* (2008), Barbour and Gnedin (2009)

optimal stopping problem: Bruss (2000), Gnedin (2004), Belomestny *et al* (2009), Bruss and Yor (2012)

oscillating Lévy exponent: Farkas *et al* (2001)

paradoxical local martingales: Elworthy *et al* (1999)

passage times: Bondesson *et al* (1996)

perturbed random walk: Alsmeyer *et al* (2013)

phase type distributions: He and Zhang (2005)

Poisson-Dirichlet distribution: Feng (2012)

Poisson Process: Gnedin (2004, 2008)

positive dependence: Hu *et al* (2004)

probability: Shields (1993), Bélisle *et al* (1997), Dudley (2002), Vershik (2004), Roussas (2004), Roussas (2005), Athreya and Lahiri (2006), Tanner (2006), Rosenthal (2006), Beam (2007), Cuesta-Albertos *et al* (2007), Berti and Regazzini (2007), Jacobsen *et al* (2008), Boman and Lindskog (2009), Kahn and Neiman (2009), DasGupta (2010), Björklund (2010), Holtz (2011), Bogachev *et al* (2011), Stroock (2011), Misiewicz and Wesolowski (2012), Shiryaev (2012)

probability inequalities: Bisgaard (1990), Galambos and Simonelli (1996), Sato (1997), Geiss (1999), Jiang and Chen (2004), Kartashov and Mishura (2004), Abadir (2005), Junge *et al* (2007), Wagner (2008), Jiand and Peng (2010)

projections of probability distributions: Remapla and Nguyen (1996), Bélisle *et al* (1997), Ushakov and Ushakov (2001), Cuesta-Albertos *et al* (2007)

quadratic forms: Christoph *et al* (2001)

quantum information channels: Maasen (2010), Collins and Nechita (2011)

quantum probability: Chebotarev (2000), Rao (2005), Maasen (2010)

quasi-concave density: Kibzun and Matveev (2010)

quasi-stationarity: Collet *et al* (2012)

queueing systems: Buitenhek and Van Houtum (1997), Cheng (1997)

R**-symmetric distribution**: Jones and Arnold (2008)

random elements in abstract spaces: Dudek and Zieba (1997), Geiss (1999), Guan and Yang (2001), Talagrand (2005), Rosalsky and Thanh (2009), Stroock (2011), Ledoux and Talagrand (2012)

random events: Hombas (1997), Stoyanov (1998, 2005), Klesov and Rosalsky (1999), Charalambides (2005), Stoyanov (2008), Grzenda and Ziemba (2008), Stepniak and Owsiany (2009)

random fields: Houdré and Lacey (1996), Bryc (2001), El Machkouri and Volný (2003), Rue and Held (2005), Bulinskii and Shashkin (2009)

random forests: Le Gall (1999), Pavlov (2000), Pitman (2006)

random functions: Ushakov and Ushakov (2001)

random partitions: Pitman (2006), Schweinberg (2010), Gnedin *et al* (2010)

random permutations: Gnedin and Olshanski (2012)

random projections: Cuesta-Albertos *et al* (2007)

random series: Nam and Rosalsky (1996), Kruglov and Bo (2001), Stoica (2008)

random sets: Kendall (2000), Gnedin and Pitman (2005), Molchanov (2005), Schneider and Weil (2008)

random trees: Le Gall (1999), Konsowa (2002)

random variables: Schief (1989), Móri (1992), Brown and Xia (1995), Suppes *et al* (1996), Shimura (1997), Pemantle and Rosenthal (1999), Edwardes (2000), Ostrovska (2002), Hardy (2003), Kaluszka and Okolewski (2004), Hu *et al* (2004), Roussas (2004), Stoyanov (2005), Bisgaard and Sasvári (2006), Marksström (2007), Oliveira (2011), Cacoullos and Papadatos (2013)

random walks: Bramson (1991), Norris (1997), Sznitman (2002), Komorowski (2003), Löwe and Matzinger (2003), Betracchi and Zucca (2004), Foss *et al* (2005), Bramson *et al* (2006), Dungey (2007), Merkl and Rolles (2007), Blanchet and Glynn (2008), Alsmeyer *et al* (2013)

reliability analysis: Harris (1991), Cheng (1997), Lin and Hu (2000), Müller and Stoyan (2002), Yu (2011), Anis (2012)

self-decomposability: Sato (1999), Bose *et al* (2002), Iksanov and Jurek (2003), Kozubowski (2005), Bertoin (2007), Sapatinas and Shanbhag (2010), Maejima and Ueda (2013)

self-similar processes: Embrechts and Maejima (2002)

semimartingales: Jacod and Shiryaev (2003), Bojdecki *et al* (2007), Schnurr (2011)

space-time Brownian motion: Burdzy and Salisbury (1999)

stable processes: Samorodnitsky and Taqqu (1994), Bondesson *et al* (1996), Kozubowski (2005)

stationary processes: Wood (1995)

Stieltjes classes: Stoyanov (2004), Ostrovska and Stoyanov (2005), Stoyanov and Tolmatz (2005), Pakes (2007), Lin and Stoyanov (2009), Penson *et al* (2011), Kleiber (2012, 2013)

stochastic control: Buckdahn *et al* (2008), Dai Pra *et al* (2010)

stochastic financial models: Duffie and Protter (1992), Shiryaev and Kabanov (1994), Monteiro and Page (1997), Delbaen and Schachermayer (1997, 1998), Becherer (2001), Kramkov (2003), Jourdain (2004), Hamza and Klebaner (2006), Roux and Zastawniak (2006), Lasserre *et al* (2006), Barndorff-Nielsen and Shiryaev (2011)

stochastic flows: Elworthy (1978, 1982), Li and Scheutzow (2011)

stochastic geometry: Bianchi (2009)

stochastic integrals and SDEs: Föllmer and Imkeller (1993), Ahn and Protter (1994), Constantin (1998), Swart (2001), Lejya (2002), Strasser (2003), Bass *et al* (2004), Večer and Xu (2004), Appelbaum (2009), Elworthy *et al* (2010), Czichowsky and Schweizer (2011)

stochastic monotonicity: Denardo *et al* (1997)

stochastic orderings: Scarsini (1998), Müller and Stoyan (2002), Hu *et al* (2004), Navarro and Shaked (2006), Shaked and Shanthikumar (2007), Zhao and Li (2009), Franco-Pereira *et al* (2010), Oliveira (2011)

stochastic processes: Kôno (1979), Pflug (1996), Feldman (1996), Perrin and Senoussi (1999), Kallenberg (2002), Bass and Chen (2010), Collet et al (2012), Ayache (2013)

stochastic processes in infinite dimensional space: Elworthy (1978, 1982), Föllmer and Gantert (1997), Geiss (1999), Lukić and Beder (2001), Brzezniak *et al* (2001), Talagrand (2005), Brzezniak *et al* (2010), Ledoux and Talagrand (2010), Feng (2012)

stochastic stability: Fragoso and Baczynski (2002)

stopping times: Kallenberg (2002), Jacod and Shiryaev (2003), Mansuy and Yor (2006), Jeanblanc *et al* (2009)

subexponential distributions: Asmussen *et al* (1999), Schmidli (1999), Foss and Richards (2010), Lin (2011), Embrechts *et al* (2012), Foss *et al* (2013)

superdiffusions: Kuznetsov (2000), Engländer (2004)

supermartingales: Choulli and Stricker (1998), Kallenberg (2002), Jacod and Shiryaev (2003), Jeanblanc *et al* (2009)

symbols of Feller processes: Bjöttcher and Schilling (2009)

symmetric density: Jones (2013)

time change: Pitman (2006), Barndorff-Nielsen and Shiryaev (2011)

transition probabilities: Feinberg and Piunovsky (2009), Jaroszewska (2013)

unimodal distributions: Wefelmeyer (1985), DasGupta (1993), Sato (1993), Basu and DasGupta (1996), Bertin *et al* (1997), Lefévre and Utev (1998), Bose *et al* (2002)

variance: Lin and Huang (1989), Edwardes (2000), Chen *et al* (2010)

weak dependence: Dedecker *et al* (2007), Bradley (2007), Doukhan *et al* (2010)

New References

Abadir, K.M. (2005) The mean-median-mode inequality: counterexample. *Econometric Theory* **21**, 477–482.

Ahn, H. and Protter, P. (1994) A remark on stochastic integration. *LMN* **1583**, 312–315.

Albin, J.M.P. (2008a) A continuous non-Brownian motion martingale with Brownian motion marginal distributions. *SPL* **78**, 682–686.

Albin, J.M.P. (2008b) A note on the closure of convolution power mixtures (random sums) of exponential distributions. *J. Austral. Math. Soc.* **84**, 1–7.

Aldous, D. (2010) More uses of exchangeability: representations of complex random structures. (LMS Series **378**). 35–63. Cambr. Univ. Press, Cambridge.

Alsmeyer, G. (1996) Nonnegativity of odd functional moments of positive r.v.s with decreasing density. *SPL* **26**, 75–82.

Alsmeyer, G. (2010) Branching processes in stationary random environment: The extinction problem revisited. *LNS* **197**, 21–36.

Alsmeyer, G., Iksanov, A. and Meiners, M. (2013) Power and exponential moments of the number of visits and related quantities for perturbed random walks. *JTP*. Available online: `doi:10.1007/s10959-012-0475-7`

Anis, M.Z. (2012) On some properties of the IDMRL class of life distributions. *JSPI* **142**, 3047–3955.

Appelbaum, D. (2009) *Lévy Processes and Stochastic Calculus*. 2nd ed. Cambr. Univ. Press, Cambridge.

Arnold, B.C., Castillo, E. and Sarabia, J.-M. (1999) *Conditional Specification of Statistical Models*. Springer, New York.

Arnold, B.C. and Villasenor, J.A. (2013) On orthogonality of $(X+Y)$ and $X/(X+Y)$ rather than independence. *SPL* **83**, 584–587.

Arnold, B.C. and Wesolowski, J. (1997) Multivariate distributions with Gaussian conditional structure. In: *Stochastic Processes and Functional Analysis*. Eds J.A. Goldstein, N. E. Gretsky and J.J. Uhl. Lecture Notes in Pure and Applied Mathematics **186**. 45–59. Marcel Dekker, New York.

Asmussen, S. (2002) Large deviations in rare events simulation: examples, counterexamples and alternatives. In: *Hong Kong Conf. Monte Carlo and Quasi-Monte Carlo Methods 2000*, pp. 1-9. Springer, Berlin.

Asmussen, S., Fuckerieder, P., Jobmann, M. and Schwefel, H.-P. (2002) Large deviations and fast simulation in the presence of boundaries. *SPA* **102**, 1–23.

Asmussen, S., Klüppelberg, C. and Sigman, K. (1999) Sampling at subexponential times, with queueing applications. *SPA* **79**, 265–286.

Athreya, K.B and Dai, J.J. (2002) On the nonuniqueness of the invariant probability for i.i.d. random logistic maps. *AP* **30**, 437–442.

Athreya, K.B. and Lahiri, S.N. (2006) *Measure Theory and Probability Theory*. Springer, New York.

Ayache, A. (2013) Continuous Gaussian multifractional processes with random pointwise Hölder regularity. *JTP* **26**, 72–93.

Babu, G.D. and Manstavichyus, E. (2002) Infinitely divisible limit processes for the Ewens sampling formula. *Lithuanian Math. J.* **42**, 232–242.

Backhausz, A. (2011) Local degree distributions: examples and counterexamples. *Period. Math. Hungar.* **63**, 153–171.

Bai, Z.B. and Shepp, L.A. (1994) A note on the conditional distribution of X when $|X-y|$ is given. *SPL* **19**, 217–219.

Baldi, P. (1986) Limit set of inhomogeneous O.-U. processes, destabilization and annealing. *SPA* **23**, 153–167.

Banuelos, R., Pang, M. and Pascu, M. (2004) Brownian motion with killing and reflection and the "hot-spots" problem. *PTRF* **130**, 56–68.

Baral, J., Fournier, N., Jaffard, S. and Seuret, S. (2010) A pure jump Markov process with a random singularity spectrum. *AP* **38**, 1924–1946.

Barbour, A.D. and Gnedin, A.V. (2009) Small counts in the infinite occupancy scheme. *Electronic J. in Probab.* **14**, 365–384.

Barndorff-Nielsen, O.E. and Shiryaev, A.N. (2011) *Change of Time and Change of Measure*. World Scientific Publ., Singapore.

Bass, R.F., Burdzy, K. and Chen, Z.-Q. (2004) SDEs driven by stable processes for which pathwise uniqueness fails. *SPA* **111**, 1–15.

Bass, R., Burdzy, K. and Chen, Z.-Q. (2005) Uniqueness for reflected Brownian motion in lip domains. *Ann. Inst. H. Poincaré Probab. Statist.* **41**, 197–235.

Bass, R.F. and Chen, Z.-Q. (2010) Regularity of harmonic functions for a class of singular stable-like processes. *Math. Z.* **266**, 489–503.

Basu, S. and DasGupta, A. (1996). The mean-median-mode of unimodal distributions. *TPA* **41**, 210–223.
Baxendale, P., Chigansky, P. and Liptser, R. (2004) Asymptotic stability of the Wonham filter: ergodic and nonergodic signals. *SIAM J. Control Optimiz.* **43**, 643–669.
Baxter, J.R., Jain, N.C. and Varadhan, S.R.S. (1991) Some familiar examples for which the large deviation principle does not hold. *Comm. Pure Appl. Math.* **44**, 911–923.
Beam, J. (2007) Unfair gambles in probability. *SPL* **77**, 681–686.
Becherer, D. (2001) The numeraire portfolio for unbounded semimartingales. *Finance Stoch.* **5**, 327–341.
Bélisle, C., Masseé, J.-C. and Ransford, Th. (1997) When is a probability measure determined by infinitely many projections? *AP* **25**, 767–786.
Bélisle, C. and Melfi, V. (2008) Independence after adaptive allocation. *SPL* **78**, 214–224.
Belomestny, D.V., Rüschendorf, L. and Urusov, M.A. (2009) Optimal stopping of integral functionals and a "no-loss" free boundary formulation. *TPA 54*, 14–28.
Benaych-Georges, F. (2005) Failure of the Raikov theorem for free r.v.s. *LNM* **1857**, 313–319.
Bent, M. (2005) Personal communication.
Bercovici, H. and Voiculescu, D. (1995) Superconvergence to the central limit and failure of the Cramér theorem for free r.v.s. *PTRF* **103**, 215–222.
Berg, C. (2000) On infinite divisible solutions to indeterminate moment problems. In: *Special Functions. Conference in Hong Kong 1999.* 31–41. World Sci. Publ., River Edge (NJ).
Berg, C. (2005) On powers of Stieltjes moment sequences. I. *JTP* **18**, 871–889.
Berg, C. and Durán, A.J. (2004) A transformation from Hausdorff to Stieltjes moment sequences. *Arkiv Mat.* **42**, 239–257.
Berg, C. and Valent, G. (1994) The Nevanlinna parametrization for some indeterminate Stieltjes moment problems associated with birth and death processes. *Methods Appl. Analysis* **1**, 169–209.
Berti, P. and Regazzini, E. (2007) Modes of convergence in the coherent framework. *Sankhya* **69**, 314–329.
Berschneider, G. and Sasvári, Z. (2012) On a theorem of Karhunen and related moment problems and quadrature formulae. *Operator Theory: Adv. Appl.* **221**, 171–185.
Bertin, E., Cuculescu, I. and Theodorescu, R. (1997) *Unimodality of Probability Measures.* Kluwer, Dordrecht.
Bertoin, J. (2006) *Random Fragmentation and Coagulation Processes.* Cambr. Univ. Press, Cambridge.
Bertoin, J. (2007) *Lévy Processes.* Cambr. Univ. Press, Cambridge.
Bertoin, J. and Le Gall, J.-F. (2000) The Bolthausen-Sznitman coalescent and the genealogy of continuous-state branching processes. *PTRF* **117**, 249–266.
Bertoin, J. and Le Gall, J.-F. (2003) Stochastic flows associated to coalescent processes. *PTRF* **126**, 261–288.
Bertoin, J. and Yor, M. (2002) On the entire moments of self-similar Markov processes and exponential functionals of Lévy processes. *Ann. Fac. Sci. Toulouse Math.* **11**, 33–45.
Betracchi, D. and Zucca, F. (2004) Classification on the average of random walks. *J. Statist. Phys.* **114**, 947–975.
Biagini, F., Hu, Y., Oksendal, B. and Zhang, T. (2008) *Stochastic Calculus for Fractal Brownian Motion and Applications*. Springer, London.
Bianchi, G. (2009) The covariogram determines three-dimensional convex polytops. *Adv. in Math.* **220**, 1771–1808.
Biane, Ph. (2003) Free probability for probabilists. In: *Quantum Probability Communications.* Vol. **XI** (Grenoble 1998). 55–71. World Sci. Publ., River Edge (NJ).

Bielecki, T.R., Jakubowski, J. and Nieweglowski, M. (2013) Intricacies of dependence between components of multivariate Markov chains: weak Markov consistency and weak Markov copulae. *Electronic J. Probab.* **18**:45, 1–21.

Bingham, N.H., Goldie, C.M. and Teugels, J.L. (1989) *Regular Variation*. (Encyclopedia of Mathematics and Its Applications). Cambr. Univ. Press, Cambridge.

Bisgaard, T.M. (1990) Hoeffding's inequalities: counterexample. *JTP* **3**, 71–80.

Bisgaard, T.M. and Sasvári, Z. (2000) *Characteristic Functions and Moment Sequences. Positive Definiteness in Probability*. Nova Sci. Publ., Huntington (NY).

Bisgaard, T.M. and Sasvári, Z. (2006) When does $E[X^k \cdot Y^l] = E[X^k]E[Y^l]$ imply independence? *SPL* **76**, 1111–1116.

Björklund, M. (2010) CLTs for Gromov hyperbolic groups. *JTP* **23**, 871–887.

Blanchet, J. and Glynn, P. (2008) Efficient rare-event simulation for the maximum of heavy-tailed random walks. *Ann. Appl. Probab.* **18**, 1351–1378.

Bogachev, L, Gnedin, A. and Yakubovich, Y. (2008) On the variance of the number of occupied boxes. *Adv. in Appl. Math.* **40**, 401–432

Bogachev, V.I. (1998) *Gaussian Measures*. Amer. Math. Soc., Providence (RI).

Bogachev, V.I., Röckner, M. and Shaposhnikov, S.V. (2011) On uniqueness problems related to elliptic equations for measures. *J. Math. Sci. (New York)* **176**, 759–773.

Bojdecki, T., Gorostiza, L.G. and Talarczyk, A. (2007) Some extensions of fBM and sub-fBM related to particle systems. *Electronic Commun. in Probab.* **12**, 161–172.

Boman, J. and Lindskog, F. (2009) Support theorems for the Radon transform and Cramér-Wold Theorems. *JTP* **22**, 683–710.

Bondesson, L., Kristiansen, G.K. and Steutel, F.W. (1996) Infinite divisibility of r.v.s and their integer parts. *SPL* **28**, 271–278.

Bose, A., DasGupta, A. and Rubin, H. (2002) A contemporary review and bibliography of infinitely divisible distributions and processes. *Sankhya A* **64**, 763–819.

Böttcher, B. (2011) An overshoot approach to recurrence and transience of Markov processes. *SPA* **121**, 1962-1981.

Böttcher, B. and Schilling, R. (2009) Approximation of Feller processes by Markov chains with Lévy increments. *Stoch. & Dynamics* **9**, 71-80.

Bourguin, S. and Tudor, C. (2011) Cramér theorem for gamma r.v.s. *Electronic Commun. in Probab.* **16**, 365–378.

Boyarsky, A. and Gora, P. (1997) *Laws of Chaos*. Birkhäuser, Boston.

Bradley, R.C. (2007) On a stationary, triple-wise independent, absolutely regular counterexample to the CLT. *Rocky Mountain J. Math.* **37**, 25–44.

Bradley, R.C. (2007) *Introduction to Strong Mixing Conditions*. Vol. **3**. Kendrick Press, Heber City (UT).

Bradley, R.C. (2010) A strictly stationary, "causal", 5-tuplewise independent counterexample in the CLT. *ALEA Latin Amer. J. Probab. Math. Stat.* **7**, 377–450.

Bradley, R.C. and Pruss, A.R. (2009) A strictly stationary, N-tuplewise independent counterexamples to the CLT. *SPA* **119**, 3300–3318.

Bramson, M. (1991) Random walk in random environment: a counterexample without potential. *J. Statist. Physics* **62**, 863–875.

Bramson, M., Zeitouni, O. and Zerner, M.P.W. (2006) Shortest spanning trees and a counterexample for random walks in random environments. *AP* **34**, 821–856.

Brzezniak, Z., Goldys, G., Imkeller, P., Peszat, S., Priola, E. and Zabczyk, J. (2010) Time irregularity of generalized Ornstein-Uhlenbeck processes. *C.R. Acad. Sci. Paris Ser. Math.* **348**, 273–276.

Brzezniak, Z., Peszat, S. and Zabczyk, J. (2001) Continuity of stochastic evolutions. *Czechoslovak Math. J.* **51**(126), 679–684.

Brown, T.C. and Xia, A. (1995) On Stein-Chen factors for Poisson approximation. *SPL* **23**, 327–332.
Bruss, F.T. (2000) Sum the odds to one and stop. *AP* **28**, 1384–1391.
Bruss, F.T. and Yor, M. (2012) Stochastic processes with proportional increments and the last-arrival problem. *SPA* **122**, 3239–3261.
Bryc, W. (2001) Stationary random fields with linear regression. *AP* **29**, 504–519.
Buckdahn, R., Ma, J. and Rainer, C. (2008) Stochastic control problems for systems driven by normal martingales. *Ann. Appl. Probab.* **18**, 632–663.
Buitenhek, R. and Van Houtum, G.-J. (1997) On first-come first-served random service discipline in multiclass closed queuing networks. *Probab. Engrg. Inform. Sci.* **7**, 326–339.
Bulinskii, A. and Shashkin, A. (2007) *Limit Theorems for Associated Random Fields and Related Systems*. World Scientific, New Jersey.
Burdzy, K. and Kendall, W.S. (2000) Efficient Markovian couplings: examples and counterexamples. *Ann. Appl. Probab.* **10**, 362–409.
Burdzy, K. and Salisbury, T.S. (1999) On minimal parabolic functions and time-homogeneous parabolic h-transforms. *TAMS* **351**, 3499–3531
Cacoullos, T. and Papadatos, N. (2013) Self-inverse and exchangeable r.v.s. *SPL* **83**, 9–12.
Campbell, J. (2012) Personal communication.
Carmichael, J.P., Massé, J.C. and Theodorescu, R. (1984) Remarks on multiple Markov dependence. *Rend. Semin. Mat. Univ. Politechn. Torino* **42**, 65–75.
Chalender, I. and Partington, J. (2007) Applications of moment problems to the overcomleteness of sequences. *Math. Scand.* **101**, 249–260.
Chandra, T. (2012) *The Borel-Cantelli Lemma*. Springer, Heidelberg.
Charalambides, Ch. (2005) *Combinatorial Methods in Discrete Distributions*. Wiley, Chichester.
Chaumont, L. and Yor, M. (2012) *Exercises in Probability: A Guided Tour from Measure Theory to Random Processes*. 2nd ed. Cambr. Univ. Press, Cambridge.
Chebotarev, A. (2000) *Lectures on Quantum Probability*. Sociedad Matematica Mexicana, Mexico City.
Chen, J., Van Eeden, C. and Zidek, J. (2010) Uncertainty and the conditional variance. *SPL* **80**, 1764–1770.
Chen, Z.-Q. (2012) Uniform integrability of exponential martingales and special bounds of non-local Feynman-Kac semigroups. Preprint available at `arXiv:1105.3020`.
Chen, Z.-Q. and Fukushima, M. (2012) *Symmetric Markov Processes, Time Change, and Boundary Theory*. Princeton Univ. Press, Princeton and Oxford.
Cheng, D.W. (1997) Line reversibility of multiserver systems. *Probab. Engrg. Inform. Sci.* **11**, 177–188.
Cheridito, P. (2001) Mixed fractional Brownian motion. *Bernoulli* **7**, 913–934.
Černý, A. and Kallsen, J. (2008) A counterexample concerning the variance-optimal martingale measure. *Math. Finance* **18**, 305–316.
Cherny, A. (2006) Some particular problems of martingale theory. In: *From Stochastic Calculus to Mathematical Finance. The Shiryaev Festschrift*. Eds Yu. Kabanov, R. Liptser and J. Stoyanov. Springer, Berlin. 109–124.
Chigansky, P. (2010) Personal communication.
Chlebowska, K. (2013) Personal communication.
Choi, C. (1997) A norm inequality for Itô processes. *J. Math. Kyoto Univ.* **37**, 229–240.
Choulli, T. and Stricker, C. (1998) Separation of a super- and submartingale by a martingale. *LNM* **1686**, 67–72.
Christiansen, J.S. (2003) The moment problem associated with the Stieltjes-Wigert polynomials. *J. Math. Anal. Appl.* **277**, 218–245.

Christoph, G., Prohorov, Yu. and Ulyanov, V. (2001) Characterization and stability problems for finite quadratic forms. In: *Asymptotic Methods in Probability and Statistics with Applications (St. Petersburg, 1998)*. Eds N. Balakrishnan, I.A. Ibragimov and V.B. Nevzorov. 39–50. Birkhäuser, Boston.

Collet, P., Martinez, S. and San Martin, J. (2012) *Quasi-Stationary Distributions: Markov Chains, Diffusions and Dynamical Systems*. Springer, New York.

Collins, B. and Nechita, I. (2011) Gaussianization and eigenvalue statistics for random quantum channels (III). *Ann. Appl. Probab.* **21**, 1136–1179.

Constantin, A. (1998) On the pathwise uniqueness of solutions of SDEs. *SAA* **16**, 231–232.

Conze, J.-P. and Le Borgne, S. (2011) Limit law for some modified ergodic sums. *Stoch. & Dynamics* **11**, 107–133.

Coquet, F., Memin, J. and Mackevičius, V. (2000) Some examples and counterexamples of convergences of tribes and filtrations. *Lithuanian Math. J.* **40**, 228–235.

Crepey, S. and Matoussi, A. (2008) Reflected and doubly reflected BSDEs with jumps: a priori estimates and comparison. *Ann. Appl. Probab.* **18**, 2041–2069.

Cuadras, C.M. (2005) Continuous canonical correlation analysis. *Res. Lett. Inf. Math. Sci.* (Massey Univ., NZ) **8**, 97–103.

Cuadras, C.M. (2013) Some quirks in simple, multiple and canonical correlation. Preprint, Department of Statistics, University of Barcelona.

Cuculescu, I. and Theodorescu, R. (2003) Are copulas unimodal? *JMVA* **86**, 48–71.

Cuesta-Albertos, J.A. (1998) Personal communication.

Cuesta-Albertos, J.A., Fraiman, R. and Ransford, T. (2007) A sharp form of the Cramér-Wold theorem. *JTP* **20**, 201–209.

Cuesta-Albertos, J.A., Gordaliza, A. and Matraán, C. (1998) On the geometric behavior of multidimensional centralization measures. *JSPI* **998**, 191–208.

Cuesta-Albertos, J.A. and Matrán, C. (1991) On the asymptotic behavior of sums of pairwise independent random variables. *SPL* **11**, 201–210.

Czichowsky, C. and Schweizer, M. (2011) Closedness in the semimartingale topology for spaces of stochastic integrals with constrained integrands. *LNM* **2006**, 413–436.

Daduna, H. and Szekli, R. (2006) Dependence ordering for Markov processes on partially ordered spaces. *JAP* **43**, 793–814.

Dai Pra, P., Louis, P.-Y. and Minelli, I.G. (2010) Realizable monotonicity for continuous-time Markov processes. *SPA* **120**, 959–982.

DasGupta, A. (1994) Distributions which are Gaussian convolutions. In: *Statistical Decision Theory and Related Topics* (Purdue Symp. V, West Lafayette (IN) 1992). Eds S.S. Gupta and J. Berger. 391–400. Springer, New York.

DasGupta, A. (2008) *Asymptotic Theory of Statistics and Probability*. Springer, New York.

DasGupta, A. (2010) *Fundamentals of Probability: A First Course*. Springer, New York.

De Jeu, M. (2003) Determinate multidimensional measures, the extended Carleman theorem and quasi-analytic weights. *AP* **31**, 1205–1227.

Davis, M.H.A. (2011) Personal communication.

De La Rue, T. (2002) Examples and counterexamples to a.s. convergence of bilateral martingales. *New York J. Math.* **8**, 133–144.

Dedecker, J., Doukhan, P., Lang, G., León, J.R., Louchichi, S. and Prieur, C. (2007) *Weak Dependence: With Examples and Applications*. (LNS **190**). Springer, New York.

Del Tenno, I. (2009) Cut points and diffusions in random environment. *JTP* **22**, 891–933.

Delbaen, F. and Schachermayer, W. (1997) The Banach space of workable contingent claims in arbitrage theory. *Ann. Inst. H. Poincaré Probab. Statist.* **33**, 113–144.

Delbaen, F. and Schachermayer, W. (1998) A simple counterexample to several problems in the theory of asset pricing. *Math. Finance* **8**, 1–11.

Delong, L. and Imkeller, P. (2010) Backward SDEs with time delayed generators—results and counterexamples. *Ann. Appl. Probab.* **20**, 1512–1536.

Dembo, A. and Zeitouni, O. (2010) *Large Deviations Techniques and Applications*. 2nd ed. Jones & Bartlett, Boston (MA).

Denardo, E.V., Feinberg, E.A. and Kella, O. (1997) Stochastic monotonicity for stationary recurrence times of first passage heights. *Ann. Appl. Probab.* **7**, 326–339.

Denisov, D. and Zwart, B. (2007) On a theorem of Breiman and a class of random differential equations. *JAP* **44**, 1031–1046.

Derriennik, Y. and Klopotowski, A. (2000) On Bernstein's example of three pairwise independent random variables. *Sankhya Ser. A* **62**, 318–330.

Devroye, L. and Györfi, L. (1990) No empirical probability measure can converge in the total variation sense for all distributions. *AS* **18**, 1496–1499.

Dewan, I. and Prakasa Rao, B.L.S. (2000) Explicit bounds on Lévy-Prohorov distance for a class of multidimensional distributions. *SPL* **48**, 105–119.

Diaconis, P. and Freedman, D. (1999) Iterated random functions. *SIAM Review* **41**, 45–76.

Dinwoodie, I.H. (1991) A note on the upper bound for i.i.d. large deviations. *AP* **19**, 1732–1736.

Doney, R.A. (2007) *Fluctuation Theory for Lévy Processes*. (LNM **1897**). Springer, Berlin.

Dorogovtsev, A.A. (2007) Conditioning of Gaussian functionals and orthogonal expansions. *Theory Stoch. Process.* **13**, 29–37.

Doukhan, P., Lang, G., Surgailis, D. and Teyssiere, G. (2010) *Dependence in Probability and Statistics*. (LNS **200**). Springer, Berlin.

Doukhan, P., Mayo, N. and Truquet, L. (2009) Weak dependence, models and some applications. *Metrika* **69**, 199–225.

Drouet Mari, D. and Kotz, S. (2001) *Correlation and Dependence*. Imperial College Press, London.

Dudek, D. and Zieba, W. (1997) A note on the strong tightness in $C[0,1]$. *Ann. Univ. M. Curie-Sklodowska (Lublin)* **51**, 227–233.

Dudley, R.M. (2002) *Real Analysis and Probability*. 2nd ed. Cambr. Univ. Press, Cambridge.

Duffie, D. and Protter, P. (1992) From discrete to continuous-time finance: weak convergence of the financial gain process. *Math. Finance* **2**, 1–15.

Dungey, N. (2007) Time regularity for random walks on locally compact groups. *PTRF* **137**, 429–442.

Dupuis, P. and Wang, H. (2005) On the convergence from discrete to continuous time in an optimal stopping problem. *Ann. Appl. Probab.* **15**, 1339–1366.

Eberle, A. (1999) *Uniqueness and Non-uniqueness of Semigroups Generated by Singular Diffusion Operators* (*LNM* **1718**). Springer, Berlin.

Edwardes, M.D.B. (2000) Is variance larger if and only if tails are larger? *SPL* **47**, 141–147.

El Machkouri, M. and Volný, D. (2003) Counterexample to the functional CLT for real random fields. *Ann. Inst. H. Poincaré Probab. Statist.* **39**, 325–337.

Elworthy, K.D. (1978) Stochastic dynamical systems and their flows. In: *Proc. Internat. Conf. Northwestern Univ., Evanston (IL), 1978*. 79–95. Acad. Press, New York.

Elworthy, K.D. (1982) *Stochastic Differential Equations on Manifolds*. (LMS Lecture Notes Series **70**). Cambr. Univ. Press, Cambridge.

Elworthy, K.D., Le Yan, Y. and Li, X.-M. (2010) *The Geometry of Filtering*. (Frontiers in Mathematics). Birkhäuser, Basel.

Elworthy, K.D., Li, X.-M. and Yor, M. (1999) The importance of strictly local martingales: applications to radial O.-U. processes. *PTRF* **115**, 325–355.

Embrechts, P., Lambrigger, D.D. and Wüthrich, M.V. (2009) Multivariate extremes and the aggregation of dependent risks: examples and counterexamples. *Extremes* **12**, 107–127.

Embrechts, P. and Maejima, M. (2002) *Selfsimilar Processes*. Princeton Univ. Press, Princeton.

Embrechts, P., Klüppelberg, C. and Mikosch, T. (2012) *Modelling Extremal Events: For Insurance and Finance*. Springer, Berlin.

Engelbert, H.-J. (2002) On uniqueness of solutions to SDEs: a counterexample. *AP* **30**, 1039–1043.

Engländer, J. (2004) An example and a conjecture concerning scaling limits of superdiffusions. *SPL* **66**, 363–366.

Fan, S.-J. and Jiang, L. (2010) A generalized comparison theorem for BSDEs and its applications. *JTP* **23**, 1–12.

Farkas, W., Jacob, N. and Schilling, R.L. (2001) *Function Spaces Related to Continuous Negative Definite Functions: ψ-Bessel Potential Spaces*. (Dissertationes Mathematicae **393**). Polish Acad. Sci., Warsaw.

Feinberg, E.A. and Piunovsky, A.B. (2009) On strong equivalent nonrandomized transition probabilities. *TPA* **54**, 300–307.

Feldman, J. (1996) ε-close measures producing nonisomorphic filtrations. *AP* **24**, 912–915.

Feldman, J. and Smorodinsky, G. (1997) Simple examples of non-generating Girsanov processes. *LNM* **1655**, 247–251.

Feldman, G.M. and Graczyk, P. (2000) On the Skitovich-Darmois theorem for compact abelian groups. *JTP* **13**, 859–869.

Feng, S. (2012) *The Poisson-Dirichlet Distribution and Related Topics*. Springer, New York.

Föllmer, H. and Imkeller, P. (1993) Anticipation canceled by a Girsanov transformation: a paradox on Wiener space. *Ann. Inst. H. Poincaré Probab. Statist.* **29**, 569–586.

Föllmer, H. and Gantert, N. (1997) Entropy minimization and Schrödinger processes in infinite dimensions. *AP* **25**, 901–926.

Föllmer, H., Wu, C.-T. and Yor, M. (2000) On weak Brownian motion of arbitrary order. *Ann. Inst. H. Poincaré Probab. Statist.* **36**, 447–487.

Foss, S., Palmowski, Z. and Zachary, S. (2005) The probability of exceeding a high boundary on a random time interval for a heavy-tailed random walk. *Ann. Appl. Probab.* **15**, 1936–1957.

Foss, S., Korshunov, D. and Zachary, S. (2013) *An Introduction to Heavy-Tailed and Subexponential Distributions*. 2nd ed., Springer, Berlin.

Foss, S. and Richards, A. (2010) On sums of conditionally independent r.v.s. *Math. Oper. Research* **35**, 102–119.

Fragoso, M.D. and Baczynski, J. (2002) Stochastic versus mean square stability in continuous time linear infinite Markov jump parameter systems. *SAA* **20**, 347–356.

Fragoso, M.D. and Rocha, N.C.S. (2005) Stationary filter for continuous-time Markovian jump linear systems. *SIAM J. Control Optim.* **44**, 801–815.

Franco-Pereira, A.M., Lillo, R.E., Romo, J. and Shaked, M. (2010) Percentile residual life orders. *Appl. Stoch. Models Business Industry* **27**, 235–252.

Fujikoshi, Y., Ulyanov, V.V. and Shimizu, R. (2010) *Multivariate Statistics: High-Dimensional and Large-Sample Approximations*. Wiley, Hoboken (NJ).

Fukushima, M., Oshima, Y. and Takeda, M. (2012) *Dirichlet Forms and Symmetric Markov Processes*. 2nd ed. Walter De Gruyter, Berlin and New York.

Fulman, J. and Wilmer, E.L. (1999) Comparing eigenvalue bounds for Markov chains: when does Poincaré beat Cheeger? *Ann. Appl. Probab.* **9**, 1–13.

Galambos, J. and Simonelli, I. (1996) *Bonferroni-Type Inequalities with Applications*. Springer, New York.

Galambos, J. and Simonelli, I. (2003) Comments on a recent limit theorem of Quine. *SPL* **63**, 89–95.

Gamkrelidze, N.G. (1968) The connection between the local and the integral theorem for lattice distributions. *TPA* **13**, 175–179.

Gamkrelidze, N.G. (1995) On a probabilistic property of the Fibonacci sequence. *Fibonacci Quarterly* **33**, 147–152.

Garsia, J. (2004) An extension of the contraction principle. *JTP* **17**, 403–434.

Geiss, S. (1999) A counterexample concerning the relation between decoupling constants and UMD-constants. *TAMS* **351**, 1355–1375.

Gilat, D. (1994) On a curious property of the exponential distribution. In: *Trans. 12th Prague Conf. Inform. Theory*. 77–80. Academia, Prague.

Giné, E. and Zinn, J. (1982) CLTs and weak LLNs in certain Banach spaces. *PTRF* **62**, 323–354.

Giné, E. and Mason, D.M. (1998) On the LIL for self-normalized sums of i.i.d. r.v.s. *JTP* **11**, 351–370.

Glasserman, P. and Wang, Y. (1997) Counterexamples in importance sampling for large deviations probabilities. *Ann. Appl. Probab.* **7**, 731–746.

Gnedin, A. (2004) Best choice from the planar Poisson process *SPA* **111**, 317–354.

Gnedin, A. (2008) Corners and records of the Poisson process in quadrant, *Electronic Commun. in Probab.* **13**, 187–193.

Gnedin, A., Haulk, C. and Pitman, J. (2010) Characterizations of exchangeable partitions and random discrete distributions by deletion properties. In: *Probability and Mathematical Genetics. Papers in Honour of Sir John Kingman*. Eds N.H. Bingham and C.M. Goldie. LMS Lecture Notes Series **378**. 264–298. Cambr. Univ. Press, Cambridge.

Gnedin, A. and Pitman, J. (2005) Regenerative composition structures. *AP* **33**, 445–479.

Gnedin, A. and Pitman, J. (2007) Poisson Representation of a Ewens Fragmentation Process. *Combinatorics Probab. Computing* **16**, 819–827.

Gnedin, A. and Olshanski, G. (2012) The two-sided infinite extension of the Mallows model for random permutations. *Adv. in Appl. Math.* **48**, 615–639.

Grimmett, G. and Stirzaker, D. (2001a) *Probability and Random Processes*. 3rd ed. Oxford Univ. Press, Oxford.

Grimmett, G. and Stirzaker, D. (2001b) *One Thousand Exercises in Probability*. 2nd ed. Oxford Univ. Press, Oxford.

Goldberg, L.A. and Jerrum, M. (2012) A counterexample to rapid mixing of the Ge-Ŝtefankoviĉ process. *Electronic Commun. in Probab.* **17**, no. 5, 1–6.

Grigoryan, A. (2009) *Heat Kernel and Analysis on Manifolds*. (AMS/IP Studies in Appl. Math. **47**). Amer. Math. Soc., Providence (RI).

Grzenda, W. and Zieba, W. (2008) Conditional CLT. *Internat. Math. Forum* **3**, 1521–1528.

Guan, Q.Y. and Yang, X.Y. (2001) A counterexample for which the LIL for the weighted sums of independent B-valued random elements is not true. *Chinese J. Appl. Probab. Statist.* **17**, 73–80. (In Chinese)

Gubner, J.A. (2005) Theorems and fallacies in the theory of long-range-dependent processes. *IEEE Trans. Inform. Theory* **51**, 1234–1239.

Häggström, O. (2005) On the CLT for geometrically ergodic Markov chains. *PTRF* **132**, 74–82. (See also a correction: **135** (2006), p. 470.)

Hall, P. and Wightwick, J.C.H. (1986) Convergence determining sets in the CLT. *PTRF* **71**, 1–17.

Hamdan, D. (2002) An ergodic Markov chain is not determined by any p-marginals. *Indag. Math.* **13**, 487–498.

Hamedani, G.G., Volkmer, H. and Wesolowski, J. (2007) Characterization problems related to the Shepp property. *Commun. Statist. Theory Methods* **36**, 1049–1057.

Hamza, K. and Klebaner, F.C. (2006) On nonexistence of non-constant volatility in the Black-Scholes formula. *Discrete Contin. Dynam. Systems* **6**, 829–834.

Hanson, D.L. (1998) Limiting σ-algebras – some counterexamples. In: *Asymptotic Methods in Probability and Statistics*. Ed. B. Szyszkowicz. 383–385. Elsevier, Amsterdam.

Hardy, M. (2003) An illuminating counterexample. *AMM* **110**, 234–238.

Harenbrock, M. and Schmitz, N. (1992) Optional sampling of submartingales with scanned index sets. *JTP* **5**, 309–326.

Harris, B. (1991) Theory and counterexamples for confidence limits on system reliability. *SPL* **11**, 411–417.

He, Q.-M. and Zhang, H. (2005) A note on unicyclic representations of phase type distributions. *Stoch. Models* **21**, 465–483.

Heinkel, B (1996) On the Kolmogorov quasimartingale property. *Probab. Math. Statist.* **16**, 113–126.

Hiai, F. and Petz, D. (2000) *The Semicircle Law, Free Random Variables and Entropy*. (Math. Surveys Monogr. **77**). Amer. Math. Soc., Providence (RI).

Holtz, O., Nazarov, F. and Peres, Y. (2011) New coins from old, smoothly. *Constr. Approx.* **33**, 331–361.

Hombas, V.C. (1997) Waiting time and expected waiting time – paradoxical situations. *AmS* **51**, 130–133.

Houdré, C. and Lacey, M.T. (1996) Spectral criteria, SLLNs and a.s. convergence of series of stationary variables. *AP* **24**, 838–856.

Hu, C.-Y., Iksanov, A.M., Lin, G.D. and Zakysylo, O.K. (2006) The Hurwitz zeta distribution. *Austral. & New Zealand J. Statist.* **48**, 1–6.

Hu, T., Müller, A. and Scarsini, M. (2004) Some counterexamples in positive dependence. *JSPI* **124**, 153–158.

Hu, T.C. (1997) On pairwise independent and independent exchangeable r.v.s. *SAA* **15**, 51–57.

Hu, Y. and Peng, S. (1995) Solution of forward-backward SDEs. *PTRF* **103**, 273–283.

Hult, H. and Lindskog, F. (2006) On Kesten's counterexample to the Cramér-Wold device for regular variation. *Bernoulli* **12**, 133–142.

Ichihara, K. (1982) Curvature, geodesics and the BM on a Riemannian manifold. II. Explosion properties. *Nagoya Math. J.* **87**, 115–125.

Ilinskii, A.I. and Chistyakov, G.P. (1997) On the problem of characterizing probability distributions by absolute moments of partial sums. *TPA* **42**, 426–432.

Iksanov, A.M. and Jurek, Z.J. (2003) Shot noise distributions and selfdecomposability. *SAA* **21**, 593–609.

Iksanov, A. and Marynych, A. (2008) A note on non-regular martingales. *SPL* **78**, 3014–3017.

Imkeller, P. (1986) On changing time for two-parameter strong martingales: a counterexample. *AP* **14**, 1080–1084.

Isaac, R. (1999) On equitable ratios of Dubins-Freedman type. *SPL* **42**, 1–6.

Jacobsen, M., Mikosch, T., Rosinski, J. and Samorodnitsky, G. (2009) Inverse problems for regular variation of linear filters, a cancellation property for σ-finite measures and identification of stable laws. *Ann. Appl. Probab.* **19**, 210–242.

Jacod, J. and Shiryaev, A.N. (2003) *Limit Theorems for Stochastic Processes*. 2nd ed. Springer, Berlin.

Jaroszewska, J. (2013) On asymptotic equicontinuity of Markov transition functions. *SPL*. Available at: `doi:10.1016/j.spl.2012.10.033`.

Jeanblanc, M., Yor, M. and Chesney, M. (2009) *Mathematical Methods for Financial Markets*. Springer, Berlin.

Jerrum, M. (2006) On the approximation of one Markov chain by another. *PTRF* **135**, 1–14.

Jia, G. (2008) A class of backward SDEs with discontinuous coefficients. *SPL* **78**, 231–237.

Jiand, G. and Peng, S. (2010) Jensen's inequality for q-convex function under g-expectation. *PTRF* **147**, 217–239.

Jiang, L. and Chen, Z. (2004) On Jensen's inequality for g-expectation. *Chinese Ann. Math. Ser. B* **25**, 401–412.

Jones, M.C. (2008) The distribution of the ratio X/Y for all centred elliptically symmetric distributions. *JMVA* **99**, 572–573.

Jones, M.C. (2013) On distributions generated by transformation of scale. *Statistica Sinica*. Accepted in 2012.

Jones, M.C. and Arnold, B. (2008) Distributions that are both R- and log-symmetric. *Electronic J. in Statist.* **2**, 1300–1308.

Jourdain, B. (2004) Loss of martingality in asset price models with lognormal stochastic volatility. *Internat. J. Theoret. Appl. Finance* **13**, 767–787.

Junge, M., Parcet, J. and Xu, Q. (2007) Rosenthal type inequalities for free chaos. *AP* **35**, 1374–1337.

Kaas, R., Govaerts, M. and Tang, Q. (2004) Some useful counterexamples regarding comonotonicity. *Belgian Actuar. Bull.* **4**, 1–4.

Kabanov, Yu. M. (2011) Personal communication.

Kagan, A., Laha, R.C. and Rohatgi, V. (1997) Independence of the sum and absolute difference of independent r.v.s does not imply their normality. *Math. Methods Statist.* **6**, 263–265.

Kahn, J. and Neiman, M. (2009) Negative correlation and log-concavity. *Random Struct. Algorithms* **37**, 367–388.

Kallenberg, O. (2002) *Foundations of Modern Probability*. 2nd ed. Springer, Berlin.

Kaluszka, M and Okolewski, A. (2004) On Fatou-type lemma for monotone moments of weakly convergent r.v.s. *SPL* **66**, 45–50.

Kantorovitz, M.R. (2007) An example of a stationary, triplewise independent triangular array for which the CLT fails. *SPL* **77**, 539–542.

Kargin, V. (2007a) On superconvergence of sums of free r.v.s. *AP* **35**, 1931–1945.

Kargin, V. (2007b) A proof of a non-commutative CLT by the Lindeberg method. *Electronic Commun. in Probab.* **12**, 36–50.

Kartashov, N. and Mishura, Yu. (2004) What random variable generates a bounded potential? *J. Appl. Math. Stoch. Anal.* **2004**, 97–106.

Kendall, W.S. (2000) Stationary countable dense random sets. *AAP* **32**, 86–100.

Kendall, W.S. and Montana, G. (2002) Small sets and Markov transition densities. *SPA* **99**, 177–194.

Khuri, M.A. and Khuri, A.I. (2003) Corrections to a well-known proposition concerning the two-dimensional density function. *Internat. J. Math. Educ. Sci. Technol.* **34**, 787–792.

Kibzun, A.I. and Matveev, E.L. (2010) Sufficient conditions for quasiconcavity of the probability function. *Automation Remote Control* **71**, 413–430.

Kifer, Yu. (1986) *Ergodic Theory of Random Transformations*. Birkhäuser, Boston.

Kingman, J.F.C. (1982) The coalescent. *SPA* **13**, 235–248.

Klaassen, C.A.J., Mokveld, P.J. and Van Es, B. (2000) Squared skewness minus kurtosis bounded by 186/125 for unimodal distributions. *SPL* **50**, 131-135.

Kleiber, C. (2012) On moment indeterminacy of the Benini income distribution. *Statist. Papers* **54**, in print.

Kleiber, C. (2013) The generalized lognormal distribution and the Stieltjes moment problem. *JTP*. Available online: `doi 10.1007/s10959-013-0477-0`.

Kleiber, C. and Stoyanov, J. (2013) Multivariate distributions and the moment problem. *JMVA* **113**, 7–18.

Klenke, A. (2008) *Probability Theory: A Comprehensive Course*. Springer, Berlin.

Klesov, O. and Rosalsky, A. (1997) On independence and conditional independence. *J. Appl. Statist. Sci.* **6**, 121–124.

Kochar, S. and Xu, M. (2011) The tail behavior of the convolutions of gamma r.v.s. *JSPI* **141**, 418–428.

Komorowski, T. (2003) A note on the CLT for two-fold stochastic random walks in random environment. *Bull. Polish Acad. Sci. Math.* **51**, 217–232.

Kôno, N. (1979) Real variable lemmas and their applications to sample properties of stochastic processes. *J. Math. Kyoto Univ.* **19**, 413–433.

Konsowa, M.H. (2002) Conductivity of random trees. *Probab. Engrg. Inform. Sci.* **16**, 233–240.

Koole, G. (1993) Marked point processes as limits of Markovian arrival streams. *JAP* **30**, 365–372.

Kopanov, P. (2012) Personal communication.

Koralov, L.B. and Sinai, Y.G. (2007) *Theory of Probability and Random Processes*. 2nd ed. Springer, Berlin.

Koshnevisan, D. (2007) *Probability*. Amer. Math. Soc., Providence (RI).

Kozubowski, T. (2005) A note on self-decomposability of stable process subordinated to self-decomposable subordinator. *SPL* **73**, 343–345.

Kramkov, D. (2003) Necessary and sufficient conditions in the problem of optimal investment in incomplete markets. *Ann. Appl. Probab.* **13**, 1504–1516.

Kristiansen, G.K. (1995) A counterexample to a conjecture concerning infinite divisible distributions. *Scand. J. Statist.* **22**, 139–141.

Kruglov, V. and Bo, C. (2001) Weak convergence of random sums. *TPA* **46**, 39–57.

Krylov, N.V. (2002) A Simple Proof of a Result of A. Novikov. University of Minnesota. Preprint available at `arXiv:math.PR/0207013v1`.

Kundu, C. and Nanda, A.K. (2010) Some reliability properties of the inactive time. *Commun. Statist. – Theory Methods* **39**, 899–911.

Kuo, W.-C., Lubaschagne, C.C.A. and Watson, B.A. (2006) Convergence of Riesz martingales. *Indag. Math. (N.S.)* **17**, 271–283.

Kutoyants, Yu. M. (2010) Personal communication.

Kuwada, K. and Sturm, K.-T. (2007) A counterexample for the optimality of Kendall-Cranston coupling. *Electronic Commun. in Probab.* **12**, 66–72.

Kuznetsov, E.E. (2000) On uniqueness of a solution of $Lu = u^{\alpha}$ with given trace. *Electronic Commun. in Probab.* **5**, 137–147.

Kyprianou, A.E. (2006) *Introductory Lectures on Fluctuations of Lévy Processes with Applications*. Springer, Berlin.

Lasserre, J.B., Prieto-Rumeau, T. and Zervos, M. (2006) Pricing a class of exotic options via moments and SDP relaxations. *Math. Finance* **16**, 469–494.

Le Gall, J.-F. (1999) *Spatial Branching Processes, Random Snakes and PDEs*. Birkhäser, Basel.

Ledoux, M. and Talagrand, M. (2012) *Probability in Banach Spaces: Isoperimetry and Processes*. Springer, New York.

Lefévre, C. and Utev, S. (1998) On order-preserving properties of random metrics. *JTP* **11**, 907–920.

Lejay, A. (2002) On the convergence of stochastic integrals driven by processes converging on account of a homogenization property. *Electron. J. Probab.* **7**, no. 18, 18 pp.

Levental, S. and Erickson, R.V. (2003) On a.s. convergence of the quadratic variation of Brownian motion. *SPA* **106**, 317–333.

Li, D., Qi, Y. and Rosalsky, A. (2010) On the set of limit points of normed sums of geometrically weighted i.i.d. bounded r.v.s. *SAA* **28**, 86–102.

Li, X.-M. and Scheutzow, M. (2011) Lack of strong completeness for stochastic flows. *AP* **39**, 1407–1421.

Lin, G.D. (1997) On the moment problem. *SPL* **35**, 85–90. Erratum: **50** (2000), p. 205.

Lin, G.D. and Hu, C.-Y. (2000) A note on the L-class of life distributions. *Sankhya* 267–272.

Lin, G.D. and Huang, J.S. (1989) Variances of sample medians. *SPL* **8**, 143–146.

Lin, G.D. and Stoyanov, J. (2009) The logarithmic skew-normal distributions are moment-indeterminate. *JAP* **46**, 909–916.

Lin, J.X. (2011) A counterexample on subexponentiality. *Xiamen Daxue Xuebao Ziran Kexue Ban* **50**, 963–965. (In Chinese)

Liu, W. and Wu, L. (2009) Identification of the rate function for large deviations of an irreducible Markov chain. *Electronic Commun. in Probab.* **14**, 540–551.

Lowther, G. (2009) Limits of one-dimensional diffusions. *AP* **37**, 78–106.

Löwe, M. and Matzinger, H. III (2003) Reconstruction of sceneries with correlated colors. *SPA* **105**, 175–210.

Luczak, A. (1997) Exponent and symmetry of operator Lévy's probability measures on finite-dimensional vector spaces. *JTP* **10**, 117–129.

Lukić, M.N. and Beder, J.H. (2001) Stochastic processes with sample oaths in reproducing kernel Hilbert spaces. *TAMS* **353**, 3945–3969.

Lyasoff, A. (2012) Personal communication.

Ma, J. and Zhang, J. (2011) On weak solutions of forward-backward SDEs. *PTRF* **151**, 475–507.

Maassen, H. (2010) Quantum probability and quantum information theory. *Lecture Notes in Physics* (Springer) **808**, 65–108.

Maejima, M. and Ueda, Y. (2013) Examples of α-selfdecomposable distributions. *SPL* **83**, 286–291.

Maes, C. Redig, F. and Van Moffaert, A. (1999) Almost Gibbsian versus weakly Gibbsian measures. *SPA* **79**, 1–15.

Mahmoud, H.M. (2008) *Pólya Urn Models*. CRC, Boca Raton (FL) and Chapman & Hall, London.

Majer, P. and Mancino, M.E. (1997) A counterexample concerning a condition of Ogawa integrability. *LNM* **1655**, 198–206.

Majumdar, S. (1999) A metric for convergence in distribution. *SPL* **42**, 239–247.

Mansuy R. and Yor, M. (2006) *Random Times and Enlargements of Filtrations in Brownian Setting*. (LNM **1873**). Springer, Berlin.

Markström, K. (2007) Negative association does not imply log-concavity of the risk sequence. *JAP* **44**, 1119–1121.

Markus, M.B. and Rosen, J. (2006) *Markov Processes, Gaussian Processes, and Local Times*. Cambr. Univ. Press, Cambridge.

Merkl, F. and Rolles, S.W.W. (2007) A random environment for linearly edge-reinforced random walks on infinite graphs. *PTRF* **138**, 157–176.

Michel, R. (2008) Some notes on multivariate generalized Pareto distributions. *JMVA* **99**, 1288–1301.

Mishura, Yu.S. (2008) *Stochastic Calculus for Fractional Brownian Motion and Related Processes*. (LNM **1929**). Springer, Berlin.

Misiewicz, J. and Wesolowski, J. (2012) Winding planar probabilities. *Metrika* **75**, 507–519.

Molchanov, I. (2005) *Theory of Random Sets*. Springer, London.

Monteiro, P.K. and Page, F.H. (1997) Arbitrage, equilibrium, and gains from trade: a counterexample. *J. Math. Economics* **28**, 481–501.

Móri, T.F. (1992) Essential correlatedness and almost independence. *SPL* **15**, 169–172.

Móri, T.F. (2009) Deviations of discrete distributions – positive and negative results. *SPL* **79**, 1089–1096.

Morters, P. and Peres, Y. (2010) *Brownian Motion*. Cambr. Univ. Press, Cambridge.

Müller, A. and Stoyan, D. (2002) *Comparison Methods for Stochastic Models and Risks*. 2nd ed. Wiley, Chichester.

Nagaev, S.V. (2002) Lower bounds for probabilities of large deviations of sums of independent r.v.s. *TPA* **46**, 728–735.

Nam, E. and Rosalsky, A. (1996) On the convergence rate of series of independent r.v.s. In: *Research Developments in Probability and Statistics. Madan Puri Festschrift*. Eds E. Brunner and M. Denker. 33–44. VSP, Utrecht (NL).

Nelsen, R.B. and Úbeda-Flores, M. (2012) How close are pairwise and mutual independence? *SPL* **82**, 1823–1828.

Nica, A. and Speicher, R. (2006) *Lectures on the Combinatorics of Free Probability*. (LMS Series **335**). Cambr. Univ. Press, Cambridge.

Norris, J.R. (1997) *Markov Chains*. Cambr. Univ. Press, Cambridge.

Oliveira, P.E. (2011) *Asymptotics of Associated Random Variables*. Springer, Berlin.

Ostrovska, S. (2002) On k-independence of r.v.s with given distributions. In: *Limit Theorems in Probability and Statistics*. Vol. **II**. 455–464. J. Bolyai Math. Soc., Budapest.

Ostrovska, S. (2006) Weak uncorrelatedness of random variables. *Acta Math. Scientia* **26B**, 379–384.

Ostrovska, S. and Stoyanov, J. (2005) Stieltjes classes for M-indeterminate powers of inverse Gaussian distributions. *SPL* **71**, 165–171.

Ouchti, L. and Volný, D. (2008) A conditional CLT which fails for ergodic components. *JTP* **21**, 687–703.

Pakes, A.G. (2001). Remarks on converse Carleman and Krein criteria for the classical moment problem. *J. Austral. Math. Soc.* **71**, 81–104.

Pakes, A.G. (2007) Structure of Stieltjes classes of moment-equivalent probability laws. *J. Math. Anal. Appl.* **326**, 1268–1290.

Pakes, A.G., Hung, W.-L. and Wu, J.-W. (2001) Criteria for the unique determination of probability distributions by moments. *Austral. & N.Z. J. Statist.* **43**, 101–111.

Păltănea, R. and Zbăganu, G. (2011) On the moments of iterated tail. *Math. Rep. (Bucur.)* **13**(63), 65–74.

Pavlov, I.V. (1980) A counterexample to the conjecture that H^∞ is dense in BMO. *TPA* **25**, 154–158.

Pavlov, Y.L. (2000) *Random Forests*. VSP, Leiden (NL).

Pedersen, H.L. (1998) On Krein's theorem for indeterminacy of the classical moment problem. *J. Approx. Theory* **95**, 90–100.

Peligrad, M. and Utev, S. (2006) Invariance principle for stochastic processes with short memory. *IMS Lecture Notes on High Dimensional Probability* **51**, 18–32.

Pemantle, R. and Rosenthal, J. (1999) Moment conditions for a sequence with negative drift to be uniformly bounded in L^r. *SPA* **82**, 143–155.

Penson, K.A., Blasiak, P., Duchamp, G.H.E., Horzela, A. and Solomon, A.I. (2010). On certain non-unique solutions of the Stieltjes moment problem. *Discrete Math. Theor. Comput. Sci.* **12**, 295–306.

Perrin, O. and Senoussi, R. (1999) Reducing non-stationary stochastic process to stationarity by a time deformation. *SPL* **43**, 393–397.

Pflug, G. (1996) *Optimization of Stochastic Models*. Kluwer, Dordrecht.

Pipiras, V. and Taqqu, M. (2001) Are classes of deterministic integrands for fBM on an interval complete? *Bernoulli* **7**, 873–897.

Pirinsky, Ch. (2009) Personal communication.

Pitman, J. (2006) *Combinatorial Stochastic Processes*. (LNM **1875**). Springer, Berlin.

Prokhorov, A.V. and Ushakov, N.G. (2003) On the problem of reconstructing a summands distribution by the distribution of their sum. *TPA* **46**, 420–430.

Rachev, S.T. (1990) A counterexample to a.s. constructions. *SPL* **9**, 307–309.

Rásonyi, M. and Stettner, L. (2005) On utility maximization in discrete-time financial market models. *Ann. Appl. Probab.* **15**, 1367–1395.

Rao, M.M. (2005) *Conditional Measures and Applications*. 2nd ed. Chapman & Hall/CRC, London/Boca Raton (FL).

Rao, M.M. and Swift, R. (2006) *Probability Theory with Applications*. 2nd ed. Springer, Berlin.

Remapla, G. and Nguyen, T.T. (1996) Non-Gaussian measures with Gaussian structure. *Probab. Math. Statist.* **16**, 287–298.

Revuz, D. and Yor, M. (2003) *Continuous Martingales and Brownian Motion*. 3rd ed. Springer, Berlin.

Ridder, A. and Schwartz, A. (2005) Large deviation without principle: join the shortest queue. *Math. Methods Oper. Research* **62**, 467–483.

Roberts, G.O. and Rosenthal, J.S. (2006) Harris recurrence of Metropolis-within-Gibbs and trans-dimensional Markov chains. *Ann. Appl. Probab.* **16**, 2123–2139.

Rogers, L.C.G. (1997) Arbitrage with fractional Brownian motion. *Math. Finance* **7**, 95–105.

Rokhlin, D.B. (2008) Equivalent supermartingale densities and measures in discrete-time infinite-horizon market models. *TPA* **53**, 626–647.

Rokhlin, D.B. (2010) Lower bounds for the densities of martingale measure in the Dalang-Morton-Willinger theorem. *TPA* **54**, 447–465.

Romano, J.P. and Siegel, A.F. (1986) *Counterexamples in Probability and Statistics*. Wadsworth & Brooks/Cole, Monterey (CA).

Rosalsky, A. and Thanh, L.V. (2007) On the strong LLN for sequences of blockwise independent and blockwise p-orthogonal random elements in Rademacher type p Banach spaces. *Probab. Math. Statist.* **27**, 205–222.

Rosalsky, A. and Thanh, L.V. (2009) Weak LLN of double sums of independent random elements in Rademacher type p and stable type p Banach spaces. *Nonlinear Anal.* **71**, 1065–1074.

Rosenthal, J.S. (2006) *A First Look at Rigorous Probability Theory*. 2nd ed. World Sci. Publ., Singapore.

Rosinski, J. and Zak, T. (1997) The equivalence of ergodicity and weak mixing for inf.div. processes. *JTP* **10**, 73–86.

Roussas, G.G. (2005) *An Introduction to Measure-Theoretic Probability*. Elsevier, Boston.

Roux, A. and Zastawniak, T. (2006) A counterexample to an option pricing formula under transaction costs. *Finance Stoch.* **10**, 575–578.

Rozovskii, L.V. (1999) On an inequality of J. Doob for ch.f.s. *TPA* **44**, 588–590.

Rue, H. and Held, L. (2005) *Gaussian Markov Random Fields: Theory and Applications*. Chapman & Hall, London and CRC, Boca Raton (FL).

Ruf, J. (2012) A new proof for the conditions of Novikov and Kazamaki. *SPA* **123**, 404–421.

Salisbury, T.S. (1986) On the Itô excursion process. *PTRF* **73**, 319–350.

Salisbury, T.S. (1986) A Martin boundary in the plane. *TAMS* **293**, 623–642.

Saloff-Coste, L. and Züniga (2009) Merging for time inhomogeneous Markov chains. I. Singular values and stability. *Electronic J. in Probab.* **14**, 1456–1494.

Samorodnitsky, G. and Taqqu, M. (1994) *Stable Non-Gaussian Random Processes*. Chapman & Hall, New York.

Sapatinas, T. (1995) Identifiability of mixtures of power-series distributions and related characterizations. *Ann. Inst. Statist. Math.* **47**, 447–459.

Sapatinas, T. and Shanbhag, D.N. (2010) Moment properties of multivariate infinitely divisible laws and criteria for multivariate self-decomposability. *JMVA* **101**, 500–511.

Sasvári, Z. (2013) *Multivariate Characteristic and Correlation Functions*. Walter de Gruyter, Berlin.

Sato, K.-I. (1993) Convolution of unimodal distributions can produce any number of modes. *AP* **21**, 1543–1549.

Sato, K.-I. (1999) *Lévy Processes and Infinitely Divisible Distributions*. Cambr. Univ. Press, Cambridge.

Sato, M. (1997) Some remark on the mean, median, mode and skewness. *Austral. J. Statist.* **39**, 219–224.

Scarsini, M. (1985) Lower bounds for the d.f. of a k-dimensional n-extendible exchangeable process. *SPL* **3**, 57–62.

Scarsini, M. (1998) Multivariate convex orderings, dependence, and stochastic equality. *JAP* **35**, 93–103.

Schief, A. (1989) Almost sure convergent r.v.s with given law. *PTRF* **81**, 559–567.

Schmidli, H. (1999) Compound sums and subexponentiality. *Bernoulli* **5**, 999–1012.

Schneider, R. and Weil, W. (2008) *Stochastic and Integral Geometry*. Springer, Berlin.

Schnurr, A. (2011) A classification of deterministic Hunt processes with some applications. *Markov Process. Related Fields* **17**, 259–276.

Scholz, F. (2012) Personal communication.

Schweinberg, J. (2010) The number of small blocks in exchangeable random partitions. *ALEA Latin Amer. J. Probab. Math. Stat.* **7**, 217–242.

Schweizer, M. (1999) A minimality property of the minimal martingale measure. *SPL* **42**, 27–31.

Shaked, M. and Shanthikumar, J.G. (2007) *Stochastic Orders*. Springer, Berlin.

Shields, P.C. (1993) Two divergence-rate counterexamples. *JTP* **6**, 521–545.

Shimura, T. (1997) The product of independent r.v.s with slowly varying truncated moments. *J. Austral. Math. Soc. Ser. A* **62**, 186–197.

Shirikyan, A. (2006) LLNs and CLT for randomly forced PDEs. *PTRF* **134**, 215–247.

Shiryaev, A.N. (2012) *Problems in Probability*. Springer, Berlin.

Shiryaev, A.N., Erlich, I.G. and Yaskov, P.A. (2013) *Probability by Theorems and Problems: With Proofs and Solutions*. Vol. **1** and **2**. MCCME, Moscow. (In Russian)

Shiryaev, A.N., Kabanov, Yu.M., Kramkov, D.O. and Melnikov, A.V. (1994) Toward a theory of pricing options of European and American types. II. Continuous time. *TPA* **39**, 61–102.

Shorack, G.R. (2000) *Probability for Statisticians*. Springer, New York.

Sokol, A. (2012) An extended Novikov-type criteria for local martingales with jumps. Available on `arXiv:1210.2866`.

Sokol, A. (2013) Optimal Novikov-type criteria for local martingales with jumps. *Electronic Commun. Probability* **18**:39, 1–6.

Song, L. (1999) A counterexample to the CLT. *Bull. London Math. Soc.* **31**, 222–230.

Stenflo, Ö. (2001) A note on a theorem of Karlin. *SPL* **54**, 183–187.

Stepniak, C. and Owsiany, T. (2009) Are Bernstein examples on independent events paradoxical? In: *Statistical Inference, Econometric Analysis and Matrix Algebra, VII.* Eds B. Schipp and W. Kräer. 411–414. Physica-Verlag, Heidelberg.

Steutel, F. and Van Harn, K. (2003) *Infinite Divisibility of Probability Distributions on the Real Line*. Marcel Dekker, Hoboken (NJ).

Stoica, G. (2008) A counterexample concerning a series of T. L. Lai. *Adv. Appl. Statist.* **9**, 141–144.

Stoyanov, J. (1998) Global dependency measure for sets of random elements: "The Italian problem" and some consequences. In: *Stochastic Processes and Related Topics. In Memory of S. Cambanis 1943-1995*. Eds I. Karatzas, B. Rajput and M. Taqqu. 357–375. Birkhäuser, Boston.

Stoyanov, J. (2000) Krein condition in probabilistic moment problems. *Bernoulli* **6**, 939–949.

Stoyanov, J. (2004) Stieltjes classes for moment-indeterminate probability distributions. *JAP* **41A**, 281–294.

Stoyanov, J. (2005) Sets of binary random variables with a prescribed independence dependence structure. *Math. Scientist* **28**, 19–27.

Stoyanov, J. and Tolmatz, L. (2005) Method for constructing Stieltjes classes for M-indeterminate probability distributions. *Appl. Math. Comput.* **165**, 669–685.

Stoyanov, J. and Lin, G.D. (2012) Hardy's condition in the moment problem for probability distributions. *Theory Probab. Appl.* **57**, 811–820.

Strasser, E. (2003) Necessary and sufficient conditions for the supermartingale property of a stochastic integral w.r.t. a local martingale. *LNM* **1832**, 385–393.

Stroock, D.W. (2011) *Probability Theory: An Analytic View*. 2nd ed. Cambr. Univ. Press, Cambridge.

Suhov, Yu. and Kelbert, M. (2005) *Probability and Statistics by Example. 1: Basic Probability and Statistics*. Cambr. Univ. Press, Cambridge.

Suhov, Yu. and Kelbert, M. (2008) *Probability and Statistics by Example. 2: Markov Chains*. Cambr. Univ. Press, Cambridge.

Sung, S.H. (1997) A note on Spătaru's complete convergence theorem. *J. Math. Anal. Appl.* **206**, 606–610.

Suppes, P., De Barros, J.A. and Oas, G. (1996) A collection of probabilistic hidden-variable theorems and counterexamples. Available at `arXiv:quant-ph/9610010v2`.

Swart, J.M. (2001) A 2-dimensional SDE whose solutions are not unique. *Electronic Commun. in Probab.* **6**, 67–71.

Székely, G.J. and Rao, C.R. (2000) Identifiability of distributions of independent r.v.s by linear combinations and moments. *Sankhya, Ser. A* **62**, 193–202.

Székely, G. J., Rizzo, M. L. and Bakirov, N. K. (2007) Measuring and testing independence by correlation of distances. *AS* **35**, 2769–2794.

Székely, G. J. and Rizzo, M. L. (2009) Brownian distance covariance. *Ann. Appl. Statist.* **3**, 1233–1303.

Sznitman, A.-S. (2002) An effective criterion for ballistic behavior of random walks in random environment. *PTRF* **122**, 509-544.

Szynal, D. (1993) On complete convergence for some classes of dependent r.v.s. *Ann. Univ. M. Curie-Sklodowska (Lublin)* **47**, 145–150.

Talagrand, M. (2005) *The Generic Chaining: Upper and Lower Bounds of Stochastic Processes*. Springer, Berlin.

Tanner, S. (2006) Nontangential and probabilistic boundary behavior pluriharmonic functions. *AP* **34**, 1623–1634.

Tian, J.P. and Kannan, D. (2006) Lumpability and commutativity of Markov processes. *SPA* **24**, 685–702.

Tsirelson, B. (1997) Triple points: from non-Brownian filtrations to harmonic measures. *Geometric Funct. Anal.* **7**, 1096–1142.

Ushakov, N.G. (1999) *Selected Topics in Characteristic Functions*. VSP, Utrecht (NL).

Ushakov, V.G. and Ushakov, N.G. (2001) The reconstruction of the probability characteristics of multivariate random functions from projections. *Moscow Univ. Comput. Math. Cybernet.* **2001**, 31–38.

Van Eeden, C. (1991) On a conjecture concerning a characterization of the exponential distribution. *CWI Quarterly* **4**, 205–211.

Van Handel, R. (2009) The stability of conditional Markov processes and Markov chains in random environments. *AP* **37**, 1876–1925.

Van Handel, R. (2012) On the exchange of integration and supremum of σ-fields in filtering theory. *Israel J. Math.* **192**, 763–764.

Večer, J. and Xu, M. (2004) The mean comparison theorem cannot be extended to the Poisson case. *JAP* **41**, 1199–1202.

Veretennikov, A. Yu. (2006) On lower bounds for mixing coefficients of Markov diffusions. In: *From Stochastic Calculus to Mathematical Finance. The Shiryaev Festschrift*. Eds Yu. Kabanov, R. Liptser and J. Stoyanov. 623–633. Springer, Berlin.

Vershik, A.M. (2004) Random metric spaces and universality. *Russian Math. Surveys* **59**, 259–295.

Voiculescu, D. (2000) *Lectures on Free Probability Theory*. (LNM **1738**). 279–349. Springer, Berlin.

Vostrikova, L.V. (2012) Personal communication.

Wagner, D.G. (2008) Negatively correlated r.v.s and Mason's conjecture for independent sets in matroids. *Ann. Combinatorics* **12**, 211–239.

Watanabe, T. (2008) Convolution equivalence and distributions of random sums. *PTRF* **142**, 367–397.
Wefelmeyer, W. (1985) A counterexample concerning monotone unimodality. *SPL* **3**, 87–88.
Wen, L. (2001) A counterexample for the two-dimensional density function. *AMM* **108**, 367–368.
Wesolowski, J. (1992) A characterization of the bivariate elliptically contoured distribution. *Statist. Papers* **33**, 143–149.
Wesolowski, J. (1996) Are continuous mappings preserving normality necessarily linear? *Appl. Math. (Warsaw)* **24**, 109–112.
Wood, A.T. (1995) When is a truncated covariance function on the line a covariance function on the circle? *SPL* **24**, 157–164.
Wulfsohn, A. (2005) Measure convolution semigroups and noninfinitely divisible probability distributions. *J. Math. Sci. (New York)* **131**, 5682–5696.
Yarho, A. (2012) Personal communication.
Yu, Y. (2011) Concave renewal functions do not imply DFR interrenewal times. *JAP* **48**, 583–588.
Zalesskii, B.A., Sazonov, V.V. and Ulyanov, V.V. (1991) A precise estimate for the rate of convergence in the CLT in a Hilbert space. *Math. USSR-Sbornik* **68**, 453–482.
Zbăganu, G. (2000) On the existence of an optimal covering strategy. *Rev. Roumaine Math. Pures Appl.* **45**, 1047–1062.
Zempléni, A. (2002) Decompositions and antiirreducibles in max-semigroups of bivariate r.v.s. *J. Math. Sci. (New York)* **106**, 2765–2768.
Zhao, P. and Li, X. (2009) Stochastic order of sample range from heterogeneous exponential r.v.s. *Probab. Engrg. Inform. Sci.* **23**, 17–29.
Zvonkin, A.K. (2012) Personal communication.

Index

The Index lists notions and results discussed in the main body of the book. The numbers after each item refer to the numbers of the counterexamples. The numbers in bold font indicate the section where a definition of the listing can be found. For recent counterexamples, which appeared after 1997, the reader is advised to see the APPENDIX.